Building and Surveying Series

Accounting and Finance for Building and Surveying A. R. Jennings
Advanced Valuation Diane Butler and David Richmond
Applied Valuation, second edition Diane Butler
Asset Valuation Michael Rayner
Auctioning Real Property R. M. Courtenay Lord
Building Economics, fourth edition Ivor H. Seeley
Building Maintenance, second edition Ivor H. Seeley
Building Procurement second edition Alan E. Turner
dition Ivor H. Seeley

ations Ivor H. Seeley
*r H. Seeley
tion Ivor H. Seeley and George P. Murray
tical Guide Philip Freedman

eering Projects Alan Griffith

*omas
tion Stephen L. Gruneberg
or Success Richard Pettinger
and Control Brian Cooke

*J. Cooke
urth edition R. McMullan
* Alan Park
act of Property

nd edition

*n David Richmond
ct 1998 edition

*orge P. Murray
Pricing, second edition

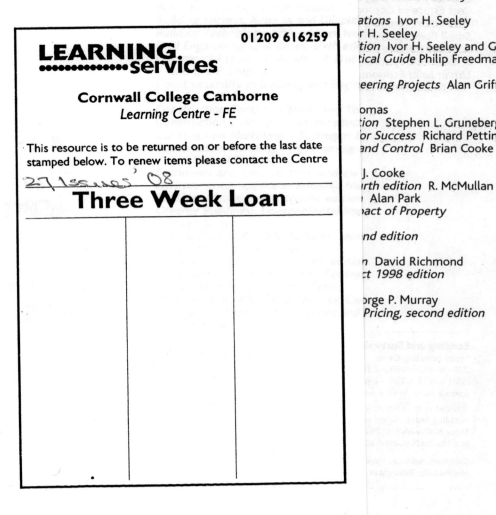

Building and Surveying Series
Series Standing Order
ISBN 0–333–71692–2 hardcover
ISBN 0–333–69333–7 paperback
(outside North America only)

You can receive future titles in this series as they are published by placing a
standing order. Please contact your bookseller or, in the case of difficulty, write
to us at the address below with your name and address, the title of the series
and the ISBN quoted above.

Customer Services Department, Macmillan Distribution Ltd
Houndmills, Basingstoke, Hampshire RG21 6XS, England

BUILDING TECHNOLOGY

IVOR H. SEELEY

BSc (Est Man), MA, PhD, FRICS,
CEng, FICE, FCIOB, FCIH

Formerly Emeritus Professor of Nottingham Trent University

Fifth Edition

palgrave

First edition 1974
Reprinted five times
Second edition 1980
Reprinted four times
Third edition 1986
Reprinted six times
Fourth edition 1993
Fifth edition 1995

Published by
PALGRAVE
Houndmills, Basingstoke, Hampshire RG21 6XS and
175 Fifth Avenue, New York, N.Y. 10010
Companies and representatives throughout the world

PALGRAVE is the new global academic imprint of
St. Martin's Press LLC Scholarly and Reference Division and
Palgrave Publishers Ltd (formerly Macmillan Press Ltd).

ISBN 0–333–62096–8

This book is printed on paper suitable for recycling and made from fully managed and sustained forest sources.

A catalogue record for this book is available from the British Library.

12   11   10   9   8   7   6
07   06   05   04   03   02   01

Printed in Malaysia

When we build, Let us think that we build forever
John Ruskin

This book is dedicated to my wife for her continual interest in building matters and for producing such a good design for our bungalow, where some of the constructional methods described in the book have been put into practice

CONTENTS

LIST OF ILLUSTRATIONS

LIST OF TABLES

PREFACE TO THE FIFTH EDITION

This book examines the general principles of building construction and applies them to practical examples of constructional work throughout all parts of simple domestic buildings. Consideration is also given to the choice of materials and implications of Building Regulations, British Standards, Codes of Practice and current government research.

The principal aim is to produce a comprehensive, simply explained and well-illustrated text which will form a basic construction textbook for first- and second-year students in a variety of disciplines, including architects, quantity surveyors, valuers, auctioneers, estate agents, building and mineral surveyors, builders, housing managers, environmental health officers, building control officers and clerks of works. It should also prove of value to those working for BTEC certificates and diplomas in construction; and degrees in architecture, building, quantity surveying, building surveying and estate management.

All drawings and building data are fully metricated, and it was decided to omit 'm' and 'mm' symbols from all descriptions as well as dimensions on the drawings, to avoid constant repetition of units of measurement. It is unlikely that any confusion will arise as a result of this. A metric conversion table is produced in an appendix to help readers who wish to make comparisons between imperial and metric measures. The abbreviations BRE, for Building Research Establishment, and DOE, for Department of the Environment, have been used extensively throughout the book.

The fourth edition was considerably extended and updated to include latest developments, including the latest British Standard requirements and BRE recommendations and the requirements of the Building Regulations 1991 and the supporting Approved Documents issued in 1989, 1991 and 1994 (1990, 1992 and 1995 editions) and their application. A new and detailed chapter was added on the important and complex scientific and technological aspects of sound and thermal insulation, dampness, ventilation and condensation and their wide ranging implications, as they cannot be covered satisfactorily on an elemental basis. The section on timber frame construction was extended and a new section added on lighting in the chapter on External Works, as well as many other additions. Overseas readers, in particular, continue to welcome the chapter on Building in Warm Climates which was introduced in the third edition.

This fifth edition has been further updated and considerably extended to include details of radon diagnosis and protection, freestanding walls, more extensive cavity wall details and their insulation, more solid and joisted floor detailing, warm and inverted warm deck roofing details, ventilated cold deck roof details, aluminium/timber and PVC-U windows, plastic doors, GRP petrol/oil interceptors, septic tanks, cesspools and sewage treatment plant, sound insulation of partitions and condensation prevention in pitched roofs. Extensive improvement in the quality of supporting drawings has also been effected.

IVOR H. SEELEY

ACKNOWLEDGEMENTS

THE author acknowledges with gratitude the willing co-operation and assistance received from many organisations connected with the building industry – manufacturing, research and contracting.

Extracts from *The Building Regulations* 1991 and supporting Approved Documents are reproduced with the permission of the Controller of HMSO. Tables and diagrams from BRE publications are reproduced by permission of the Controller of Her Majesty's Stationery Office.

Extracts from British Standards and Codes of Practice are reproduced by permission of the British Standards Institution, Linford Wood, Milton Keynes, MK14 6LE. The Institution allows a discount on sales of publications to students and educational bodies and BS Handbook 3 contains useful summaries of British Standards for building, which are updated periodically.

Drawings forming figures 4.9, 4.10, 4.11 and 4.12 from *Timber Frame Construction* are reproduced with the permission of the Timber Research and Development Association.

Grateful acknowledgement is made to the following organisations who so willingly provided advice and technical data which has done much to increase the value of the book to the reader by detailing latest developments and techniques:

Rockwool Ltd, Pencoed, Bridgend (insulating materials:

figures 4.5, 4.6, 6.2, 6.3, 7.6, 7.9 and 15.2); British Flat Roofing Council and CIRIA, joint copyright holders and publishers of 'Flat Roofing Design and Good Practice', from which figures 7.7 and 7.8 are taken; Schüco International, Milton Keynes (PVC-U windows and doors: figure 8.6); Sampson Modul Windows, Needham Market (figure 8.5); L.B. Plastics Ltd, Nether Heage, Derby (Sheerframe PVC-U windows and doors: figure 8.7); Klargester Environmental Engineering Ltd, Aylesbury (pollution control equipment: figures 13.2.4, 13.5 and 13.6.2); Entec (pollution control) Ltd, Andover (figure 13.6.1); Catnic Ltd, Caerphilly (steel lintels); and Cavity Trays, Yeovil (ventilation products: figure 15.3).

Among others who provided valuable information were Rehau (PVC-U windows), Pilkington Insulation, Kingspan Insulation, Gyproc Insulation, British Gypsum, Conder and Spel (pollution control products).

The quality and usefulness of the book has been very much enhanced by the large number of first-class drawings which do so much to bring the subject to life. Ronald Sears devoted many hundreds of hours to the painstaking preparation of the initial drawings, for which he deserves special appreciation. The author also expresses his gratitude to the publishers, for their continual assistance throughout the production of the book, and to various academics for their helpful suggestions.

1 THE BUILDING PROCESS AND SITEWORKS

This book describes and illustrates the constructional processes, materials and components used in the erection of fairly simple domestic and associated buildings, and examines the principles and philosophy underlying their choice. Its primary aim is to meet the needs of students of a variety of disciplines whose studies embrace building technology.

This is a wide-ranging subject and no single textbook could cover every aspect. The student is advised to widen his reading by perusing technical journals, Building Research Establishment and other relevant government publications and trade catalogues. He will also obtain a better understanding of the processes, materials and site organisation by visiting buildings under erection. All building work of any significance is subject to control under the Building Regulations. In addition, restrictions on the siting and appearance of buildings stem from the operation of the Town and Country Planning Acts.

All dimensions are in metric terms, and in this connection the metric conversion table in the appendix (page 327) may prove useful.

CHOICE OF SITE

Various matters should be considered when selecting a building site. The high cost of land and planning controls frequently combine to prevent the acquisition of the ideal site. Indeed many sites are now being developed which in years gone by would have been considered unsuitable. Some of the more important factors are now investigated.

Climate. This varies widely throughout the country being broadly warmer in the south, colder in the east and north, and wetter in the west. There are however local variations with some areas being more susceptible to mist and fog. The general climate of the UK, the influence of microclimate and improving microclimate through design are well described in BRE Digest 350.[1]

Aspect. The aspect of a site is important as it will determine the amount of sun received on the various elevations of the building. The ideal site is generally throught to be one below the summit on a gentle southerly slope, thus securing maximum sunshine and some protection from northerly winds. A house on a site with a north frontage should ideally have its main rooms at the rear. A further point to consider is the altitude of the sun varying from 61° 57' at the summer solstice to 15° 03' at the winter solstice. All habitable rooms should be planned, wherever practicable, to receive sunlight during their normal period of use.

Elevation. Elevated sites are generally preferable to low-lying ones, being drier and easier to drain, while hollows are likely to be cold and damp. Nevertheless, a hilltop site may need protection from winds by a tree belt. Slopes steeper than 1 in 10 with clay soils can suffer from long-term creep of the slope.

Prospect. Ideally a site should command pleasant views and the adjoining land uses should be compatible. The site itself will be more attractive if it is gently undulating and contains some mature trees, but if the trees are too close to the building they will restrict light and air and may cause settlement problems as described in chapter 3.

Available facilities. A housing site should ideally have ready access to schools and shops and other facilities such as parks, sports facilities, swimming pools, libraries and community centres, and good public transport services.

Services. Adequate and accessible gas and water mains, electricity cables and sewers are generally essential, although in rural areas it may be necessary to provide septic tank installations, of the type described in chapter 13.

Subsoils. Subsoils merit special consideration because of their effect on the building work.

(1) Hard rock provides a good foundation but increases excavation costs and may cause difficulties in the disposal of sewage effluent on isolated sites.

(2) Gravel is probably the ideal subsoil, being strong and easily drained.

(3) Sand drains well but if loose could be subject to movement; in practice it is usually combined with clay or gravel.

(4) Clay often has a good bearing capacity but does not drain well; with the shrinkable varieties it is necessary to take the foundations down to at least one metre below ground level because of the variations that occur with differing climatic conditions.

(5) Chalk provides a stable and easily drained subsoil of good bearing capacity.

(6) Made ground is where soil or other fill has been deposited to make up levels and lengthy periods are needed for settlement; it may be necessary to use raft or piled foundations.

Contamination. This can arise with landfill and former industrial sites which are best avoided as they could involve expensive site works to remove potential hazards, as provided for in part C2 of Schedule 1 to the Building Regulations 1991.

Water table. It is essential that the building should be erected well above the highest groundwater level. This level will vary according to the extent of the underground flow, evaporation and rainfall, and is normally highest in winter.

Subsidence. In areas liable to mining subsidence special and costly precautions of the type described in chapter 3 should be taken.

SITE INVESTIGATIONS

All potential building sites need to be investigated to determine their suitability for building and the nature and extent of the preliminary work that will be needed. Particular attention should be given to the nature of the soil and its probable load-bearing capacity usually by means of trial holes or borings, as there may be variations over the site. The level of the water table should also be established as a high water table may necessitate subsoil drainage and could cause flooding in winter. A study of Ordnance Survey maps could show the presence of disused mines or former ponds or tips on the site which have since been filled.

The position and size of main services should be determined and it is advisable to take a grid or framework of levels over the site to indicate the amount of earthwork and the ease or otherwise of drainage. The nature and condition of site boundaries should be noted together with the extent of site clearance work, such as old buildings to be demolished and trees and shrubs to be grubbed up and removed.

By way of contrast there may be some attractive mature trees on the site which ought to be retained and they may even be the subject of tree preservation orders or sited within a conservation area. Consideration should also be given to such matters as access, storage space and working conditions, which will all affect the cost of the project. Indeed, it is good practice to assess the approximate cost of site clearance work in building up an estimate of the likely cost of the complete scheme.

The local planning authority should be approached to ascertain whether there are any special or significant restrictions which could adversely affect the development of the site, and the position of the building line, so that the location of any new buildings can be established with certainty. The survey should include details of neighbouring development and the position with regard to facilities in the area. Enquiries will reveal the existence of any restrictive covenants such as rights of way, light and drainage, which may restrict the development.

BRE Digest 318[2] advocates the implementation of desk studies to collect as much material as possible about the site. This information can be obtained more cheaply and quickly than information derived, for example, from boreholes and trial pits. In the UK sources include geological maps, Ordnance Survey maps (both old and new), air photographs, geological books and journals, mining records and reports of previous site investigations. Table 1.1 illustrates a typical desk study checklist for low-rise building.

The walk-over survey, as described in BRE Digest 348,[3] is an integral and important part of the site investigation process which should always be undertaken. When used in conjunction with an effective desk study, it provides valuable information which cannot be obtained in any other way, by carrying out a thorough examination of the site and the surrounding area on foot. BRE Digest 348[3] describes how local authorities, local inhabitants and people working in the area such as builders and public utilities (gas, water and electricity undertakings) should be questioned, and local libraries and archives visited for relevant in-

Table 1.1 Desk study checklist for low-rise building
(Source: BRE Digest 318[2])

___

1. *Topography, vegetation and drainage*
(a) does the site lie on sloping ground, and if so what is the maximum slope angle?
(b) are there springs, ponds, or water courses on or near the site?
(c) are, or were there, trees or hedges growing in the area of proposed construction?
(d) is there evidence of changes in ground level (e.g. by placement of fill), or of the demolition of old structures?

2. *Ground conditions*
(a) what geological strata lie below the site, and how thick are they?
(b) what problems are known to be associated with this geological context?
(c) is the site covered by Alluvium, Glacial Till (Boulder Clay) or any other possibly soft deposits?
(d) is there available information on the strength and compressibility of the ground?
(e) is the subsoil a shrinkable clay?
(f) does experience suggest that groundwater in these soil conditions may attack concrete?
(g) is there evidence of landslipping either on or adjacent to this site or on similar ground nearby?
(h) is there, or has there ever been, mining or quarrying activity in this area?
(i) are there coal seams under the site?

3. *The proposed structure*
(a) what area will the buildings occupy?
(b) what foundation loading is expected?
(c) how sensitive is the structure likely to be to differential foundation movements?
(d) what soils information is required for the design of every likely type of foundation?
(e) is specialist geotechnical skill required?

___

formation on past features which can affect the building work. A structured report is then produced from the information gathered at the site and from local enquiries.

Methods of Soil Investigation

A number of methods can be used to determine the soil conditions. For simple residential buildings on reasonable sites digging several holes about three spits deep, drilling holes up to 2 or 2.5 m deep with a hand auger, or driving a pointed steel bar about 1.5 m into the ground, are generally sufficient. With larger buildings or more difficult sites, the following methods are applicable

(1) Excavating trial holes about 1.5 m deep outside the perimeter of the building.

(2) Drilling boreholes by percussion or rotary methods. The percussion method uses a steel bit with a chisel point screwed to a steel rod. The rotary method employs a hollow rod with a rotating bit and a core of strata is forced back up the hollow rod.

(3) Load testing such as by applying a compressive load on a platform which exerts a recorded pressure on a steel plate between 300 and 1000 mm diameter placed on the surface of the bottom of a trial pit, as described in BS 5930.[4]

More information on trial pits is given in BRE Digest 381 (1993).

Classification of Soils

Soils can conveniently be classified under five main headings

(1) rocks, which include igneous rocks, limestones and sandstones;
(2) cohesive soils, such as clays where the constituent fine-grained particles are closely integrated and stick together;
(3) non-cohesive soils, such as gravels and sands, whose strength is largely dependent on the grading and closeness of the coarse-grained particles;
(4) peat, which is decayed vegetable matter of low strength with a high moisture and acidic content;
(5) made ground which may contain waste of one kind or another and can cause settlement problems.

Soils are defined in BS 1377[5] and this British Standard also prescribes methods of testing soils.

Serious problems can arise from a variety of ground conditions on the site as illustrated in table 1.2.

GROUNDWATER DRAINAGE

On low-lying or damp sites, it is advisable to provide an effective system of groundwater drainage, as described in the BS Code of Practice for building drainage.[6] The Building Regulations 1991 (paragraph C3 of Schedule 1) require subsoil drainage to be provided if it is needed to avoid (a) the passage of ground moisture to the interior of the building, or (b) damage to the fabric of the building. This would apply, for example, where the water table can rise to within 0.25 m of the lowest floor of the building.

Table 1.2   Ground problems and low-rise building
(Source: BRE Digest 318[2])

Differential settlement or heave of foundations or floorslabs
- soft spots under spread footings on clays
- growth or removal of vegetation on shrinkable clays
- collapse settlements on pre-existing made ground
- mining subsidence
- self-settlement of poorly compacted fill
- floorslab heave on unsuitable fill material

Soil failure
- failure of foundations on very soft subsoil
- instability of temporary or permanent slopes

Chemical processes
- groundwater attack on foundation concrete
- reactions due to chemical waste or household refuse

Variations during construction
- removal of soft spots to increase depth of footings
- dewatering problems
- piling problems

Groundwater is the portion of the rainwater which is absorbed into the ground and drainage of it may be necessary for the following reasons

(1) to increase the stability of the ground;
(2) to avoid surface flooding;
(3) to alleviate or avoid dampness in basements;
(4) to reduce humidity in the immediate vicinity of the building.

The standing level of the groundwater (water table) will vary with the season, the amount of rainfall and the proximity and level of watercourses. Main groundwater drains should follow natural falls wherever possible and should be sited so as not to endanger the stability of buildings or earthworks.

Systems of Groundwater Drainage

The most commonly used systems embrace porous or perforated pipes or gravel-filled trenches (French drains) laid to one of the following arrangements

(1) *Natural*. The drains follow natural depressions or valleys on the site with branches discharging into the main pipe (figure 1.1.1).
(2) *Herringbone*. There are a number of main drains into which small subsidiary drains discharge from both sides. The subsidiaries run parallel to each other but at an angle to the main drains, and should not exceed 30 m in length (figure 1.1.2). The spacing of the subsidiary drains varies with the type of soil and where the main drains are 600 to 900 mm deep, the

spacing of the subsidiaries might be within the following ranges

| Sand | 30 to 45 m | Sandy clay | 11 to 12 m |
| Loam | 23 to 26 m | Clay | 8 to 9 m |

(3) *Grid*. A main drain or drains are laid near the boundaries of a site into which branches discharge from one side only (figure 1.1.3).
(4) *Fan*. The drains converge to a single outlet at one point on the boundary of the site without the use of a main drain (figure 1.1.4).
(5) *Moat or cut-off system*. Drains are laid on one or more sides of a building to intercept the flow of groundwater and thereby protect the foundations (figure 1.1.5).

The choice of system will depend on site conditions. A common arrangement is 100 mm main drains with 75 mm branch drains, 600 to 900 mm deep with gradients determined by the fall of the land. The outlet of the groundwater drainage system will discharge into a soakaway or through a catchpit into the nearest ditch or watercourse (figures 1.1.6 and 1.1.7). Other forms of soakaway construction including trench type and the method of determining soil infiltration characteristics are detailed in BRE Digest 365.[7] Where these are not available groundwater drains may be connected, with the approval of the local authority, through a silt trap to the surface water drainage system. If connected to a foul drain a reverse-action intercepting trap would be required.

In clay or heavy soils, *mole drains* provide an alternative to pipes. These are formed by drawing a steel cartridge through the ground by a strong, thin blade attached to a mole plough driven by a tractor. Mole diameters range from 50 to 150 mm, at depths varying from 300 to 750 mm and spacings from 3 to 5 m. Their life is rather limited.

Groundwater Drains

A variety of materials is available as follows

(1) Clayware field pipes to BS 1196[8] of unglazed, porous pipes with square ends (butt jointed) – 65 to 300 mm nominal bores.
(2) Clay surface water drainpipes to BS 65,[9] non-porous and laid with open joints. Fully and half-perforated pipes are also available – 75 to 900 mm nominal bores.

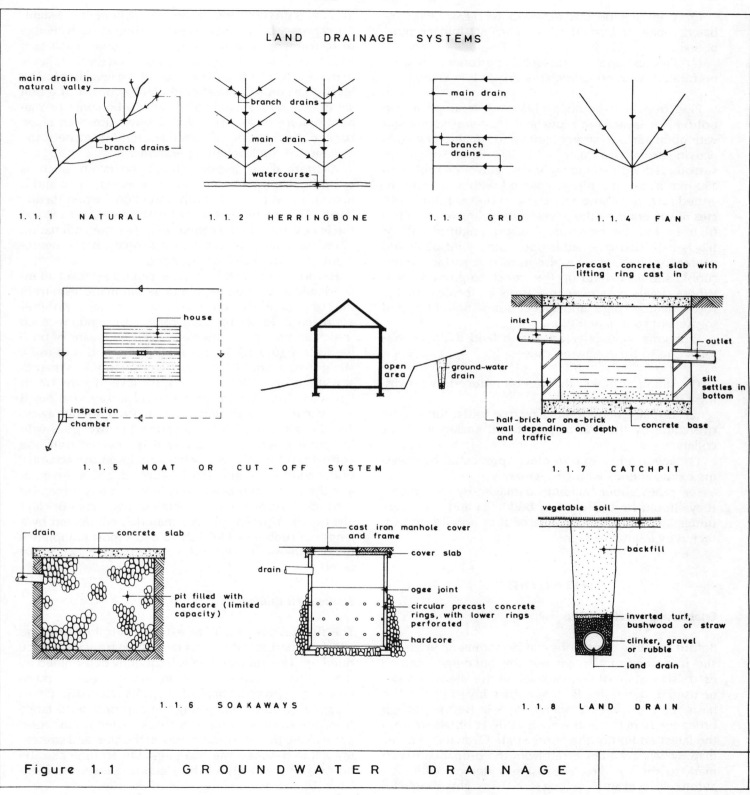

LAND DRAINAGE SYSTEMS

1.1.1 NATURAL

main drain in natural valley
branch drains

1.1.2 HERRINGBONE

branch drains
main drain
watercourse

1.1.3 GRID

main drain
branch drains

1.1.4 FAN

1.1.5 MOAT OR CUT-OFF SYSTEM

house
inspection chamber

open area
ground-water drain

1.1.7 CATCHPIT

precast concrete slab with lifting ring cast in
inlet
outlet
silt settles in bottom
half-brick or one-brick wall depending on depth and traffic
concrete base

1.1.6 SOAKAWAYS

drain
concrete slab
pit filled with hardcore (limited capacity)

cast iron manhole cover and frame
cover slab
drain
ogee joint
circular precast concrete rings, with lower rings perforated
hardcore

1.1.8 LAND DRAIN

vegetable soil
backfill
inverted turf, bushwood or straw
clinker, gravel or rubble
land drain

Figure 1.1 GROUNDWATER DRAINAGE

(3) Concrete porous pipes to BS 1194[10] with rebated, ogee or butt joints – 75 to 900 mm nominal bores.

(4) Plastics pipes to BS 4962,[11] perforated or non-perforated, with outside diameters up to 400 mm.

Pipe trenches should be just wide enough at the bottom for laying the pipes and they are mainly laid with open joints to straight lines and suitable gradients (varying from 1 in 80 to 1 in 300). Pipes should be surrounded with suitable clinker, gravel or rubble to 150 mm above the pipes, covered with a layer of inverted turf, brushwood or straw to prevent fine particles of soil entering the pores of the filling or the pipes (figure 1.1.8). On occasions, a short length of slate or tile is provided over each open pipe joint as added protection. If the drain is also to receive surface water, rubble should extend up the trench to ground level. Where drains pass near tree roots or through hedge roots, jointed spigot and socket pipes should be used to prevent root penetration.

Problems sometimes arise with land drainage systems due to the following causes

(1) silting of pipes: this can be reduced by providing catchpits (figure 1.1.7);

(2) displacement of pipes by tree roots: this can be overcome by the use of spigot and socket joints or collars;

(3) access of vermin to pipes: prevented by covering exposed ends with wire gratings;

(4) pipes under buildings damaged by settlement: they should be laid clear of buildings and if passage under a building is unavoidable they should be protected by a lintel or arch.

SETTING OUT

Establishing Levels on Site

Before building operations can be commenced on the site it is necessary to set out the building(s) and to establish a point of known level on the site which can be used to determine floor and drain invert levels. The basis for the levelling operations will be the nearest ordnance bench mark, whose value is obtained from the latest edition of the appropriate Ordnance Survey map. Levels are transferred from the ordnance bench mark to the building site using a dumpy, tilting or automatic level and levelling staff. The sights should be as long as possible and preferably be kept equidistant. A temporary bench mark is established at each change of instrument on a permanent fixed point such as a road kerb or top of a boundary wall, and this is suitably marked with a knife and detailed in the level book. The level point on the site should ideally consist of a bolt set in concrete or could alternatively be some permanent feature on the site. Flying levels are then taken back to the ordnance bench mark (OBM) to check the accuracy of the levelling operations.

All normal precautions should be taken, such as ensuring that the levelling staff is fully extended and is held truly vertical in both directions when taking levels. The instrument must be accurately set up and the levels properly recorded on either the collimation or rise and fall method. It is essential for the instrument to be in proper adjustment.

Foundation trench level pegs can be established individually with a dumpy, tilting or automatic level from the temporary bench mark on the site. Another approach is to set up sight rails at the ends of each trench and to fix intermediate levels by means of boning rods (figure 1.2.1). Yet another method is to use a straight-edge about 2.5 to 3 m long and not less than 25 × 150 mm in section, and a spirit level (figure 1.2.2). Peg A is driven in to the correct level and another peg B is driven in almost the length of the straight-edge away. The straight-edge is placed across the two pegs with the spirit level on it and peg B is lowered until the correct level is achieved. After checks on the accuracy of the spirit level and straight-edge, peg C is driven in and the levelling process repeated. A quicker method is to use a water level consisting of two tubes of clear glass or transparent plastics material, connected by a length of rubber or PVC tubing and almost completely filled with water. The water surfaces give two equal levels.

Setting Out Buildings

The first task is usually to establish the building line which demarcates the outer face of the front wall of the building. The building line is frequently determined by the highway authority and in urban areas is often around 8 m from the back of the public footpath. Often it can be fixed by measuring the appropriate distance from the highway boundary (back of the public footpath) along the side boundaries of the site, and stretching a line between the two pegs. On an infill site the line may be established by sighting between the existing buildings on either side of the site.

The next step will be to position the front of the building on the building line by checking the dimensions between the new building and the side boundaries. Flank walls will then be set out at right-angles to the building line often using a large builder's or timber square or the 3:4:5 method (figure 1.2.3). The builder's square is a right-angled triangular timber frame with sides varying in length from 1.50 to 3.00 m. The square is placed against the building line and two pegs are driven in on the return side. By sighting across the two pegs a third peg can be driven in the same straight line, and a bricklayer's line stretched between them.

If a builder's square is not available, a right-angle can be set out based on a right-angled triangle whose sides are in the ratio of 3:4:5 (derived from the theorem of Pythagoras). A peg is first driven in at the corner of the building and a distance of 3 m is measured back along the building line. A peg is driven in at this point and the ring of the tape is placed over a nail driven into the top of the peg. The tape is held at the 12 m mark against the ring on the first peg and with the tape around the corner peg, the tape is stretched out to give the position of the third peg at the 7 m mark. The line extended through the third peg is at right-angles to the building line.

To secure permanent line markers, profiles are established at the corners of the building and at wall intersections (figure 1.2.4). They often consist of boards about 25 × 150 mm or 25 × 100 mm nailed to 50 × 50 mm posts or pegs driven firmly into the ground. Saw cuts or nails demarcate the width of walls and the spread of foundations and the bricklayer sets his lines to these and, when working below ground, plumbs down from them at the corners of the building. A radius rod worked from a centre pivot can be used to set out curved bays, whilst a framed profile would be made up for a cant bay.

The setting out of a steel-framed building requires extreme accuracy as the stanchions and beams are cut to length at the fabricator's works, and any error in setting out can involve expensive alterations on site. It is usual to erect continuous profiles around the building and to set out the column spacings along them. Wires stretched between the profiles will give the centres of stanchions around which templets are formed.

Building Research Establishment Digest 234[12] emphasises the need to attain predictable standards of accuracy in the setting out and erection processes, resulting from the increase in the partial or complete prefabrication of building structures, and methods for their achievement are described in BS 5606.[13] Errors in linear measurement may arise through reading, calibration or tension errors. However, if corrections are made for slope, the maximum error is likely to be about 25 mm in a 30 m length. Basic precautions include keeping the tape level to reduce wind effect, pulling tight enough to remove kinks but not to stretch, keeping the tape clean, checking periodically against an engineer's steel tape and obtaining a refill if damage occurs.[14]

CONTROL OF BUILDING WORK

Proposals for most new buildings, extensions and material alterations or changes of use of existing buildings are subject to the requirements of the Building Regulations 1991; as is also the installation of cavity wall insulation, underpinning, sanitary equipment, drainage, unvented hot water systems and fixed heating appliances in which fuel is burnt. Unvented hot water systems are under pressure and eliminate the need for a storage tank and overflow, and provide greater flexibility in the siting and operation of the system. In addition, approval is often required under the Town and Country Planning Acts, with particular reference to appearance, layout, access and use, as described later in the chapter.

Building Regulations

Application of the Building Regulations means that the appropriate local authority has to be notified and the building work will have to comply with the Regulations. The main purpose of the Regulations is to ensure the health and safety of people in or about the building, and they are also concerned with energy conservation and access to buildings for the disabled. The client may choose either the local authority or an approved inspector to supervise the work.

Where the client opts for local authority supervision, he has a further choice of depositing full plans or submitting a much less detailed building notice, and a fee is payable to the local authority. The local authority can prosecute if work is started without taking either course of action. Where an approved inspector is selected, the client and the inspector must jointly give the local authority an initial notice accompanied by a site plan. Work must not be commenced before the notice has been accepted by the local authority. The inspector's fee is negotiable.

The two alternative procedures are now considered in more detail.

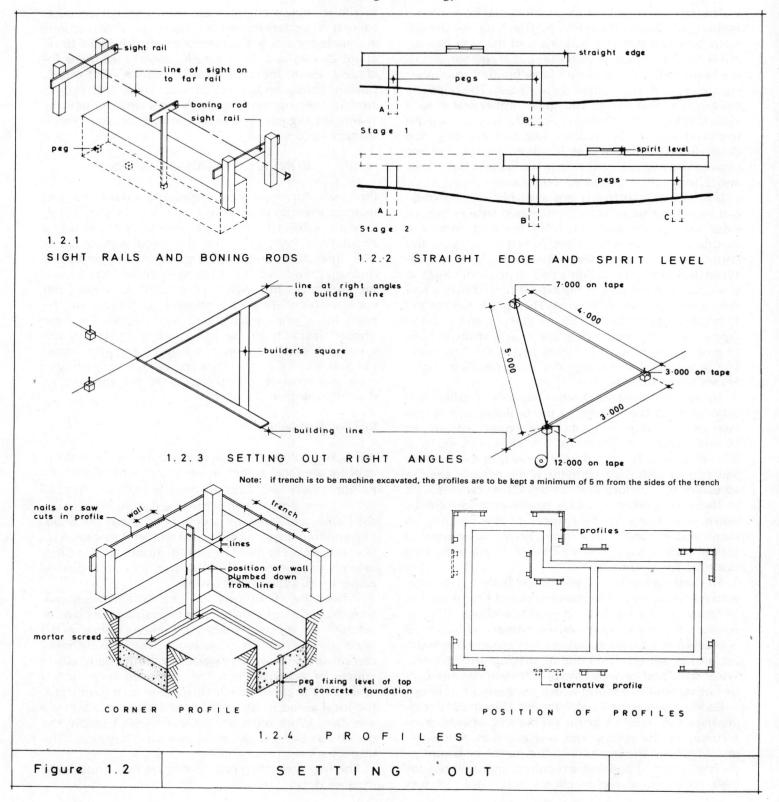

1.2.1
SIGHT RAILS AND BONING RODS

sight rail

line of sight on to far rail

boning rod

sight rail

peg

1.2.2 STRAIGHT EDGE AND SPIRIT LEVEL

straight edge

pegs

A

B

Stage 1

spirit level

pegs

A

B

C

Stage 2

1.2.3 SETTING OUT RIGHT ANGLES

line at right angles to building line

builder's square

building line

7·000 on tape

4·000

5·000

3·000 on tape

3·000

12·000 on tape

Note: if trench is to be machine excavated, the profiles are to be kept a minimum of 5 m from the sides of the trench

nails or saw cuts in profile

wall

trench

lines

position of wall plumbed down from line

mortar screed

peg fixing level of top of concrete foundation

CORNER PROFILE

profiles

alternative profile

POSITION OF PROFILES

1.2.4 PROFILES

Figure 1.2　　　　　　SETTING OUT

(1) *Local authority control*

Where the proposal encompasses the erection of offices or shops, full plans must be deposited. Where full plans are deposited, the local authority may pass or reject them within 5 weeks, or 2 months if the client agrees. The plans must be accompanied by a certificate that the plans show compliance with the structural stability and/or energy conservation requirements of the Regulations. The local authority has to consult the fire authority about proposed means of escape in the case of certain factories, offices, shops, railway premises, hotels and boarding houses.

The local authority may pass plans subject to either or both of the following conditions

 (i) modifications in the deposited plans;
 (ii) the depositing of further plans.

Work may begin at any time after the submission of a building notice or deposited plans, provided the local authority is given 48 hours' notice. If the local authority considers that any work contravenes the requirements of the Regulations, it may serve a notice requiring demolition or alteration within 28 days.

(2) *Supervision by approved inspector*

The approved inspector and the client should jointly give the local authority an initial notice together with a declaration that an approved scheme of insurance applies to the work, which must be signed by the insurer. The initial notice must contain a description of the work and, in the case of a new building or extension, a site plan and information about drainage.

The local authority have 10 working days in which to consider the notice and may only reject it on prescribed grounds. On acceptance, the local authority may impose conditions. It is a contravention of the Regulations to start work before the notice has been accepted. As a general rule the approved inspector must be independent of the designer or builder, but he need not be if the work consists of alterations or extensions to one or two storey houses. The National House-Building Council is a major provider of private building control services, alongside practising professionals.

Where a client wishes to have detailed plans of work certified as complying with the Building Regulations, he should ask the approved inspector to supply a plans certificate and the local authority also receives a copy. When the work is complete the approved inspector should give the client and the local authority a final certificate.

Unlike a local authority, an approved inspector has no direct power to enforce the Building Regulations. He is, however, required to inform the client if he believes that any work being carried out under his supervision contravenes the Regulations. If the client fails to remedy the alleged contravention within 3 months he is obliged to cancel the initial notice. He must also inform the local authority of the contravention, unless a second approved inspector is taking over responsibility.

Approved Documents

The Building Regulations are supported by Approved Documents which give guidance about some of the ways of meeting the requirements of the Regulations, and they will be referred to individually in appropriate chapters of the book. The client can choose whether or not to use the Approved Documents.

The technical solutions in the Approved Documents describe some of the more widely used forms of construction which achieve an acceptable level of performance and yet, at the same time, permit adequate flexibility. Alternative approaches are usually based on the recommendations of a British Standard or by using a product with a British Board of Agrément Certificate or other European Organisation for Technical Approvals (EOTA) certification.

Inspection

Where the building work is subject to supervision by the local authority, it must receive notice of commencement and completion of certain stages of the work. The builder is required to supply the local authority with not less than 48 hours' notice in writing of the commencement of the work, and 24 hours' written notice before the covering up of any excavation for a foundation, any foundation, damp-proof course, concrete or other material laid over a site; and before any drain or private sewer is haunched or covered; and not more than five days after the laying of any drain or private sewer. If the builder neglects to give any of these notices, he may be required to cut into, lay open or pull down so much of the building work as prevents the local authority from ascertaining whether any of these regulations have been contravened. Finally, the builder is required to give the local authority written notice not more than five days after completion of the building (excluding any Saturday, Sunday, Bank holiday or public holiday).

British Standards

It should be noted that many building materials and components and constructional techniques are the subject of British Standards. These documents are issued by the British Standards Institution, which is a private body representing all interests with financial support from the government. The work of the Institution is mainly performed through technical committees. The use of these standards ensures a good standard of materials and workmanship, and they embrace the results of latest research. Apart from satisfying Building Regulations, the use of British Standards reduces the work involved in specification writing and makes for greater uniformity in requirements, which helps the builder. Where a British Standard covers several grades or classes of material, it is essential that the required grade or class shall be stated. British Standard Codes of Practice describe in detail the best practical methods to be used and designs to be adopted. They are, however, being replaced with British Standards as and when they are revised. In the years ahead British Standards will probably be replaced by Eurocodes.

Agrément Certificates

Many new materials and products have been produced in the last three decades and it became vitally necessary to provide an effective means of assessing their likely performance. In 1966, the government set up the Agrément Board to assess each product and its performance for the prescribed use, and to devise and carry out appropriate tests. On satisfactory completion of its investigations, the Board issues an agrément certificate which contains details of the proprietary product, the design data appropriate to its use, site handling and subsequent maintenance. The British Board of Agrément (BBA) is a member body of the European Organisation for Technical Approvals (EOTA).

Improvement Lines and Building Lines

Section 73 of the Highways Act 1980 provides for the highway authority to prescribe an improvement line on either or both sides of a street to which the street is to be widened, in front of which no new building shall be erected, and limits compensation for buildings erected after the prescription of the line. Section 74 of the same Act provides similarly for the prescription of a frontage line for new buildings on either or both sides of a highway.

Planning Consent

Planning applications must be submitted where it is proposed to carry out development which requires permission under the Town and Country Planning Acts. Such applications are usually required to be submitted in triplicate accompanied by plans, and also accompanied by a fee.

'Development' is defined in the Town and Country Planning Act 1990 as 'the carrying out of building, engineering, mining or other operations in, on, over or under land, or the making of any material change in the use of any buildings or other land.'

Certain works do not constitute development and do not therefore require planning consent. Typical examples are works of maintenance, improvement or other alteration of a building which affect only the interior of the building, or which do not materially affect its external appearance, the change of use of a building to another use within the same class as listed in the Town and Country Planning (Use Classes) Order 1987, and maintenance or improvement works to roads by a local highway authority. On the other hand, the conversion of a single dwelling into two or more separate dwellings, the deposit of refuse or waste material and the display of advertisements generally require planning permission. Limited extensions to dwelling houses, boundary fences and walls within certain limits of height, painting the exteriors of buildings and erection of contractors' site huts are classified as 'permitted development' and no planning permission is needed, but this list is not exclusive.

A prospective developer often wishes to know whether the development he has in mind is likely to receive planning permission, before he purchases the land or incurs the cost of preparation of detailed plans. In these circumstances, the developer is able to submit an *outline application*, which entails merely the submission of an application form and a site plan. Consent may be given subject to subsequent approval by the local planning authority of any matters relating to siting, design, external appearance, means of access and landscaping of the site. In this way the applicant can obtain officially the local planning authority's reaction to his proposals, without the necessity of preparing comprehensive plans.

All planning applications are recorded in a planning register, which is available for inspection by the public. The entries in the register will include details of the applicant and agent (if any), nature and location of proposed development, planning decision and result of appeal (where applicable). The local planning authority has three choices

(1) to grant permission unconditionally;
(2) to grant permission subject to conditions;
(3) to refuse permission.

Decisions must include full reasons for refusal or for any conditions attached to a permission, and an aggrieved applicant has the right of appeal to the Secretary of State.

Readers requiring more detailed information on planning control are referred to Telling[15] and Heap.[16]

Other Statutory Requirements

New factory buildings are subject to various statutory controls, apart from Building Regulations, by virtue of the provisions of the Factories Act 1961 and the Clean Air Acts 1956 and 1968, as amended by the Local Government, Planning and Land Act 1980. New offices and shops have to comply with the Offices, Shops and Railway Premises Act 1963. A number of statutes prescribe minimum provision in relation to means of escape in case of fire and fire-fighting appliances, including the Fire Precautions Act 1971, while hotels and restaurants are examined under the Food Hygiene Regulations.

Other statutory provisions affecting building work include the Local Government, Planning and Land Act 1980, the Defective Premises Act 1972, the Health and Safety at Work Act 1974, the Construction Regulations and the CDM new construction (design and management) regulations which would extend the responsibility for health and safety from contractors to designers and clients, probably from 1995 onwards, although there were concerns over the high cost of operation in 1994.

SEQUENCE OF BUILDING OPERATIONS

The building client may be a public authority, a private organisation or even an individual. Where an architect is engaged, the client supplies him with a brief of his basic requirements which should be as comprehensive as possible. The architect will inspect the site and consult with the local authorities having planning and building control functions, and with the various statutory authorities that will provide services to the site. He will then prepare preliminary sketch drawings for consideration by the client and from which a quantity surveyor can prepare an approximate estimate of cost. On the larger projects, the quantity surveyor is likely to produce a cost plan as a basis for the effective control of the cost of the project throughout the design stage. The architect will proceed with the working drawings, and quotations will be obtained from sub-contractors and statutory undertakers, in addition to the various consents and approvals necessary. In the traditional contract approach, the specification and bill of quantities will be formulated and together with the drawings and conditions of contract will be sent to building contractors for the submission of tenders. The tenders are examined by the quantity surveyor who submits a report to the client through the architect and a contractor is selected. In some circumstances it is preferable to negotiate a price with a single contractor. Many other forms of building procurement are now available as detailed by Turner,[17] and include design and build, construction management, management contracting, and design and manage.

Following acceptance of his tender the contractor prepares a programme, orders materials and starts work on the site. His first task is to install certain vital temporary works, such as access roads, hoardings, site huts, storage compound and temporary services. Most of the temporary works are removed by the time the permanent work is complete. The contractor is responsible for the organisation of all the work on the site including that of nominated sub-contractors.

The building will be set out on the site in accordance with the dimensions and levels supplied by the architect. Topsoil is stripped and probably stockpiled on the site. Foundation trenches are excavated and levelled, following which pegs are driven into the bottoms of the trenches with their tops delineating the finished surface of the foundation concrete, which is then poured. Bricklaying follows starting at the quoins and stringing lines between them, and concrete oversite laid. Brickwork is levelled to receive the dampproof course and the erection of walls continues with the fixing of external door frames and windows as the work proceeds.

The carpenter builds in first floor joists and follows with the roof timbers. The roofer covers the roof to complete the carcassing. Partitions are erected and floor finishes follow after the insertion of services. Then comes the internal joinery in doors and staircases, followed by plastering to walls and ceilings, fireplaces and joinery fittings. The final stages of the project embrace the completion of services, decorations and cleaning of floors and windows, external works including landscaping, and removal of surplus materials from the site.

In winter it is necessary to take certain precautions in order to maintain an acceptable level of productivity. Important precautions include good internal access roads, protection of areas to be excavated when frost is anticipated, protection of water services from frost, care of plant, erection of temporary shelters or framed enclosure, normally covered with suitable polythene sheeting to working areas and heated where necessary, and adequate protection of all new work during exceptionally adverse weather conditions.

The architect is responsible for ensuring that the work is built in accordance with the contract documents, and on larger projects will be assisted by a clerk of works who stays on the site. The quantity surveyor will make periodic valuations of the completed work and materials brought onto the site and the architect will then issue certificates authorising the client to make payments to the contractor. After completion of the work, the quantity surveyor will agree the final account with the contractor. At the end of the defects liability period, usually six months after completion, the architect will prepare a schedule of defects for which he considers the contractor is responsible. The contractor will not receive payment of the outstanding balance of retention money until these defects have been remedied.

PROBLEMS IN DESIGN AND CONSTRUCTION OF BUILDINGS

Principles of Design

Before we consider design and constructional problems it would probably be helpful to examine the elements or principles of design. A prime objective in building design is to secure an attractive building – one of high aesthetic value. It will also be appreciated that buildings serving different purposes tend to assume different forms. Hence public buildings often take a symmetrical form to emphasise formality, order and dignity, whilst individual houses are often quite informal. Character is derived from the composition as a whole.

The relationship of the various units (such as doors, windows, plinths and pilasters) to each other and to the whole building is termed 'proportion', and their relationship in size is described as 'scale'. A well-designed building will always be well proportioned. Much of the beauty of older buildings stems from the use of local materials which generally weather to attractive colours. An effort should be made to harmonise with adjoining buildings by sympathetic choice of colours and to obtain an attractive street picture. The texture of the materials should also be considered; in general rougher surfaces have a more interesting texture with variations in colour which mellow over a period of time.

Designing a building is essentially a matter of making a long series of choices – choices about ends and choices about means.[18] Both client and designer are concerned with the choice of ends, embracing the purpose and character of the building and the client's special requirements. The means are solely the responsibility of the designer and will include the detailed planning, structural form, services and finishes; the completed design represents a set of instructions for the erection of the building.

The architect is concerned with providing a building which will satisfy the client's needs. In performing this task he will, however, be obliged to have regard to such factors as maximum use of land, cost, availability of labour and materials, technological constraints, planning and building regulations, relationship with other buildings, landscaping, services, circulation networks, and fire and noise prevention, and desirably be environmentally friendly.

Even the design of a dwelling house can be quite complex if it is to effectively meet the changing needs of the occupants over a long period of time. In the last three decades there has been a change of approach to the design of housing accommodation, stemming from the government report *Homes for Today and Tomorrow*.[19] Accommodation standards generally cease to be based on minimum room sizes, and depend on functional requirements and levels of performance, with minimum overall sizes for the dwelling related to the size of family. There should for instance be space for activities requiring privacy and quiet, for satisfactory circulation, for adequate storage and to accommodate new household equipment, in addition to a kitchen arranged for easy housework and with sufficient room in which to take at least some meals. The report includes recommendations for the provision of sanitary appliances, kitchen fitments, bedroom cupboards, electric-socket outlets and minimum heating standards. In 1980 the Government proposed modifications to the Parker Morris standards, described earlier, which were considered unduly restrictive.

Design Problems

Until the 1930s architects were seldom tempted to depart from traditional construction, and they tended to rely on inherited specifications which were based on the known effects of time, wear and weathering on local materials. Today the designer is faced with a wide range of both old and new materials and components, necessitating careful thought to design details to avoid unsatisfactory results.

The Building Research Establishment[20] has instanced cases where accelerated deterioration and/or unsightly appearance has resulted from the unsuitable placing of incompatible materials or inadequate attention to design details. Streaking by rainwater washings over walling, resulting from lack of suitable projecting features with adequate drips at the head of walls to buildings with flat roofs is a typical example.

Changes in appearance of materials used externally in buildings may result from any of the following three causes

(1) Preventable faults in design and/or construction where there is sufficient information available at the time to avoid them.

(2) Unforeseeable behaviour of a material in a given situation where it could reasonably be argued that there was insufficient information generally available at the time, and where the change may not be wholly explicable in the light of existing knowledge.

(3) The process of 'natural' weathering, often as the result of microclimatic influences, frequently unpredictable in detail.

Some materials are commonly believed to be maintenance-free but this is rarely so in practice. For instance, untreated teak and western red cedar often become unsightly, particularly in urban situations. An annual application of a linseed oil/paraffin wax mixture containing a fungicide or a suitable wood stain is the minimum necessary to preserve appearance. Aluminium used externally requires anodic treatment, followed by periodic washing, to maintain a satisfactory appearance.[21]

The extensive use of system building and timber frame houses generated a new set of problems, some of which will be described in chapter 4.

Building Regulations Approved Document M (1991) covers access and facilities for disabled people to which more attention is now being paid in the design of buildings.

Building Maintenance

Building maintenance work uses extensive resources of labour and materials and in 1985 the backlog on the maintenance of the public housing stock in England was estimated by the Urban Housing Renewal Unit[22] at £18.8b (approximately £5000 per house). By 1994 the total national housing repair needs were probably in the order of £50b. Hence it is vitally important that the probable maintenance and running costs of a building should be considered at the design stage, and adequate attention directed towards the maintenance implications of alternative designs. A reduction in initial constructional costs often leads to higher maintenance and running costs. The cheapest heating system in installation costs is often the most expensive to operate.[23]

Ransom[24] shows the findings of a Building Research Advisory Service survey of defects in 1975 which gave a breakdown of 24 per cent rain penetration, 16 per cent condensation, 5 per cent entrapped moisture, 5 per cent other causes of damp, 18 per cent cracking, 15 per cent detachment and 17 per cent miscellaneous items.

Performance Standards

New buildings have to meet the performance standards prescribed by the Building Regulations 1991.[25] For instance the strength and stability of a building is covered by part A of the Building Regulations, and site preparation and resistance to moisture in part C, as it is vital that any building shall be wind and weather-tight. With the greater provision of central heating, thermal insulation assumes greater importance and provisions relating to the conservation of fuel and power are contained in part L of the Building Regulations, whilst as the volume of noise increases and occupiers of buildings become more susceptible to noise, so the provision of adequate sound insulation (part E) becomes important. The inclusion of minimum standards of resistance to spread of fire, adequate means of escape in case of fire and adequate facilities for the fire service are essential and these are detailed in part B of the

Building Regulations. Occupants of buildings want to secure ample hygiene and effective drainage and waste disposal and heat-producing appliances, and minimum standards for these are scheduled in parts G, H and J of the Building Regulations. More detailed reference will be made to many of these regulations in later chapters.

Quality Assurance

Griffith[26] has described how in recent years, increasing concern has been expressed at the frequent low standards of performance and quality achieved in UK building work, which highlighted the need for structured and formal systems of construction management to improve performance, workmanship and quality.

BS 5750,[27] with its origins in manufacturing industry, prescribes the national standard for quality assurance systems in the UK. This standard is presented in six parts: part 1 covers the specification for design/development, procedure, installation and servicing; part 2 deals with the specification for production and installation; part 3 encompasses the specification for final inspection and test; while parts 4, 5 and 6 provide guidance for the practical implementation of the quality assurance systems specified in parts 1, 2 and 3.

The CIOB saw quality assurance as an objective demonstration of the builder's ability to produce building work in a cost-effective way to meet the customer's requirements,[28] while the RICS considered that it involved a management process designed to give confidence to the client by consistently meeting stated objectives.[29]

Griffith[26] has aptly described how quality assurance is concerned with developing a formal structure, organisational and operational procedure to ensure good quality throughout the total building process. Quality is a measure of fitness for purpose and assurance relates to the assessment and recognition of an organisation's quality management system by an independent assessor, termed the certification body. Individual companies develop their own quality systems to the guidelines prescribed by BS 5750.[27]

The basic requirement of quality assurance within the construction industry is to give clients confidence in the ability of designers, contractors, suppliers and other parties to meet their requirements. Hence quality assurance certification must include an assessment of the scope of management experience and efficiency and the technical and financial competence of all parties involved in the building process.[26]

Quality assurance can be applied to the following five groups of participants in the construction industry

(1) client in the project brief;
(2) designer in the design and specification;
(3) manufacturers in the supply of materials, products and components;
(4) contractors and subcontractors in construction, supervision and management;
(5) user in the use, upkeep and repair of the new building.

Supervision of the work in progress is a vital part of quality planning and implementation and BS 8000,[30] covering most of the traditional crafts, aims to encourage good workmanship by providing:

(1) most frequently required recommendations on workmanship for building work in a readily available and convenient form to those working on site;
(2) assistance in the efficient preparation and administration of contracts;
(3) recommendations on how designers' requirements for workmanship may be realised satisfactorily;
(4) definitions of good practice on building sites for supervision and for training purposes, although this is not intended to supplant normal training in craft skills;
(5) reference for quality of workmanship on building sites.

However, criticisms have been levelled at this British Standard, with some justification, on the code being unduly costly, cumbersome and repetitive.

Dimensional Co-ordination

BS 4011[31] recommends that the first selection of basic sizes for the co-ordinating dimensions of components should be in the order of preference, n by 300 mm; n by 100 mm; n by 50 mm; and n by 25 mm, where n is equal to any natural number including unity. It is essential to relate components to a grid if they are to fit into the space planned for them within a building, without the need for trimming and cutting. Also, by using them in accordance with a grid, it is possible to reduce the variety of components and increase standardisation and rationalisation in building. The grid lines can be based on the centre lines of walls and columns or on wall and column faces. Axial grids are most suitable for framed buildings and face grids are better suited for buildings with load-bearing walls.

Key reference planes constitute load-bearing walls, columns and upper and lower surfaces of floors and roofs. The spaces between these key reference planes are termed *controlling zones* and the distances between these zones are called *controlling dimensions*. The recommended sizes of controlling dimensions are prescribed in BS 4330.[32] Intermediate key reference planes control the position of the heads and sills of windows and heads of door sets. BS 6100[33] distinguishes building reference systems and defines dimensions and dimensional and modular co-ordination.

ENVIRONMENTALLY FRIENDLY BUILDINGS

Apart from statutory requirements for buildings, there was in the 1990s a strong and increasing awareness of the urgent and vital need to produce more environmentally friendly buildings, and this received further momentum from the publication of the government White Paper on the Environment (This Our Inheritance) in 1990. A strong case can be made for developing energy efficient and environmentally friendly products suitably coded.

Chlorofluorocarbons (CFCs) are man-made chemicals mostly used in the building industry and they constitute a main cause in the thinning of the ozone layer and the 'greenhouse' warming of the earth, culminating in the Montreal Protocol in 1987.[34] The UK government expressed its intention to eliminate all sources of CFCs by year 2000 and to reduce significantly the proportion of halcon. BRE information paper IP 23/89[35] detailed how the use of CFCs can be minimised in the short term and completely replaced in the long term. The implications for users and specifiers of air conditioning equipment, and of using alternative insulation materials in roofing, walls, flooring and timber frame construction, are given, including technical risk issues.

The most important greenhouse gas is carbon dioxide (CO_2), the levels of which have risen by 25 to 30 per cent since 1840. The largest source of CO_2 results from burning fossil fuels to supply energy needs and, in the UK, over half of all energy production is consumed in buildings. Levinson *et al.*[36] identified the following three straightforward techniques which could assist significantly in producing environmentally friendly buildings

(1) to produce energy efficient buildings with a minimal need to burn fossil fuels or use expensive electricity, and generating large cost savings;

(2) to ensure that materials used in buildings are environmentally friendly, such as ceasing to use tropical hardwoods;

(3) to ensure that buildings are managed so that they continue to have a low environmental impact.

Developers of new buildings designed to be more environmentally friendly can seek recognition through a BRE environmental assessment method (BREEAM), which was introduced in 1990, starting with office buildings and extended to homes in 1991, to cover a range of issues affecting the global, neighbourhood and internal environments, and gives credits for each aspect of design where specific targets are met. Global issues encompass global warming, ozone depletion, rain forest destruction and resource depletion. Neighbourhood issues comprise Legionnaires' disease (from cooling towers), local wind effects and re-use of existing site. Indoor issues embrace Legionnaires' disease (from water supplies), lighting, indoor air quality and hazardous materials.[37]

Radon

Radon is a colourless, odourless gas which is radioactive. It is formed where uranium and radium are present and can move through cracks and fissures in the subsoil, and so into the atmosphere or into spaces under and in dwellings. Where it occurs in high concentrations it can be a risk to health.

The Building Regulations Approved Document C states that 'where a house or extension is to be erected in Cornwall or Devon, or parts of Somerset, Northamptonshire or Derbyshire there may be radon contamination of the site and precautions against radon may be necessary'. The BRE Report 211[38] gives detailed guidance on where protection is necessary and also contains practical constructional details.

There are two main methods of achieving radon protection in new dwellings: passive and active.

(1) The *passive* system consists of an airtight and therefore substantially radon-proof barrier, generally of 1200 gauge polyethylene (polythene) sheet, across the whole of the building including the floor and walls. This is the preferred method but may need to be supplemented by secondary protection, probably involving underfloor ventilation or subfloor depressurisation.

(2) The *active* system consists of a powered radon extraction system as an integral part of the services of the house, comprising a sump(s) which is usually ducted

internally with a fan outlet through the roof. It will incur running and maintenance costs for the life of the building.

Further guidance on the provision of radon sumps in existing dwellings is given in BRE report BR 227[39] and on radon surveys and their implementation in existing dwellings in BRE Report BR 250.[40]

REFERENCES

1. *BRE Digest 350*: Climate and site development (1990)
2. *BRE Digest 318*: Site investigation for low-rise building: desk studies (1987)
3. *BRE Digest 348*: Site investigation for low-rise building: the walk-over survey (1989)
4. *BS 5930: 1981* Code of practice for site investigations
5. *BS 1377: 1975* Methods of test for soil for civil engineering purposes
6. *BS 8301: 1985* Code of practice for building drainage
7. *BRE Digest 365:* Soakaway design (1991)
8. *BS 1196: 1989* Clayware field drains and junctions
9. *BS 65: 1988* Specification for vitrified clay pipes, fittings, joints and ducts
10. *BS 1194: 1969* Concrete porous pipes for under-drainage
11. *BS 4962: 1989* Plastics pipes and fittings for use as subsoil field drains
12. *BRE Digest 234*: Accuracy in setting-out (1980)
13. *BS 5606: 1978* Code of practice for accuracy in building
14. G. Hedley and C. Garrett. *Practical Site Management*. Godwin (1983)
15. A.E. Telling. *Planning Law and Procedure*. Butterworths (1993)
16. D. Heap. *An Outline of Planning Law*. Sweet and Maxwell (1991)
17. A. Turner. *Building Procurement*. Macmillan (1990)
18. *BRE Digest 12*: Structural design in architecture (1969)
19. Ministry of Housing and Local Government. *Homes for today and tomorrow* (Parker Morris report). HMSO (1963)
20. *BRE Digests 45* and *46*: Design and appearance (1964)
21. I.H. Seeley. *Building Maintenance*. Macmillan (1987)
22. Department of the Environment. *An Inquiry into the Condition of the Local Authority Housing Stock in England*. HMSO (1985)
23. I.H. Seeley. *Building Economics: appraisal and control of building design cost and efficiency*. Macmillan (1995)
24. W.H. Ransom. *Building Failures: diagnosis and avoidance*. Spon (1987)
25. *The Building Regulations 1991 and Approved Documents (1989, 1991 and 1995)*. HMSO
26. A. Griffith. *Quality Assurance in Building*. Macmillan (1990)
27. *BS 5750: 1987* Quality systems
28. CIOB. *Quality Assurance in Building* (1987)
29. RICS. *Quality Assurance: Introductory Guidance* (1989)
30. *BS 8000: 1990*. Workmanship on building sites
31. *BS 4011: 1966*. Recommendations for the co-ordination of dimensions in building. Co-ordinating sizes for building components and assemblies
32. *BS 4330: 1974*. Recommendations for the co-ordination of dimensions in building. Controlling dimensions
33. *BS 6100: 1984*. Co-ordination of dimensions, tolerances and accuracy
34. United Nations Environment Programme. *Montreal Protocol on substances that deplete the ozone layer* (1987)
35. BRE Information Paper IP 23/89. *CFCs and the Building Industry* (1989)/*BRE Digest 358*; CFCs in buildings (1992)
36. Levinson, DSSR and Gleeds. *A Guide to Environmentally Friendly Buildings* (1990)
37. BRE. First green assessment for buildings. *BRE News of Construction Research, August 1990*
38. BRE. *BR Report 211: Radon: guidance on protective measures for new dwellings* (1991, revised 1992)
39. BRE. *BR Report 227: Radon sumps: a BRE guide to radon remedial measures in existing dwellings* (1992)
40. BRE. *BR Report 250: Surveying dwellings with high indoor radon levels: a BRE guide to radon remedial measures in existing dwellings* (1993)

2 BUILDING DRAWING

Building drawing is important as it is often easier to explain building details by drawings or sketches than by written descriptions. Drawing thus forms an effective means of communication and drawings constitute an essential working basis for any building project. It is desirable to achieve maximum uniformity in the presentation of building drawings and this is assisted by implementing the recommendations contained in BS 1192: Construction drawing practice Part 1: 1984 Recommendations for general principles, supported by Part 2: 1987 Recommendations for architectural and engineering drawings and Part 3: 1987 Recommendations for symbols and other graphic conventions. This chapter is concerned with the various drawing materials in use and the basic principles to be observed in the preparation of drawings and sketches.

DRAWING INSTRUMENTS AND MATERIALS

Students are advised to purchase good-quality drawing instruments which can quite easily last a lifetime. It is possible to purchase complete sets with varying ranges of instruments or to buy instruments singly. A popular and useful set of instruments is illustrated in figure 2.1. *Compasses*. These are usually 125 or 150 mm in length and a lengthening bar (figure 2.1.6) permits the drawing of large radius curves. They should be designed to take a pencil point (figure 2.1.1), pen point (figure 2.1.4) or divider point (figure 2.1.5). Ideally they should be self-centring so that even pressure can be maintained on both points. Pen or ink points are often made from tungsten steel for use with plastic tracing film which is highly abrasive. In some cases the instruments are double knee-jointed and the points fitted with fine adjusting devices. The pencil should be sharpened to a chisel point and fitted tangentially to the circle.

Beam compasses. These are used for drawing circles and curves of extra large radius. They consist of attachments which are fastened to a wood lath or metal bar and have needle points and interchangeable pen and pencil points. They can be useful for plotting land surveys but are rarely needed for building drawings.
Spring bow compasses. These are used for drawing small radius circles and curves. They are obtainable as separate dividers, and pen and pencil compasses (figures 2.1.7, 2.1.8 and 2.1.9).

Many sets of drawing instruments also include the larger *divider compasses* shown in figure 2.1.2. These are useful for setting out larger divisions of equal length.
Ruling pens. These consist of a handle and two blades of equal length connected by an adjusting screw for varying the thickness of line (figure 2.1.3). One blade is normally hinged or pivoted for ease of cleaning. The ink is fed in between the blades and the pen is held in a vertical position and drawn along the edge of a tee square or set square. The pen must be carefully cleaned after use. These pens are, however, little used nowadays.

In recent years many draughtspersons have used 'Graphos', 'Rapidograph', 'Rotring–Isograph' and similar type pens which are a form of fountain pen that take Indian ink and have interchangeable heads or nibs for producing lines of different thicknesses. A Rotring–Isograph pen is illustrated in figure 2.1.10. They are also useful for sketching and writing. Some manufacturers have produced compasses which will take inter-changeable ink heads with ink containers.
Freehand pens. These are useful for producing small lettering such as explanatory or descriptive notes on drawings of building details. Larger lettering can be obtained with a Gillott Nr 303 or similar nib attached to a standard holder, while mapping pens with fixed or interchangeable nibs are useful for small lettering.

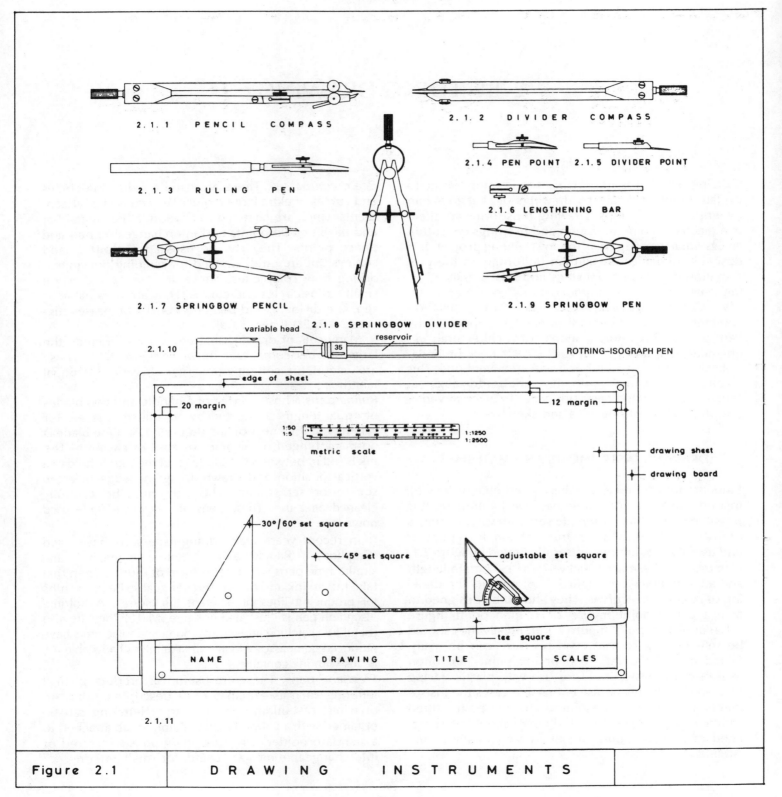

2.1.1   PENCIL   COMPASS

2.1.2   DIVIDER   COMPASS

2.1.3   RULING   PEN

2.1.4 PEN POINT   2.1.5 DIVIDER POINT

2.1.6 LENGTHENING BAR

2.1.7 SPRINGBOW   PENCIL

2.1.9 SPRINGBOW   PEN

2.1.8 SPRINGBOW   DIVIDER

2.1.10   variable head   reservoir   ROTRING–ISOGRAPH PEN

edge of sheet

20 margin   12 margin

1:50
1:5   1:1250
1:2500

metric scale

drawing sheet

drawing board

30°/60° set square

45° set square

adjustable set square

tee square

NAME   DRAWING   TITLE   SCALES

2.1.11

Figure 2.1   DRAWING   INSTRUMENTS

Drawing ink. This is usually black in colour and water-proof, although a wide range of coloured inks is also available. Indian ink is sold in bottles or tubes which must be kept sealed.

Stencils. These are used by draughtspersons, especially for main headings. They are available in a wide range of sizes and in a number of different styles of letters and figures. Special pens are used with the stencils and have ink feeds. They are not suitable for examination purposes.

Pencils. They are produced in a variety of grades and HB is generally found to be most suitable for small scale drawings and B pencils for large scale work and sketches. It is advisable to use good-quality pencils sharpened to a round point, preferably with a pen-knife.

Erasers. The erasing of pencil work is generally done with soft rubber erasers, although art gum may be used for large surfaces. Ink work can be removed with green ink erasers but much quicker results can be obtained with electrically operated erasers. Erasing work is facilitated by the use of erasing shields which consist of thin pieces of metal or plastic incorporating perforations of various shapes, which mask the parts of the drawing to be retained. Electrically operated erasers are often used in large drawing offices.

Drawing boards. These are made in a variety of sizes and to various designs. BS 6381: 1983 (specification for drawing boards) gives the recommended sizes for drawing boards and tee squares. The most common sizes are A2: 650 × 470 mm; A1: 920 × 650 mm; and A0: 1270 × 920 mm. Good-quality boards may be made of silver spruce with a projecting edge of black plastic to take the head of a tee square. The dovetail jointed boards have slotted hardwood battens screwed to their back face. Other drawing boards consist of laminboard faced on both sides with mahogany veneer and with beech capping pieces on all edges. Boards for student use include clamped softwood boards and poplar-faced blockboard, possibly fitted with aluminium moulded ends. Most draughtspersons use drawing tables fitted with a parallel motion straight-edge operating on a continuous wire and pulley principle.

Tee squares. These are used to produce horizontal lines and as a working platform for set squares. The most common variety is made of mahogany with black plastic working edges. The usual lengths are 650, 920 and 1270 mm (see figure 2.1.10).

Set squares. These are used to draw vertical or inclined lines. One type of set square has two angles of 45° while the other has angles of 60° and 30° respectively. For other angles adjustable set squares or protractors may be used. Set squares and protractors are usually made of transparent plastic and set squares may have plain or bevelled edges with sizes varying from 150 to 350 mm (see figure 2.1.11). Rotring make an adjustable set square with an ink edge, comprising a rebate to prevent ink spreading under the square by capillary action, and this has proved popular.

French curves. These are made of transparent plastic and to a variety of forms as aids to drawing architectural curves with pens, and in the form of railway curves ranging from 50 to 6000 mm radius.

Scales. These are used for plotting dimensions prior to drawing and for scaling from finished drawings. Most scales are now manufactured from PVC, which is dimensionally stable, and are either of 150 or 300 mm nominal length of flat or oval section; the majority are fully divided. A very popular and useful scale of oval section with scales on four edges has 1:10 and 1:100, 1:20 and 1:200, 1:5 and 1:50, and 1:1250 and 1:2500 scales.

Commonly used scales are shown in table 2.1.

Choice of scale is influenced by the need to communicate information accurately and adequately to secure maximum output, the nature of the subject and the desirability of producing all drawings for a particular project of uniform size. Further guidance is given in CPI (Building Project Information Committee) *Production Drawings: a code of procedure for building works* (1987) on the production and arrangement of drawings.

A scale of 1:100 is a very popular scale of plans, sections and elevations of new buildings. One millimetre on the drawing represents one hundred millimetres on the site. It is very close to the old $\frac{1}{8}$ inch to 1 foot scale, which had a ratio of 1:96. Scale drawings enable building details to be drawn in direct proportion to the work to be undertaken on site and in a convenient form for extraction of relevant information.

Drawing paper. This is produced in the form of cartridge paper for general use and handmade paper for fine line work and to receive colour washes. Cartridge paper is made in several qualities and weights and either in rolls 25 m long or in sheets. Trimmed sizes of drawing sheets as specified in BS 3429:1975 (Sizes of drawing sheets) are: 841 × 1189 mm (A0); 594 × 841 mm (A1); 420 × 594 mm (A2); 297 × 420 mm (A3); and 210 × 297 mm (A4). It is desirable to use standard size sheets wherever possible. Detail paper, which resembles thick tracing paper, is used on occasions for

Table 2.1 Recommended scales

Type of drawing	Function of drawing	Scales for use with metric system	Notes
DESIGN STAGE			
Sketch Drawings	preliminary drawings, sketches or diagrams to show designer's general intentions		Scales will vary but it is recommended that preference be given to those used in production stage
PRODUCTION STAGE			
Location drawings			
Block plan	to identify site and locate outline of building in relation to town plan or other wider context	1:2500* 1:1250*	*Ordnance Survey indicate that due to large costs and labour involved these scales will probably have to be retained for a transitional period
Site plan	to locate position of buildings in relation to setting out point, means of access, general layout and drainage	1:500 1:200	
General arrangement	to show position occupied by various spaces in building, general construction and location of principal elements, components and assembly details	1:200 1:100 1:50	
Component drawings			
Ranges	to show basic sizes, system of reference and performance data on a set of standard components of a given type	1:100 1:50 1:20	
Details	to show all the information necessary for the manufacture and application of components	1:10 1:5 1:1 (full size)	
Assembly	to show in detail the construction of buildings; junctions in and between elements, between elements and components, and between components	1:20 1:10 1:5	

preliminary drawings and is often suitable for photo-copying.

Tracing paper. This has either a smooth or matt surface and can be obtained in either roll or sheets in several weights, suitable for ink and pencil drawing. For greater durability tracing cloth and polyester film are available. It is advisable to hang tracing cloth before use to permit expansion to take place and to dust the working surface with French chalk to remove grease and permit the ink to run evenly.

Reproduction of Drawings

This can be done in various ways as described below. Reproduction is important as it is rarely satisfactory to use the original drawing on the site.

Diazo. Material (paper, cloth or film) is exposed in contact with the original to an ultra-violet light source and developed with liquid ammonia vapour or heat. It normally produces dark brown to black lines and is very suitable for working drawings.

True-to-scale (TTS). A manual system of offset printing using gelatine plate, with ink lines on a variety of materials, and produces dark black permanent lines.

Blueprint. This consists of white lines on blue paper or opaque cloth and is now almost entirely superseded by diazo. The material (paper or cloth) is exposed in contact with the original to a suitable light source and developed in water.

Contact copying or photostats. This produces a black line on paper, translucent cloth or film. A direct positive is made from the translucent original by transmitted light and a negative positive made by reflex copying of an opaque original to give a laterally reversed first copy and right reading subsequent copies.

Optical copying. A black line is produced on silver sensitised materials from a photographic negative, using a camera and film.

For smaller drawings of size A3 and below, three other reproduction processes are available: namely, diffusion transfer, electrophotography and offset litho.

LAYOUT AND PRESENTATION OF DRAWINGS

It is important that drawings shall be logically and neatly arranged to give a balanced layout. BS 1192 Part 2 recommends that every drawing sheet shall have a filing margin, title and information panel. The title and information panel is located at the bottom right-hand corner of the sheet and incorporates the project title, subject of drawing, scale, date of drawing, project number and revision suffix, and name and possibly address and telephone number of architect or surveyor responsible for the project. Often the names or initials of the persons drawing and checking the drawing are also added. Above this information is provision for the legend and notes. The student will not need to give so many particulars and can reduce the size of the title and information panel on the lines adopted for the figures in this book. He/she particularly needs to incorporate his/her name, the subject of the drawing and the operative scales, as indicated in figure 2.1.11.

BS 1192 Part 2 contains a variety of illustrated examples relating to location/general arrangement drawings, including a block plan, site plan, floor plans, section and elevation, and assembly/detail drawings encompassing a window opening and component range/schedule and component detail for doorsets. There is also a hardware schedule, reinforced concrete and steelwork details, drainage plan, hot and cold water services plan and isometric projection, lighting and electric power plans and details of other services.

Setting out the various parts of a drawing may prove difficult to the student in his/her early days. It may therefore be helpful to draw rough outlines on tracing paper in the first instance. Ensure that the drawing paper has its smoothest surface upwards and pin the sheet onto the drawing board with a drawing pin at the top left-hand corner. Manipulate the sheet so that the top edge is parallel to the tee square and a second pin can then be inserted at the top right-hand corner of the sheet. With small sheets it is not advisable to use pins at the bottom corners of the sheet as they obstruct the tee square. Another alternative is to fix the sheets with Sellotape.

The first step is to draw the border lines using a tee square in both directions. Thereafter all vertical lines are drawn with set squares. The title and information panel should follow, and the space remaining is available for drawing.

Apart from the desirability of securing a balanced arrangement and a nice-looking drawing, it is also possible to save a considerable amount of time by carefully locating the various parts in a logical relationship one with another. Hence when preparing plans, sections and elevations of buildings or components parts, such as windows and doors, it is good practice to draw the plan(s) at the bottom of the sheet and the section(s) on the right-hand side. The elevation(s) can then be drawn by projecting many of the lines upwards from the plan(s) and horizontally across from the section(s). Many examples of this arrangement appear in figures throughout this book.

GENERAL ASPECTS RELATING TO DRAWINGS

Some of the more important general aspects relating to building drawings are now considered. Firstly it is vital that all instruments, scales, tee squares and set squares are kept clean with a suitable cloth or duster. Scale and plot a number of points before connecting them with lines in order to accelerate the drawing process. Always use a scale which is sufficiently large to permit the drawing of details that can easily be read, with adequate space for annotated notes. Draw all lines lightly in the first instance and work from centre lines wherever practicable.

Lines

BS 1192 Part 2 recommends the use of lines of different thicknesses for the following specific purposes.
Thick lines. These should be used for site outline of buildings on block and site drawings; primary functional elements, such as loadbearing walls and structural slabs, on general location and assembly drawings; outlines requiring emphasis on component ranges and horizontal and vertical profiles on component details. Services are shown by thick broken lines.
Medium lines. These should be used for general details on site drawings; secondary elements and components, such as non-loadbearing partitions, windows and doors, on general location and assembly drawings; outlines of components, on component ranges; and general details on component details.
Thin lines will be used for reference grids, dimension lines, leader lines and hatching on all types of drawing; centre lines are shown by thin chain lines.

Dimensions

Dimension lines for figured dimensions should always be drawn in positions where they cannot be confused with other information on the drawing. The terminal points to which dimension lines refer must be clearly shown and BS 1192 makes certain recommendations. Open arrows of the type in figure 2.2.1 should be used for basic or modular dimensions or for the sizes of

spaces or components taken to grid lines, centre lines or unfinished carcase surfaces. Tolerances or gaps, and the work sizes of components which are specified for manufacture (so that allowing for the tolerances, the actual size lies between the required limits) shall be indicated by solid arrows of the type shown in figure 2.2.2 and 2.2.3.
Dimension figures. These should be written immediately above and along the line to which they relate, as shown in figure 2.2.1. Running dimensions should take the form illustrated in figure 2.2.4. In all cases dimension figures, when not written for viewing from the bottom, should be written only for viewing from the right-hand edge (see figure 2.2.5).

Linear dimensions may be expressed in both metres and millimetres on the same drawing. In order to avoid confusion metres should be given to three places of decimals, for example, 2.1 m should be entered as 2.100. Millimetres will be entered as the actual number involved, for example, a wall thickness of 215. In this way the need for symbols (m and mm) is largely avoided. Summing up, the approach to be adopted for metric dimensions is

(1) whole numbers indicate millimetres;
(2) decimalised expressions to three places of decimals indicate metres;
(3) all other dimensions should be followed by the unit symbol.

The sequence of dimensions of components must be in the order of (a) length, (b) width and (c) depth or height. New levels, such as floor levels, should be distinguished from existing ground levels by inserting them in boxes; for example, a finished floor level could be shown as FFL 150.750.

Graphical Symbols

Materials in section on plans and vertical sections are best hatched to assist in interpreting the drawings. Hatching is preferable to colouring which is costly, laborious and conducive to error. Hatchings representing the more commonly used materials are contained in BS 1192 Part 3 and some of these are reproduced in figure 2.2.6. The use of nationally recognised hatchings leads to uniformity in the presentation of drawings and enables them to be more readily and easily understood.

A large number of graphical symbols are in common use representing components in connection with services, installations and fixtures and fittings, and a selected sample is shown in figure 2.2.7. BS 1192 Part 3 shows many more examples of graphic symbols, including a wide range of sanitary fittings, manholes, gullies, piped and electrical services. Recognised abbreviations may also be used and these can result in a reduction in the length of explanatory notes on drawings. A selected sample of officially recognised abbreviations follows.

Aggregate	Agg	Hardboard	hdb
Air brick	AB	Hardcore	hc
Aluminium	al	Hardwood	hwd
Asbestos	abs	Inspection chamber	IC
Asphalt	asph	Insulation	insul
Boarding	bdg	Joist	jst
Brickwork	bwk	Mild steel	MS
Cast iron	CI	Pitch fibre	PF
Cement	ct	Plasterboard	pbd
Concrete	conc	Rainwater pipe	RWP
Copper	copp	Reinforced concrete	RC
Cupboard	cpd	Softwood	swd
Damp-proof course	DPC	Stainless steel	SS
Damp-proof membrane	DPM	Tongue and groove	T&G
Foundation	fdn	Vent pipe	VP
Glazed pipe	GP	Wrought iron	WI

Lettering

Lettering is needed on drawings to provide information which could not otherwise be obtained. The aim should be to produce uniform, neat and easily legible lettering. General notes may with advantage be collected in groups but more specific particulars should be inserted as near as possible to the items to which they relate. Care must however be taken not to obscure any part of the drawing with lettering or lines linking lettering and details.

Lettering can take a variety of forms from block letters to small case (italics), or even neatly handwritten notes. It should be appreciated that the lettering on a drawing can quite easily occupy one-third of the total time required to prepare the drawing. Hence if time was short in the examination, a candidate would be justified in completing drawings to later questions with neatly handwritten notes, provided that he had earlier lettered up a drawing adopting the more orthodox approach. Each individual adopts his own style of lettering – vertical or sloping, thin or broad, simple or more flamboyant. Generally, a student would be well advised to adopt a fairly simple style of lettering and to gain adequate practice to achieve a fair standard of competence. Small-case lettering is best suited for notes and block letters for headings.

In any event, in the early stages, the student should produce his lettering between faint parallel lines. The letters and figures should be well formed with uniform spacing between the letters, using a fairly soft pencil, such as a B grade, kept continuously sharpened. BS 1192 recommends heights of 4 to 7 mm for headings and 1.5 to 2.5 mm for notes. Three styles of lettering are illustrated in figure 2.3.1. The first is a simple broad, vertical style for headings, the second a thinner, sloping and more elaborate style also suitable for headings, and the third is small case for notes.

SKETCHES

In recent years many examining bodies have introduced sketches into technology questions as a substitute for scale drawings. This approach is based on the philosophy that more drawings can be produced in a given period by sketching than by scale drawing. Hence the examination candidate has a greater opportunity to demonstrate his knowledge of the subject and of his ability to apply this knowledge to practical construction problems. Furthermore, the drawing process is simply a means of communication and the main aim of a building technology examination is not to test the ability of the candidate in producing scale drawings. Sketches can often be used to advantage in an examination to supplement a written description.

Some students have expressed concern at the move towards sketches, using the argument that they never have been and never will be artists. There is something of a fallacy here. Sketching requires practice in the same way as scale drawing does and the majority of students can acquire a reasonable skill over a period of time. There are several basic rules to be observed

(1) All sketches should be fairly large: it is difficult to show sufficient details on small sketches.

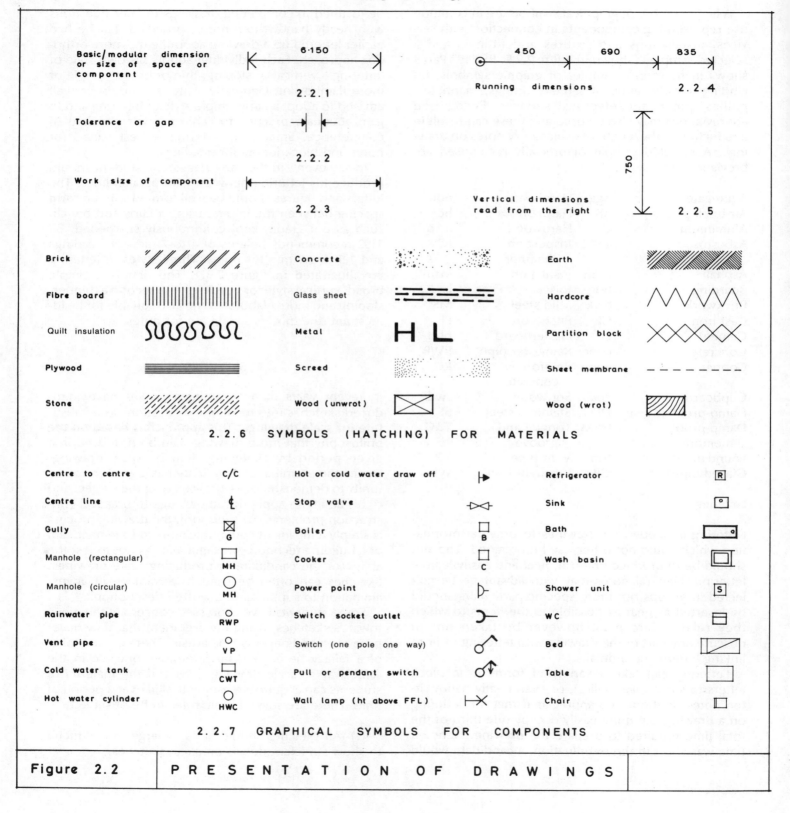

Basic/modular dimension or size of space or component   6·150   2.2.1

Running dimensions   450   690   835   2.2.4

Tolerance or gap   2.2.2

Work size of component   2.2.3

Vertical dimensions read from the right   750   2.2.5

Brick		Concrete		Earth	
Fibre board		Glass sheet		Hardcore	
Quilt insulation		Metal	H L	Partition block	
Plywood		Screed		Sheet membrane	
Stone		Wood (unwrot)		Wood (wrot)	

2.2.6   SYMBOLS (HATCHING) FOR MATERIALS

Centre to centre	C/C	Hot or cold water draw off		Refrigerator	R
Centre line	₵	Stop valve		Sink	
Gully	G	Boiler	B	Bath	
Manhole (rectangular)	MH	Cooker	C	Wash basin	
Manhole (circular)	MH	Power point		Shower unit	S
Rainwater pipe	RWP	Switch socket outlet		WC	
Vent pipe	VP	Switch (one pole one way)		Bed	
Cold water tank	CWT	Pull or pendant switch		Table	
Hot water cylinder	HWC	Wall lamp (ht above FFL)		Chair	

2.2.7   GRAPHICAL SYMBOLS FOR COMPONENTS

| Figure 2.2 | PRESENTATION OF DRAWINGS |

ABCDEFGHI
JKLMNOPQR
STUVWXYZ

ABCDEFGHIJ

KLMNOPQRST

UVWXYZ

abcdefghijklmn

opqrstuvwxyz

1234567890

STYLES OF LETTERING

2.3.1

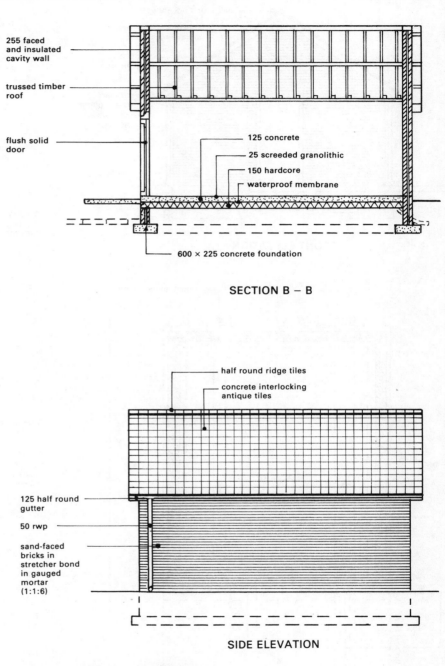

255 faced
and insulated
cavity wall

trussed timber
roof

flush solid
door

125 concrete

25 screeded granolithic

150 hardcore

waterproof membrane

600 × 225 concrete foundation

SECTION B – B

half round ridge tiles

concrete interlocking
antique tiles

125 half round
gutter

50 rwp

sand-faced
bricks in
stretcher bond
in gauged
mortar
(1:1:6)

SIDE ELEVATION

2.3.2 ELECTRICITY SWITCH HOUSE

(See figure 2.4 for further working
drawings of electricity switch house)

Figure 2.3	STYLES OF LETTERING AND WORKING DRAWINGS	Scale: 1 : 100

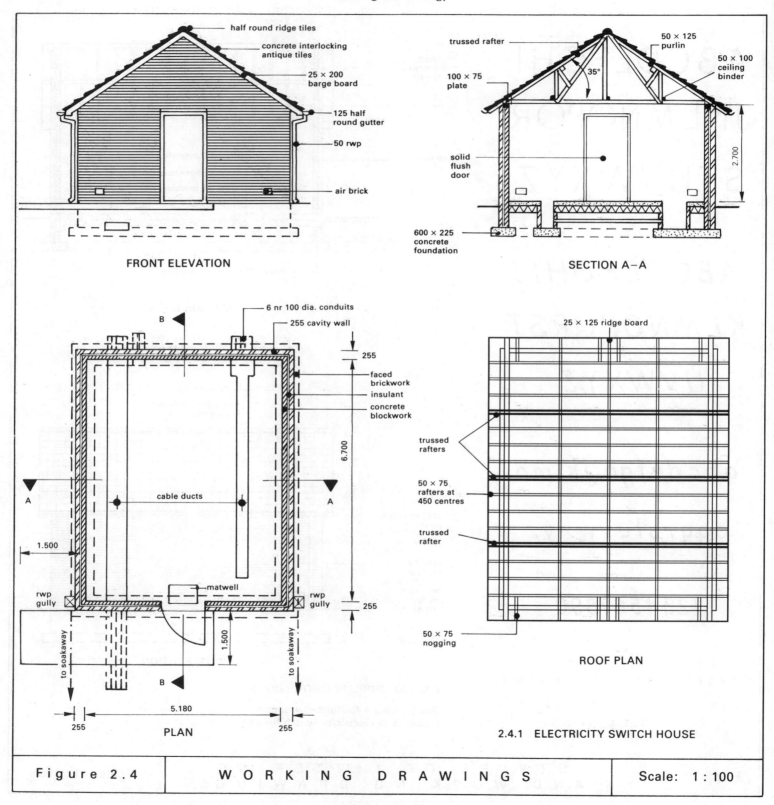

half round ridge tiles

concrete interlocking
antique tiles

25 × 200
barge board

125 half
round gutter

50 rwp

air brick

FRONT ELEVATION

trussed rafter

50 × 125
purlin

50 × 100
ceiling
binder

100 × 75
plate

35°

solid
flush
door

2.700

600 × 225
concrete
foundation

SECTION A–A

B

6 nr 100 dia. conduits

255 cavity wall

255

faced
brickwork

insulant

concrete
blockwork

6.700

A   A

cable ducts

1.500

rwp
gully   rwp
gully

to soakaway   matwell   to soakaway

1.500

255

B

255   255

5.180

PLAN

25 × 125 ridge board

trussed
rafters

50 × 75
rafters at
450 centres

trussed
rafter

50 × 75
nogging

ROOF PLAN

2.4.1 ELECTRICITY SWITCH HOUSE

Figure 2.4	W O R K I N G D R A W I N G S	Scale: 1 : 100

(2) Sketches should be in correct proportions and one useful approach is to plot a few leading dimensions to a suitable scale before starting to sketch.

(3) Use a soft pencil (a B grade pencil is often used); faint lines can be thickened up subsequently.

(4) Develop the ability to move the hand down or across the paper working from the shoulder or elbow, and not merely the fingers.

(5) Start by sketching simple objects like garden sheds, summerhouses and garages: valuable experience can be obtained by taking a notebook on one's travels and sketching buildings or building details.

(6) To give the sketches a measure of solidity, establish a source of light and shade all faces which are hidden from it.

Some typical sketches are shown in figures 7.5, 9.6, 12.1 and 13.3.

WORKING DRAWINGS

The majority of working drawings consist of plans, sections and elevations drawn by orthographic projection, whereby they are all in flat planes. The working drawings form one of the most effective ways of conveying the designer's requirements to the contractor, to assist him in constructing the work on the site.

Plan. This represents a view from above of an object on a horizontal plane. With buildings, plans are normally drawn of each floor at about one metre above floor level, looking down at the floor and cutting through walls, doors and windows.

The drawing represents a sectional plan of the walls, doors and windows, as if they were cut open. A roof plan looks down on the roof which is opened up to show the roof members.

Sections. These are taken vertically through a building cutting through the foundations, walls, partitions, floors, roof, windows and doors, to show the form of construction. Some features such as internal doors, wall tiling, skirtings and picture rails will appear in elevation where they are seen in the background.

Elevations. These represent external faces of a building including windows and doors.

BS 1192 describes how, in theory, with orthographic projection one or more views of an object are obtained by dropping perpendiculars from all significant points on the object to one or more planes of projection.

These perpendiculars are known as projectors and the principal planes of projection are assumed to be at right angles to one another.

Figures 2.3.2 and 2.4.1 illustrate plans, sections and elevations of an electricity switch house and are a good example of orthographic projection.

In practice all the components on figures 2.3 and 2.4 would be on a single sheet to permit projections from all components horizontally and vertically using a tee square, and set square, but the restriction of the page size prevents this in its entirety. An alternative scale is 1:50 which permits the inclusion of greater detail and provides more space for descriptive notes and dimensions.

Other forms of projection are used on occasions to make drawings more easily understood.

Isometric projection. This is a common form of projection in which length, breadth and height of the object are shown on the one drawing. An example of a house drawn to isometric projection is illustrated in figure 2.5.3 from the plan in figure 2.5.1 and elevation in figure 2.5.2. All vertical lines are drawn as verticals and horizontal lines at 30° to the horizontal. A layman can far more readily appreciate and comprehend the nature and appearance of the building from the isometric sketch than he can from the plan and elevation.

Axonometric projection. This projection is another way to draw the building, as shown in figure 2.5.4. This is similar to isometric projection, except that horizontal lines are drawn at 45° to the horizontal. Both projections are useful for preparing drawings of components and of service layouts in buildings. However, BS 1192, Part 1: 1984 adopts different criteria when describing projections.

Pictorial projection. There are in fact two methods of drawing in perspective: angular perspective for external views of buildings and parallel perspective which is normally used for interiors. Both methods provide projections from a spectator situated some distance from the building or room being drawn. These processes are not considered in detail as they are thought to be outside the scope of this book.

Building Research Establishment Current Paper 18/73 (Working drawings in use) drew attention to the need for the proper classification of drawings with appropriate titles and references to other drawings for further information where applicable, to assist in their dissemination. Coded referencing of materials and components with the corresponding items in the bill of quantities and specification can be very helpful to the contractor.

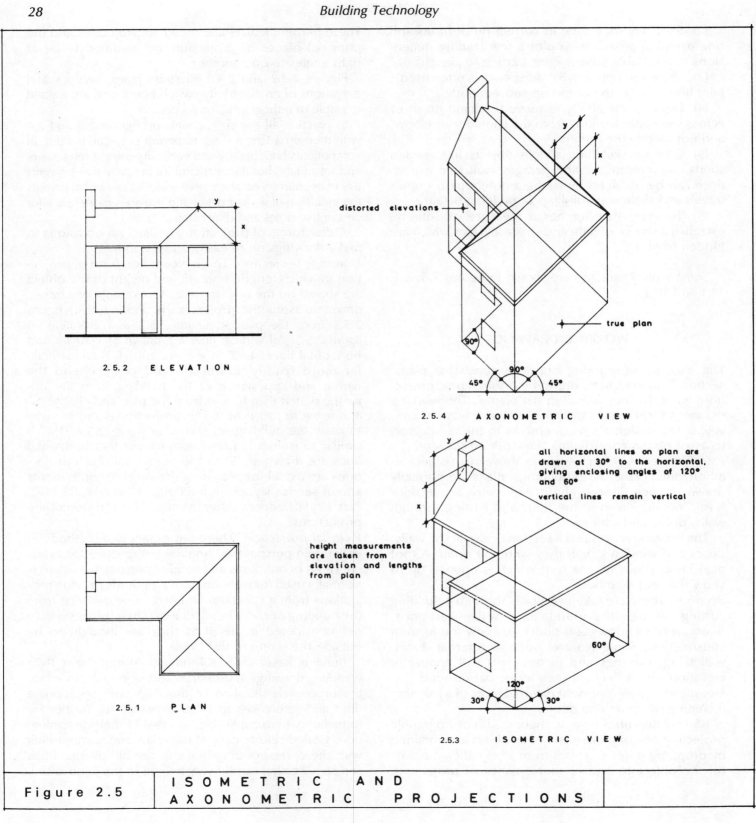

2. 5. 2   E L E V A T I O N

distorted elevations ←

true plan ←

90°

90°

45°   45°

2. 5. 4   A X O N O M E T R I C   V I E W

2. 5. 1   P L A N

all horizontal lines on plan are
drawn at 30° to the horizontal,
giving enclosing angles of 120°
and 60°

vertical lines remain vertical

height measurements
are taken from
elevation and lengths
from plan

60°

120°

30°   30°

2.5.3   I S O M E T R I C   V I E W

| Figure 2.5 | I S O M E T R I C   A N D   A X O N O M E T R I C   P R O J E C T I O N S |

3 FOUNDATIONS, EARTHWORK SUPPORT AND CONCRETE

It is imperative that the foundations of a building be properly designed to spread the dead and superimposed loads over a sufficient area of soil, after the removal of all vegetable matter. In this context 'soil' is that part of the earth which lies below the topsoil and above the rock, having been formed by the erosion of the earth's crust by water, atmospheric means, and intense pressure over many thousands of years. This involves an understanding of soil types and their characteristics, and awareness of the different types of foundation that are available. Chapter 1 dealt with site investigations and these included an examination of soil conditions and the level of the water table. These aspects are becoming even more important as some building sites now occupy land which has been avoided in the past. Building Research Establishment Digest 64[1] suggests an initial approach to the local authority with its intimate knowledge of soil and general conditions in the area. Older editions of Ordnance Survey maps may provide useful information on features that cause difficulty, such as infilled ponds, ditches and streams, disused pipes and sites of old buildings and tips. A polygonal pattern of cracks about 25 mm wide on the ground surface during a dry summer indicates a shrinkable soil. Whereas larger cracks, roughly parallel to one another generally indicate deeper-seated movements and may be caused by mining, brine pumping, or landslips.[1]

When building on filled sites, such as inner city redevelopment and reclaiming industrial waste lands, it is necessary to establish whether the fill is able to support the building without excessive settlement and whether the fill contains materials which are hazardous to health and harmful to the environment or building. Building Regulation Approved Document C2 (1991) describes the measures needed to remove solid and liquid contaminants, to fill and seal the surface and to deal with gaseous contaminants such as radon and landfill gas and methane. Further information on the provision of protection against radon is given in chapter 1.

IDENTIFICATION AND CHARACTERISTICS OF SOIL

Table 3.1 is extracted from Building Research Establishment Digest 64[1] and classifies soil types; it shows how they are identifiable in the field, and details possible foundation difficulties. Quantitative tests in the field and laboratory will be necessary where comprehensive information on soil conditions is required. In this connection reference to BS 5930[2] and BS 1377[3] should prove useful.

BS 8004 and 8103[4] provide a basis for field identification of soil in terms of the predominant size of soil particle and the strength features which have an important influence on foundation behaviour; the following descriptions are mainly extracted from these Standards.

Non-cohesive Soils

Gravel. A natural deposit consisting of rock fragments in a matrix of finer and usually sandy material. Many of the particles are larger than 2 mm in size.

Sand. A natural sediment consisting of the granular and mainly siliceous products of rock weathering. It is gritty with no real plasticity. The particles normally range between 0.06 and 2.00 mm in size.

Well-graded sand. A sand containing a proportion of all sizes of sand particles with a predominance of the coarser grades.

Compact gravel and sand. Deposits require a pick for removal and offer high resistance to penetration by excavating tools.

Table 3.1   Soil identification

Soil type	Field identification	Field assessment of structures and strength	Possible foundation difficulties
Gravels	Up to 76.2 mm (retained on No. 7 BS sieve) Some dry strength indicates presence of clay	Loose – easily removed by shovel. 50 mm stakes can be driven well in	Loss of fine particles in water-bearing ground
Sands	Pass No. 7 and retained on No. 200 BS sieve Clean sands break down completely when dry Individual particles visible to the naked eye and gritty to fingers	Compact – requires pick for excavation. Stakes will penetrate only a little way	Frost heave, especially on fine sands Excavation below water table causes runs and local collapse, especially in fine sands
Silts	Pass No. 200 BS sieve. Particles not normally distinguishable with naked eye Slightly gritty; moist lumps can be moulded with the fingers but not rolled into threads Shaking a small moist lump in the hand brings water to the surface Silts dry rapidly; fairly easily powdered	Soft – easily moulded with the fingers Firm – can be moulded with strong finger pressure	As for fine sands
Clays	Smooth, plastic to the touch, sticky when moist. Hold together when dry. Wet lumps immersed in water soften without disintegrating Soft clays either uniform or show horizontal laminations Harder clays frequently fissured, the fissures opening slightly when the overburden is removed or a vertical surface is revealed by a trial pit	Very soft – exudes between fingers when squeezed Soft – easily moulded with the fingers Firm – can be moulded with strong finger pressure Stiff – cannot be moulded with fingers Hard – brittle or tough	Shrinkage and swelling caused by vegetation Long-term settlement by consolidation Sulphate-bearing clays attack concrete and corrode pipes Poor drainage Movement down slopes; most soft clays lose strength when disturbed
Peat	Fibrous, black or brown Often smelly Very compressible and water retentive	Soft – very compressible and spongy Firm – compact	Very low-bearing capacity; large settlement caused by high compressibility Shrinkage and swelling – foundations should be on firm strata below
Chalk	White – readily identified	Plastic – shattered, damp and slightly compressible or crumbly Solid – needing a pick for removal	Frost heave Floor slabs on chalk fill particularly vulnerable during construction in cold weather Swallow holes
Fill	Miscellaneous material – for instance rubble, mineral, waste, decaying wood		To be avoided unless carefully compacted in thin layers and well consolidated May ignite or contain injurious chemicals

Source: BRE Digest 64[1]

Loose gravel and sand. Deposits readily removable by hand-shovelling only.

Uniform or poorly graded sand. The majority of particles lie within a fairly restricted size range.

Cohesive Soils

Clay. A natural deposit consisting mainly of the finest siliceous and aluminous products of rock weathering. It has a smooth, greasy touch, sticks to the fingers and dries slowly. It shrinks appreciably on drying and has considerable strength when dry.

Stiff clay. A clay which requires a pick or pneumatic spade for its removal and cannot be moulded with the fingers at its natural moisture content.

Firm clay. A clay which can be excavated with a spade and can be moulded by substantial pressure with the fingers at its natural moisture content.

Very soft clay. Extruded between fingers when squeezed in fist at its natural moisture content.

Soft clay. A clay which can be readily excavated and can be easily moulded with the fingers at its natural moisture content.

Boulder clay. A deposit of unstratified clay or sandy clay containing subangular stones of various sizes.

Silt. A natural sediment of material of finer grades than sand. Most of the grains will pass a 75 micrometre test sieve. It shows some plasticity, is not very gritty and has appreciable cohesion when dry.

Soil Identification and Classification Tests

Soils may be subjected to a number of tests to establish their identity and classify them. Some of the more important tests will now be briefly described.

Particle size distribution. This test determines the proportion of gravel, sand, silt and clay in a particular soil. The soil is dried and sieved through a nest of sieves, and the weight retained of each soil is recorded. The results of the grading test are plotted on a graph. Sand has particles between 0.060 and 2.000 mm, silt between 0.060 and 0.002 mm and clay is less than 0.002 mm.

Liquid limit test. The object is to determine the moisture content at which the soil passes from plastic to liquid state. The moisture content is expressed as a percentage of the dry weight of the soil.

Plastic limit test. This determines the moisture content at which the soil ceases to be plastic (soil sample can be rolled into a thread 3 mm diameter without breaking).

Plasticity index. This refers to the difference between liquid and plastic limits.

Casagrande devised a soil classification chart and a group symbol of two letters for each soil, the first letter representing the size of the soil particles and the second letter its main characteristic or property, as indicated below.

First letter		Second letter	
Gravel	G	Fines	F
Sand	S	High compressibility	H
Silt	M	Intermediate compressibility	I
Clay	C	Low compressibility	L
Organic silts and clays	O	Poorly graded	P
		Well graded	W
		Clay	C

A poorly graded sand would have a group symbol of SP, if it were well graded it would be SW, and if it were a sandy clay it would be SC.

Other tests include a dry density test to determine the density of the dry soil in its natural position, the standard penetration and consolidation tests to determine the compressibility and the shear vane, unconfined compression and triaxial compression tests to determine the shear strength parameters; both the compressibility and shear strength parameters are needed to ascertain the allowable bearing pressure.

DESIGN OF FOUNDATIONS

The primary aim must always be to spread the loads from the building over a sufficient area of soil to avoid undue settlement, particularly unequal settlement. The Building Regulations 1991[5] Schedule I, Part A lays down the following requirements for the foundations of a building.

Loading

A1 (1) The building shall be constructed so that the combined dead, imposed and wind loads are sustained and transmitted to the ground –

(a) safely, and

(b) without causing such deflection or deformation of

any part of the building, or such movement of the ground, as will impair the stability of any part of another building

(2) In assessing whether a building complies with sub-paragraph (1), regard shall be had to the imposed and wind loads to which it is likely to be subjected in the ordinary course of its use for the purpose for which it is intended.

Ground Movement

A2 The building shall be constructed so that ground movement caused by (a) swelling, shrinkage or freezing of the subsoil; or (b) land-slip or subsidence (other than subsidence arising from shrinkage), in so far as the risk can be reasonably foreseen, will not impair the stability of any part of the building.

The further structural requirements in the Building Regulations 1991 on disproportionate collapse in A3 applies only to buildings having five or more storeys and A4 to public buildings, shops and shopping malls with a clear span exceeding 9 m between supports.

BS 8004[4] includes a table of presumed allowable bearing values under static loading for different soils and these are incorporated in table 3.2. The dead, imposed and wind loads of a building are calculated in accordance with the principles laid down in BS 6399.[6] For instance the imposed load on a floor to a house is likely to be about 1.50 kN/m^2. In this way the total load in kilonewtons per linear metre of wall can be calculated, and dividing this by the safe bearing capacity of the soil in kilonewtons per square metre will give the required width of foundation in metres. The Building Regulations Approved Document A1/2 (1991) also contains a schedule (table 12) giving the minimum width of strip foundation for varying total loads expressed in kilonewtons per linear metre of wall for use on various soils. For example, with gravel and sand the minimum width ranges from 250 to 650 mm for loads varying from 20 to 70 kN/m. For firm clay or sandy clay the corresponding widths are 300 and 850 mm. With soft silt, clay, sandy clay or silty clay, the foundation widths are 450 mm for a loading of 20 kN/m and 650 mm for 30 kN/m. For greater loadings it would be advisable to pile or use a raft foundation. Table 12 in Building Regulations Approved Document A1/2 (1991) is illustrated in diagrammatic form in figure 3.1.

Building Research Establishment Digest 67[7] describes how a typical two-storey semi-detached house of about 85 m^2 floor area, in cavity brickwork, with lightweight concrete or clay block partitions, timber floors and tiled roof, exerts a combined load approaching 1000 kN, excluding the weight of foundations. The loads at ground level are, in kilonewtons per linear metre of wall, approximately: party wall 50, gable end wall 40, front and back walls 25 and internal partitions less than 15. When the foundations have been designed, their weight must be added to the loadings already calculated so as to obtain the total bearing pressure on the soil beneath.

A load applied through a foundation always causes settlement as it compresses the soil beneath it. However, not even uniform ground uniformly loaded settles evenly, and the complex properties of soil make it difficult to assess the settlement of individual foundations or to predict the distortion of complete buildings.

Clays which shrink on drying and swell again when wetted often cause movement in shallow foundations. Shrinkage of clays occurs both horizontally and vertically, and so there is a tendency for walls to be drawn outwards, in addition to settling, and for cracks to open between the clay and the sides of the foundations. Water may enter the cracks during the following winter and soften the clay against or below the foundations.

Trees which are close to buildings on a clay soil can cause extensive damage, particularly fast-growing trees with wide-spread roots like poplars and also oak, willow, cherry, plum and whitebeam. The tree roots extract water from the soil causing the clay to shrink because of the reduction in moisture content. Buildings on shallow foundations should not be closer to single trees than their height on maturity and not closer to clumps or rows of trees than one-and-a-half times the mature height of the trees. BRE Digest 298[8] describes how the minimum distance between buildings and trees varies with the species of tree and that it can be half the height of the tree with lime, ash, elm, sycamore, hawthorn, beech, birch and cypress. Constant pruning of growing trees is often necessary to restrict their height and avoid risk of damage to buildings. Adequate time should be allowed after felling trees on new building sites to allow time for the clay to regain its water content, accompanied by ground heave. Cutler and Richardson[9] have provided a useful source of reference on the effect of roots of different species of trees on buildings. Further guidance is available in BS 5837.[10]

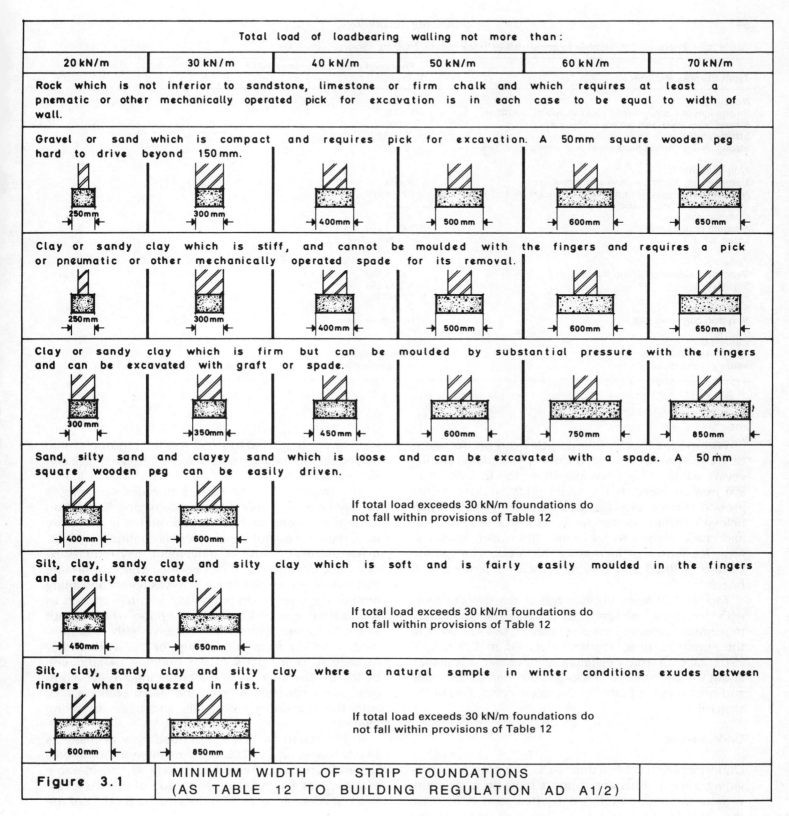

Total load of loadbearing walling not more than:					
20 kN/m	30 kN/m	40 kN/m	50 kN/m	60 kN/m	70 kN/m

Rock which is not inferior to sandstone, limestone or firm chalk and which requires at least a pnematic or other mechanically operated pick for excavation is in each case to be equal to width of wall.

Gravel or sand which is compact and requires pick for excavation. A 50mm square wooden peg hard to drive beyond 150mm.

| 250mm | 300mm | 400mm | 500mm | 600mm | 650mm |

Clay or sandy clay which is stiff, and cannot be moulded with the fingers and requires a pick or pneumatic or other mechanically operated spade for its removal.

| 250mm | 300mm | 400mm | 500mm | 600mm | 650mm |

Clay or sandy clay which is firm but can be moulded by substantial pressure with the fingers and can be excavated with graft or spade.

| 300mm | 350mm | 450mm | 600mm | 750mm | 850mm |

Sand, silty sand and clayey sand which is loose and can be excavated with a spade. A 50mm square wooden peg can be easily driven.

| 400mm | 600mm | If total load exceeds 30 kN/m foundations do not fall within provisions of Table 12 | | | |

Silt, clay, sandy clay and silty clay which is soft and is fairly easily moulded in the fingers and readily excavated.

| 450mm | 650mm | If total load exceeds 30 kN/m foundations do not fall within provisions of Table 12 | | | |

Silt, clay, sandy clay and silty clay where a natural sample in winter conditions exudes between fingers when squeezed in fist.

| 600mm | 850mm | If total load exceeds 30 kN/m foundations do not fall within provisions of Table 12 | | | |

| Figure 3.1 | MINIMUM WIDTH OF STRIP FOUNDATIONS (AS TABLE 12 TO BUILDING REGULATION AD A1/2) | |

Table 3.2 *Presumed allowable bearing values under static loading* (Source: BS 8004[4])

Types of rocks and soils	Presumed allowable bearing value (kN/m^2)
Rocks	
Strong igneous and gneissic rocks in sound condition	10 000
Strong limestones and strong sandstones	4 000
Schists and slates	3 000
Strong shales, strong mudstones and strong siltstones	2 000
Non-cohesive	
Dense gravel, or dense sand and gravel	> 600
Medium dense gravel, or medium dense gravel and sand	200 to 600
Loose sand, or loose sand and gravel	< 200
Compact sand	> 300
Medium dense sand	100 to 300
Loose sand	< 100
Cohesive	
Very stiff boulder clays and hard clays	300 to 600
Stiff clays	150 to 300
Firm clays	75 to 150
Soft clays and silts	< 75
Very soft clays and silts	Not applicable
Other soils	
Peat and organic soils	Not applicable
Made ground or fill	Not applicable

With beds of sand, fine particles can on occasions be washed out of the bed, reducing its stability. During severe winters, frost may penetrate soil to a depth of 600 mm or more. If the water table is close to the ground surface and the spaces between the soil particles are within a certain range, as with fine sands, silts and chalk, water can move into the frozen zone and form ice lenses of increasing thickness. The ground surface is lifted and this movement is termed *frost heave*.[11]

Methods of assessing damage in low-rise buildings with particular reference to progressive foundation movement, causing a range of crack types and sizes to the superstructure, are well detailed in BRE Digest 251[12], ranging from negligible hairline cracks less than about 0.1 mm wide to cracks greater than 25 mm wide and necessitating partial or complete rebuilding of the structure.

Underpinning

Underpinning has been defined as 'the process of providing new, permanent support beneath a structure without the need to remove it, often so as to increase the capacity of the structure'. Remedial underpinning is the process applied to structures that, through some inadequacy in their support, have suffered damage and distortion.

BRE Digest 352[13] emphasises that underpinning is usually necessary only if further damaging foundation movements need to be stopped, and it is therefore important to find out if movement is continuing. Identifying the exact cause of movement may require expensive exploration and soil testing, while confirming that movement is continuing involves time–consuming monitoring. Hence underpinning is often chosen as the easiest, quickest and safest option where doubt exists. However, difficulties can arise with underpinning, resulting from poor design of the work and/or damage to other parts of the building or adjoining buildings. The majority of underpinning projects comprise mass concrete or pier and beam construction, with the remainder using pile and beam or piling methods.

A BRE report in 1991[14] described how as many as 14 000 homes in the UK, most of them in south-east England, are underpinned each year at considerable expense, yet a significant proportion of the work carried out is not necessary for the structural safety of the house.

FOUNDATION TYPES AND THEIR SELECTION

There are a number of different foundation types available for use with domestic and small industrial and commercial buildings. Their selection is influenced by the type of building and the nature of the loadings, and the site conditions, and Barnbrook[15] provides some useful guidelines. The more common types of foundation are now described.

Strip Foundations

The majority of buildings up to four storeys in height have strip foundations, in which a continuous strip of concrete provides a continuous ground bearing under the loadbearing walls. A typical strip foundation is illustrated in figure 3.2.1. This type of foundation is placed centrally under the walls and is generally composed of plain concrete often to a mix of 1:3:6 by volume (1 part cement, 3 parts sand and 6 parts coarse aggregate, usually gravel), with the thickness being not less than the projection of the foundation and in no case less than 150 mm. The Building Regulations Approved Document A1/2 1991 (Section IE) recommends a concrete mix of 50 kg of cement to not more than 0.1 m^3 of fine aggregate (sand) and 0.2 m^3 of coarse aggregate. Alternatively, grade STI concrete to BS 5328: Part 2[16] or grade 15 concrete to BS 8110: Part I[17] would be acceptable. Internal walls may be supported by separate strip foundations or by thickening the concrete oversite.

On a sloping site the most economic procedure is to use a *stepped foundation* (figure 3.3.4), thus reducing the amount of excavation, backfill, surplus soil removal and trench timbering/support. The foundation is stepped to follow the line of the ground and the depth of each step is usually 150 or 225 mm (multiple of brick courses) and the lap of concrete at the step should not be less than the depth of the concrete foundation and in no case less than 300 mm. The damp-proof course may also be stepped in a like manner. Where the slope exceeds 1 in 10, it is desirable to use short bored pile foundations to overcome the sliding tendency.

Figure 3.2.1 also shows the linking of the damp-proof course in the external wall with the damp-proof membrane under the floor by means of a short vertical connecting damp-proof course, so that the ground floor is completely tanked. Topsoil is removed from over the whole area of the building and in figure 3.2.1 is replaced by suitable hardcore which forms a base for the solid concrete ground floor. Alternatively the damp-proof membrane may be laid under the floor screed, where it will be less prone to damage and will probably eliminate the need for a vertical membrane.

Wide-strip Foundations

Where the loadbearing capacity of the ground is low, as for example with marshy ground, soft clay silt and 'made' ground, wide strip foundations may be used to spread the load over a larger area of soil. It is usual to provide transverse reinforcement to withstand the tensions that will arise (see figure 3.2.2). The depth below ground level should be the same as for orthodox strip foundations. All reinforcement should be lapped at corners and junctions. If there is any danger of the foundation failing as a beam in the longitudinal direction, it may be necessary to use a reinforced inverted 'T' beam of the type illustrated in figure 3.2.3. In the construction shown in figure 3.2.3, some engineers prefer to omit the hooked ends to the reinforcing bars and to obtain sufficient bond with straight bars.

Deep-strip/Trench Fill Foundations

This type of foundation was first introduced to reduce the expense entailed in constructing orthodox strip foundations to depths of 900 mm or more in shrinkable clay soils, to counteract the variable soil conditions at different seasons. In reducing the width of the foundation trench, the quantity of excavation, backfill and surplus soil removal are also reduced. The deeper foundation also provides greater resistance to fracture from unequal settlement.

In more recent times concrete trench fill has been advocated as a more economical substitute for the traditional strip foundation with brickwork below ground in many situations. There can also be time savings because of the quicker completion with concrete trench fill. A typical deep strip or trench fill foundation is illustrated in figure 3.3.1 and a minimum width of 425 mm is advocated for use with a cavity wall.[18]

Raft Foundations

These cover the whole area of the building and usually extend beyond it. They consist primarily of a reinforced concrete slab up to 300 mm thick which is often thickened under loadbearing walls. The level of the base of the raft is usually within 300 mm of the surface

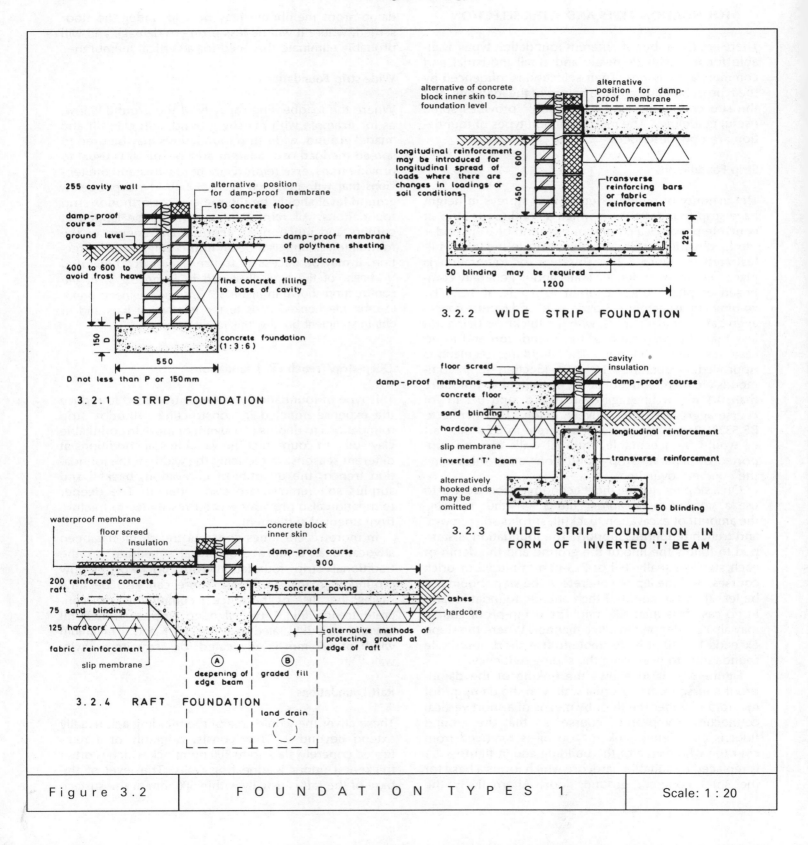

255 cavity wall

damp-proof course

ground level

400 to 600 to avoid frost heave

alternative position for damp-proof membrane

150 concrete floor

damp-proof membrane of polythene sheeting

150 hardcore

fine concrete filling to base of cavity

concrete foundation (1:3:6)

P

150 D

550

D not less than P or 150mm

3.2.1   STRIP FOUNDATION

alternative of concrete block inner skin to foundation level

alternative position for damp-proof membrane

longitudinal reinforcement may be introduced for longitudinal spread of loads where there are changes in loadings or soil conditions

transverse reinforcing bars or fabric reinforcement

450 to 600

225

50 blinding may be required

1200

3.2.2   WIDE STRIP FOUNDATION

floor screed

damp-proof membrane

concrete floor

sand blinding

hardcore

slip membrane

inverted 'T' beam

alternatively hooked ends may be omitted

cavity insulation

damp-proof course

longitudinal reinforcement

transverse reinforcement

50 blinding

3.2.3   WIDE STRIP FOUNDATION IN FORM OF INVERTED 'T' BEAM

waterproof membrane

floor screed

insulation

concrete block inner skin

damp-proof course

900

200 reinforced concrete raft

75 concrete paving

75 sand blinding

125 hardcore

fabric reinforcement

slip membrane

Ⓐ deepening of edge beam

Ⓑ graded fill

land drain

ashes

hardcore

alternative methods of protecting ground at edge of raft

3.2.4   RAFT FOUNDATION

| Figure 3.2 | FOUNDATION TYPES 1 | Scale: 1:20 |

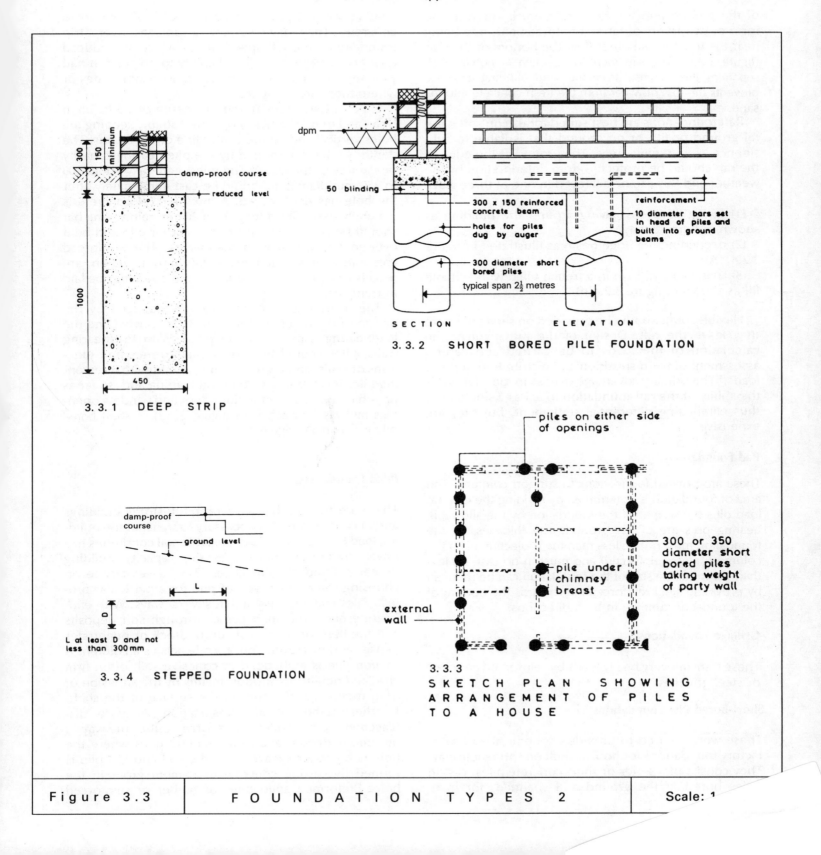

300

150 minimum

damp-proof course

reduced level

1000

450

3.3.1 DEEP STRIP

dpm

50 blinding

300 x 150 reinforced
concrete beam

holes for piles
dug by auger

reinforcement

10 diameter bars set
in head of piles and
built into ground
beams

300 diameter short
bored piles

typical span 2½ metres

SECTION   ELEVATION

3.3.2 SHORT BORED PILE FOUNDATION

piles on either side
of openings

300 or 350
diameter short
bored piles
taking weight
of party wall

pile under
chimney
breast

external
wall

3.3.3
SKETCH PLAN SHOWING
ARRANGEMENT OF PILES
TO A HOUSE

damp-proof
course

ground level

L

D

L at least D and not
less than 300 mm

3.3.4 STEPPED FOUNDATION

| Figure 3.3 | FOUNDATION TYPES 2 | Scale: ¹ |

of the ground and the reinforcement is often in the form of two layers of fabric reinforcement, one being near the top and the other near the bottom of the slab (figure 3.2.4). The slip membrane, often of polythylene, separates the concrete from the sand blinding and also prevents the loss of water from the fresh concrete into the sand.

Raft foundations are best suited for use on soft natural ground or fill, or on ground that is liable to subsidence as in mining areas. The ground at the edge of the raft should be protected from deterioration by the weather and this can be achieved in one of three ways

(1) laying concrete paving around the building as shown in figure 3.2.4;

(2) deepening the edge beam as illustrated in figure 3.2.4. (A);

(3) laying a field drain in a trench filled with suitable fill as shown in figure 3.2.4.(B).

Flexible joints should be provided on services where they leave the raft. Design of the raft involves the calculation of the loads to be carried and careful assessment of the disposition and distribution of these loads.[4] The primary advantage over strip foundations is the ability of the raft foundation to act as a single unit, thus eliminating differential settlement, but they are expensive.

Pad Foundations

These are isolated foundations to support columns. The area of foundation is determined by dividing the column load plus the weight of the foundation by the allowable bearing pressure of the ground. The thickness of the foundation must not be less than the projection from the column (unless reinforced) and must in no case be less than 150 mm. The size of the foundation can be reduced by providing steel reinforcement towards the bottom of the foundation running in both directions.

Grillage Foundations

These transmit very heavy loads by reinforced concrete or steel grillages.

Short-bored Pile Foundations

These were devised to provide economical and satisfactory foundations for houses built on shrinkable clay. They consist of a series of short concrete piles, cast in holes bored in the ground and spanned, for load-

bearing walls, by light beams usually of reinforced concrete. They have several advantages over strip foundations through speed of construction, reduced quantity of surplus spoil and ability to proceed in bad weather. Problems do however arise on stony sites or where there are many tree roots.

Holes are normally bored to a depth of 2.5 to 3.5 m by a hand or mechanically operated augur, keeping the holes vertical and on the centre line of the beams. The depth will be determined by the pile-bearing capacity or stability of the clay.[18] The piles, generally about 300 to 350 mm diameter, should be cast immediately after the hole has been bored. A mix of 1:2:4 concrete is generally used. Short lengths of 20 mm reinforcing bar should be set in the top of each corner pile and bent over to be cast in with the beams. The reinforced concrete beams, often 300 × 150 mm in section, are usually cast in formwork but in some cases may be laid in trenches.

The distribution of the piles is influenced by the design of the building, the loads to be carried and the loadbearing capacity of the piles. With loadbearing walls, piles should be provided at corners and junctions of walls, and under chimney stacks with intervening piles located to give uniform loading and, so far as possible, to keep ground floor door and window openings midway between piles (figure 3.3.3). A short bored pile is detailed in figure 3.3.2.

Piled Foundations

These are frequently used with multi-storey buildings and in cases where it is necessary to transmit the building load through weak and unstable soil conditions to a lower stratum of sufficient bearing capacity. Building Research Establishment Digest 315[19] gives advice on choosing the correct type of pile for a particular project. Piles may be classified in several ways. With *end-bearing* piles, the shaft passes through soft deposits and the base or point rests on bedrock or penetrates dense sand or gravel, and the pile acts as a column. A *friction* pile is embedded in cohesive soil, often firm clay, and obtains its support mainly by the adhesion or 'skin friction' of the soil on the surface of the shaft. Another method of pile classification relates to 'displacement' piles where soil is forced out of the way as the pile is driven, and 'replacement' piles where the hole is bored or excavated in the soil and the pile is formed by casting concrete or cement grout in the hole. Preformed solid piles of timber or reinforced concrete, and concrete or steel tubes or 'shells' with the lower end closed are examples of 'displacement' piles. The choice of pile depends on site and soil conditions, economic considerations and structural requirements. Sometimes piles are linked by beams to carry loadbearing walls.

Table 3.3. gives guidance on the choice of foundation for a variety of site conditions (extracted from Building Research Establishment Digest 67[7]).

EXCAVATION AND TRENCH TIMBERING/SUPPORT

Preliminary Siteworks

BS 8103[4] recommends that bulk excavation and filling should if possible be carried out as soon as the site is cleared. The filling should be placed in thin layers, preferably not exceeding 150 mm deep, and consolidated at a moisture content which will ensure adequate compaction. As indicated in chapter 1, if the contours of the site are such that surface water will drain towards the building, land or other drains should be laid to divert the water from the vicinity of the building. The Building Regulations 1991[5], Part C in Schedule 1, require the ground to be covered by the building to be reasonably free from vegetable matter, subsoil drainage if needed to avoid the passage of ground moisture to the interior of the building or damage to its fabric, as described in chapter 1, and precautions taken to avoid danger to health and safety caused by substances found on or in the ground to be covered by the building.

Excavation

Excavations should be cleared of water before concrete is deposited and those in clay, soft chalk or other soils likely to be affected by exposure to the atmosphere should, wherever practicable, be concreted as soon as they are dug. When this is not possible it is advisable to protect the bottom of the excavation with a 75 mm layer of lean concrete blinding, or to leave the last 50 to 75 mm of excavation until the commencement of concreting. Any excess excavation should be refilled with lean concrete; a mix of 1:12 is suitable. The method of excavation will depend on the nature and size of job and the soil conditions. The bulk of excavation work is now performed by machine. The principal machines are now briefly described.

Dragline. A bucket is dragged towards the machine, and it generally excavates below its own level.
Face shovel. This digs in deep faces above its own level.
Drag shovel or backactor. This digs below its own level and toward itself and is primarily used for trench excavation.
Skimmer. This is for shallow excavation up to 1.5 m deep and is particularly useful for levelling and roadwork.
Grab and clamshell. This is for moving loose materials.

Other machines include the *scraper* which operates like an earth plane and carries its scrapings with it; the *bulldozer* and *angledozer* for bulk excavation and grading; and the *rooter*, which is a tractor-drawn toothed scarifier for breaking up hard surfaces.

Problems with Wet Ground

Various methods have been employed to excavate below the water table. A common method is to excavate under the continuous protection of tightly closed sheeting and to form one or more sumps from which pumping proceeds continuously. For wide excavation, the strutting is expensive and inconvenient and difficulty arises in maintaining the floor of the excavation. There is also a danger of fine particles being extracted from the soil by the continuous pumping and possibly causing settlement of neighbouring ground.

Another method is the *wellpoint* system whereby cylindrical metal tubes of about 50 mm diameter are sunk into the ground until they have penetrated below the lowest level of the proposed excavation at about 1.5 m intervals. A riser pipe connects each wellpoint with a header pipe on the surface, which in turn is coupled to a suction pump. The pump creates a vacuum in the system which draws the water from the ground and so lowers the water table. A wellpoint is capable of extracting large quantities of water with a minimum of air and fine particles.

Where the ground is impervious it may be possible to use *electro-osmosis*, whereby water is extracted from the ground and the ground is stabilised. Steel tubing is used for cathodes with smaller diameter tubing as the anode. A potential of 40 to 180 volts is applied and the groundwater flows to the cathode (negative pole) from which it is extracted.

Where the ground is unstable because of groundwater, it may be necessary to enclose the excavation

Table 3.3   Choice of foundation

Soil type and site condition	Foundation	Details	Remarks
Rock, solid chalk, sands and gravels or sands and gravels with only small proportions of clay, dense silty sands	Shallow strip or pad foundations as appropriate to the loadbearing members of the building	Breadth of strip foundations to be related to soil density and loading (see figure 3.1). Pad foundations should be designed for bearing pressures tabled in BS 8004: 1986.[4] For higher pressures the depth should be increased and BS 8004: 1986[4] consulted	Keep above water wherever possible. Slopes on sand liable to erosion. Foundations 0.5 m deep should be adequate on ground susceptible to frost heave although in cold areas or in unheated buildings the depth may have to be increased. Beware of swallow holes in chalk
Uniform, firm and stiff clays: (1) Where vegetation is insignificant	Bored piles and ground beams, or strip foundations at least 1.0 m deep	Deep strip foundations of the narrow width shown in figure 3.3.1 can conveniently be formed of concrete up to the ground surface	
(2) Where trees and shrubs are growing or to be planted close to the site	Bored piles and ground beams	Bored pile dimensions as in table 3 of Building Research Establishment Digest 67[7]	Downhill creep may occur on slopes greater than 1 in 10. Unreinforced piles have been broken by slowly moving slopes
(3) Where trees are felled to clear the site and construction is due to start soon afterwards	Reinforced bored piles of sufficient length with the top 3 m sleeved from the surrounding ground and with suspended floors, or thin reinforced rafts supporting flexible buildings, or basement rafts		
Soft clays, soft silty clays	Strip foundations 1.0 m wide if bearing capacity is sufficient, or rafts	See figure 3.1 and BS 8004: 1986[4]	Settlement of strips or rafts must be expected. Services entering building must be sufficiently flexible. In soft soils of variable thickness it is better to pile to firmer strata (see Peat and fill below)
Peat, fill	Bored piles with temporary steel lining or precast or *in situ* piles driven to firm strata below	Design with large safety factor on end resistance of piles only as peat or fill consolidating may cause a downward load on pile (see Building Research Establishment Digest 63).[11] Field tests for bearing capacity of deep strata or pile loading tests will be required	If fill is sound, carefully placed and compacted in thin layers, strip foundations are adequate. Fills containing combustible or chemical wastes should be avoided
Mining and other subsidence areas	Thin reinforced rafts for individual houses with loadbearing walls and for flexible buildings	Rafts must be designed to resist tensile forces as the ground surface stretches in front of a subsidence. A layer of granular material should be placed between the ground surface and the raft to permit relative horizontal movement	Building dimensions at right angles to the front of longwall mining should be as small as possible

Source: BRE Digest 67[7]

with *steel sheet piling*. The piling is made in a variety of sections of differing strengths to resist a range of pressures. The pile sections are usually driven by a double-acting steam-operated pile hammer.

Some work in waterlogged ground has been made possible by *freezing* which has both solidified the loose ground and prevented water flowing into the working area. Freezing is normally undertaken by drilling a series of vertical boreholes of about 150 to 175 mm diameter at approximately one metre intervals around the perimeter of the work. The boreholes are lined with 100 to 150 mm diameter freezing tubes closed at the bottom end. An inner tube is then inserted with the bottom left open. Cooled brine solution is fed into the inner tube and returned to the next cooling tube or the refrigeration plant from the outer tube. The constant passage of the brine solution with a temperature lower than the freezing point of water gradually freezes the groundwater.

Other processes include the use of compressed air, soil stabilisation and grouting of the soil. All these methods are very costly and waterlogged sites should only be used as a last resort.[20]

Timbering of Excavations

The amount of timbering required to the sides of excavations is largely dependent upon two main factors – the depth of the excavation and the nature of the soil to be upheld. Vibration and loads from traffic or other causes, position of water table, climatic conditions and the time for which the excavation is to remain open also affect the decision. Figure 3.4.2 lays down broad guidelines for timbering requirements under different site conditions.

In relatively shallow trenches in firm soil it may be possible to dispense with timbering or, as it is sometimes termed, planking and strutting. The most that would be required would be pairs of 175 × 38 mm poling boards, spaced at about 1.80 m centres, and strutted with a single 100 × 100 mm strut. Alternatively adjustable steel struts may be used as illustrated in figure 3.4.1.

Most of the timber used is softwood, often red or yellow deal, possibly using pitch pine for heavy struts. The various members and their uses are now described.

Poling boards. These are boards 1.00 to 1.50 m in length (depending on the depth of excavation) and they vary in cross-section from 175 × 38 to 225 × 50 mm. The boards are placed vertically and abut the soil at the sides of the excavation.

Walings. These are longitudinal members running the length of the trench or other excavation and they support poling boards. They vary in size from 175 × 50 to 225 × 75 mm.

Struts. These are usually square timbers, either 100 × 100 or 150 × 150 mm in size. They are generally used to support the walings which, in turn, hold the poling boards in position. Struts are usually spaced at about 1.80 m centres to allow adequate working space between them.

Sheeting. This consists of horizontal boards abutting one another to provide a continuous barrier when excavating in loose soils, and so permitting the timbering to closely follow the excavation. A common size for the sheeting is 175 × 50 mm.

Runners. These are poling boards in continuous formation with tapered bottom edges, possibly shod with metal shoes as shown in figure 3.4.3. A common size for these members is 225 × 50 mm. They are particularly suitable for use in loose or waterlogged soils which will not stand unsupported.

In *moderately firm ground* the timbering is likely to take the form illustrated in figure 3.4.4. It consists of a series of poling boards which are quite widely spaced at about 600 m centres, supported by walings and struts in the normal way. In shallow trenches the poling boards would probably only be needed at about 1.80 m centres, with each pair of poling boards individually strutted with a single 100 × 100 mm strut, and no walings, on the lines indicated in figure 3.4.1.

Timbering in *loose soil* may take one of two main forms. One method uses continuous horizontal sheeting supported by pairs of poling boards and struts at about 1.80 m centres. Another method employs a continuous length of poling boards or runners supported by walings and struts. If the trench exceeds 1.50 m in depth, it is necessary to step the timbering so that the timbering to the lower stage fits inside the timbering to the upper section of the trench, and there is an overlap of about 150 mm at the point of connection between the two stages as shown in figure 3.4.5. It will be noticed that the effective width of the trench is reduced at each step. Vertical members called puncheons may be inserted between the walings on the line of the struts, to give the timbering greater rigidity.

Timbering for *shafts* can take a number of different

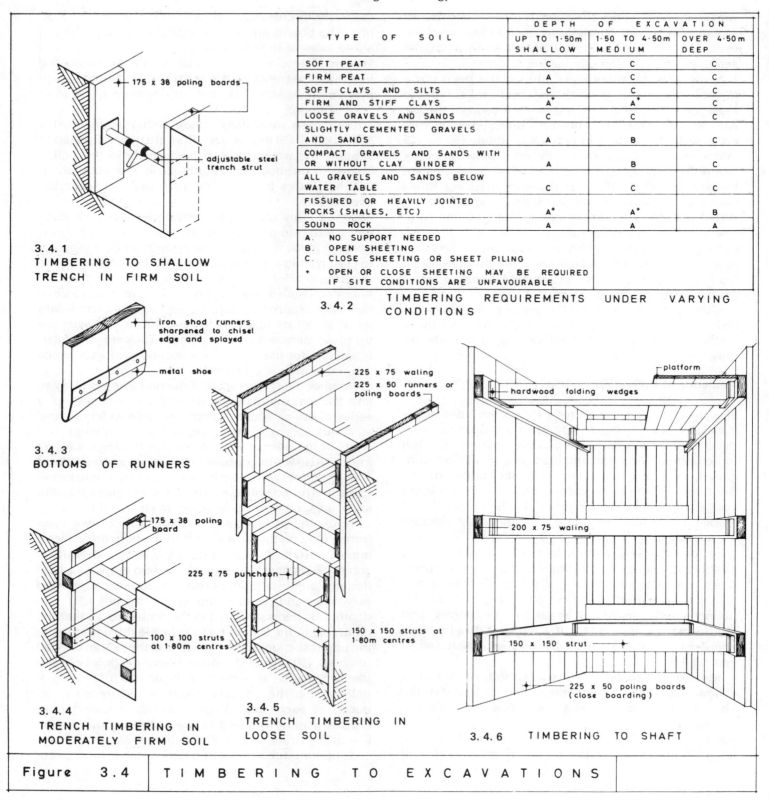

3.4.1
TIMBERING TO SHALLOW
TRENCH IN FIRM SOIL

175 x 38 poling boards

adjustable steel
trench strut

TYPE OF SOIL	DEPTH OF EXCAVATION		
	UP TO 1·50m SHALLOW	1·50 TO 4·50m MEDIUM	OVER 4·50m DEEP
SOFT PEAT	C	C	C
FIRM PEAT	A	C	C
SOFT CLAYS AND SILTS	C	C	C
FIRM AND STIFF CLAYS	A⁺	A⁺	C
LOOSE GRAVELS AND SANDS	C	C	C
SLIGHTLY CEMENTED GRAVELS AND SANDS	A	B	C
COMPACT GRAVELS AND SANDS WITH OR WITHOUT CLAY BINDER	A	B	C
ALL GRAVELS AND SANDS BELOW WATER TABLE	C	C	C
FISSURED OR HEAVILY JOINTED ROCKS (SHALES, ETC)	A⁺	A⁺	B
SOUND ROCK	A	A	A

A.   NO SUPPORT NEEDED
B.   OPEN SHEETING
C.   CLOSE SHEETING OR SHEET PILING
⁺   OPEN OR CLOSE SHEETING MAY BE REQUIRED
     IF SITE CONDITIONS ARE UNFAVOURABLE

3.4.2   TIMBERING REQUIREMENTS UNDER VARYING
        CONDITIONS

iron shod runners
sharpened to chisel
edge and splayed

metal shoe

3.4.3
BOTTOMS OF RUNNERS

225 x 75 waling
225 x 50 runners or
poling boards

platform

hardwood folding wedges

175 x 38 poling
board

225 x 75 puncheon

100 x 100 struts
at 1·80m centres

150 x 150 struts at
1·80m centres

200 x 75 waling

150 x 150 strut

225 x 50 poling boards
(close boarding)

3.4.4
TRENCH TIMBERING IN
MODERATELY FIRM SOIL

3.4.5
TRENCH TIMBERING IN
LOOSE SOIL

3.4.6   TIMBERING TO SHAFT

Figure   3.4   TIMBERING TO EXCAVATIONS

forms. One arrangement is to use 225 × 38 or 50 mm vertical sheeting or poling boards supported by 150 × 150 walings which are further strengthened by beams (possibly 250 × 150 to 300 × 225 mm in size) and vertical hangers (probably 300 × 75 mm). The beams, hangers and walings are usually connected with 20 mm mild steel bolts.

Another and rather simpler arrangement for timbering a shaft or pit, such as might be necessary for a heating chamber, is illustrated in figure 3.4.6. The timbering is made up of 225 × 50 mm poling boards supported by 200 × 75 mm walings and 150 × 150 mm struts.

Readers can obtain further information on timbering to excavations in *Timber in Excavations*.[21]

Earthwork Support with Steel Sheeting or Piles

For larger projects steel sheeting and interlocking steel sheet piling are often fixed vertically to support the sides of excavations, with trench sheeting normally in the 2–6 m depth range while piles may be up to 13 m long. The effective pile size may be influenced by the difficulty of driving, depending on the type of ground and type of hammer.[22]

The framing consists of horizontal or vertical members supporting the sheeting and may be of timber, rolled steel sections, box piles and occasionally precast or *in situ* concrete beams. Alternatively, concrete diaphragm walls can be used, in lengths not exceeding 6 m, and they will also form part of the permanent structure, comprising both framing and sheeting in a single structural unit.[22]

Another approach is to drive a series of soldier piles, usually of steel and often placed in prebored holes to eliminate the noise of pile driving. Horizontal members are then placed between them as the excavation proceeds. Soldier piles can also take the form of concrete bored piles.[22]

The loads created by the ground are taken through the sheeting and framing to a reaction member composed of any of the materials previously described. One modern development is the ground anchor drilled into either soft ground or rock and providing significant holding power, thus causing no interference with subsequent operations as the excavation support is external to the trench. Another alternative is to bolt steel tie rods above ground to steel channels supported by vertical or raking steel members driven into the ground. Further information concerning the use of steel sheet piling can be obtained from *Piling Handbook*.[23]

CONCRETE

The principal material used in foundations is concrete and this section of the chapter examines the basic materials used in concreting, selection of mixes, site operations and the use of reinforced and prestressed concrete.

Cements

Cements are substances which bind together the particles of aggregates (usually sand and gravel) to form a mass of high compressive strength (concrete). The most commonly used cement is Portland cement which may be of the ordinary variety or be rapid-hardening. There are several other cements that will produce concretes with more specialised properties.[24] The properties of concrete can also be modified by the addition of admixtures.[24,25]

Portland cement. This is made by mixing together substances containing calcium carbonate, such as chalk or limestone, with substances containing silica and alumina, such as clay or shale, heating them to a clinker and grinding them to a powder. The basic requirements for Portland cement, as fully detailed in BS 12,[26] cover composition, sampling procedures and tests for fineness, chemical composition, compressive strength, setting time and soundness. The cement combines with water to form hydrated calcium silicate and hydrated calcium aluminate. The initial set takes place in about 45 minutes and the final set within ten hours, and develops strength sufficiently rapidly for most concrete work. The setting time of rapid-hardening cement is similar to that of ordinary Portland cement, but after setting it develops strength more rapidly, enabling formwork to be struck earlier. It also has advantages in cold weather.[24]

Low heat Portland cement. This type of cement is manufactured to comply with the requirements of BS 1370[27], and it sets, hardens and evolves heat more slowly than ordinary Portland cement. It is used primarily in large structures such as dams and massive bridge abutments and retaining walls which use large volumes of concrete, and where the generated heat cannot easily be dissipated and high early strength is not usually required.

White and coloured Portland cements. White cement is produced by reducing the content of iron compounds in the cement through careful selection of the raw materials and using special manufacturing processes. Coloured cements are obtained by adding suitable pigments to white cement. These cements are mainly used for decorative purposes and have been introduced to good effect in floors and pavings.

Sulphate-resisting cement. This cement should comply with BS 4027.[28] It is more resistant than ordinary Portland cement to the effect of sulphates, which are found in some soils.

Hydrophobic Portland cement. This cement has been developed to prevent partial hydration of cement during storage in humid conditions, resulting in a reduction in early strength. Substances added during the grinding process form a water-repellent film around each grain of cement and so prevent deterioration during storage. During mixing the protective acid film is lost by abrasion and behaves as an air-entraining agent, but mixing time is normally about 25 per cent longer than for ordinary Portland cement.[29]

Portland blastfurnace cement. This is made by grinding a mixture of ordinary Portland cement clinker and granulated blastfurnace slag, complying with BS 146.[30] It is similar to ordinary Portland cement but evolves less heat and is useful when concreting in large masses.[24] It is rather more resistant to chemical attack by sulphates when the slag content is between 70 and 90 per cent.[31]

Supersulphated cement. This consists of granulated blastfurnace slag, calcium sulphate and a small percentage of Portland cement or lime, complying with BS 4248.[32] Its prime advantage is the high resistance to chemical attack by sulphate-bearing waters and weak acids. It deteriorates rapidly if stored under damp conditions. Concrete based on this cement requires a longer mixing period and the surface of the finished concrete needs to be kept moist during curing.[24]

High alumina cement. This cement differs in method of manufacture, composition and properties from Portland cements and should comply with BS 915.[33] Its main advantages stem from its very high early strength and resistance to chemical attack. Heat evolution is rapid, permitting the concrete to be placed at lower temperatures than ordinary Portland cement concrete. When mixed with a suitable aggregate, such as crushed firebrick, it makes an excellent refractory concrete to withstand high temperatures. It should not be used in structural concrete,[24] as it suffers from chemical conversion, which is accelerated by hot and damp conditions.

Pozzolanic cements. These cements are mixtures of a Portland cement and a pozzolanic material (one combining with lime to form a hard mass). They offer good resistance to chemical attack, but the rate of heat evolution and strength attainment is reduced.[24]

Aggregates

Aggregates are gravels, crushed stones and sand which are mixed with cement and water to make concrete. The two most essential characteristics for aggregates are durability and cleanliness; cleanliness includes freedom from organic impurities.

Fine aggregate. This consists of natural sand, or crushed stone or crushed gravel sand that mainly passed through a 5 mm British Standard sieve with a good proportion of the larger particles.

Coarse aggregate. This is primarily natural gravel, or crushed gravel or stone that is mainly retained on a 5 mm British Standard sieve. Both types of aggregate should comply with the grading requirements of BS 882.[34] Artificial coarse aggregates such as clinker and slag are used for lightweight concrete.[35] The maximum size of coarse aggregate is determined by the class of work. With reinforced concrete the aggregate must be able to pass readily between the reinforcement and it rarely exceeds 20 mm. For foundations and mass concrete work the size can be increased possibly up to 40 mm. The type of aggregate used directly influences the fire protection and thermal insulation qualities of the concrete.[24]

BRE Information Paper IP 16/89[36] describes how indigenous resources of porous carboniferous sandstone in north-west England could make a significant contribution to the supply of aggregates for concrete for less demanding applications such as foundations and floors. However, there is a considerable variation in the quality of materials between different quarries and between different layers in some quarries, and they therefore require careful assessment to determine whether

(1) higher cement content may be required to achieve sufficient strength;

(2) drying shrinkage is too high for the proposed end use and exposure conditions;

(3) embedded metal may need extra protection in some exposure conditions;

(4) design changes may be needed to accommodate

low elastic moduli of the concretes.

A BRE survey[37] showed that some igneous rocks of the basalt and dolerite types and certain sedimentary rocks (especially greywacke and mudstone) can be shrinkable, and these are encountered extensively in the industrial belt of Scotland. Aggregate shrinkage produces increased shrinkage of concrete despite adopting good practice, although the shrinkage can be minimised by taking normal precautions, such as the careful control of mix proportions. As a general rule, the smaller the proportion of cement in the concrete or the larger the maximum particle size of aggregate, the less its shrinkage. However, even when such precautions are taken, the use of shrinkable aggregates may still have significant practical consequences for design.

SITE TESTING OF MATERIALS

It is sometimes necessary to carry out site tests on materials to determine their suitability. The following tests relating to cement and sand serve to illustrate the approach.

Cement

(1) Examine to determine whether it is free from lumps and is of a flour-like consistency (free from dampness and reasonably fresh).

(2) Place hand in cement and if of blood heat then it is in satisfactory condition.

(3) Settle with water as paste in a closed jar to see whether it will expand or contract.

Sand

(1) Handle the sand; it should not stain hands excessively, ball readily or be deficient in coarse or fine particles.

(2) Use a standard sieve test – if more than 20 per cent is retained on a 1.25 mm sieve, it is unsuitable for use.

(3) Apply a silt or organic test – a jar half filled with sand and made up to the three-quarters mark with water; shake vigorously and leave for three hours; the amount of silt on top of the sand is then measured and this should not exceed six per cent.

CONCRETE MIXES

Concrete must be strong enough, when it has hardened, to resist the various stresses to which it will be subjected and it often has to withstand weathering action. When freshly mixed it must be of such a consistency that it can be readily handled without segregation and easily compacted in the formwork. The fine aggregate (sand) fills the interstices between the coarse aggregate, and both aggregates need to be carefully proportioned and graded. The strength of concrete is influenced by a number of factors

(1) proportion and type of cement;
(2) type, proportions, gradings and quality of aggregates;
(3) water content;
(4) method and adequacy of batching, mixing, transporting, placing, compacting and curing the concrete.

Concrete mixes can be specified by the volume or weight of the constituent materials or by the minimum strength of the concrete. BRE Digest 326[38] outlines the disadvantages of nominal mix proportions by volume (prescribed mixes) such as 1:3:6 for plain concrete foundations, 1:2:4 for normal reinforced concrete and 1:1½:3 for strong reinforced concrete work. These mixes fail to specify adequately the cement content of the mix, as the actual cement content of a cubic metre of concrete made to a particular nominal mix varies with different aggregates and different water contents. Hence there is considerable merit in specifying mixes in terms of the cement content in kilogrammes per cubic metre of fresh concrete.

Nevertheless, the strength of concrete produced under site conditions varies widely. A more realistic approach is to specify a minimum strength of the concrete and for the proportions of cement, sand, coarse aggregate and water then to be selected to achieve it (designed mixes). This trend away from the specification of nominal mixes and towards the design of mixes on a strength basis is reflected in BS 8110[39] and BS 5328.[40] Table 3.4 gives details of the more common grades of concrete for designed mixes. BS 8110[39] recommends grade 7 to 10 plain concrete (7 to 10 N/mm^2) and grade 15 to 25 for reinforced concrete. BS 5075[25] details five categories of admixture which may be used to modify one or more of the properties of cement concrete, such as accelerating and retarding qualities.

Table 3.4　Grades of concretes

Grade	Characteristic compressive strength at 28 days (N/mm²)	Lowest grade permitted for use in	Minimum cement content for use in	
			Plain concrete (kg/m³)	Reinforced or prestressed concrete (kg/m³)
C7.5	7.5	Concrete not containing embedded metal	120	–
C10	10.0		150	–
C15	15.0	Reinforced concrete with lightweight aggregate	180	240
C20	20.0	Reinforced concrete with dense aggregate	220	240
C25	25.0			300
C30	30.0	Concrete with post-tensioned tendons		
C35	35.0			300
C40	40.0	Concrete with pre-tensioned tendons		
C45	45.0			
C50	50.0			
C55	55.0			
C60	60.0			

Source: BRE Digest 326[38]

The water/cement ratio is a most important factor in concrete quality. It should be kept as low as possible consistent with sufficient workability to secure fully compacted concrete with the equipment available on the site. The higher the proportion of water, the weaker will be the concrete. Water/cement ratios are usually in the range of 0.40 to 0.60 (weight of water divided by weight of cement). Allowance has to be made for absorption by dry and porous aggregates and the surface moisture of wet aggregates. Badly proportioned aggregates require an excessive amount of water to give adequate workability, and this results in low strength and poor durability. A common test for measuring workability on the site is the slump test, although for greater accuracy the compacting factor test or consistometer test should be used.[41]

The *slump test* consists of filling a 300 mm high open-ended metal cone with four consolidated layers of concrete and then lifting the cone and measuring the amount of slump or drop of the cone-shaped section of concrete. Slumps vary from 25 mm for vibrated mass concrete to 150 mm for heavily reinforced non-vibrated concrete.

The apparatus for the *compacting factor test* consists of two conical hoppers mounted above a cylindrical container. The top hopper is filled with concrete and this is allowed to fall through a hinged trapdoor at the base into the second hopper below. The trapdoor of the second hopper is then released and the concrete allowed to fall into the container, when the top surface of the concrete is levelled off and the combined weight of the cylinder and partially compacted concrete is taken. Finally the cylinder is filled with concrete in 50 mm layers which are each compacted and the cylinder is weighed once again. The compacting factor is the ratio of the partially compacted to the fully compacted weight.

CONCRETING OPERATIONS

Concreting operations comprise batching, mixing, transporting, placing, compacting and curing the concrete. However, most concrete is now supplied ready-mixed, when it is important to check that the correct material has been delivered to the site and that it is placed in its final position in the right condition. Even on small sites, concrete is often pumped from the mixer to the trenches or formwork and this assists in ensuring that a concrete of suitable consistency has been supplied.

Batching

Prior to use the cement needs to be stored in a damp-proof and draught-proof structure, while aggregates should be stored on clean, hard surfaces. For very small projects the aggregates may be measured by volume using a timber gauge box, with its size related to the quantity of aggregate required for a bag of cement. Weigh batching gives much more accurate results. It is necessary to check constantly that the gauging apparatus is being used correctly and that allowance is made for the moisture content of the sand.

Mixing

Concrete can be mixed by hand or machine on the site or purchased ready-mixed, when it is delivered to the site from a central batching plant in lorries with mixing drums or with revolving containers. Hand mixing should only be used for limited quantities of concrete on the smallest projects, with the proportion of cement increased by ten per cent. The materials are mixed dry on a clean, hard surface, water added through a rose and further mixing continued until a uniform colour is obtained. Concrete mixers are designed by their capacity of concrete and type (tilting, non-tilting and reversing.)[42] The normal mixing time is at least one minute after all materials, including water, have been placed in the mixer. The mixer driver should monitor visibly the workability, homogeneity and cohesiveness of each mix, as well as the consistency of production.[43]

Transportation

The type of plant used to transport concrete from the mixer to the point of deposition depends on the size of the project and the height above ground at which the concrete is to be placed. Barrows, lorries, dumpers and mechanical skips are used for this purpose, and they must be kept clean, and on very large schemes pumping may prove to be easier, quicker and cheaper. Outputs claimed vary from 14 to 24 m^3/h per cylinder with a range of 500 mm horizontally and 50 m vertically through a 180 mm pipeline. Concrete should generally be placed in its final position within 30 minutes of leaving the mixer.

Placing

All formwork should be checked, cleaned and oiled before concrete is placed against it. Concrete should not be permitted to fall freely more than one metre, to prevent air pockets forming and segregation of the materials.

Compaction

The concrete must be adequately compacted to secure maximum density. It can be done by hand or by using vibrators. Vibrators are more efficient and can be used with drier mixes. There are three main types: external, internal and surface vibrators. Care must be taken to ensure that concrete is well compacted against forms and at corners and junctions. Construction joints, at the end of a day's work, should be carefully formed

and be so positioned as not to be too conspicuous in the finished work. The 'laitance' film of cement and sand at the joint is removed to ensure a good bond between the new and old work. In large areas of concrete expansion joints should be incorporated to permit movement of the concrete with differences in temperature.

In cold weather the rate of setting and hardening of cement slows down and practically ceases at freezing point. Concrete should be at least 5°C when placed and should not fall below 2°C until it has hardened. Special precautions should be taken such as keeping aggregates and mixing plant under cover; covering exposed concrete surfaces with insulating material; using a richer mix of concrete and/or rapid-hardening cement; heating water and aggregate; placing concrete quickly and leaving the formwork in position for longer periods.

Curing

The chemical reaction which accompanies the setting of cement and hardening of concrete is dependent on the presence of water; so exposed concrete should be covered with bubble plastic sheets or quilts of plastic with fibres, glass wool, straw or other suitable material, to protect it from the sun and drying winds for at least seven days.[43]

Formwork

Formwork or shuttering is used to retain concrete in a specific location until the concrete has developed sufficient strength to stay in position without support. The general requirements for formwork are as follows

(1) it should be sufficiently rigid to prevent significant movement during the placing of the concrete;
(2) it should be of sufficient strength to withstand working loads and weight of wet concrete;
(3) it must have sufficiently tight joints to prevent loss of fine material from concrete;
(4) erection and stripping shall be easily performed with units of a manageable size;
(5) it should produce concrete face of good appearance; and
(6) it must permit removal of side forms prior to striking of soffit shuttering.

Formwork may be of timber, with possibly six or seven uses, or of sheet steel or metal-faced plywood. The use of telescopic units, adjustable props and special beam and column clamps, permits the easy erection and dismantling of steel formwork.

The period which should elapse before formwork is struck depends on the type of cement, nature of component and the temperature. For example, formwork to beam sides, walls and columns could be removed after one day with rapid hardening Portland cement and normal weather conditions (15°C), whereas six days would be required for ordinary Portland cement in cold weather (just above freezing). Similarly with slabs the periods would range from two to ten days, and with beam soffits from four to fourteen days, with props remaining for longer periods in both cases.

CHEMICAL FAILURES IN CONCRETE

Concrete is subject to two main forms of chemical failure, namely carbonation and alkali–silica reaction, and these have become more prevalent in recent years. Carbonation is a form of deterioration which attacks exposed concrete. All Portland cement contains a proportion of calcium hydroxide (free lime). The hydration, or hardening of the cement in concrete, does not change the state of the free lime. Subsequently, concrete is in contact with carbon dioxide in the atmosphere, and this reacts with the free lime to form calcium carbonate, thereby reducing the alkalinity of the concrete. The alkalinity is necessary to protect mild steel reinforcement in concrete from oxidation or rusting. If the alkalinity of the concrete is destroyed by carbonation, the steel is liable to rust in the presence of moisture and air. The same concrete which is vulnerable to carbonation is the type of concrete which is inclined to be porous and therefore most likely to expose the reinforcement to these two elements. The resulting rust can cause progressive cracking and spalling of the concrete.[44]

Carbonation starts at the surface of the concrete and proceeds inwards at a decreasing rate determined by the type of concrete, its quality and its density. It was originally thought that there was little risk of carbonation becoming a problem if the reinforcement had sufficient cover of concrete. In practice, the cover has often proved to be inadequate and the concrete too permeable, producing a fatal combination. This adverse condition is sometimes aggravated by the addition of chloride additives to the concrete mixes.[44]

Where the carbonation has not reached the steel reinforcement, overcladding the superstructure to prevent extensive rain penetration or treatment with a coating which is resistant to water ingress and carbon dioxide diffusion can reinstate its life expectancy. If carbonation had reached the steel, the area of carbonated concrete requires cutting out, then the rusting steel should be cleaned and treated and the wall reinstated with a water-repellent mix before the whole wall surface is overclad or resurfaced.[44]

The other major concrete problem is alkali–silica reaction, which has been termed 'concrete cancer'. This defect is caused by a chemical reaction between the alkalis normally present in concrete and certain forms of aggregate quarried in various parts of the United Kingdom. For this defect to occur, moisture has to be present. Alkali–silica reaction is most likely to occur in exposed concrete and the best means of protection is to shield it from excessive wetting by rain or condensation.[44]

REINFORCED CONCRETE

Concrete is strong in compression but weak in tension, and where tension occurs it is usual to introduce steel bars to provide the tensile strength which the concrete lacks. For example, with a concrete beam or lintel, compression occurs at the top and tension at the bottom, so the reinforcement is placed about 25 mm up from the bottom of the beam and the ends are often hooked to provide a grip. The 25 mm cover prevents rusting of the reinforcement.

The steel must be free from loose mill scale, loose rust, grease, oil, paint, mud and other deleterious substances which impair the bond between the steel and concrete. The most common form of reinforcement is the mild steel bar to BS 4449[45] or BS 4482.[46] Medium and high tensile bars are also available, and deformed bars which are twisted and/or ribbed provide a better bond and greater frictional resistance than round bars and obviate the need for hooked ends. It is important that the reinforcement is fixed securely to avoid displacement during the placing of the concrete. The bars are tied together with soft wire at intersections, and spacers of small precast concrete blocks, concrete rings or plastic fittings ensure the correct cover of

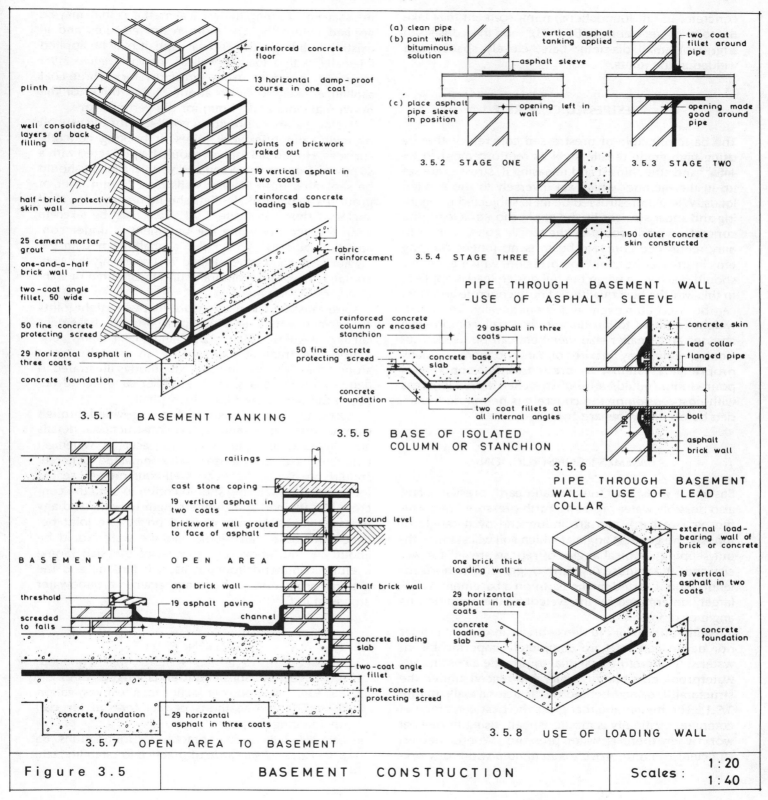

reinforced concrete floor

plinth

13 horizontal damp-proof course in one coat

well consolidated layers of back filling

joints of brickwork raked out

19 vertical asphalt in two coats

half-brick protective skin wall

reinforced concrete loading slab

25 cement mortar grout

one-and-a-half brick wall

two-coat angle fillet, 50 wide

fabric reinforcement

50 fine concrete protecting screed

29 horizontal asphalt in three coats

concrete foundation

3.5.1 BASEMENT TANKING

(a) clean pipe
(b) paint with bituminous solution

vertical asphalt tanking applied

two coat fillet around pipe

asphalt sleeve

(c) place asphalt pipe sleeve in position

opening left in wall

opening made good around pipe

3.5.2 STAGE ONE

3.5.3 STAGE TWO

150 outer concrete skin constructed

3.5.4 STAGE THREE

PIPE THROUGH BASEMENT WALL - USE OF ASPHALT SLEEVE

reinforced concrete column or encased stanchion

29 asphalt in three coats

50 fine concrete protecting screed

concrete base slab

concrete foundation

two coat fillets at all internal angles

3.5.5 BASE OF ISOLATED COLUMN OR STANCHION

concrete skin

lead collar

flanged pipe

bolt

asphalt brick wall

3.5.6 PIPE THROUGH BASEMENT WALL - USE OF LEAD COLLAR

railings

cast stone coping

19 vertical asphalt in two coats

brickwork well grouted to face of asphalt

ground level

BASEMENT   OPEN AREA

one brick wall

half brick wall

threshold

19 asphalt paving

channel

screeded to falls

concrete loading slab

two-coat angle fillet

fine concrete protecting screed

concrete foundation

29 horizontal asphalt in three coats

3.5.7 OPEN AREA TO BASEMENT

external load-bearing wall of brick or concrete

one brick thick loading wall

19 vertical asphalt in two coats

29 horizontal asphalt in three coats

concrete loading slab

concrete foundation

3.5.8 USE OF LOADING WALL

| Figure 3.5 | BASEMENT CONSTRUCTION | Scales: | 1:20 1:40 |

concrete. In raft foundations, reinforcement may take the form of steel fabric to BS 4483,[47] and this consists of a grid of small diameter bars, closely spaced and welded at the joints.

PRESTRESSED CONCRETE

The basic principle of prestressed concrete is that by using high-grade tensile steel as reinforcement to BS 4486[48] and stretching it and releasing it, stresses are set up in the member which act inversely to the applied loads. When a prestressed beam is subjected to bending and shear stresses by the application of a load, the concrete in what would be the tensile zone of an ordinary reinforced concrete beam is no longer carrying tensile stresses but only compression, which generally should reach zero when the full design load is applied. In this way the best qualities of both concrete and steel can be used effectively and economically. The dead-weight is likely to produce only small compressive stresses and smaller size members can be used. Tensioning cables may be wires or bars and there are many proprietary systems. In *pre-tensioning* concrete is poured into moulds around stretched wires, whereas with *post-tensioning* the concrete is permitted to harden before the wires are stretched.

BASEMENT CONSTRUCTION

Basement walls have to withstand earth pressures and also possibly water pressure. Earth pressures vary with the type of earth and are influenced by its angle of repose (the greatest angle at which soil will stand without slipping). Typical angles of repose are 16° for wet clay and 30° for loose dry earth or gravel. The horizontal pressure exerted by the earth on a basement wall is largely determined by the weight of the earth and its angle of repose.

Neither brickwork in dense bricks in cement mortar nor dense concrete are completely impermeable to water. It is therefore essential to provide a continuous waterproof membrane, preferably placed under the structural floor and behind the structural walls (figure 3.5.1). The membrane that gives the best guarantee of complete continuity is mastic asphalt, using three-coat work when there is water pressure. Where there is sufficient room to provide a minimum width of work-

ing space of 600 mm, the vertical mastic asphalt may be applied externally. On more restricted sites and in existing buildings, the vertical asphalt may be applied internally, with a loading wall as shown in figure 3.5.8 to resist pressure from groundwater. Three-coat asphalt is likely to finish about 29 mm thick for the horizontal work and 19 mm for the vertical work.

It is necessary to provide a good key for the asphalt by raking out brickwork joints or hacking concrete surfaces. Horizontal asphalt should be protected with a 50 mm fine concrete screed and vertical asphalt should be separated from loading walls by a 25 mm layer of grout. Two-coat asphalt fillets should be provided at all angles as shown in figure 3.5.1. Care must be taken to keep the site dry while the basement is under construction and to protect the asphalt from damage. Figure 3.5.5 shows the method of continuing the horizontal damp-proof membrane under a column or stanchion in the basement.

Care must be taken to ensure that watertight joints are formed wherever pipes pass through basement tanking. Figures 3.5.2, 3.5.3 and 3.5.4 show one method of treating pipes using an asphalt sleeve around the pipe. An alternative method is illustrated in figure 3.5.6 using a sheet lead collar sandwiched between the flanges and into the asphalt.

Access to basements is sometimes provided through external open areas and typical constructional details are illustrated in figure 3.5.7. To secure a watertight concrete basement dense, well-compacted concrete should be placed between well-constructed, sealed formwork. De-watering should continue until the concrete has attained sufficient strength to withstand any pressures subsequently exerted on it. The joint between the base slab (floor) and the wall should be about 250 mm above the surface of the slab. Some useful recommendations on asphalt tanking and the general protection of buildings against groundwater are given in CP 102.[49]

REFERENCES

1. *BRE Digest 64*: Soils and foundations: 2 (1972)
2. *BS 5930: 1981* Code of practice for site investigation
3. *BS 1377: 1975* Methods of test for soil for civil engineering purposes
4. *BS 8004: 1986* Code of practice for foundations. *BS 8103 Part 1: 1986* Code of practice for stability, site

investigation, foundations and ground floor slabs for housing
5. *The Building Regulations 1991 and Approved Documents A (1991) and C (1991)*. HMSO
6. *BS 6399* Design loading for buildings. *Part 1: 1984* Code of practice for dead and imposed loads
7. *BRE Digest 67*: Soils and foundations: 3 (1980)
8. *BRE Digest 298*: The influence of trees on house foundations in clay soils (1987)
9. D.F. Cutler and I.B.K. Richardson. *Tree Roots and Buildings*. Construction Press (1989)
10. *BS 5837: 1980* Code of practice for trees in relation to construction
11. *BRE Digest 63*: Soils and foundations: 1 (1979)
12. *BRE Digest 251*: Assessment of damage in low-rise buildings (1990)
13. *BRE Digest 352*: Underpinning (1990)
14. R. Hunt, R.H. Dyer and R. Driscoll. *Foundation Movement and Remedial Underpinning in Low-rise Buildings*. BRE (1991)
15. G. Barnbrook. *House Foundations for the Builder and Building Designer*. Cement and Concrete Association (1981)
16. *BS 5328: 1981* Methods for specifying concrete, including ready-mixed concrete
17. *BS 8110* Structural use of concrete. *Part 1: 1985* Code of practice for design and construction
18. *BRE Digest 241*: Low-rise buildings on shrinkable clay soils: Part 2 (1990)
19. *BRE Digest 315*: Choosing piles for new construction (1986)
20. *BS 6031: 1981* Code of practice for earthworks
21. Timber Research and Development Association (TRADA). *Timber in Excavations* (1984)
22. C.J. Wilshere. Temporary works. *Civil Engineer's Reference Book*. Butterworths (1989)
23. British Steel Corporation. *Piling Handbook* (1984)
24. *BRE Digest 325*: Concrete—Part 1: Materials (1988)
25. *BS 5075* Concrete admixtures. *Part 1: 1982* Specification for accelerating admixtures, retarding admixtures and water-reducing admixtures
26. *BS 12: 1978* Ordinary and rapid-hardening Portland cement
27. *BS 1370: 1979* Specification for low-heat Portland cement
28. *BS 4027: 1980* Specification for sulphate-resisting Portland cement
29. G.D. Taylor. *Materials of Construction*. Construction Press (1984)
30. *BS 146: 1991* Portland blastfurnace cements
31. *BRE Digest 363*: Sulphate and acid resistance of concrete in the ground (1991)
32. *BS 4248: 1974* Specification for supersulphated cement
33. *BS 915, Part 2: 1972* High alumina cement
34. *BS 882: 1983* Specification for aggregates from natural sources for concrete
35. *BS 877, Part 2: 1977* Foamed or expanded blastfurnace slag lightweight aggregate for concrete; *BS 1165: 1985* Specification for clinker and furnace bottom ash aggregates for concrete; *BS 3797, Part 2: 1982* Specification for lightweight aggregates for concrete
36. *BRE Information Paper IP16/89* Porous aggregates in concrete: sandstones from NW England
37. *BRE Digest 357*: Shrinkage of natural aggregates in concrete (1991)
38. *BRE Digest 326*: Concrete, Part 2: Specification, design and quality control (1988)
39. *BS 8110* Structural use of concrete, *Part 1: 1985* Code of practice for design and construction
40. *BS 5328: 1981* Methods of specifying concrete, including ready-mixed concrete
41. *BS 1881* Testing concrete, *Part 103: 1983* Method for determination of compacting factor
42. *BS 1305: 1974* Batch type concrete mixers
43. G. Taylor. *Concrete site work*. Telford (1984)
44. I.H. Seeley. *Building Maintenance*. Macmillan (1987)
45. *BS 4449: 1988* Specification for carbon steel bars for the reinforcement of concrete
46. *BS 4482: 1985* Specification for cold reduced steel fabric for the reinforcement of concrete
47. *BS 4483: 1985* Specification for steel fabric for the reinforcement of concrete
48. *BS 4486: 1980* Specification for hot rolled and hot rolled and processed high tensile alloy steel bars for the prestressing of concrete
49. *CP 102: 1973* Code of practice for protection of buildings against water from the ground

4 WALLS AND PARTITIONS

Walls to buildings can be constructed in various ways using a variety of materials. In order to appreciate the different constructional techniques and their relative merits, it is necessary to know the functions of walls in various locations. The principal methods used in the construction of both external and internal walls and partitions are examined in some detail.

FUNCTIONS OF WALLS

External walls should perform a number of functions

(1) support upper floors and roofs together with their superimposed loads;

(2) resist damp penetration (mainly detailed in chapter 15);

(3) provide adequate thermal insulation (as detailed in chapter 15);

(4) provide sufficient sound insulation (as detailed in chapter 15);

(5) offer adequate resistance to fire;

(6) look attractive and satisfactorily accommodate windows and doors.

Strength of Walls

Walls must be of sufficient thickness to keep the stresses within the limits of the safe compressive stresses of the materials in the wall, for example bricks and mortar. Tensile stresses may be induced by unequal loadings on the wall and the bonding and jointing of the bricks or blocks must be able to withstand them. The thickness-to-height ratio must be sufficient to prevent buckling and there must be adequate lateral support to resist overturning. The materials in the wall must be durable and able to withstand soluble salts, atmospheric pollution and other adverse conditions.

BS 5628[1] recommends permissible stresses for load-bearing walls, and from these stresses the thickness of walls can be determined in relation to the loads to be carried. To utilise the full capacity of high-strength bricks (70 N/mm^2 or more) a 1:3 cement-sand mix is needed. For lower strength bricks, mortars with an increased proportion of lime can be used without any great loss in brickwork strength. Suitable mortar mixes for various unit strengths are listed in table 4.1. The strength of a 255 mm cavity wall with loads spread over both leaves is 20 per cent less than that of a one-brick wall. Wall ties built into cavity walls are intended to share lateral forces and deflections between the two leaves. Another useful British Standard concerned with the design of low-rise buildings is BS 8103.[2]

Table 4.1 Suitable mortar mixes for various unit strengths

Unit strength (N/mm^2)	Mortar mix (cement: lime: sand by volume)	Designation in BS 5628[1]
10	1:2:9	(iv)
20 to 35	1:1:6	(iii)
48 to 62	1:¼:3	(i)
70 or more	1:0:3	(i)

Source: Building Research Establishment Digest 246[3]

The Building Regulations 1991[4] prescribe ways of calculating wall thicknesses for residential buildings up to 3 storeys high and small single storey non-residential buildings. Solid walls constructed of coursed brickwork or blockwork should be at least as thick as $\frac{1}{16}$ of the storey height. Approved Document A1/2 (1991) of the Building Regulations[4] prescribes minimum thicknesses of certain external, compartment and separating walls in table 5 of the Document. Typical examples are 190 mm thickness for a wall of a length not exceeding 12 m and a height not exceeding 3.5 m, and for walls with a height between 3.5 m and 9 m and with a length not exceeding 9 m. For walls exceeding 9 m and not exceeding 12 m long of a similar height, the thickness shall be 290 mm from the base for the height of one storey and 190 mm for the rest of its height. The thicknesses of walls for smaller buildings are however determined by considerations other than strength, such as weather resistance, thermal insulation and fire resistance.

BRE has developed the 'brench' (bond wrench), an *in situ* tool for testing the bond of masonry units (bricks and blocks) to mortar. It consists of a long lever which is clamped to a brick or concrete block at one end and the other end is free. An increasing weight is gradually applied at the free end until the brick is rotated free from the mortar joint immediately below it. The load and the associated moment at which it occurs is a measure of the strength of the masonry.[5]

Resistance to Dampness

Damp penetration is one of the most serious defects in buildings. Apart from causing deterioration of the structure, it can also result in damage to finishings and contents and can in severe cases adversely affect the health of occupants. Dampness may enter a building by a number of different routes and these are now examined.

Water introduced during construction. In building a traditional house several tonnes of water are introduced into the walls during bricklaying and plastering. The walls often remain damp until a summer season has passed. As the moisture dries out from inner and outer surfaces it is liable to leave deposits of soluble salts or translucent crystals. A porous wall with an impervious coating on one surface will cause drying out on the other surface. Typical moisture contents of some of the more common building materials are plaster: 0.2 to 1.0 per cent; lightweight concrete: up to 5 per cent; and timber: 10 to 20 per cent.

Penetration through roofs, parapets and chimneys. Tiled roofs may admit fine blown snow and fine rain, particularly in exposed situations. Both tiles and slates must be laid to an adequate pitch and be securely fixed. It is wise to provide a generous overhang at eaves. Parapets and chimneys can collect and deliver water to parts of the building below roof level, unless they have adequate damp-proof courses and flashings. Leakage through flat roofs is more difficult to trace and needs to be distinguished from condensation.

Penetration through walls. Penetration occurs most commonly through walls exposed to the prevailing wet wind or where evaporation is retarded as in light wells. On occasions the fault stems from excessive wetting from a leaking gutter or downpipe. There must be a limit to the amount of rain that a solid wall can exclude. In the wetter parts of the country (for example SW, W and NW England rising to a maximum in NW Scotland) and on exposed sites, one-brick thick solid walls may permit penetration of water. Hence Approved Document C4 (1991) of the Building Regulations[4] recommends thicker solid brick walls at least 328 mm thick, dense aggregate concrete blockwork at least 250 mm thick, or lightweight aggregate or aerated autoclaved concrete blockwork at least 215 mm thick, or walls rendered on the external face in conditions of very severe exposure, as an alternative to cavity walls. In this connection Lacy has devised indices of exposure to driving rain.[6] The greatest penetration is likely to occur through the capillaries between the mortar joints and the walling units. The more impervious the mortar and the denser the bricks or blocks, the more serious is the penetration likely to be. Dense renderings can prevent moisture drying out more effectively than preventing its entry, and this tendency is accentuated in cracked renderings with moisture penetrating the cracks by capillary action, becoming trapped behind the rendering and subsequently drying out on the inner face of the wall.

Rain penetration normally occurs in localised patches, with well-defined edges, but not in any particular position. The patches will increase in wet weather, especially after driving rain, and fade away in prolonged dry spells.[7]

Cavity walls when properly detailed and soundly constructed will not permit penetration of rain. Penetration when it occurs is usually the direct result of faulty detailing at openings or mortar droppings on wall ties. Finally, disintegration of brickwork may be caused by the action of sulphates or frost when the bricks are saturated.

Rising damp. In older buildings damp may rise up walls because of the lack of damp-proof courses. In newer buildings rising damp may occur through a defective damp-proof course, the bridging of the damp-proof course by a floor screed internally, or by an external rendering, path or earth outside the building, or mortar droppings in the cavity. Damp may also penetrate a solid floor in the absence of a damp-proof membrane.

It usually results in a horizontal tidemark about 600 to 900 mm above ground level on external walls.

Other causes. Dampness may result from leaks in a plumbing system, although this must not be confused with condensation on cold pipes. The amount of water vapour that air can contain is limited and when this is reached the air becomes saturated. The saturation point varies with the temperature – the higher the temperature of the air, the greater the weight of water vapour it can contain.[8]

When warm, damp weather follows a period of cold, the fabric of a building which has not been fully centrally heated may remain cold for some time. The warm, moist air will tend to condense on the cold wall and floor surfaces. Furthermore, the humidity inside a building is usually higher than outside. The occupants and many of their activities, such as cooking and washing, produce moisture vapour. Changing living habits and constructional methods, such as hard plasters and solid floors, accentuate the condensation problem. Water vapour from a kitchen or bathroom may circulate through a house and condense on colder surfaces of stairwells and unheated bedrooms. Remedial measures include the provision of background heating, thermal insulation and adequate ventilation. Further information on dampness and condensation is given in chapter 15.

Thermal Insulation

Thermal insulation serves a number of purposes

(1) to prevent excessive loss of heat from within a building;
(2) to prevent a large heat gain from the outside in hot weather;
(3) to prevent condensation;
(4) to reduce expansion and contraction of the structure.

The thermal transmittance or *U-value* of a wall, roof or floor of a building is a measure of its ability to conduct heat out of a building; the greater the *U*-value, the greater the heat loss through the structure. The total heat loss through the building fabric is found by multiplying *U*-values and areas of the externally exposed parts of the building, and then multiplying the result by the temperature difference between inside and outside. *U*-values are expressed in $W/m^2 K$ (watts per square metre for 1° Celsius difference between internal and external temperatures).

The statutory requirements (Building Regulations 1991, part L[4]) and their implementation are examined in detail in chapter 15, because of the various structural components that are affected.

Sound Insulation

Noise levels both inside and outside buildings have increased significantly in recent years. The Building Regulations 1991 in Part E of Schedule 1 require party walls and walls between habitable rooms of a dwelling to provide reasonable resistance to the transmission of airborne sound, and means of determining suitable construction are given in Approved Document E/1/2/3 (1991) and these are detailed in chapter 15, as they involve other components in addition to walls.

Fire Resistance

Internal fire spread (linings)

The spread of fire over a surface can be restricted by provisions for the surface material to have low rates of surface spread of flame, and in some cases to restrict the rate of heat produced (paragraph B2 of Schedule 1 to the Building Regulations 1991). The provisions in Approved Document B2 (1991) prescribe class 3 materials for wall and ceiling linings of small rooms not exceeding 4 m^2 floor area in residential buildings and 30 m^2 in non-residential buildings and class 1 materials in other rooms and circulation spaces within dwellings.

Internal fire spread (structure)

Premature failure of the structure can be prevented, by provisions for loadbearing elements of the structure to have a minimum standard of fire resistance as prescribed in Approved Document B3 (1991) of the Building Regulations. The minimum periods of fire resistance for all loadbearing elements of the structures of dwellinghouses are prescribed in table A2 of Approved Document B (1991) as 30 mins in basement storeys and upper storeys with the height of the top floor above ground in building or separating part of building not more than 5 m and 60 mins where the height exceeds 5 m and does not exceed 20 m.

A BRE report[9] gives guidance on the notional

periods of fire resistance for a wide range of constructions. For example, loadbearing timber framing members at least 44 mm wide and spaced at not more than 600 mm apart, with lining both sides of 12.5 mm plasterboard with all joints taped and filled; and 100 mm reinforced concrete wall with 25 mm minimum cover to reinforcement, both provide half-hour fire resistance. A solid masonry loadbearing wall, with or without plaster finish, at least 90 mm thick, will provide one hour fire resistance. For masonry cavity walls, the fire resistance may be taken as that for a single wall of the same construction, whichever leaf is exposed to fire.

In general, domestic loadbearing external walls of normal materials satisfying the conditions of strength, stability and weather resistance will usually provide sufficient resistance to fire. Methods of upgrading existing partitions are described in BRE Digest 230.[10]

Compartmentation

The spread of fire can also be restricted by provisions for sub-dividing the building into compartments of restricted floor area and cubic capacity, by means of compartment walls and compartment floors (Approved Document B3: 1991) to the Building Regulations). This provision applies to flats and maisonettes.

Other forms of compartmentation apply in the case of a division between adjoining buildings by means of a separating wall, such as walls between terraced and semi-detached houses carried above roof level or suitably fire stopped or between a house and an attached garage (Section 8 of Approved Document B3 (1991) of the Building Regulations). The wall and any floor between a garage and a house are to have 30 mins fire resistance and any opening in the wall is to be at least 100 mm above garage floor level and be fitted with a half-hour fire resisting door (Approved Document B3: 1991).

Concealed spaces and fire stopping

Concealed spaces or cavities in the construction of a building provide a ready route for smoke and flame spread. This is particularly so in the case of voids above spaces in a building as, for example, above a suspended ceiling or in a roof space. As any spread is concealed it presents a greater danger than would a more obvious fire weakness in the building fabric. Provision is therefore made in section 9 of Approved Document B3 (1991) to the Building Regulations to restrict the hidden spread of fire in concealed spaces by interrupting cavities which could form a pathway around a barrier to fire and subdividing extensive cavities. It is also necessary to effectively seal the openings around all penetrating pipes, cables and other services.

Constructional techniques are illustrated in Approved Document B3 (1991). Ends of cavities should be effectively sealed to provide fire stops.

External fire spread

The requirements with regard to external fire spread in respect of dwellinghouses as detailed in Approved Document B4 (1991) to the Building Regulations are more rigorous if the distance of the wall from the boundary is less than 1 m, when class 0 materials, such as brickwork, blockwork and concrete are acceptable, as shown in diagram 36 in Approved Document B4 (1991).

Stairways

An internal stairway in a house which has more than two storeys, excluding a basement, may need to be enclosed and protected in accordance with Approved Document B1 (1991), in support of the requirements of B1 of Schedule 1 to the Building Regulations (1991), whereby the building shall be designed and constructed so that there are means of escape in case of fire from the building to a place of safety outside the building capable of being safely and effectively used at all material times.

Access and facilities for the fire service

Part B5 of Schedule 1 to the Building Regulations 1991 requires that

(1) the building shall be designed and constructed so as to provide facilities to assist fire fighters in the protection of life;

(2) provision shall be made within the site of the building to enable fire appliances to gain access to the building.

Other Defects in Brickwork

Defects in brickwork sometimes result from the use of unsuitable materials or an unsatisfactory combination of materials. The following examples will serve to illustrate some of these defects, and further information can be obtained from BRE Digest 359.[11]

Sulphate attack on mortar. This comprises a chemical action between sulphate salts in clay bricks and the aluminate constituent of Portland cement in the mortar. A new compound is formed with resultant expansion. The most vulnerable situations are parapet, boundary and retaining walls and chimney stacks, where deterioration and expansion of the mortar and cracking and spalling of the brickwork may occur. It is advisable to use bricks with a low sulphate content in these situations and to pay particular attention to damp-proofing arrangements. Another alternative is to use mortar containing sulphate-resisting Portland cement.[12]

Unsuitable mortars. Pitting and cracking of mortar may occur where the pointing mortar is excessively strong or there are unslaked particles of quicklime in the mortar.

Work in adverse weather conditions. Damage to brickwork by frost is usually restricted to new work or walling which remains wet throughout the winter. Strong pointing mortar, porous bedding mortar and soft or weak bricks are particularly vulnerable. Precautions to be taken when bricklaying in bad weather are described in BRE Defect Action Sheet 64.[13]

These include ceasing work when the air temperature is below 2°C, in strong winds or steady rainfall, checking that aggregates are not frost-bound, ensuring that new work is protected from rain and possible overnight frost and obtaining local weather forecasts.

Crystallisation of salts in brickwork. The crystallisation of water-soluble salts may result in a white powder or feathery crystals on the surface of the bricks (efflorescence), surface disintegration of the brickwork or decay of individual bricks in severe cases. The salts may originate from the bricks themselves, soil in contact with the bricks or sea water. Surface efflorescence is normally removed by periodic dry-wire brushing of the surface of the brickwork. In more severe cases it may be necessary to cut out and replace individual defective bricks or to apply a rendering to more extensive areas. Nevertheless, efflorescence is usually a temporary springtime ocurrence on new brickwork and is relatively harmless.

Movements with changes in moisture content of brickwork. Shrinkage movements may cause stepped cracks following the horizontal and vertical joints or vertical joints through both bricks and mortar. This type of defect is most likely to occur with sandlime and concrete bricks, and is remedied by cutting out defective bricks and joints and making good. Preventive measures include using comparatively dry bricks, protecting work from rain and snow during erection and using a gauged mortar not stronger than 1:2:9.[14]

Corrosion of iron and steel in the brickwork. The corrosion may be caused by the action of moisture, acids, sulphates or chlorides on the metals, resulting in expansion and opening of brick joints, cracking of bricks and staining of brickwork by rust. Preventive measures include priming the metal, applying a coat of bituminous paint and then encasing it with cement mortar 25 mm thick, and the use of zinc-coated steel wall ties to BS 1243[15] in cavity walls. In 1981 the amount of zinc on wall ties was raised to 940 g/m^2 in the British Standard to withstand a 60 year building life.

Movement. Long runs of walling should be provided with 10 mm wide vertical expansion joints about 12 mm apart.[16]

Cracks occur in masonry buildings for a variety of reasons as discussed in BRE Digest 251[17] and procedures for the simple measuring and monitoring of movement in low-rise buildings are described and illustrated in BRE Digests 343[18] and 344.[19]

BRICKS AND BLOCKS

Brick Types

Bricks are still the most popular form of walling unit for domestic construction. Their limited size and variety of colours and textures make them an attractive proposition. There is a wide range of bricks available, varying in the materials used, method of manufacture and form of brick, and these are now considered.

BS 3921[20] classifies *clay bricks* in three different ways
Varieties and functions. (a) Common: suitable for general building work but generally of poor appearance; (b) facing: specially made or selected to give an attractive appearance; (c) engineering: dense and strong semi-vitreous to defined limits for absorption and strength.

Qualities. (a) Internal: suitable for internal use only; (b) ordinary: normally sufficiently durable for external use; (c) special: durable in situations of extreme exposure.

Types. Solid: shall not have holes, cavities or depressions; perforated: holes not exceeding 25 per cent of gross volume of brick; cellular: may have frogs or cavities exceeding 20 per cent of gross volume of brick; frogged: depressions in one or more bed faces not exceeding 20 per cent of gross volume of brick.

The usual brick size is 215 × 102.5 × 65 mm. Allowing for a 10 mm mortar joint, this gives an overall walling unit size of 225 × 112.5 × 75 mm. Some modular bricks are being produced, ranging from 288 or 190 mm long × 90 mm wide × 90 or 65 mm high.

BS 3921[20] prescribes minimum requirements as to dimensional deviations, durability, compressive strength and water absorption. Requirements are also prescribed for soluble salt content and frost resistance. Bricks of ordinary quality shall be well-fired and reasonably free from deep or extensive cracks and from damage to edges and corners, from pebbles and expansive particles of lime. They shall also, when a cut surface is examined, show a reasonably uniform texture.

Engineering bricks are classified into two groups

Class	Average compressive strength in N/mm^2 not less than	Average absorption (boiling or vacuum) per cent weight not greater than
A	70	4.5
B	50	7.0

Manufacture of Clay Bricks

These bricks are made from clays composed mainly of silica and alumina, with small amounts of lime, iron, manganese and other substances. The different types of clay produce a wide variety of colours and textures. Most clay bricks are kiln-burnt, either in a *Hoffman* kiln in which the fire passes through a series of chambers whilst others are used for heating and cooling, or a *tunnel* kiln where the bricks pass through a long chamber with the firing zone in the centre.

Hand-made bricks are rather irregular in shape and size with uneven arrises; they are used solely for facing work and weather to attractive shades. The majority of clay bricks are either machine-pressed or wirecut. With *machine-pressed* bricks the clay is fed into steel moulds and shaped under heavy pressure with a single or double frog, whereas with *wirecut* bricks, the clay is extruded from a pugmill in a continuous band and then cut into bricks by wires attached to a frame. The Build-

ing Research Establishment has designed vertically perforated 'V' bricks to provide the equivalent of a cavity wall in a single leaf with a reduction in labour, mortar and weight. Standard return bricks ('L' bricks) are available for bonding quoins and stopped ends. Solid and perforated 'Calculon' clay bricks are designed for highly stressed brickwork and are particularly well suited for crosswall construction. BRE Digest 273[21] describes how perforated bricks have good thermal insulation properties and have satisfactory strength, fire resistance and waterproofing qualities. They also conserve energy and clay in their manufacture and are becoming increasingly popular.

Other Types of Brick

There are types of brick available apart from clay. *Calcium silicate bricks* (sandlime and flintlime) are made from lime and sand or siliceous gravel moulded under heavy pressure and then subjected to steam pressure in an autoclave. BS 187,[22] as amended, prescribes five acceptable classes of brick (3 to 7) with minimum compressive strengths ranging from 20.5 to 48.5 N/mm^2. Class 1 and 2 bricks are only really suitable for internal walls, whereas the stronger classes of brick can be used for external work, including exposed positions and below damp-proof courses. The drying shrinkage of calcium silicate bricks calls for care in design, in storage and during building, with vertical movement joints provided in external walls above damp-proof course at intervals of 7.5 to 9 m.[23] The requirements for *concrete bricks* and fixing bricks are detailed in BS 6073.[24] *Fire bricks* are made from refractory clay with a high fusing point and they are laid in refractory mortar with tight joints.

Choice of Bricks

The location in which the bricks are to be used has an appreciable effect on the choice of brick. Special quality bricks only should be used in positions of severe exposure, such as parapet, free standing and retaining walls. If a building has a parapet, the choice of brick for the whole wall will probably be restricted to those suitable for severe exposure as it is rarely acceptable architecturally to use one facing brick up to roof level and another for the parapets. Strength is not necessarily an index of durability, nor does high water absorption necessarily indicate a brick of low standard.

Bricks of ordinary quality are suitable for normal exposure (in an external wall between eaves and damp-proof course).

The student should examine and handle bricks used in his/her own locality and note their particular characteristics. He/she should also be aware of the main properties of some of the more commonly used bricks.

Flettons. These are made from Oxford clay, primarily in Bedfordshire and Northamptonshire. These common bricks are produced in very large quantities and have single frogs, sharp arrises, and faces crossed with bands of dark and light pink. They have a hard exterior skin but are relatively soft inside. Because of their unattractive appearance they are used primarily for internal walls and the inner leaf of cavity walls.

Stocks. These are mainly yellow in colour and made from London clay. They are made in several grades and are good all-purpose bricks. They are suitable for facings, and have the unusual property of increasing in strength with age.

Staffordshire blues. These are bluish-grey engineering bricks made from clays containing iron oxides and are burnt at high temperatures. They are very heavy and dense, with a high crushing strength and low water absorption. These bricks are ideal for positions requiring great strength, such as bridge abutments and retaining walls, and for damp-proof courses.

Southwater reds. These come from Sussex and have similar properties to Staffordshire blues but are red in colour.

Facing bricks. They may be machine-made or hand-made to a variety of colours and textures. Some are multi-coloured and embrace a range of colours, such as red and purple to brown. Sandfaced bricks have a rough textured surface obtained by applying coloured sand prior to burning.

Concrete Blocks

BS 6073[24] prescribes the requirements for precast concrete blocks made from cement and one of a number of different aggregates from natural aggregates to lightweight materials. The British Standard recognises three types of block: solid, hollow and cellular. They are made to various sizes, a common size being 440 × 215 mm with a wide range of thicknesses from 60 to 250 mm. The Standard gives minimum compressive strengths for all block types. Where severe exposure or pollution is likely, blocks should have an average compressive strength of not

less than 7 N/mm^2, but blocks of lower strength can be protected by rendering.

Autoclaved aerated concrete, commonly known as AAC, is a microcellular construction product made from a siliceous base material, such as fine sand or pulverised fuel ash (PFA), or mixtures and a binder of lime and ordinary Portland cement (OPC). The cellular structure, which gives good thermal properties and a high strength:density ratio, results from the reaction between the alkaline lime/OPC component and added aluminium powder to give bubbles of hydrogen. The properties and uses are well described in BRE Digest 342.[25]

Leca energysaver concrete blocks combine two highly efficient insulating materials, lightweight expanded clay aggregate (Leca) to BS 3797 and polyurethane foam. The Leca concrete shell is manufactured to BS 6073, and is cured prior to the addition of the polyurethane. The compressive strength of the concrete combines with the tensile strength of the polyurethane giving the blocks improved working characteristics and robustness. For example, a 125 mm thick block will achieve a *U*-value of 0.45 when used in cavity wall construction without the provision of secondary insulation or the need for 'trade offs', thereby increasing significantly standards of occupier comfort in relation to fuel expenditure.

Clay blocks

These are hollow with keyed surfaces and have a high thermal insulation value.

MORTARS

Brick and blocks are bedded in and jointed with mortar. A good mortar spreads readily, remains plastic while bricks are being laid to provide a good bond between bricks and mortar, resists frost and acquires early strength, particularly in winter. Mortars should not be stronger than necessary, as an excessively strong mortar concentrates the effects of any differential movement in fewer and wider cracks.

There are a number of different types of mortars in use.[26]

Lime mortar. In the past lime mortars composed of one part of lime to three parts of sand were widely used, although they developed strength slowly. With the advent of cement their use diminished rapidly.

Cement mortar. A mix of 1:3 (cement to sand) is workable but is too strong for everyday use. It would be suitable for heavily loaded brickwork or in extremely wet situations. The sand should comply with BS 1200[27] and be clean and well graded.

Cement-lime mortar. This is sometimes described as gauged or compo-mortar and is the most useful general purpose mortar. The best properties of both cement and lime are utilised to produce a mortar which has good working, water-retention and bonding qualities, and also develops early strength without an excessively high mature strength. The lime should be non-hydraulic or semi-hydraulic complying with BS 890.[28] White limes (chalk lime) and magnesium limes (from dolomitic limestone) are both non-hydraulic limes and do not set under water. Greystone limes (obtained from beds of greyish chalk or limestone) are semi-hydraulic limes and harden under water in a few weeks.

Air-entrained (plasticised) mortar. Mortar plasticisers, which entrain air in the mix, provide an alternative to lime for improving the working qualities of lean cement-sand mixes. Hence a 1:6 cement-sand mortar gauged with plasticiser is a good alternative to a 1:1:6 cement-lime-sand mix.

Masonry cement mortar. This consists of a mixture of Portland cement with a very fine mineral filler and an air-entraining agent, complying with the requirements of BS 5224.[29]

Mortars containing special cements. High alumina cement may be used where high early strength or resistance to chemical attack is required. Sulphate-resisting cement may be used where sulphate attack is possible.

Choice of Mortars

It is advisable to restrict the cement content to a minimum consistent with obtaining mortar of adequate strength. In winter, richer mixes are needed to obtain early strength and so resist the effects of frost. Weak mortars should always be used with bricks or blocks of high-drying shrinkage. Tables 4.2 and 4.3 provide useful guidelines for the selection of mortars for different situations and using alternative types of brick or block. These tables have been extracted from BRE Digest 362,[26] with some amendments introduced by BS 5628, Part 1,[1] and the guiding principle is that the mortar should contain no more cement than is necessary to give adequate strength in the brickwork.

Pointing and Jointing

Mortar joints may be finished in a number of ways after the bricks have been laid. Where the work is carried out while the mortar is still fresh, the process is called *jointing*. When the mortar is allowed to harden and then some is removed and replaced with fresh mortar, the process is termed *pointing*. The mortar used in pointing is generally weaker than that used in the joints, to ensure that the load is carried by the bed joint. The five main types of finish are illustrated in figure 4.1.

1. *Struck flush* (figure 4.1.1) gives maximum strength and weather resistance to brickwork but may appear irregular with uneven bricks.

2. *Curved recessed* (bucket handle) (figure 4.1.2) gives a superior finish to struck flush but with little difference in strength or weather resistance.

3. *Struck or weathered* (figure 4.1.3) produces interplay of light and shadow on brickwork but has less strength and weather resistance than the previous two finishes.

4. *Overhung struck* (figure 4.1.4) produces variation of light and shade but weakens brickwork and provides a surface on which rainwater may lodge, and this may result in discolouration or frost damage.

5. *Square recessed* (figure 4.1.5) should only be used with very durable bricks and preferably in sheltered locations, as it offers less strength and weather resistance, but has good appearance.

Jointing is quicker and cheaper than pointing and, as the surface finish is part of the bedding mortar, there is less risk of the face joints failing through frost action or insufficient adhesion. On the other hand it may not look as attractive as pointing and it is difficult to keep the work clean and of uniform colour. With pointing, the preparatory work is important – the joints must be properly raked out and brushed to remove loose mortar and dust, and wetted to provide good adhesion.

BUILDING BRICK AND BLOCK WALLS

Bricks and blocks are bonded to give maximum strength and adequate distribution of loads over the wall. Unbonded walls contain continuous vertical joints with the accompanying risk of failure as shown in figure 4.1.6. Bonded walls provide lateral stability and resistance to side thrust (figure 4.1.7), and the bond can be selected to give an attractive appearance to the

Table 4.2   Mortar mixes (proportions by volume) (Source: BS 5628 Part 1[1])

Mortar designation	Type of mortar (proportion by volume)				Mean compressive strength at 28 days	
	Cement: lime: sand	Masonry cement: sand		Cement: sand with plasticiser	Preliminary (laboratory tests) (N/mm^2)	Site tests (N/mm^2)
(i)	1:0 to ¼:3	—		—	16.0	11.0
(ii)	1:½:4 to 4½	1:2½ to 3½		1:3 to 4	6.5	4.5
(iii)	1:1:5 to 6	1:4 to 5		1:5 to 6	3.6	2.5
(iv)	1:2:8 to 9	1:5½ to 6½		1:7 to 8	1.5	1.0

Increasing resistance to frost attack during construction
——————————————————————►

Improvement in bond and consequent resistance to rain penetration
◄——————————————————————

Note:
- Mortar designation (i) is the strongest; (iv) is the weakest.
- The weaker the mortar the more it can accommodate movement, for example, due to settlement or temperature and moisture changes.
- Where a range of sand contents is given, the larger quantity should be used for sand that is well graded and the smaller for coarse or uniformly fine sand.
- Because damp sands bulk, the volume of damp sand used may need to be increased.

Table 4.3   Selection of mortar groups

Type of brick:	Clay		Concrete and calcium silicate	
Early frost hazard[a]	no	yes	no	yes
Internal walls	(iv)	(iii) or (iv)[b]	(iv)	(iii) or plast (iv)[b]
Inner leaf of cavity walls	(iv)	(iii) or (iv)[b]	(iv)	(iii) or plast (iv)[b]
Backing to external solid walls	(iv)	(iii) or (iv)[b]	(iv)	(iii) or plast (iv)[b]
External walls; outer leaf of cavity walls:				
–above damp-proof course	(iv)[c]	(iii)[c]	(iv)	(iii)
–below damp-proof course	(iii)[d]	(iii)[b, d]	(iii)[d]	(iii)[d]
Parapet walls; domestic chimneys:				
–rendered	(iii)[e, f]	(iii)[e, f]	(iv)	(iii)
–not rendered	(ii)[g] or (iii)	(i)	(iii)	(iii)
External freestanding walls	(iii)	(iii)[b]	(iii)	(iii)
Sills; copings	(i)	(i)	(ii)	(ii)
Earth-retaining walls (backfilled with free-draining material)	(i)	(i)	(ii)[d]	(ii)[d]

[a]during construction, before mortar has hardened (say 7 days after laying) or before the wall is completed and protected against the entry of rain at the top

[b]If the bricks are to be laid wet, see 'Cold weather bricklaying'

[c]if to be rendered, use group (iii) mortar made with sulphate-resisting cement

[d]if sulphates are present in the groundwater, use sulphate-resisting cement

[e]parapet walls of clay units should not be rendered on both sides; if this is unavoidable, select mortar as though *not* rendered

[f]use sulphate-resisting cement

[g]with 'special' quality bricks, or with bricks that contain appreciable quantities of soluble sulphates

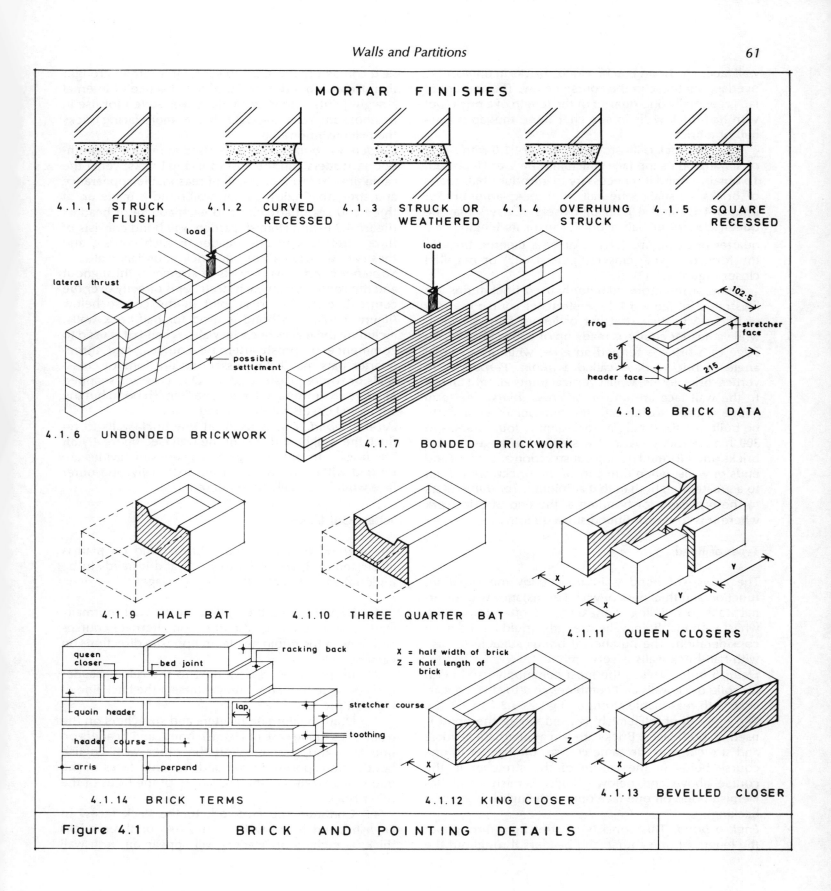

MORTAR FINISHES

4.1.1 STRUCK FLUSH

4.1.2 CURVED RECESSED

4.1.3 STRUCK OR WEATHERED

4.1.4 OVERHUNG STRUCK

4.1.5 SQUARE RECESSED

4.1.6 UNBONDED BRICKWORK

load

lateral thrust

possible settlement

4.1.7 BONDED BRICKWORK

load

4.1.8 BRICK DATA

frog
stretcher face
header face
102·5
65
215

4.1.9 HALF BAT

4.1.10 THREE QUARTER BAT

4.1.11 QUEEN CLOSERS

X   Y
X   Y

4.1.14 BRICK TERMS

queen closer
bed joint
racking back
quoin header
lap
stretcher course
header course
toothing
arris
perpend

X = half width of brick
Z = half length of brick

4.1.12 KING CLOSER

X   Z

4.1.13 BEVELLED CLOSER

X

Figure 4.1   BRICK AND POINTING DETAILS

wall face. The bond is achieved by bricks in one course overlapping those in the course below. The amount of lap is generally one-quarter of the length of a brick, but with half-brick walls in stretcher bond the lap is one-half of a brick.

A typical brick is illustrated in figure 4.1.8 with a frog or sinking on its top face which may be V or U-shaped. To obtain a bond it is necessary to introduce bats (parts of bricks – usually one-half or three-quarter bricks: figures 4.1.9 and 4.1.10) or closers. Closers may be queen closers of half a brick, cut in its length, or a quarter brick (figure 4.1.11), or be a tapered brick in the form of a king closer (figure 4.1.12) or bevelled closer (figure 4.1.13).

Some of the more commonly used brick terms are illustrated in figure 4.1.14. A stretcher course is a row of bricks with their stretcher or longer faces showing, while a header course is made up of ends of bricks. An edge of a brick is termed an *arris*, while an external angle of brickwork is called a *quoin*. *Perpends* are vertical brick joints, whilst vertical joints at right angles to the wall face are known as *cross joints*. The term *gauge* is used to indicate the number of courses to be built in a fixed height, for example four courses to 300 mm. *Racking back* is a stepped arrangement of bricks often formed during construction at corners and ends of walls, when one part of the brickwork is built to a greater height than that adjoining. *Toothing* refers to the provision for bonding at the end of brickwork where it is to be continued subsequently.

Types of Bond

The choice of bond is influenced by the situation, function and thickness of wall. For instance with external walls to buildings appearance is often important, whilst with manhole walls strength would be the main consideration. The number of bonds suitable for use with half-brick walls is very limited.

Flemish bond. This is the most commonly used bond for solid brick walls as it combines an attractive appearance with reasonable strength (figures 4.2.7 to 4.2.12). The term 'double Flemish' is used where the bond is used on both faces. Bricks are laid as alternate headers and stretchers in the same course, the header in one course being in the centre of the stretcher in the course above and below. Single Flemish bond has Flemish bond on one face only with English bond as a backing.

English bond. This consists of stretchers throughout the length of one course and headers throughout the

next course (figures 4.2.1 to 4.2.6). It is rather stronger than Flemish bond because of the absence of internal straight joints, and is particularly well suited for use in manhole and retaining walls. It uses more facing bricks than Flemish bond.

Garden wall bonds. These are used to reduce the number of headers and the cost of facing bricks, yet at the same time to produce a wall of reasonable appearance and strength. English garden wall bond is made up of three courses of stretchers to each course of headers (figure 4.3.6) and Flemish garden wall bond consists of three stretchers to one header in each course, and these are well suited for one-brick boundary walls.

Stretcher bond. This consists of stretchers throughout and the centre line of each stretcher is directly over the centre line of the cross joint in the course below (figure 4.3.7). This bond is used for half-brick walls, including the leaves of cavity walls. A half-bat is used to commence or finish alternate courses, and three-quarter bats may be necessary at junctions of cross-walls or where pilasters occur. On occasions snapped headers are introduced to imitate Flemish bond but the additional cost is rarely justified.

Rat-trap bond. This is used on one-brick walls to reduce the weight and cost of the wall (see figure 4.3.8). The bricks are laid on edge and a series of cavities are formed within the wall. If used externally, the outer face would normally be rendered.

Setting Out Bonds

The method of setting out a brick bond on plan is illustrated in figures 4.3.1 to 4.3.5, and it needs to be undertaken systematically in logical stages.

1. Draw to scale the outline of the wall, normally starting at a right-angled corner, on alternate courses and draw in the quoin header in opposite directions on each course.

2. Insert queen closers next to the quoin header and continue them until they intersect the back line of the wall produced.

3. Draw in alternate headers and stretchers on the front faces in the case of Flemish bond, and headers on one face and stretchers on the other with English bond. The exposed header and stretcher faces of the quoin header determine the nature of the faces of the other bricks.

4. Continue the facework to the back faces. In English bond if the wall is one, two or three-bricks thick, stretchers or headers will appear on both wall

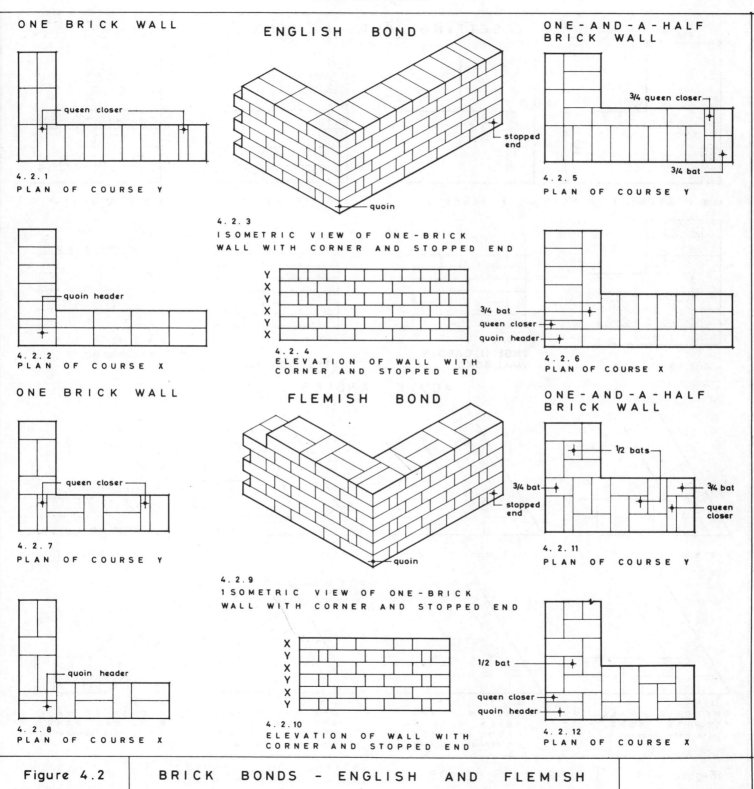

ONE BRICK WALL

4.2.1 PLAN OF COURSE Y
queen closer

4.2.2 PLAN OF COURSE X
quoin header

ENGLISH BOND

4.2.3 ISOMETRIC VIEW OF ONE-BRICK WALL WITH CORNER AND STOPPED END
stopped end
quoin

4.2.4 ELEVATION OF WALL WITH CORNER AND STOPPED END
Y X Y X Y X

ONE-AND-A-HALF BRICK WALL

4.2.5 PLAN OF COURSE Y
3/4 queen closer
3/4 bat

4.2.6 PLAN OF COURSE X
3/4 bat
queen closer
quoin header

ONE BRICK WALL

4.2.7 PLAN OF COURSE Y
queen closer

4.2.8 PLAN OF COURSE X
quoin header

FLEMISH BOND

4.2.9 ISOMETRIC VIEW OF ONE-BRICK WALL WITH CORNER AND STOPPED END
stopped end
quoin

4.2.10 ELEVATION OF WALL WITH CORNER AND STOPPED END
X Y X Y X Y

ONE-AND-A-HALF BRICK WALL

4.2.11 PLAN OF COURSE Y
1/2 bats
3/4 bat
3/4 bat
queen closer

4.2.12 PLAN OF COURSE X
1/2 bat
queen closer
quoin header

| Figure 4.2 | BRICK BONDS - ENGLISH AND FLEMISH |

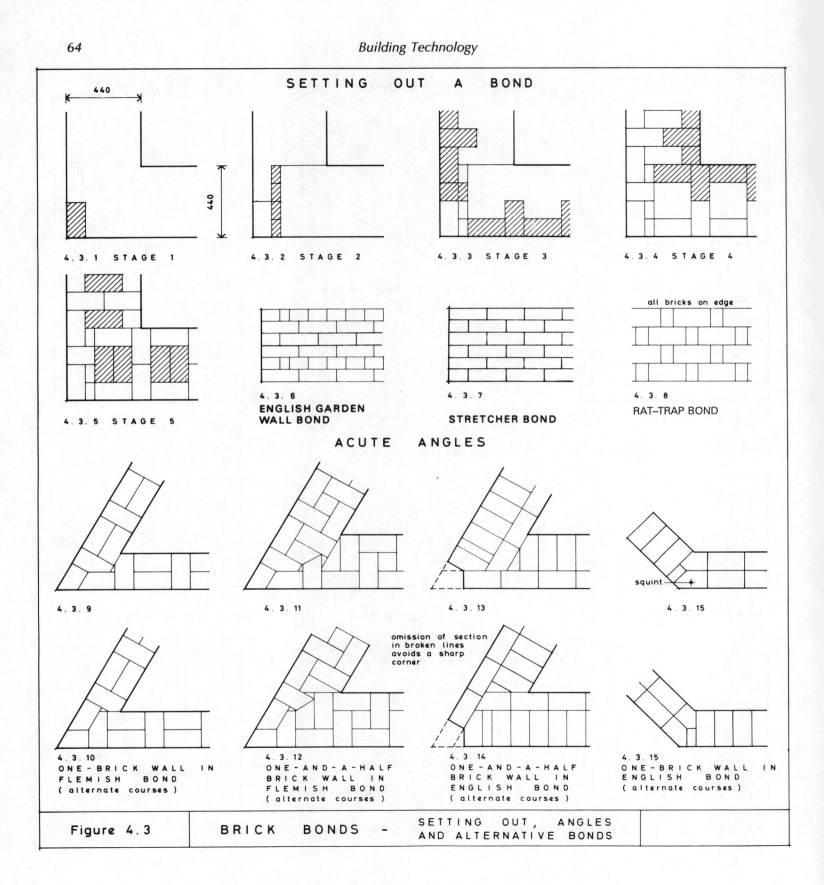

SETTING   OUT   A   BOND

440

440

4.3.1 STAGE 1   4.3.2 STAGE 2   4.3.3 STAGE 3   4.3.4 STAGE 4

4.3.5 STAGE 5

4.3.6
ENGLISH GARDEN
WALL BOND

4.3.7
STRETCHER BOND

all bricks on edge

4.3.8
RAT–TRAP BOND

ACUTE   ANGLES

4.3.9

4.3.11

4.3.13

squint

4.3.15

4.3.10
ONE–BRICK WALL IN
FLEMISH   BOND
(alternate courses)

4.3.12
ONE–AND–A–HALF
BRICK WALL IN
FLEMISH   BOND
(alternate courses)

omission of section
in broken lines
avoids a sharp
corner

4.3.14
ONE–AND–A–HALF
BRICK WALL IN
ENGLISH   BOND
(alternate courses)

4.3.15
ONE–BRICK WALL IN
ENGLISH   BOND
(alternate courses)

| Figure 4.3 | BRICK BONDS – | SETTING OUT , ANGLES AND ALTERNATIVE BONDS | |

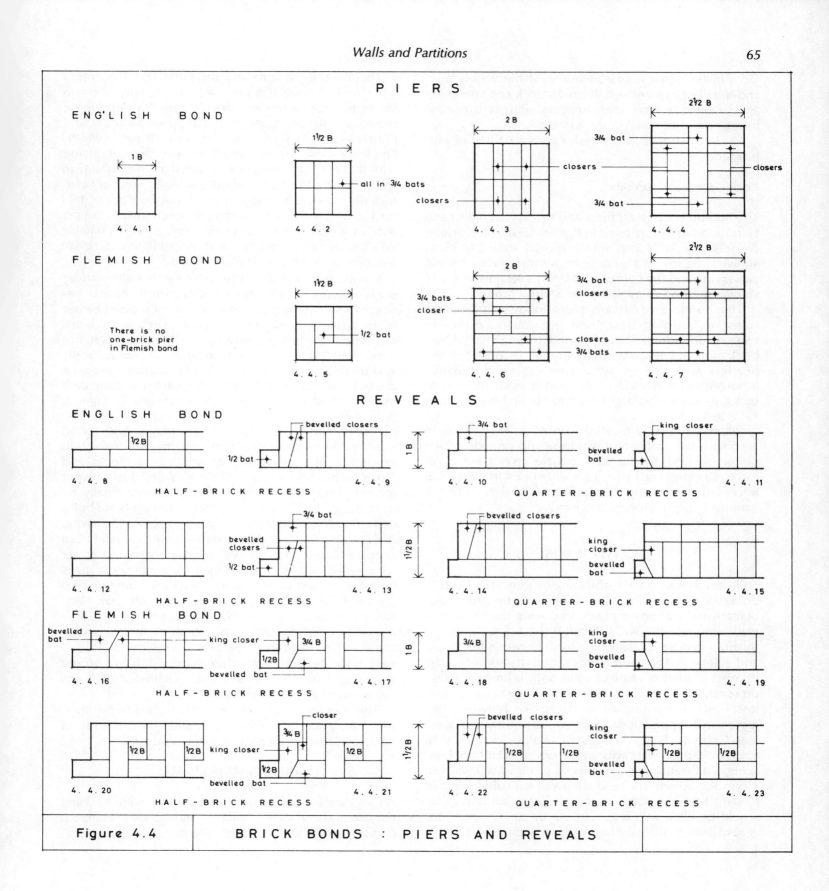

Figure 4.4   BRICK BONDS : PIERS AND REVEALS

faces in the same course, whereas if the wall is one-and-a-half or two-and-a-half-bricks thick and stretchers appear on one face, then headers will occur on the other side of the wall in the same course.

5. In thick walls, the interior gaps are filled in with headers.

Angles, Piers and Reveals

The bonding of angles, piers and reveals in English and Flemish bond is illustrated in figures 4.3 and 4.4. Squint corners usually incorporate a special squint brick as shown in figure 4.3.15. Acute angles introduce bats and closers around the corner, and figures 4.3.13 and 4.3.14 show a method of avoiding the sharp corner.

The bonding of square piers follows the normal rules of bonding described previously. Alternate courses are identical but turned through 90°. One-brick piers in English bond (figure 4.4.1) consist of two headers laid side by side; there is no alternative arrangement in Flemish bond. Note how closers and/or bats are used in the larger piers to obtain bond (figures 4.4.2 to 4.4.7).

Recessed reveals may be formed in solid walls to accommodate doors and windows and provide added protection from the weather. Square reveals are identical to stopped ends which are illustrated in figure 4.2. Recessed reveals (figures 4.4.8 to 4.4.23) require the extensive use of closers and bats.

CAVITY WALLS

Solid brick walls of the thicknesses generally used in domestic construction are not always damp-proof, particularly in locations exposed to driving rain. Moisture may penetrate the walls through the bricks, joints or through minute cracks which form between the bricks and joints as the mortar dries out. In addition, the thermal insulation value of solid walls is limited. These disadvantages can be overcome by using cavity or hollow walls, consisting of an air space between two leaves with upper floor loads supported by the inner leaf. Cavity walls often consist of a half-brick outer leaf, 52.50 to 80 mm wide cavity and inner leaf of 100 mm insulating, loadbearing concrete blocks, giving a total width of 255 to 282.50 mm. The head of the wall is usually built solid to spread the roof loads over both leaves and the cavity should be filled with fine concrete up to but not above ground level to stiffen the base of the wall (figure 4.5.3).

The Building Regulations in Approved Document A1/2 (1991), clause 1C8 and table 11, require the two leaves to each be not less than 90 mm thick and to be provided with adequate lateral support by roofs and floors. The cavity shall be not less than 50 mm in width. The leaves shall be tied together with ties complying with BS 1243,[15] or suitable equivalent to give stability to the wall. The British Standard describes vertical twist, butterfly and double triangle types of wall ties made of mild steel coated with zinc, galvanised wire, copper, copper alloys or stainless steel. Austenitic stainless steel or suitable non-ferrous ties should be used in conditions of severe exposure as defined in BS 5628, Part 3.[16]

Polypropylene wall ties provide another alternative; some failures of wall ties have occurred, mainly because of sub-standard galvanising, as described earlier in the chapter, or aggressive mortars such as the black ash type. The ties are spaced at distances apart not exceeding 900 mm horizontally and 450 mm vertically and there shall be, within 225 mm of openings, a tie every 300 mm of height if the leaves are not connected by a bonded jamb (table 6 of Approved Document A1/2: 1991).

Cavity walls offer distinct advantages with good water penetration prevention and thermal insulation properties, greater choice of bricks and blocks and fewer facing bricks required. It is essential to prevent mortar droppings falling into the cavity, ideally by bringing a wood lath up the cavity as the walls are built, as mortar droppings can form bridges across the cavity. A damp-proof course should be inserted over reinforced concrete lintels to door and window openings; under window sills (figure 4.6.1) and the ends of cavities at jambs to openings must be effectively sealed and damp-proofed (figures 4.5.4 and 4.5.5). Figures 4.5 and 4.6 show alternative methods of sealing the cavity, using mineral fibre insulation. Another approach is to apply an internal insulating skin. The satisfactory performance of cavity walls is very much dependent on close supervision and proper detailing. The cavity must not be ventilated or it will cease to act as a thermal insulator.

The statutory requirements with regard to the thermal insulation of cavity walls and the methods of achieving them are described in chapter 15.

FREESTANDING WALLS

All freestanding walls must be stable under wind load and durable under service conditions, and extensive guidance on their design and construction in both brick and blockwork is provided in BRE Good Building Guide 14 (1992): Building brick or blockwork freestanding walls.

The guide gives rule-of-thumb dimensions for maximum height above ground level and minimum foundation width for brick and blockwork freestanding walls in four exposure zones in the UK. The most exposed zone (4) covers N and NW Scotland while the least exposed (1) is SE England, with zones 2 and 3 in between.

There are also differences in exposure within each of the four exposure zones. *Sheltered locations* are typical urban areas and others where there is considerable local interruption of wind flow. *Exposed locations* are typical rural areas or others where there is a clear view of open country or little local resistance to wind flow.

Table 4.4 gives a range of maximum wall heights and minimum foundation widths for brick freestanding walls of different thicknesses in both sheltered and exposed locations in zones 1 and 4, extracted from GBG 14. It should, however, be noted that the guide lists a variety of situations where the rules of thumb may not apply, such as in areas subject to significant pressures, on the crest of a hill, supporting a large gate or door, on ground of doubtful stability or where the difference in ground level between each side of the wall exceeds twice the wall thickness.

Foundations in dry soils or wet soils with low sulphate contents should have a minimum cement content of 220 kg/m^3, while those in wet soils with higher sulphate levels should have a minimum cement content of 280–380 kg/m^3, depending on the amount of salt in the ground and cements with sulphate resisting properties may be required.

For stability, bricks should have a density of not less than 1200 kg/m^3, with a generous overhang (45 mm) to the coping, when a type MN or ML brick to BS 3921 (1985) will normally suffice, but using type F bricks below the low level dpc and above the high level dpc. The suggested mortar mix is 1 part Portland cement, $\frac{1}{2}$ part lime, $4\frac{1}{2}$ parts sand and a proprietary air entraining agent. Piers, $1\frac{1}{2}$ bricks square, are needed at all ends of half brick and one brick walls.

Brick cappings to BS 4729 (1990) or precast concrete, clayware or stone copings to BS 5642, Part 2 (1983) must be frost resistant and shed rainwater clear of the wall. Copings must have a 'soft' waterproof joint above all movement joints and copings must not be continuous over movement joints.

A high level damp-proof course must be provided beneath the coping to protect the wall from water penetration around the joints in the coping. Flexible high bond impermeable materials are suitable for this lightly-loaded situation. Where frost-susceptible materials are used, such

Table 4.4   Brick freestanding wall dimensions

Wall thickness	Location	Zone	Maximum wall height above ground (mm)	Minimum foundation width (mm)
Half brick	Sheltered	1	725	350
Half brick	Sheltered	4	525	375
One brick	Sheltered	1	1925	525
One brick	Sheltered	4	1450	575
One and a half bricks	Sheltered	1	2500	525
One and a half bricks	Sheltered	4	2450	725
Half brick	Exposed	1	525	375
Half brick	Exposed	4	375	400
One brick	Exposed	1	1450	600
One brick	Exposed	4	1075	650
One and a half bricks	Exposed	1	2400	725
One and a half bricks	Exposed	4	1825	850

as high absorptive brick masonry, a low level dpc of two courses of damp-proof bricks should be provided to resist rising water and ground salts, but avoiding the use of clay damp-proof course bricks if the wall is to be built of calcium silicate or concrete units, to prevent differential movement occurring.

DAMP-PROOF COURSES

Damp-proof courses are required in various locations in a building to prevent moisture penetration, which is a prime cause of deterioration in building structures and materials, and the presence of excess moisture encourages the growth of moulds and wood-rotting fungi. The upward movement of moisture is prevented by damp-proof courses placed in walls at not less than 150 mm above ground level and in solid floors, either sandwiched in the concrete slab or underneath the floor screed. Defective damp-proof courses permit rising damp with consequential damage to walls and flooring materials and finishes. The deficiencies can arise from

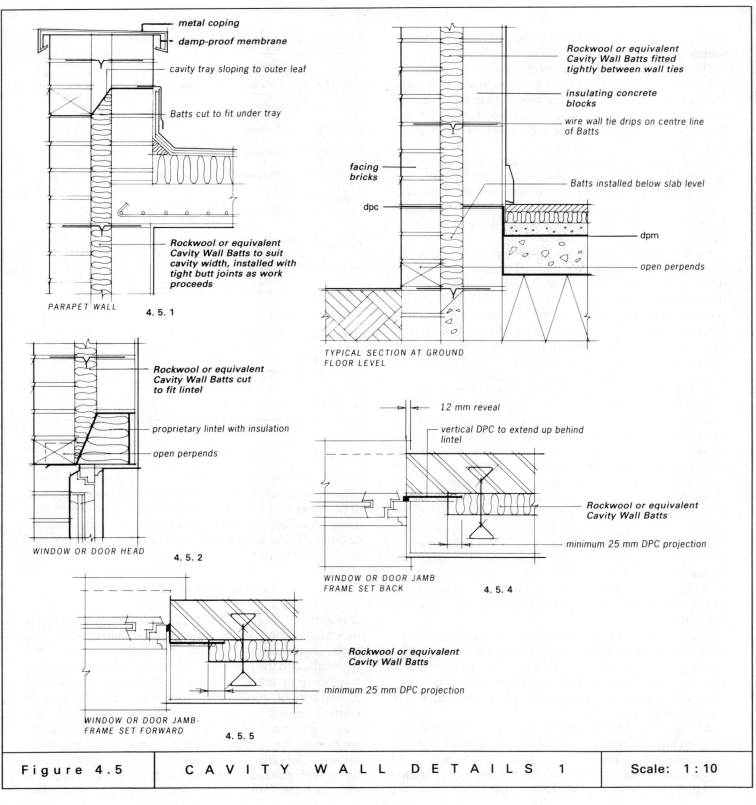

metal coping

damp-proof membrane

cavity tray sloping to outer leaf

Batts cut to fit under tray

*Rockwool or equivalent
Cavity Wall Batts to suit
cavity width, installed with
tight butt joints as work
proceeds*

PARAPET WALL   **4. 5. 1**

*Rockwool or equivalent
Cavity Wall Batts cut
to fit lintel*

proprietary lintel with insulation

open perpends

WINDOW OR DOOR HEAD   **4. 5. 2**

*Rockwool or equivalent
Cavity Wall Batts fitted
tightly between wall ties*

**insulating concrete
blocks**

wire wall tie drips on centre line
of Batts

**facing
bricks**

dpc

Batts installed below slab level

dpm

open perpends

TYPICAL SECTION AT GROUND
FLOOR LEVEL

12 mm reveal

vertical DPC to extend up behind
lintel

*Rockwool or equivalent
Cavity Wall Batts*

minimum 25 mm DPC projection

WINDOW OR DOOR JAMB
FRAME SET BACK   **4. 5. 4**

*Rockwool or equivalent
Cavity Wall Batts*

minimum 25 mm DPC projection

WINDOW OR DOOR JAMB-
FRAME SET FORWARD   **4. 5. 5**

| Figure 4.5 | CAVITY WALL DETAILS 1 | Scale: 1 : 10 |

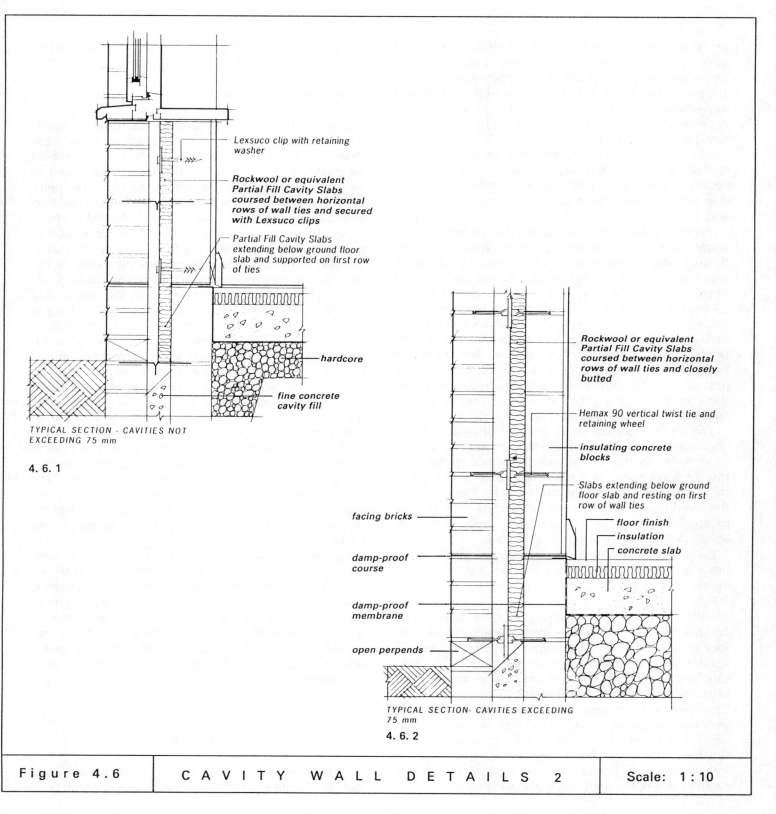

Lexsuco clip with retaining washer

Rockwool or equivalent Partial Fill Cavity Slabs coursed between horizontal rows of wall ties and secured with Lexsuco clips

Partial Fill Cavity Slabs extending below ground floor slab and supported on first row of ties

hardcore

fine concrete cavity fill

TYPICAL SECTION - CAVITIES NOT EXCEEDING 75 mm

4. 6. 1

Rockwool or equivalent Partial Fill Cavity Slabs coursed between horizontal rows of wall ties and closely butted

Hemax 90 vertical twist tie and retaining wheel

insulating concrete blocks

Slabs extending below ground floor slab and resting on first row of wall ties

floor finish

insulation

concrete slab

facing bricks

damp-proof course

damp-proof membrane

open perpends

TYPICAL SECTION- CAVITIES EXCEEDING 75 mm

4. 6. 2

| Figure 4.6 | CAVITY WALL DETAILS 2 | Scale: 1:10 |

the damp-proof courses being pointed or rendered over, bridged by mortar droppings, punctured, not lapped or though fill and paving not being kept at least 150 mm below them.[30] Barriers to resist the downward movement of moisture are inserted in parapets, chimneys and above reinforced concrete and some forms of steel lintel in cavity walls. On occasions horizontal movement has to be resisted as where the outer leaf of a cavity wall is returned at an opening to close the cavity.[31] Typical details of damp-proof course arrangements around window and door openings are shown in figures 4.5.2, 4.5.4 and 4.5.5.

Parapet walls have in the past often proved to be a source of moisture penetration into a building, either through the absence or incorrect placing of damp-proof courses, or through rendering parapet walls on both sides and the top; where the rendering cracks, rain enters and is trapped behind the rendering.[32] In low parapet walls, a damp-proof course should be inserted under the coping, and the roof covering extended up the inside face of the wall and tucked in under the damp-course (figure 4.7.1). With taller parapets two damp-proof courses are required; one under the coping and the other at or near roof level and lapped with the roof covering (figure 4.7.2).[31]

The most important characteristics of a damp-proof course are that it shall be impervious and durable, in order to last the life of the building without permitting moisture to penetrate. A reasonable degree of flexibility is important to prevent fracture with movement of the building. Other desirable properties include ease of application, ability to withstand loads, and availability in thin sheets at reasonable cost.

Building Research Establishment Digest 380[31] classifies materials for damp-proof courses into three categories

Flexible. Sheet lead and copper, bitumen, polythylene and pitch/bitumen polymer – these are suitable for most locations and they alone should be used over openings and for bridging cavities. Flexible materials are the most satisfactory for stepped damp-proof courses and they can be dressed to complex shapes without damage.
Semi-rigid. Mastic asphalt – particularly suitable for very thick walls and to withstand high water pressure.
Rigid. Dense bricks and slates – these are particularly suitable where high bond strength is required as in freestanding walls but can provide a barrier only to the capillary rise of moisture, not water under pressure.

Table 4.5 lists the requirements and principal characteristics of damp-proof course materials.

Damp-proofing Old Walls

Various methods are available for damp-proofing existing buildings which are subject to rising damp in the absence of damp-proof courses. Probably the best method is to saw a slot in a mortar bed joint and to insert a damp-proof membrane in the slot. The membrane is normally inserted in about 1 m lengths and may be of slate, bitumen-felt, copper, lead or polythylene.

Another process is called electro-osmotic damp-proofing. Damp rises in walls from soils by capillarity, osmotic diffusion within the porous substance of the wall and by evaporation from its surface. Damp walls are negatively charged in relation to the underlying soil. The wall acts as an accumulator which, as long as evaporation losses are replenished from the soil, is virtually on permanent trickle charge. The electro-osmotic installation consists of 25 mm holes drilled from the outside with strip electrodes of high conductivity copper mortared into the drillings and looped into copper strips set into bed joints at damp-course level along the wall face. Earth electrodes are also installed consisting of 15 mm diameter copper or copper-clad steel rod electrodes driven from depths of 4 to 6 m at intervals of 10 to 13 m. The object is to provide a bridge between the wall at damp-course level and the soil, thus destroying the surface tension and preventing rising moisture. However, this method is not included in BRE Digest 245,[40] and has not always proved successful in practice.

Yet another process is to inject silicone or aluminium stearate water repellants into the lowest accessible mortar bed joint through closely spaced holes along the line of the dpc at different depths to ensure penetration through the entire thickness of the wall, preferably in late summer. They are not intended to provide a damp-proof barrier against a substantial positive pressure of water or in highly alkaline conditions.[40]

SCAFFOLDING

Most brickwork is erected from scaffolding made up of steel or aluminium alloy tubes of 38 mm nominal bore

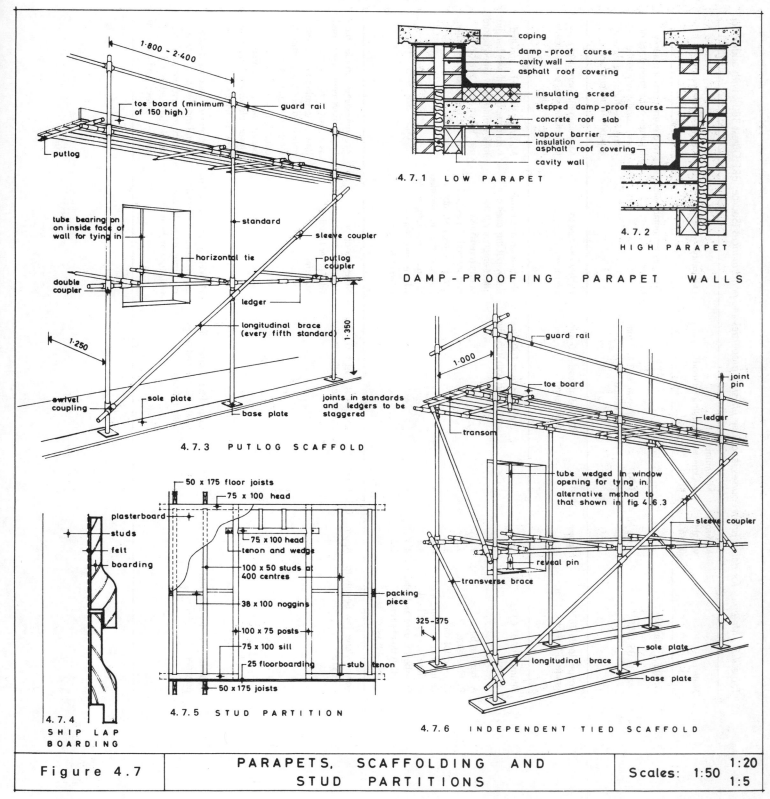

1·800 - 2·400

toe board (minimum of 150 high)

guard rail

putlog

tube bearing on on inside face of wall for tying in

standard

sleeve coupler

horizontal tie

putlog coupler

double coupler

ledger

longitudinal brace (every fifth standard)

1·250

1·350

swivel coupling

sole plate

base plate

joints in standards and ledgers to be staggered

4.7.3   PUTLOG SCAFFOLD

coping

damp-proof course

cavity wall

asphalt roof covering

insulating screed

stepped damp-proof course

concrete roof slab

vapour barrier

insulation

asphalt roof covering

cavity wall

.4.7.1   LOW PARAPET

4.7.2

HIGH PARAPET

DAMP-PROOFING   PARAPET   WALLS

guard rail

1·000

toe board

joint pin

transom

ledger

tube wedged in window opening for tying in.
alternative method to that shown in fig 4.6.3

sleeve coupler

reveal pin

transverse brace

325 - 375

longitudinal brace

sole plate

base plate

4.7.6   INDEPENDENT TIED SCAFFOLD

50 x 175 floor joists

75 x 100 head

plasterboard

studs

felt

boarding

75 x 100 head
tenon and wedge

100 x 50 studs at 400 centres

38 x 100 noggins

100 x 75 posts

75 x 100 sill

25 floorboarding

packing piece

stub tenon

50 x 175 joists

4.7.4

SHIP LAP BOARDING

4.7.5   STUD PARTITION

| Figure 4.7 | PARAPETS,   SCAFFOLDING   AND STUD   PARTITIONS | Scales: 1:50 | 1:20 1:5 |

Table 4.5 Materials for damp-proof courses, and their physical properties and performances (Source: BRE Digest 380[31] with additions)

Material	Minimum weight (kg/m²)	Minimum thickness (mm)	Jointing method	Durability	Other considerations
Group A: Flexible					
Lead to BS 1178[33]	Code Nr 4	1.80	Lapped at least 100 mm and welted.	Corrodes in contact with mortars.	To prevent corrosion apply bitumen or bitumen paint of heavy consistency to the corrosion-producing surface and both lead surfaces.
Copper to BS 2870[35] (grade C 104 or C 106)	Approx 2.28	0.25	Lapped at least 100 mm and welted.	Highly resistant to corrosion.	May stain masonry. Not easy to work on site, so unsuitable for cavity trays.
Bitumen to BS 6398[36] in class:			Lapped at least 100 mm and sealed with adhesive.	The hessian or felt may decay but this does not affect efficiency if the bitumen remains undisturbed. Types D, E and F are most suitable for buildings that are intended to have a very long life or where there is a risk of movement.	Materials should be unrolled with care. In cold weather the rolls should be warmed before use. When used as a cavity tray, the DPC should be fully supported.
A: hessian base	3.8	—			
B: fibre base	3.3	—			
C: asbestos base	3.8	—			
D: hessian base and lead	4.4	—			
E: fibre base and lead	4.4	—			
F: asbestos base and lead	4.9	—			
High bond strength asbestos base	2.2	—			
Polyethylene, low density to BS 6515[37]	Approx. 0.5	0.46	Lapped at least 100 mm and sealed with adhesive.	No evidence of deterioration in contact with other building materials.	Accommodates considerable lateral movement. When used as a cavity tray, may be difficult to hold in place and may need bedding in mastic for the full thickness of the outer leaf of walling, to prevent rain penetration. Unsuitable where compressive stress is minimal.
Bitumen polymer and pitch polymer	Approx 1.5	1.10	Lapped for distance of at least 100 mm and sealed with adhesive	Unlikely to be impaired by any movements normally occurring up to the point of failure of the wall.	Accommodates considerable lateral movement. If used as a cavity tray, preformed cloaks should be used, e.g. at changes of level and junctions.

Group B: Semi-rigid			
Mastic asphalt to BS 6925[38] or BS 6577[39]	12	No deterioration	To provide mortar key, beat up to 35% grit into asphalt immediately after application and leave proud of surface, or score surface while warm
Group C: Rigid			
DPC brick to BS 3921[20]	—	No deterioration	Particularly suitable if DPC is required to transmit tension as in freestanding walls. Does not resist downward movement of water
Slate to BS 743[34]		No deterioration	—

and a variety of fittings. Upright members are called standards; horizontal members along the length of the scaffold are termed ledgers; and the short cross members are called putlogs when one end is supported by the wall, and transoms when both ends rest on ledgers. A putlog scaffold (figure 4.7.3) has a single row of standards set about 1.25 m or 1.30 m from the wall and is partly supported by the building. An independent tied scaffold (figure 4.7.6) has a double row of standards about one metre apart, and does not rely on the building for support.

The spacing of standards depends on the load to be carried and varies from about 1.80 m to 2.40 m for bricklayers. Every scaffold should be securely tied into the building at intervals of approximately 3.60 m vertically and 6.00 m horizontally. Putlogs cannot be relied upon to act as ties. Alternative methods of fixing to window openings are shown in figures 4.7.3 and 4.7.6, as also are bracing arrangements. Substantial working platforms are required of 32 to 50 mm thick boards with butted joints and guard rails and toe boards for safety purposes, all fixed in accordance with the Construction (Working Places) Regulations.

ARCHES AND LINTELS

Openings in walls are spanned by arches or lintels, or a combination of both. An *arch* is an arrangement of wedge-shaped blocks mutually supporting one another. The superimposed load is transmitted through the blocks and exerts a pressure, both downwards and outwards, upon its supports. A *lintel* is a solid horizontal member or beam spanning an opening to carry the load and transmit it to the wall on either side.

Arches

The simpler forms of brick arch can be classified into three main types.

Flat gauged arches. Sometimes referred to as camber arches on account of the small rise or camber given to the soffit of the arch. In the Georgian period they were widely used in good-class buildings and were usually constructed of special dark red soft bricks known as 'rubbers' and 'cutters'. They are normally confined to the facework and are backed with rough brick arches or concrete lintels.

To set out the arch on elevation (figure 4.8.1), draw the centre and springing lines and then the lines of the skewbacks (maximum projections $\frac{1}{10}$ of span) and produce them to intersect the centre line of the arch. This locates the centre from which all the arch joints will radiate. The arch bricks are then set out along the flat top of the arch (usually 300 mm above springing line), ensuring that there is a central brick or key to the arch. Lines drawn from these points to the centre point give the arch joints. This type of arch would be built off a solid turning piece.

Axed arches. These can be segmental (figure 4.8.2) or semi-circular (figure 4.8.3) and are built up from centres consisting of ribs and ties (about 150 × 25 mm) supporting laggings (about 19 mm thick). The centres rest on bearers and folding wedges (for ease of removal) supported by props or posts. The arch bricks or *voussoirs* are axed or cut as closely as possible to the required shape, and they form attractive arches. To set out an axed segmental arch (figure 4.8.2) draw the springing line and set out the rise on the centre line. Join the springing line to the point so obtained and bisect this line. The centre of the arch curve is at the intersection of the bisector and the centre line. The top and bottom arch curves and skewbacks are then drawn; the bricks set out at 75 mm centres along the extrados (upper edge) with dividers and these points connected with the centre point to give the brick joints.

Rough arches. These are normally constructed of common bricks in half-brick rings without any cutting and are separated by wedge-shaped joints. Hence the arches are set out from their bottom edges in this instance. To obtain the parallel sides of the bricks, draw a 75 mm diameter circle, to the scale of the drawing, around the centre of the arch and join up the points on the soffit of each arch ring tangentially to each side of the circle (figure 4.8.4).

Lintels

Brick lintels or *soldier arches.* These are very popular, and they are superior to wood lintels which tend to decay, and concrete lintels which are of poor appearance. The bricks are usually placed on end, offer little strength in themselves and are not true arches. When bricks are laid on edge over openings, it is customary to make up the 40 mm to the next brick joint with two courses of nibless roofing tiles. The additional support

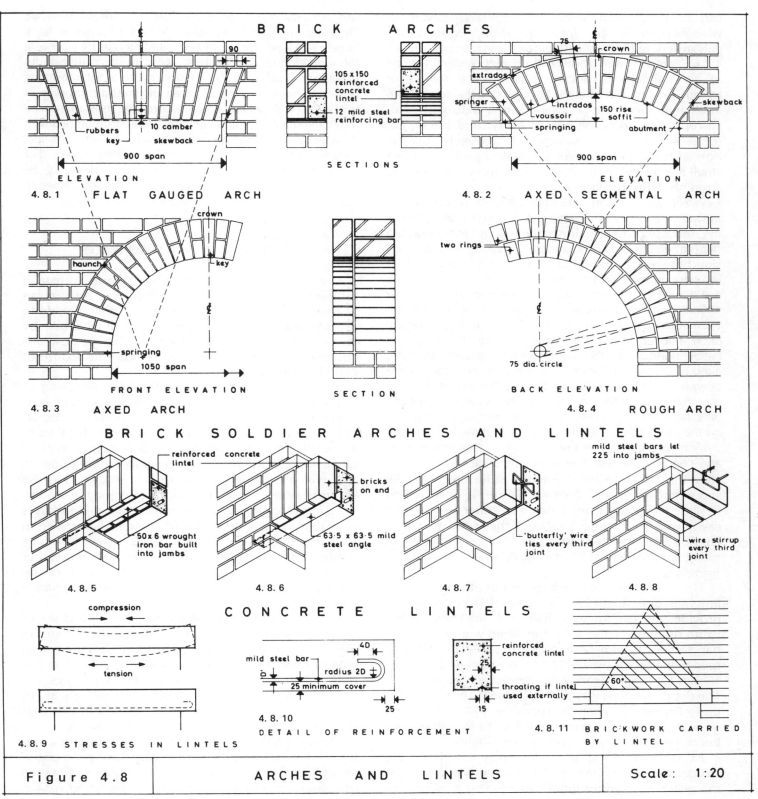

BRICK ARCHES

4.8.1 FLAT GAUGED ARCH

- 90
- rubbers
- key
- 10 camber
- skewback
- 900 span
- ELEVATION

SECTIONS

- 105 x 150 reinforced concrete lintel
- 12 mild steel reinforcing bar

4.8.2 AXED SEGMENTAL ARCH

- 75
- crown
- extrados
- springer
- intrados
- voussoir
- springing
- 150 rise soffit
- skewback
- abutment
- 900 span
- ELEVATION

4.8.3 AXED ARCH

- crown
- haunch
- key
- springing
- 1050 span
- FRONT ELEVATION

SECTION

4.8.4 ROUGH ARCH

- two rings
- 75 dia. circle
- BACK ELEVATION

BRICK SOLDIER ARCHES AND LINTELS

4.8.5
- reinforced concrete lintel
- 50 x 6 wrought iron bar built into jambs

4.8.6
- bricks on end
- 63·5 x 63·5 mild steel angle

4.8.7
- 'butterfly' wire ties every third joint

4.8.8
- mild steel bars let 225 into jambs
- wire stirrup every third joint

CONCRETE LINTELS

4.8.9 STRESSES IN LINTELS
- compression
- tension

4.8.10 DETAIL OF REINFORCEMENT
- 4D
- mild steel bar
- radius 2D
- 25 minimum cover
- 25

4.8.11 BRICKWORK CARRIED BY LINTEL
- reinforced concrete lintel
- throating if lintel used externally
- 25
- 15
- 60°

Figure 4.8	ARCHES AND LINTELS	Scale: 1:20

can be obtained in a number of ways: galvanised wrought iron flat bar for small spans (figure 4.8.5); galvanised mild steel angle for larger spans (figure 4.8.6); 'butterfly' wall ties every third joint connecting to an *in situ* reinforced concrete lintel (figure 4.8.7); or by building two mild steel bars across the top of the lintel into the brick jambs on either side with wire stirrups hung from the bars at every third brick joint (figure 4.8.8). Another alternative is to merely continue normal brick courses over the opening and to insert small mesh steel reinforcement into three bed joints to strengthen the brickwork.

Concrete lintels. Concrete lintels placed over openings in walls are subject to forces which tend to cause bending, inducing compression at the top and tension at the bottom (figure 4.8.9). Concrete is strong in compression but weak in tension and so steel reinforcement is introduced to resist the tensile stresses. The steel bars are placed 25 mm up from the bottom of the lintel, or 40 mm if fixing inserts are provided, to give protection against corrosion. The ends of reinforcing bars are often hooked as shown in figure 4.8.10 to give a better bond with the concrete and greater strength, although some engineers consider that they can secure adequate bond by using straight bars. The mix of concrete is normally 1:2:4; the smaller lintels are usually precast and only the very long and heavy lintels formed *in situ* (in place) on the job. With *in situ* lintels, timber shuttering or formwork is assembled over the opening, reinforcement fixed and concrete poured. The shuttering is struck (removed) after the concrete has hardened sufficiently to take superimposed loads. Boot lintels are 'L' shaped and designed to carry both inner and outer leaves of a cavity wall (see figure 8.2.2) but they tend to create cold bridges. The load to be carried by a lintel is the triangular area above it with angles of 60° as shown in figure 4.8.11.

The size of reinforcing bars and depth of concrete lintels is dependent on the span. It is customary to provide one bar for each half-brick thickness in the width of lintel. Hence a 105 mm lintel would have one bar and a 215 mm lintel two bars. The diameter of reinforcing bars is usually 12 mm for spans up to 1.60 m, 16 mm for longer spans or the provision of two bars. The depth of lintels varies from about 150 mm for spans up to 700 mm, 225 mm for spans from 700 mm to 1.30 m, and 300 mm for spans from 1.30 m to 2.20 m. End bearings normally vary from 100 to 150 mm.

BS 5977, Part 1[41] gives performance criteria for the structural requirements for lintels, and Part 2[42] covers the materials used for lintels. Reinforced masonry lintels are covered in BS 5628, Part 2.[43] Steel lintels, as illustrated in figure 4.5.2 and filled with insulation have largely replaced reinforced concrete lintels as they prevent the formation of cold bridges. They are normally hot dip galvanised and may be coated with a thermosetting polyester powder applied by an electrostatic process to give a tough, durable and highly corrosion resistant surface as obtained with Catnic steel lintels, which, combined with its sloping outer face, avoids the need for a separate dpc. Another alternative is to use stainless steel lintels. For windows over 1.80 wide, fixings should desirably be provided for curtain rails. Weep holes should be formed in every fourth joint to the brickwork over steel lintels.[44]

EXTERNAL TREATMENTS TO WALLS

A one-brick thick solid wall is unlikely to withstand satisfactorily severe weather conditions and, as an alternative to cavity wall construction, it is possible to provide adequate protection against driving rain by applying a suitable external finish. The greatest disadvantage is that the attractive colour and texture of good brickwork is lost.

The most common external finish is rendering or roughcast. *Rendering* may consist of two coats of Portland cement and sand (1:3 or 1:4), possibly incorporating a waterproofing compound, finished with a float to a smooth finish and often painted with two coats of emulsion or stone paint. This produces a rather dense coat which is liable to develop hair cracks. Moisture enters the cracks, becomes trapped behind the rendering and evaporates from the inner surface. Hence it is advisable to use a rendering of cement: lime: sand in the proportions of 1:1:6 or 1:2:9. These produce porous finishes which absorb water in wet weather and permit free evaporation when the weather improves.[45] External rendered finishes should comply with BS 5262.[46]

With *roughcast* or *pebbledashing* there is less risk of the external coat cracking or breaking away from the wall. In roughcast the coarse aggregate, usually gravel, is mixed into the second coat which is applied to the wall, whereas in pebbledashing the chippings are thrown onto the second coat whilst it is still 'green'. In both cases the appearance is rather unattractive but the need for periodic decoration is eliminated.

Common brick surfaces can be decorated in a variety of different ways. *Masonry paints* have good water-shedding properties and are available in a wide range of colours. In more recent times, *chlorinated rubber paints* have been applied to external brick and block walls with satisfactory results. *Bituminous paints* give an almost impervious surface coating but their use restricts future treatment owing to the bitumen bleeding through other applications. Silicones and other colourless water repellants are useful on exterior wall surfaces in good condition, when the appearance of the brickwork or stonework is to be maintained. The permanence of the protection is variable, depending on the type of treatment selected and the condition of exposure and of the masonry. Thick, textured, sprayed coatings, often based on alkyd resins with mica, perlite and sometimes fibres, are widely used and can have a life of up to 10 years.[47]

Vertical tile hanging on the external elevations of buildings can be very attractive if skilfully fixed, but it is expensive. It is extremely durable provided the fixing battens are pressure-impregnated with preservative. With plain tiling all tiles must be double nailed but a 40 mm lap is sufficient. Angles may be formed with purpose-made tiles or be close-cut and mitred with soakers, whilst vertical stepped flashings are usually introduced at abutments to form a watertight joint.

Weatherboarding provides a most attractive finish and may be either of painted softwood, or of cedar boarding treated periodically with a suitable preservative such as a high-solids exterior wood stain.[48] Feather-edge boarding is used extensively but probably the best results are obtained with ship-lap boarding (figure 4.7.4). The boarding is often nailed through a felt backing to studs (vertical timbers). Internal linings will be selected to meet thermal insulation requirements. A variety of *sheet claddings* is also available in profiled coloured steel, aluminium, fibre-reinforced cement, plastics and GRP (glass fibre reinforced polyester).

STONE WALLS

Building Stones

Building stones can be classified into three groups as follows

(1) *Sedimentary rocks*. These are formed by the weathering of land masses and the removal of the particles by water and their deposition in lakes or the sea. The deposited grains are united by some cementitious material. The principal stones in this class are limestones and sandstones.

Limestones are composed largely of calcium carbonate and small quantities of clay, silica and magnesia. They include the most widely used stones in the country with varying textures and colours ranging from white to grey, and have a natural bed. Some of the more common limestones are: *Anston*, quarried near Sheffield, cream, soft, easily carved, but unsuitable for polluted atmospheres; *Bath*, open texture and warm colour; *Clipsham*, quarried near Oakham, cream to buff, medium-grained and useful for general building work; *Hopton Wood*, from Derbyshire, hard surface and takes good polish; *Portland*, from Isle of Portland, cream or chalky white, very popular.

Sandstones are consolidated sands from grey to red in colour and have harder, rougher surfaces than limestones. The grains of sand are cemented together by siliceous, calcareous (calcium carbonate) or ferruginous (iron-bearing) substances. Typical examples are: *Bramley Fall*, quarried near Leeds, light brown, hard and weathers well; *Blue Bristol Pennant*, from Bristol, greyish blue, fine grained and hard; *Darley Dale*, from Derbyshire, light buff to dark grey, fine grained and weathers well; *Mansfield* (Notts.), yellow to red, unsuitable for polluted atmospheres; *Forest of Dean*, blue or grey, smooth and strong.

(2) *Igneous rocks*. Volcanic in origin, they have been fused by heat and then cooled and crystallised. The principal stones in this class are granites, syenites and basalt.

Granites are composed of quartz, felspar and mica, the greater part of the stone being felspar, which is normally white in colour. Granite is extremely hard and durable and often takes a high polish. Examples of some of the more common granites are: *Creetown* (S.W. Scotland), light grey, medium grain and polishes well: *Penryn*, from Cornwall, light grey, rather coarse grain, takes good polish, good all-purpose stone; *Peterhead* (N.E. Scotland), red, takes good polish, used extensively in building, civil engineering and monumental work; *Rubislaw*, from N.E. Scotland, grey in colour, fine grain, polishes well, used extensively in building work; *Shap* (Cumbria), pink to brown, fine grained.

(3) *Metamorphic rocks*. These may be sedimentary or igneous in origin and have been affected by heat or earth movement to such an extent as to assume a new structure and/or form new materials. The principal stones in this class are marbles, serpentines and slates.

Marble is a limestone which has been subjected to intense heat and pressure deep down in the earth's crust. The calcium carbonate is crystallised into calcite and a rough cleavage develops which gives the stone a veined appearance. True marbles are virtually non-existent in the British Isles, as Purbeck marble from Dorset is really a very compact limestone which has not undergone metamorphosis. Marbles are imported from Norway, Sweden, Belgium, Italy and France. They are available in a variety of colours and are mainly used as thin internal linings, owing to their high cost.[49]

Useful guidance on the selection of natural building stone is given in BRE Digest 269.[50]

Quarrying, Conversion, Working and Bedding of Stone

Sedimentary rocks are generally quarried with the aid of picks, bars and wedges and are split along their natural bed, while igneous rocks split readily in various directions. On quarrying the stone contains some moisture, known as 'quarry sap', which aids conversion (reducing to required sizes) by saws or feathers and wedges. Stone may be worked by hand or machine to a variety of finishes, some of which are illustrated.

Plain face. Consists of working the stone to an accurately finished face with a saw or chisel.

Hammer dressed. This is worked to rough hammered finish to give appearance of strength (figure 4.9.1).

Boasted. Parallel but not continuous chisel marks are formed across face of stone with a boaster (figure 4.9.2)

Tooled. This is a good-class finish with continuous chisel cuts (figure 4.9.3).

Vermiculated. Draughted margin with haphazard formation to interior; suitable for quoins (figure 4.9.4).

Furrowed. Draughted margin and base of furrows at same level, for quoins, string courses and similar features (figure 4.9.5).

Combed. A steel comb produces an irregular pattern on face of stone; also known as 'dragged' (figure 4.9.6).

Rubbed. A smooth face obtained by rubbing similar stones with sand and water.

Polished: Consists of polishing stone with putty powder.

Stones should be bedded with a mortar which matches the stone. It is usually made up of mason's putty, consisting of one part of lime putty to three parts of stone dust, and joints are often pointed with Portland cement and stone dust (1:3). In walling, stones should be laid with their natural beds horizontal, arch stones with their natural beds normal to the curve of the arch and vertical in the case of cornices and coping stones.

Walling Types

Stone was once a popular building material, but owing to the high cost, its use has diminished and it is now mainly employed as a facing to prestige buildings with a backing of brickwork, concrete or other material. The stone facework is made up of accurately dressed stones with fine joints and the surface finish can vary from a plain face to one of the finishes illustrated in figures 4.9.1 to 4.9.6. This type of stonework is called *ashlar* and is illustrated in figure 4.9.12. The stone facing can be bonded into the backing material or alternatively be fixed by non-ferrous metal angles, dowels, cramps or corbels built into slots formed in the back of the stonework. It is advisable to paint the back of the stone with bituminous paint where it adjoins brickwork or concrete to prevent cement stains appearing on the front face.

Stone members often have to be joined by special fittings or fastenings where the mortar joint offers insufficient adhesion. Coping stones can be connected by slate cramps (figure 4.9.13) about 150 × 62 × 50 mm set in chases in the adjoining stones in Portland cement, or by metal cramps (figure 4.9.14) of copper or bronze, about 250–300 × 25–50 × 6–15 mm set in sinkings in Portland cement, lead or asphalt, with the bedding material covering the cramp. Joggle joints (figure 4.9.15) help to bind adjoining stones with the jointing material filling the grooves. Dowels (figure 4.9.16) can be used for joining sections of vertical members like columns and mullions. They may be of slate or metal.

In stone-producing districts, waste stone cut to rectangular blocks about 150 mm thick (bed joint) are sometimes used to form the outer leaf of cavity walls but they are more expensive than brick. Another alternative quite widely adopted in the past, is to use rubble walling composed of roughly dressed stone built to one of a number of patterns with wide joints. *Uncoursed random rubble* (figure 4.9.7) consists of roughly dressed stones taken at random and not laid to any pre-determined pattern. It is used in boundary walls where it may be laid dry (without mortar joints), as well as in cottages and farm buildings. With *random rubble built to courses* (figure 4.9.8), the stone walling is levelled up at intervals in the height of the wall, such as at damp-proof course, window sill, window head and eaves. *Coursed squared rubble* (figure 4.9.9) consists of walling stones dressed to a roughly rectangular shape and every course is one stone in height, although the heights of the individual courses can vary. Indeed all the walling arrangements can be made up of either random rubble or squared rubble, but it is customary to build quoins, window and door jambs, and chimneys in dressed stonework. *Snecked rubble* has small sneck stones, about 100 × 100 mm, spaced at approximately one metre intervals to break the joint. Joints radiate from the sneck stone in the manner illustrated in figure 4.9.10). *Kentish rag* (figure 4.9.11) uses roughly dressed stone as a backing to brickwork or concrete. It has an attractive appearance and is used quite extensively in panels to the front elevations of modern houses.

Rubble walling is often constructed of selected facing stones on both faces with the intervening voids in the heart of the wall filled with stone chippings or 'spalls' grouted up with liquid lime mortar. Bonders or through stones (about one to the square metre) will tie the two faces together. Stone walls should be at least one-third thicker than the corresponding brick wall, as stone is apt to be rather porous. The use of preservative impregnated battens on the inside face to take plasterboard or other finishings, assists in combatting condensation. A useful finish to a window head is a stone lintel with rebated joggled joints, as illustrated in figure 4.9.17.

Reconstructed stone is widely used as a substitute for natural stone in order to reduce cost. Natural stone, sand and cement form the basic materials and reinforcement can be introduced during casting, in addition to any required sinkings. Special facing mixes permit a wide range of finishes. It eliminates the defects of natural stone but problems may arise through crazing (formation of fine hair cracks).

CROSSWALL CONSTRUCTION

In crosswall construction, all the vertical loads of the roof and upper floors are transmitted to the ground by transverse party and flank walls, as distinct from the more orthodox method whereby all enclosing and some internal walls carry the loads. The relative costs are influenced by a large number of factors. The front

and rear walls no longer carry loads and this permits flexibility in the choice of claddings (figure 4.9.18). On the other hand the uniform spacing of crosswalls, normally between 3 and 5.5 m, places constraints on layout. Some stiffening of the crosswalls by longitudinal members such as upper floor joists is advisable. One method is to stagger the joists on each side of a party wall and to tie the floor to the wall with metal anchors (figure 4.9.19) at intervals not exceeding 1.20 m. Anchor straps have one end screwed to the floor joist and the other end mortared into a joint in the crosswall and bent over a steel bar built into the wall. An alternative is to use mild steel joist hangers. The part of the load of a traditional pitched roof which is normally supported by the front and rear walls, is transferred to the crosswalls by beams. To further stabilise the walls, it is good practice to return the ends of the crosswalls in the manner shown in figure 4.9.18.

NON-TRADITIONAL HOUSING

General Approach

During the period 1945 to 1955 there was a shortage of labour and materials and the house building industry was unable to meet the demand using traditional methods. Hence non-traditional methods were employed on a substantial scale, embracing about ½ million dwellings in the public sector.

Non-traditional methods included concrete posts and infilling panels; thin concrete slabs supported on light structural steel framing; pre-assembled panels of brickwork; stressed-skin resin-bonded plywood panels; asbestos sheeting in various forms; curtain walling and the like. These were usually produced in a factory and transported to the site, requiring only to be placed and secured in position. Cupboard fittings for kitchens, items of joinery, and even complete dwellings were handled in this way.

In the nineteen sixties and early nineteen seventies, demand again exceeded supply and industrialised and system building methods were employed, with a predominance of high rise buildings. Industrialisation was primarily concerned with the rationalisation of both the type of construction and the building process itself. In building operations it implied the use of mechanical plant and the replacement of *in situ* work by prefabricated units. On the other hand, system

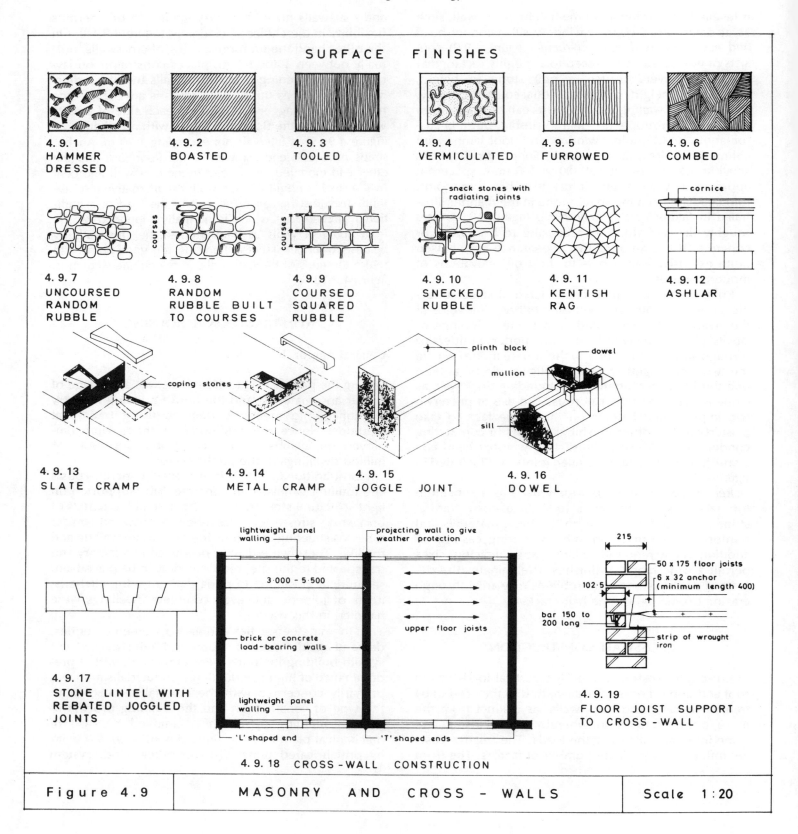

SURFACE FINISHES

4.9.1 HAMMER DRESSED

4.9.2 BOASTED

4.9.3 TOOLED

4.9.4 VERMICULATED

4.9.5 FURROWED

4.9.6 COMBED

4.9.7 UNCOURSED RANDOM RUBBLE

4.9.8 RANDOM RUBBLE BUILT TO COURSES

4.9.9 COURSED SQUARED RUBBLE

4.9.10 SNECKED RUBBLE

4.9.11 KENTISH RAG

4.9.12 ASHLAR

4.9.13 SLATE CRAMP

4.9.14 METAL CRAMP

4.9.15 JOGGLE JOINT

4.9.16 DOWEL

4.9.17 STONE LINTEL WITH REBATED JOGGLED JOINTS

4.9.18 CROSS-WALL CONSTRUCTION

4.9.19 FLOOR JOIST SUPPORT TO CROSS-WALL

| Figure 4.9 | MASONRY AND CROSS - WALLS | Scale 1:20 |

building usually relied on both non-traditional and industrialised methods. A system provided both a design and technique, often with specially made plant and components, which was available only from one design and, possibly, construction organisation.

Defects

The use of non-traditional housing has resulted in major problems, involving heavy remedial costs, much of which would not have occurred had the well tried and established traditional methods been employed. The Association of Metropolitan Authorities[51] has listed the main defects that have been identified, as follows

(1) Concrete external walls: corrosion of metal reinforcement and metal panel ties; movement of panels at joints; deterioration of concrete finish; uneven *U*-values; sound transmission problems; continuous cavities from ground floor to eaves; and the use of high alumina cement. These defects have threatened the structural stability of dwellings, destroyed weatherproofing, exacerbated condensation problems and permitted rapid fire spread.

(2) Timber framed external walls: deterioration of timber frames and cladding; fire spread risk through continuous cavities; compaction of thermal insulation material within cavity; deterioration of metal cladding; poor thermal insulation values; and inadequate tying in of internal and external leaves.

(3) Steel framed external walls: deterioration of steel frame, metal and other claddings; spalling and disintegration of asbestos cement panels; and inadequate thermal insulation.

(4) Roofs: extensive deterioration of roof sheeting; lack of insulation and condensation problems in roof spaces; need for work to party walls in roof spaces to prevent fire spread from one dwelling to another; deterioration of gable end claddings; inadequate wind bracing allowing lateral movement; and deterioration of concrete beams as a result of condensation.

The defects associated with non-traditional dwellings are very serious and relate to the main structural components. Extensive remedial works were found to be urgently necessary not only to replace defective components, but to check further decline. Demolition of some types of dwellings was found necessary as remedial costs proved prohibitive. The Association of

Metropolitan Authorities estimated in 1983 that the average cost of work to non-traditional dwellings, for both repair, or in severe cases demolition and replacement, amounted to £10 000 per unit.

Industrialised and system building declined extensively in the nineteen seventies, and in the nineteen eighties many buildings were suffering from major problems of structural instability, rainwater penetration and condensation, apart from unsatisfactory designs and layouts. Many local authorities faced enormous bills for repairs and replacements and many blocks were demolished. However, in the 1990s a significant number of system built tower blocks were being refurbished.

TIMBER FRAME CONSTRUCTION

Construction Methods

Timber frame is a method of construction as opposed to a system of building, although there are a number of systems which use timber frame as a basis. In Scandinavia, timber frame buildings started as a site built operation, but over the last fifty years most of the components have become factory prefabricated, thereby limiting the work on site to erection and fitting out. However, in the USA and Canada a significant amount of timber frame housing was still site built in the late 1980s.

In the UK, timber frame largely encompasses factory manufactured wall frames and roof trusses with only a few specialist companies making and erecting frames on site. There are variations in the extent of factory construction, with some panel prefabrication consisting of simple sheathed stud panels and others including a range of components such as joinery, insulation, cladding and internal linings. Timber frame construction is based on timber members and components forming a structural frame which transmits all vertical and horizontal loads to the foundations. The external cladding is non-load bearing and its main functions are to make the building weatherproof and provide an attractive external appearance.[52]

Most timber frame buildings in the UK use a platform frame, whereby each storey is framed up with the floor deck becoming the erection platform for the one above, as illustrated in figure 4.10. The prefabricated wall panels can be either small units approximately 3.6 m in length for ease of manhandling into position or full width panels incorporating ancillary components for placing with a crane.

External wall panels are constructed from vertical studs normally at 400 mm or 600 mm centres, nailed with simple butt joints to top and bottom rails. Stress graded timber is required and the most common size is 89 × 38 mm. Wind bracing is generally provided by a wood based sheet material nailed to the external face of the frame or possibly cross bracing coupled with internal plasterboard. External claddings can be selected from a wide range of materials including brickwork, tiling, timber boarding, cement rendering and proprietary claddings. High levels of insulation are achieved within the structural frame as illustrated in figure 4.11.1.[52]

Internal load bearing and non-load bearing walls are often constructed of a stud frame lined on each side with plasterboard or other sheet material. Separating walls are usually constructed of two separate stud frames with room linings of at least 30 mm of plasterboard applied in two or more layers with staggered joints, as described and illustrated later in the chapter.

TRADA[52] have described how the fire resistance of a timber frame structure is achieved by a combination of the internal lining material, the timber framing and the wall insulation. Thirty and sixty minutes fire resistance are the normal requirements, and the higher resistance can be obtained by using a double layer of plasterboard with staggered joints.

External walls

A typical external wall is illustrated in figure 4.11.1 and comprises

(1) Internal lining; often gypsum plasterboard which is an incombustible material, with a class 0 spread of flame rating.

(2) Vapour check (barrier): this is always incorporated either as a separate polythylene sheet – normally 500 g thick, or an alternative material forming an integral part of the plasterboard lining.

(3) Structural timber frame of timber studs, noggings, sole plate and head: the wall panel will be produced in a factory, complete with its sheathing board and usually with the breather membrane attached. The frames are of stress graded softwood timber.

(4) Insulating quilt: the space between the vapour check barrier and the sheathing board is filled with incombustible insulating quilt, often of mineral wool with a typical *U*-value of 0.39 W/m^2 K.

(5) Sheathing board: this is of plywood or other wood-based sheet material nailed to the timber studwork.

(6) Breather membrane: this is usually applied to the sheathing board on the outer face of the timber frame, to keep rain out of the structure during construction and to allow the wall to breath.

(7) Wall ties: the inner leaf of the timber structure and the outer leaf of brickwork are connected by flexible wall ties.

(8) Cavity: a nominal 50 mm wide ventilated cavity is formed to separate the wet brick outer leaf from the dry inner leaf.

(9) External finish: most houses are finished with an outer leaf of brickwork which is completed after the structure has been erected. However, timber frame accepts many other claddings such as tile hanging, timber boarding or natural stone.

Figure 4.11.2 shows an isometric detail of a typical timber frame external wall with brick cladding.

Internal walls

These usually consist of factory-produced framed panels with site applied plasterboard linings. Sometimes sheathing board is also applied to internal partitions for extra stiffness. Where point loads such as beams, trimmers or purlins are to be supported, additional vertical studs are incorporated to carry the load to the foundations.

Separating walls

These walls usually consist of two independent wall frames spaced 50 to 100 mm apart, preferably with a combined width of 250 mm for satisfactory acoustic performance. The lining, which provides fire and acoustic protection, consists of two or three layers of dense plasterboard on the room side of each leaf. Absorbent quilt is inserted in one leaf or the cavity to provide an acceptable level of thermal insulation. Light metal restraint straps of 1.6 mm metal are often provided at approximately 1.2 m centres to tie the two leaves together at ceiling level, principally to stiffen the frames during construction, as shown in figure 4.12. Plan details at a stagger or step and stagger of a timber frame separating wall are illustrated in figure 4.13.1.

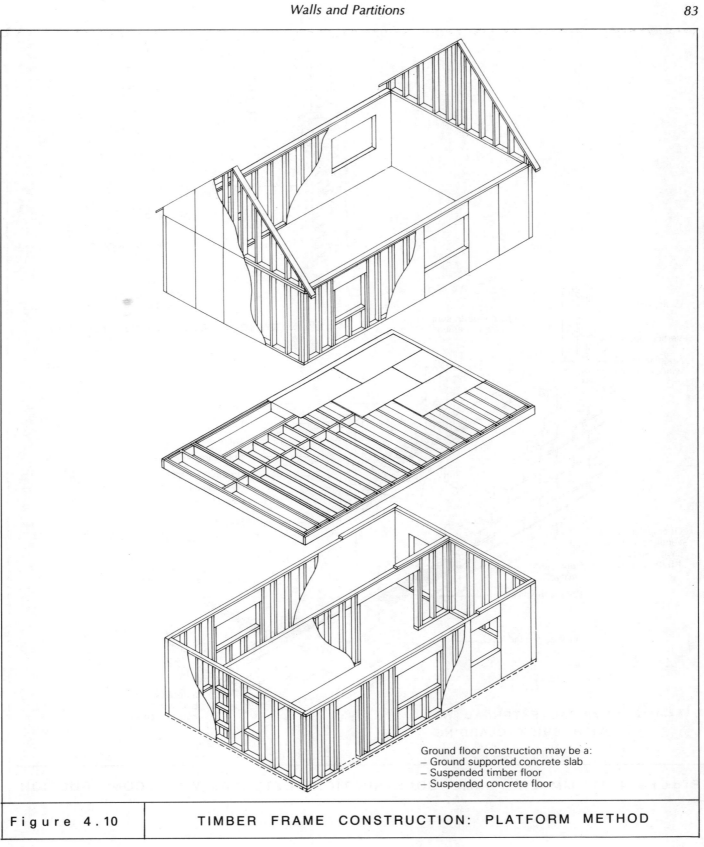

Ground floor construction may be a:
— Ground supported concrete slab
— Suspended timber floor
— Suspended concrete floor

| Figure 4.10 | TIMBER FRAME CONSTRUCTION: PLATFORM METHOD |

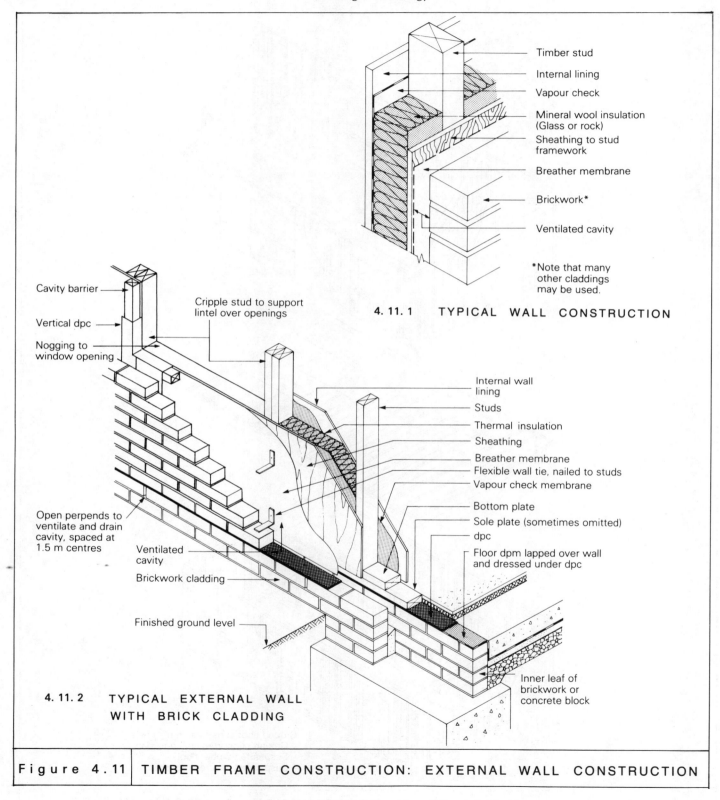

Timber stud
Internal lining
Vapour check
Mineral wool insulation (Glass or rock)
Sheathing to stud framework
Breather membrane
Brickwork*
Ventilated cavity

*Note that many other claddings may be used.

4. 11. 1   TYPICAL WALL CONSTRUCTION

Cavity barrier
Vertical dpc
Nogging to window opening
Cripple stud to support lintel over openings

Internal wall lining
Studs
Thermal insulation
Sheathing
Breather membrane
Flexible wall tie, nailed to studs
Vapour check membrane
Bottom plate
Sole plate (sometimes omitted)
dpc
Floor dpm lapped over wall and dressed under dpc

Open perpends to ventilate and drain cavity, spaced at 1.5 m centres
Ventilated cavity
Brickwork cladding
Finished ground level

Inner leaf of brickwork or concrete block

4. 11. 2   TYPICAL EXTERNAL WALL WITH BRICK CLADDING

| Figure 4.11 | TIMBER FRAME CONSTRUCTION: EXTERNAL WALL CONSTRUCTION |

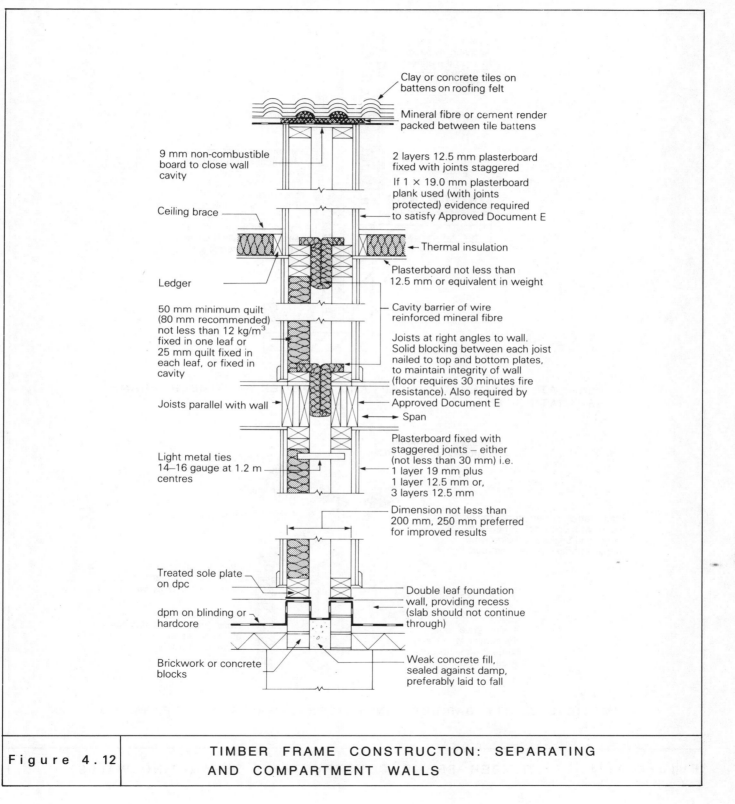

Clay or concrete tiles on battens on roofing felt

Mineral fibre or cement render packed between tile battens

9 mm non-combustible board to close wall cavity

2 layers 12.5 mm plasterboard fixed with joints staggered

If 1 × 19.0 mm plasterboard plank used (with joints protected) evidence required to satisfy Approved Document E

Ceiling brace

Thermal insulation

Plasterboard not less than 12.5 mm or equivalent in weight

Ledger

50 mm minimum quilt (80 mm recommended) not less than 12 kg/m³ fixed in one leaf or 25 mm quilt fixed in each leaf, or fixed in cavity

Cavity barrier of wire reinforced mineral fibre

Joists at right angles to wall. Solid blocking between each joist nailed to top and bottom plates, to maintain integrity of wall (floor requires 30 minutes fire resistance). Also required by Approved Document E

Joists parallel with wall

Span

Light metal ties 14–16 gauge at 1.2 m centres

Plasterboard fixed with staggered joints – either (not less than 30 mm) i.e.
1 layer 19 mm plus
1 layer 12.5 mm or,
3 layers 12.5 mm

Dimension not less than 200 mm, 250 mm preferred for improved results

Treated sole plate on dpc

Double leaf foundation wall, providing recess (slab should not continue through)

dpm on blinding or hardcore

Brickwork or concrete blocks

Weak concrete fill, sealed against damp, preferably laid to fall

| Figure 4.12 | TIMBER FRAME CONSTRUCTION: SEPARATING AND COMPARTMENT WALLS |

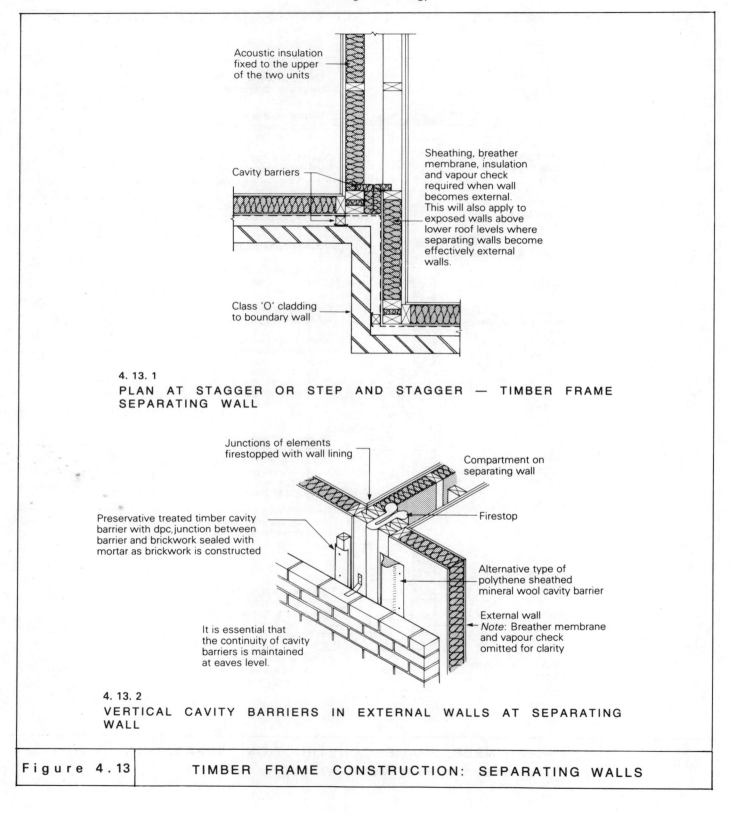

Acoustic insulation fixed to the upper of the two units

Cavity barriers

Sheathing, breather membrane, insulation and vapour check required when wall becomes external. This will also apply to exposed walls above lower roof levels where separating walls become effectively external walls.

Class 'O' cladding to boundary wall

4. 13. 1

PLAN AT STAGGER OR STEP AND STAGGER — TIMBER FRAME SEPARATING WALL

Junctions of elements firestopped with wall lining

Compartment on separating wall

Preservative treated timber cavity barrier with dpc, junction between barrier and brickwork sealed with mortar as brickwork is constructed

Firestop

Alternative type of polythene sheathed mineral wool cavity barrier

External wall
Note: Breather membrane and vapour check omitted for clarity

It is essential that the continuity of cavity barriers is maintained at eaves level.

4. 13. 2

VERTICAL CAVITY BARRIERS IN EXTERNAL WALLS AT SEPARATING WALL

| Figure 4.13 | TIMBER FRAME CONSTRUCTION: SEPARATING WALLS |

Fire stops and cavity barriers

Fire stops of suitable non-combustible material are fixed in the cavity between the outer brickwork and the inner wall panel, to prevent the spread of fire. They mainly consist of either mineral fibre or wool or possibly cementious material (but not asbestos cement), gypsum based plasterboards or intumescent mastic.

Cavity barriers prevent the penetration of smoke or flame or restrict their movement. They are required around all openings in external walls, at each floor level above ground floor in external and separating walls, at eaves level, verges, junctions of external walls with separating walls and in cavities at a maximum of 8 m centres. Cavity barriers can consist of preservative treated timber battens, size 38 mm × cavity width, mineral fibre board, wire reinforced mineral wool blanket at least 50 mm thick or polyethylene sleeved mineral wool or mineral wool slab.[52] Both fire stops and cavity barriers are illustrated in figure 4.13.2 and covered in Approved Document B3 (1991).

Problems in Construction

A BRE report[53] identified potentially dangerous faults in the design and construction of timber frame houses which the Establishment surveyed on ten sites between 1983 and 1985. A total of 433 types of faults were found, one in four potentially reducing the durability of the house and one in six which could cause reduced performance in a fire. More than half of the faults found were described as infringements of good practice as judged by the Building Regulations, the National House-Building Council and the British Standards Institution.

The report showed that faults associated with the use of brick cladding included very common lack of provision for vertical relative movement between the frame and the cladding, widespread inadequate provision for the exclusion of rainwater and universal inability to provide effective barriers against fire. It also found faults like torn or missing breather papers which together with poor construction of flashings, gutters and windows can lead to excessive water penetration, such that the construction will remain wet for long periods.

Nevertheless, the BRE emphasised that the faults are defined in the report as departures from good practice,

of a kind which are commonly found in all forms of housing construction and it is clearly desirable that they are avoided; but they only rarely lead to significant failures in service. Whilst design and construction practice needs to be improved, the evidence available to BRE indicates that the performance in service of timber frame housing is no less satisfactory than that of traditional construction.

Advantages of Timber Frame

The substantial interest in timber framed housing has occurred for many reasons, but principally because of the following advantages.

(1) The shell of the house, including the roof, can be constructed in as short a time as one day, permitting the following trades to work under cover.

(2) The dry construction can be largely prefabricated under factory conditions, thus reducing costs.

(3) Full insulation values are achieved immediately upon completion and not months later, as with wet forms of construction when the building water has dried out.

(4) The problems of drying shrinkage of internal concrete blockwork are eliminated.

(5) Security on site is less of a problem because of speedy completion.

(6) Financial outlay for materials and components can be delayed because of quick factory production.

(7) Subsequent modifications to internal walls are simplified and made less costly and the internal layout assumes greater flexibility.

(8) Intermittent heating can be adopted when the dwelling is unoccupied, thereby saving energy; the lightweight and well insulated walls warm up faster than brickwork and blockwork.

(9) The traditional outer brick face can be retained, which is still very popular in the United Kingdom.

INTERNAL PARTITIONS

There is a wide range of materials available for the construction of thin dividing walls in buildings. The choice will be influenced by a number of factors: thickness, weight, sound insulation, cost, decorative treatment and possibly fire resistance.

Brick

Half-brick walls may be used on the ground floors of domestic buildings, but they need some form of foundation on account of their weight. They have good loadbearing qualities and have a notional fire resistance period of two hours. Brick-on-edge partitions could be used to enclose small areas such as cupboards.

Clay Block

Hollow clay blocks containing cavities are available. A common size is 290 × 215 mm with thicknesses ranging from 62.5 to 150 mm. The blocks are usually laid to lap half a block and special blocks are available for stopped ends and junctions. Some blocks have interlocking or tongued joints which increase rigidity.

Concrete Block

Concrete blocks can be either loadbearing or non-loadbearing and are made from a wide range of lightweight aggregates, including sintered pulverised fuel ash from power stations, foamed slag, expanded clays and shales, clinker, expanded vermiculite and aerated concrete, as prescribed in BS 6073.[24] The loadbearing blocks range from 75 to 215 mm in thickness and non-loadbearing are 60 and 75 mm thick with a commonly used block size of 440 × 215 mm. Suitable mortars include cement: lime: sand 1:2:9 or 1:1:6; masonry cement: sand 1:5; and cement: sand 1:7 with a plasticiser. The blocks should be allowed to dry out before plastering and in long lengths of partition vertical joints filled with mastic or bridged by metal strips should be provided at 6 m intervals. Concrete block partitions can be tied to brick walls by leaving indents or recesses in the brickwork, to receive the ends of blocks on alternate courses, or by building expanded metal strips into the bed joints of the brickwork and blockwork.

Timber

Timber stud partitions (figure 4.7.5) are still used to a significant extent in the upper floors of domestic buildings. They are generally constructed of a 100 × 75 mm head and sill with vertical members or studs, ranging from 75 × 38 mm to 100 × 50 mm, framed between them at about 400 mm centres. Nogging pieces are often inserted between the studs to stiffen the parti-

tion. Plasterboard is generally nailed to either side of the partition with rose-headed galvanised nails and finished with a 3 to 5 mm skim coat of plaster. The sound insulating properties of a stud partition can be increased by staggering the studs and mounting the partition on a pad of resilient material. Stud partitions on upper floors are usually supported by floor joists. Where loads are to be carried, trussed partitions incorporating two sets of braces may be used. Sound insulation of partitions is dealt with in chapter 15 and illustrated in figure 15.2.

Plasterboard

There are several types of dry partitioning incorporating plasterboard faces which can be used to form lightweight partitions with good sound insulating properties. One quite popular variety consists of two gypsum wallboards separated by a core of cellular construction in accordance with BS 4022,[54] to thicknesses of 38 to 100 mm depending on the area of the partition. The surfaces may be prepared for immediate decoration, or to receive gypsum plaster or plastic sheeting. Another alternative is to use hot-dipped galvanised steel track and studs, faced each side with gypsum wallboard as a non-loadbearing relocatable partition. Laminated partitions are also available made up of layers of wallboard and plank to an overall thickness of 50 or 65 mm.

Other Types

There is a variety of relocatable steel partitions on the market which although expensive may be ideally suited for offices. It is also possible to use panels of wood wool or compressed strawboard which are comparatively easy to cut and erect. Glass bricks can be used where it is required to transmit light through a partition. Non-loadbearing partitions should comply with BS 5234.[55]

REFERENCES

1. *BS 5628* Code of practice for use of masonry, *Part 1: 1978* Structural use of unreinforced masonry
2. *BS 8103* Structural design of low-rise buildings, *Part 1: 1986* Code of practice for stability, site investigation, foundations and ground floor slabs for housing
3. *BRE Digest 246*: Strength of brickwork, blockwork and concrete walls: design for vertical load (1981)

4. *The Building Regulations 1991 and Approved Documents A (1991), B (1991), C (1991) and L (1995).* HMSO

5. *BRE Digest 360*: Testing bond strength of masonry (1991)

6. R.E. Lacy. An index of exposure to driving rain: *BRE Digest 127* (1971)

7. *BRE Digest 297*: Surface condensation and mould growth in traditionally-built dwellings (1990)

8. *BRE Digest 110*: Condensation (1972)

9. BRE. *Guidelines for the construction of fire resisting structural members* (1982)

10. *BRE Digest 230*: Fire performance of walls and linings (1984)

11. *BRE Digest 359*: Repairing brick and block masonry (1991)

12. *BS 4027: 1991* Sulphate-resisting Portland cement

13. *BRE Defect Action Sheet 64*: External walls – bricklaying and rendering when weather may be bad (1985)

14. *BRE Digest 200*: Repairing brickwork (1981)

15. *BS 1243: 1978 (1981)* Specification for metal ties for cavity wall construction

16. *BS 5628* Code of practice for use of masonry, *Part 3: 1985* Materials and components, design and workmanship

17. *BRE Digest 251*: Assessment of damage in low-rise buildings (1990)

18. *BRE Digest 343*: Simple measurement and monitoring of movement in low-rise buildings – Part 1: cracks (1989)

19. *BRE Digest 344*: Simple measurement and monitoring of movement in low-rise buildings – Part 2: settlement, heave and out-of-plumb (1989)

20. *BS 3921: 1985* Specification for clay bricks

21. *BRE Digest 273*: Perforated bricks (1983)

22. *BS 187: 1978* Specification for calcium silicate (sandlime and flintlime) bricks

23. *BRE Digest 157*: Calcium silicate brickwork (1992)

24. *BS 6073* Precast concrete masonry units, *Part 1: 1981* Specification for precast concrete masonry units

25. *BRE Digest 342*: Autoclaved aerated concrete (1989)

26. *BRE Digest 362*: Building mortar (1991)

27. *BS 1199* and *1200: 1976* Sands for mortar for plain and reinforced brickwork, block walls and masonry (amended 1984)

28. *BS 890: 1972* Building limes

29. *BS 5224: 1976* Masonry cement

30. *BRE Defect Action Sheets 35* and *36*: Substructure: DPCs and DPMs – specification and installation (1983)

31. *BRE Digest 380*: Damp-proof courses (1993)

32. L. Addleson. *Building failures: A guide to diagnosis, remedy and prevention.* Butterworths (1992)

33. *BS 1178: 1982* Specification for milled lead sheet for building purposes

34. *BS 743: 1970* Materials for damp-proof courses

35. *BS 2870: 1980* Specification for rolled copper and copper alloys: sheet, strip and foil

36. *BS 6398: 1983* Specification for bitumen damp-proof courses for masonry

37. *BS 6515: 1984* Specification for polyethylene damp-proof courses for masonry

38. *BS 6925: 1988* Specification for mastic asphalt for building and engineering (limestone aggregate)

39. *BS 6577: 1985* Specification for mastic asphalt for building (natural rock asphalt aggregate)

40. *BRE Digest 245*: Rising damp in walls: diagnosis and treatment (1986)

41. *BS 5977, Part 1: 1981* Lintels – method of assessment of load

42. *BS 5977, Part 2: 1983* Lintels – specification for prefabricated lintels

43. *BS 5628, Part 2: 1985* Structural use of reinforced and prestressed masonry

44. BRE. *Quality in traditional housing Vol. 2: an aid to design* (1982)

45. *BRE Digest 196*: External rendered finishes (1976)

46. *BS 5262: 1976* Code of practice for external rendered finishes

47. *BRE Digest 197*: Painting walls, Part 1: choice of paint (1982); *BRE Digest 198*: Painting walls, Part 2: failures and remedies (1984)

48. *BRE Digest 286*: Natural finishes for exterior timber (1984)

49. A. Everett. *Materials.* Batsford (1986)

50. *BRE Digest 269*: The selection of natural building stone (1989)

51. Association of Metropolitan Authorities. *Defects in housing, Part 1: Non-traditional dwellings of the 1940s and 1950s* (1983)

52. TRADA. *Timber Frame Construction* (1988)

53. BRE. *Quality in timber frame housing* (1985)

54. *BS 4022: 1970* Prefabricated gypsum wallboard panels

55. *BS 5234: 1975* Code of practice: internal non load bearing partitioning

5 FIREPLACES, HEAT PRODUCING APPLIANCES, FLUES AND CHIMNEYS

Before examining the constructional requirements and techniques associated with fireplaces, flues and chimneys, it is helpful to consider their purpose, the changing form of appliances and the general objectives, against the background of statutory requirements.

GENERAL PRINCIPLES OF DESIGN AND INSTALLATION OF HEAT PRODUCING APPLIANCES

The primary objective of an open fire is to heat the room in which it is placed, but it often fulfills a secondary objective of providing a focal point in the living room or lounge of a dwelling. Indeed so much importance may be attached to the secondary objective that it may become the deciding factor in providing an open fire in a new centrally heated dwelling, when heating efficiency considerations alone (it is probably less than 45 per cent efficient) would preclude its use. Hence the appearance of the fireplace is important and there is a wide range of materials from which to choose.

Many of the older types of open fire were inefficient and much of the obtainable heat was lost as a result of incomplete combustion. Furthermore, they tended to emit considerable quantities of dark smoke. The efficiency of modern open fires has been increased substantially by means of improved appliances, flues, air supply and fuels. A number of variants of the traditional open fire have been produced to secure greater efficiency and the more important of these will be described and illustrated later.

The principal requirements of a fireplace and flue are threefold

(1) to secure maximum heat for the benefit of occupants;

(2) to take adequate precautions against spread of fire; and

(3) to ensure effective removal of smoke and avoidance of downdraught.

All fires require sufficient air for combustion purposes, and as air is drawn into the flue from the room, further air is needed to replace it. Apart from blockage of the flue by soot or debris, there are three sets of conditions which can prevent the chimney operating satisfactorily, namely

(1) insufficient air entering the room to replace that passing up the chimney;

(2) adverse flow conditions resulting from the poor design of passages through which the smoke passes (throat, gathering and flue);

(3) downdraught caused by the build-up of pressure at the chimney top; this is influenced by the form of the building itself, neighbouring buildings, trees and the topography of the site.

The Building Regulations 1991 and Approved Document J1/2/3 (1989)[1] requirements in respect of heat producing appliances, which are designed to burn solid fuel, oil or gas, or are incinerators, are as follows.

(1) They shall be so installed that there is an adequate supply of air to them for combustion and for the efficient working of any flue, pipe or chimney.

(2) They shall have adequate provision for the discharge of the products of combustion to the outside air.

(3) The appliances and flue pipes shall be so installed, and fireplaces and chimneys so constructed, as to reduce to a reasonable level the risk of the fabric of

the building being damaged by heat or fire. In Approved Document J1 (1989) non-combustible materials are defined as those capable of satisfying the test for non-combustibility in BS 476, Part 2.[4]

Approved Document J1/2/3 (1989) details provisions which give acceptable levels of performance for fireplaces, constructional hearths, chimneys and flues, and flue outlets. The Approved Document classifies appliances into three categories, namely Section *2*: solid fuel burning appliances with a rated output up to 45 kW; Section *3* gas burning appliances with a rated input up to 60 kW operated by natural draught; and section *4* oil burning appliances with a rated output up to 45 kW. The detailed recommendations of the Approved Document will be considered later in the chapter.

An open fire in a living room or dining room may incorporate a *boiler unit* in place of the usual firebrick at the back of the fireplace, so that the fire may heat water as well as the room in which it is placed.

Convector fires are an efficient type of solid-fuel fire and they can also be fitted with a back boiler for domestic hot water supply. They emit convected warm air into the room in addition to radiant heat, and can either be inset in a normal fireplace recess (figure 5.4.5) or be freestanding, and they have a heat efficiency of 45 to 65 per cent. *Room heaters* are generally cleaner and more efficient than open fires, and can emit radiant heat, but are now less popular. They can be quite attractive in appearance and generally contain a transparent or translucent door(s) which can be left open if desired (figures 5.3.4, 5.3.5 and 5.3.6). They have a heat efficiency of 45 to 65 per cent when the doors are open, and up to 75 per cent when they are closed.

The Clean Air Act 1956 made provision for the establishment of smoke control areas in which authorised smokeless fuels only can be used. Approved Document J1/2/3 (1989) recommends that with a solid fuel burning opening appliance, there should be a permanent air entry opening(s) with a total free area of 50 per cent of the appliance throat opening.

Application of British Standards

There are a number of British Standards providing guidelines for the design and installation of a variety of heat producing appliances and their main provisions will now be outlined, with the exception of BS 1251,[3] dealing with open fireplace components, which is described later in the chapter.

BS 3376[4] contains detailed information on solid mineral fuel open fires with convection, with or without boilers. It gives the construction and performance requirements and describes the methods of test for all types of open fire, with or without a boiler, and for use with solid mineral fuels. Solid mineral fuels comprise bituminous coal, anthracite, Welsh dry steam coal, and manufactured smokeless fuels derived from these coals. Both overnight burning and intermittent burning types of appliance are included. Backboilers for use with domestic solid fuel appliances are described in BS 3377[5] and can be constructed of mild steel tubes, mild steel sheet and plate, cast iron, copper or stainless steel.

BS 4834[6] covers inset open fires without convection and with or without boilers, burning solid mineral fuels. It gives the constructional, dimensional and performance requirements for inset open fires burning solid mineral fuels, with nominal widths of 350, 400, 450 and 500 mm, designed for either fireplace components complying with BS 1251[3] and installed in accordance with BS 8303,[7] or with special fireplace components and boilers supplied with the appliance. It does not apply to inset open fires designed for space heating by convection which are covered by BS 3376,[4] as previously described.

BS 8303[7] covers the installation of domestic heating and cooking appliances burning solid mineral fuels. It encompasses the selection and installation of four main categories: (a) open fires without convection; (b) room heaters; (c) independent boilers and freestanding cookers; and (d) warm air heating appliances with natural convection.

Room heaters burning solid mineral fuels are covered by BS 3378[8] and this standard embraces room heaters with or without a boiler and with manual or thermostatic control.

There are two British Standards relating to gas fired appliances, of which BS 5258[9] encompasses gas fires, combined gas fires and back boilers, and decorative gas log and other fuel effect appliances. While BS 5871[10] deals with the installation of space heating appliances burning gas, mainly intended for heating single rooms in domestic and other premises and also gives information on the installation of gas fires when combined with back boilers. There are four basic types of space heater: radiant gas fires, radiant convector gas fires, gas convectors, and fire/back boilers. The back boiler may provide domestic hot water only or central heating with or without domestic hot water.

HEARTHS

Every fireplace must have a constructional hearth extending both under and in front of the opening. The Approved Document J1/2/3 prescribes that the hearth shall project not less than 500 mm in front of the jambs, extend not less than 150 mm beyond each side of the opening between the jambs and be not less than 125 mm thick. These requirements are illustrated in figures 5.1.1 and 5.1.2. Where the hearth adjoins a floor of combustible material, it shall be laid so that the upper surface of the hearth is not lower than the surface of the floor. No combustible material, other than timber fillets supporting the edges of a hearth, where it adjoins a floor, shall be placed under a constructional hearth, within a distance of 250 mm, measured vertically from the upper side of the hearth, unless such material is separated from the underside of the hearth by an air space of not less than 50 mm.

Hearths are almost invariably constructed of concrete and, when formed on a solid concrete ground or upper floor, become part of the normal floor construction. When constructed in association with a suspended timber ground floor, the hearth is supported on fender walls and hardcore in the manner shown in figures 5.1.1, 5.1.2 and 5.1.3. With upper timber floors, the hearth is normally supported by the brick chimney breast at the back, and by metal sheeting or temporary timber formwork as illustrated in figures 5.2.2 and 5.2.3.

Suspended concrete hearths may be formed of precast concrete, reinforced with steel fabric reinforcement, and built in as the work proceeds. They are more frequently cast *in situ* on timber formwork, which must be removed, or on metal flat sheeting which can be left in position. Some hearths in very old buildings are supported by brick trimmer arches.

Timber upper floors need to be trimmed around the hearth and chimney breast in the manner shown in figures 5.2.2 and 5.2.3. The trimmer and trimming joists are each 25 mm thicker than the normal floor joists and the joints of the intersecting timbers must be well framed together.

FIREPLACE CONSTRUCTION

Fireplace Recesses

Where the structure accommodating the fireplace opening and flue projects into a room, as in figures 5.2.1 and 5.3.1, the projection is termed a *chimney breast*. The fireplace opening is the recess housing the firegrate (figure 5.1.3); opening sizes in fireplace surrounds vary from 360 to 460 mm wide by 560 mm high.[3] The depth of the recess is usually about 338 mm (one-and-a-half bricks deep).

Approved Document J1/2/3 (1989) prescribes minimum dimensions for the backs and jambs of fireplace recesses. The jambs must be at least 200 mm thick. The back of the recess if a solid wall shall be not less than 200 mm thick, or if a cavity wall each leaf shall be not less than 100 mm thick (figure 5.4.1), extending for the full height of the recess. Where the recess is situated in an external wall and no combustible external cladding is carried across the back of the recess, the back wall of the recess may be reduced to 100 mm thick. Nevertheless in practice it would be desirable to provide a thickness of 200 mm to achieve a reasonable standard of thermal insulation. Where a wall, other than one separating dwellings, serves as the back of two recesses built on opposite sides of the wall, it may be reduced to 100 mm thick (see figure 5.4.3). Economy in construction, in addition to greater efficiency in heating, can be obtained if fireplaces are grouped together preferably on an internal wall.

To improve the air supply to a fireplace the air may be ducted up through the front hearth as shown in figure 5.4.5 or through an ash container pit as in figure 5.4.4. It is advisable that the sides and bottom of the pit shall be constructed of non-combustible material not less than 50 mm thick and that no combustible material shall be built into a wall or beside the pit within 225 mm of the inner surface of the pit.

Throats and Lintels

The throat is the part of the flue immediately above the fireplace openings (contraction at bottom of flue). It is restricted in size, often about 100 × 250 mm in cross-section, to accelerate the flow of the flue gases and to ensure adequate draught. The throat may be provided in a precast concrete throat unit of the form illustrated

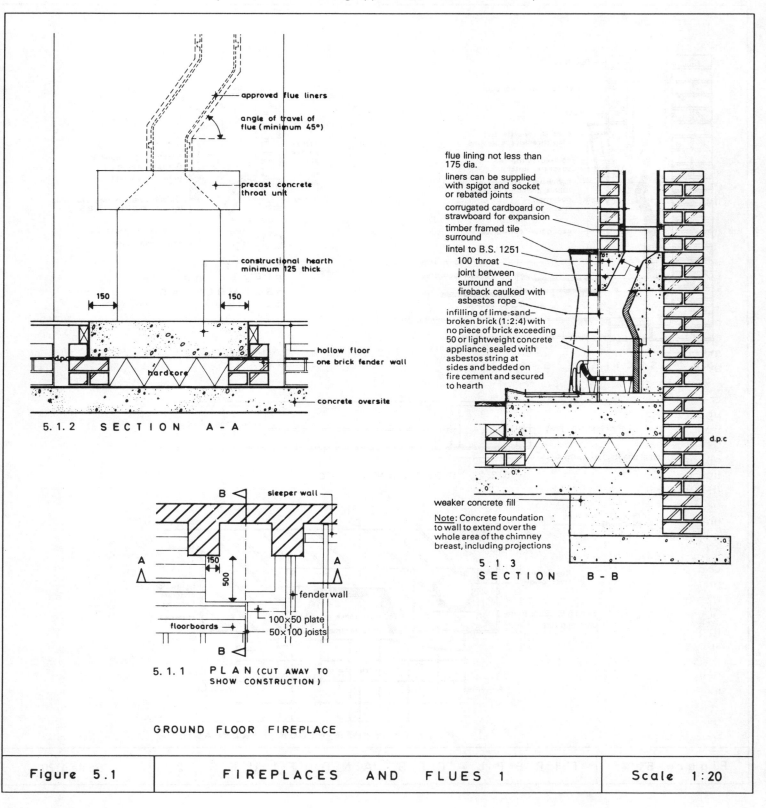

approved flue liners

angle of travel of flue (minimum 45°)

precast concrete throat unit

constructional hearth minimum 125 thick

150   150

hollow floor

one brick fender wall

hardcore

concrete oversite

5.1.2   SECTION   A - A

flue lining not less than 175 dia.

liners can be supplied with spigot and socket or rebated joints

corrugated cardboard or strawboard for expansion

timber framed tile surround

lintel to B.S. 1251

100 throat

joint between surround and fireback caulked with asbestos rope

infilling of lime-sand–broken brick (1:2:4) with no piece of brick exceeding 50 or lightweight concrete appliance sealed with asbestos string at sides and bedded on fire cement and secured to hearth

d.p.c

weaker concrete fill

Note: Concrete foundation to wall to extend over the whole area of the chimney breast, including projections

5.1.3   SECTION   B - B

B   sleeper wall

A   A

150

500

fender wall

100×50 plate

floorboards   50×100 joists

B

5.1.1   PLAN (CUT AWAY TO SHOW CONSTRUCTION)

GROUND FLOOR FIREPLACE

| Figure 5.1 | FIREPLACES AND FLUES 1 | Scale 1:20 |

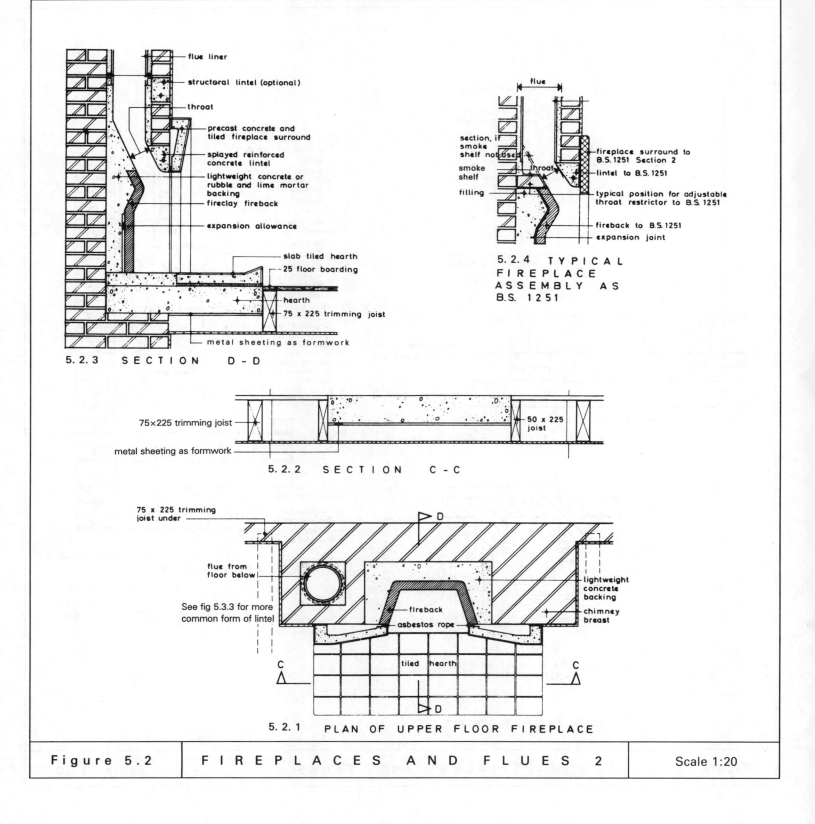

flue liner

structural lintel (optional)

throat

precast concrete and
tiled fireplace surround

splayed reinforced
concrete lintel

lightweight concrete or
rubble and lime mortar
backing

fireclay fireback

expansion allowance

slab tiled hearth

25 floor boarding

hearth

75 x 225 trimming joist

metal sheeting as formwork

5.2.3   SECTION   D-D

flue

section, if
smoke
shelf not used

smoke
shelf

filling

throat

fireplace surround to
B.S. 1251 Section 2

lintel to B.S. 1251

typical position for adjustable
throat restrictor to B.S. 1251

fireback to B.S. 1251

expansion joint

5.2.4   TYPICAL
FIREPLACE
ASSEMBLY AS
B.S. 1251

75×225 trimming joist

metal sheeting as formwork

50 x 225
joist

5.2.2   SECTION   C-C

75 x 225 trimming
joist under

flue from
floor below

See fig 5.3.3 for more
common form of lintel

fireback

asbestos rope

tiled hearth

lightweight
concrete
backing

chimney
breast

5.2.1   PLAN OF UPPER FLOOR FIREPLACE

Figure 5.2   FIREPLACES AND FLUES 2   Scale 1:20

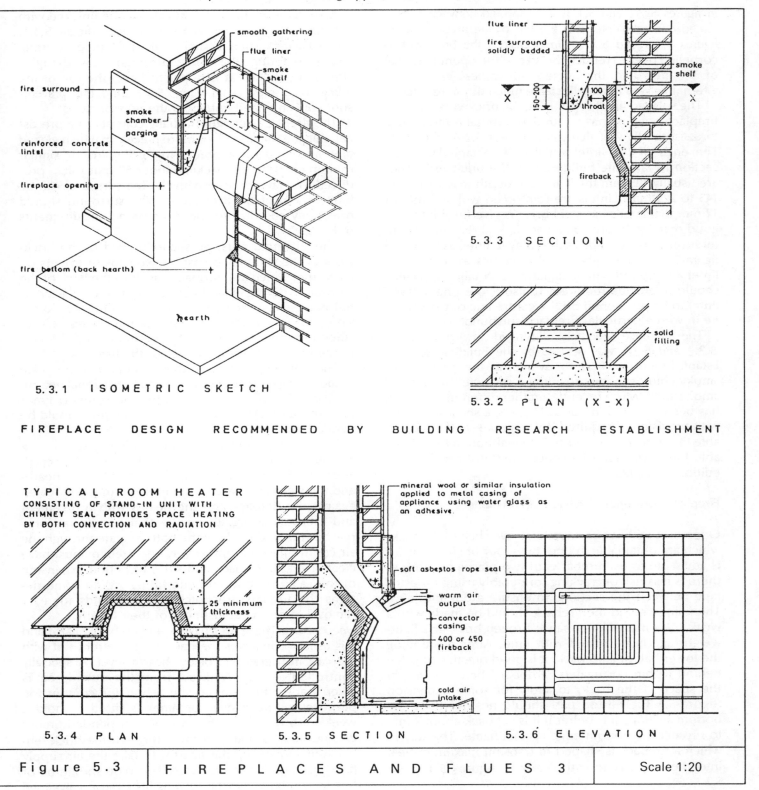

smooth gathering

flue liner

smoke shelf

fire surround

smoke chamber

parging

reinforced concrete lintel

fireplace opening

fire bottom (back hearth)

hearth

5.3.1   ISOMETRIC   SKETCH

flue liner

fire surround solidly bedded

smoke shelf

150-200

100

throat

fireback

5.3.3   SECTION

solid filling

5.3.2   PLAN   (X - X)

FIREPLACE   DESIGN   RECOMMENDED   BY   BUILDING   RESEARCH   ESTABLISHMENT

TYPICAL   ROOM   HEATER
CONSISTING OF STAND-IN UNIT WITH
CHIMNEY SEAL PROVIDES SPACE HEATING
BY BOTH CONVECTION AND RADIATION

25 minimum thickness

5.3.4   PLAN

mineral wool or similar insulation applied to metal casing of appliance using water glass as an adhesive.

soft asbestos rope seal

warm air output

convector casing

400 or 450 fireback

cold air intake

5.3.5   SECTION

5.3.6   ELEVATION

Figure 5.3   FIREPLACES   AND   FLUES   3   Scale 1:20

in figures 5.1.2 and 5.1.3 or be formed between a reinforced concrete lintel and the top of the fireback, as in figures 5.3.3 and 5.2.4. On occasions the brickwork is corbelled over to reduce the size of the opening to that of the flue and this is termed *gathering*. Another alternative is to fit an adjustable metal throat restrictor.

The splayed lintel supporting the brickwork over the fireplace opening has a rounded internal bottom edge to assist the flow of flue gases (figures 5.2.3 and 5.3.3). The ends of the lintel will be of rectangular cross-section to facilitate building into the brickwork. They are usually 125 mm thick and the depth may vary from 145 to 225 mm. Lintels are reinforced with a 8 mm or 12 mm diameter mild steel bar. To give flexibility, it is good practice to provide an additional rectangular lintel several courses above the splayed lintel as shown in figure 5.2.3. The higher lintel then acts as a structural lintel and permits the installation of a taller appliance should it later be required. It also allows the splayed lintel to be positioned off the centre line of the wall to tie in with a particular fireplace surround.

The fireplace assembly illustrated in figures 5.3.1, 5.3.2 and 5.3.3 was devised by the Building Research Establishment in 1955 to overcome the problem of smoky chimneys, by providing a restricted throat and smoke shelf. More recently a rather different approach has been adopted in BS 1251[3] using a shallower lintel with provision, if required, for a smoke shelf or adjustable throat restrictor (figure 5.2.4), although the adjustable throat restrictor has been omitted from the 1987 edition of BS 1251.[3]

Fireplace Surrounds, Firebacks and Ashpits

Open fires without back boilers have shaped firebacks which are obtainable in one, two, four or six pieces. It is advisable to use firebacks of two pieces or more as there is less risk of cracking, preferably using asbestos tape between the sections to ensure a durable joint. The filling behind the fireback should be made with a weak 1:2:4 mix of lime: sand: broken brick or a lightweight aggregate mixed with lime. After positioning the lower half of a fireback, it is good practice to place behind it thin corrugated cardboard before filling in the space at the back, to allow for some expansion (figure 5.1.3). The top part should not overhang the bottom section; it is better if it is set back about 2 mm to protect the bottom edge against flame. The way in which a fireback is shaped to transmit maximum heat into the room is shown clearly in figures 5.1.3 and 5.2.1.

The filling behind the fireback may be finished with a horizontal surface or smoke shelf as in figure 5.3.1. On occasions such a shelf will assist in preventing smoke entering the room through downdraught, although generally it is better to finish the top of the filling to a splay as in figure 5.2.3, and this prevents soot collecting and enhances the flow of gases.

The fireplace surround usually consists of a precast ceramic tile slab in one or more pieces, in which glazed tiles are fixed on a precast high alumina or similar reinforced concrete backing about 50 mm thick, produced in a factory. Alternatively, the surround can be built *in situ* of brick or stone. The surround should preferably conform to the dimensional requirements of BS 1251.[3]

The surround must be securely fixed to the brickwork of the chimney breast by two clips or eyelets on each side, which are normally cast into the back of the surround during manufacture, and the plaster is finished against the edges of the surround. An expansion joint of soft asbestos rope should always be provided between the back of the surround and the face of the lintel and leading edges of the fireback.

The matching hearth for the surround may be made of tiles laid *in situ* directly on the constructional hearth. Tiles laid *in situ* should be bedded in cement: lime: sand mortar (1:1:6) and preslabbed hearths should be bedded in a similar manner.

To ensure proper control of burning, *inset* or *continuous burning fires* are often provided, in accordance with BS 4834.[6] These are firmly fixed to the hearth and the sides and base of the firefront must be sealed against the fireback and the hearth with fire cement and asbestos string, to ensure that primary air only reaches the underside of the bottom grate through the air control in the ashpit cover. The firebars must have clearance at both back and sides to permit expansion of the cast iron bottom grate fittings. With some designs the firebars are notched at the ends for easy removal to permit adjustment of their length.

Deep ashpit. These fires have a sunken back hearth and this usually eliminates the need for a firefront fret with the cast iron grate set below hearth level. A specially controlled air supply direct from the outside air is taken under the floor to the ashpit for combustion purposes. A deep ash container is fitted to hold a week's ash refuse. In the case of a suspended timber ground floor, a hole is formed through the brick fender wall supporting the hearth, to take the air supply pipe, usually 75 mm diameter. This pipe should be of non-combustible material such as light gauge galvanised steel. With a solid floor, ducts are normally built into the floor. It is best to install a dual system of pipe ducts laid at right angles to each other from the outside walls to avoid suction. These should terminate on the face of external walls with suitable grilles or ventilators to prevent vermin penetrating the building (figure 5.4.4).

Open fires with back boilers. These kind of fires have become quite popular because of their dual function of both space and water heating. The usual fireback is replaced by a boiler unit, complying with BS 3377,[5] incorporating a specially designed boiler flue and specially shaped firebrick side cheeks. A boiler control damper is also fitted. Most approved appliances of this type contain fires 400 mm wide. A relief valve should be fitted as close to the boiler as possible together with an emptying cock for draining water from the system. Short lengths of metal pipes should be built into the surrounding brickwork to form 'sleeves' to accommodate the hot water pipes. The pipework is connected to the boiler as soon as the unit is fixed and it is then tested for leaks before installing the grate. The grate must be rigidly fixed to the hearth and the space behind the appliance and the side cheeks filled with a weak lightweight aggregate or brick rubble and weak lime mortar. The voids around pipes passing through sleeves in the brickwork are sealed by caulking with asbestos string. Such an appliance might heat up to 4 m² of heating surface, including pipework, in addition to an indirect hot water cylinder of about 136 litres capacity.

FLUES AND CHIMNEYS

Where practicable fireplaces, flues and chimneys should be located on internal walls as the flues help to warm parts of the building, heat loss is reduced and the flue gases are warmer resulting in more efficient operation. For a flue to work effectively, there must be an adequate air supply to the fire and the outlet to the flue (the chimney stack) must be suitably located in relation to external features, such as nearby buildings and trees.

It might be helpful to the reader to distinguish between *chimneys* and *flues*. A chimney is normally interpreted as any part of the structure of a building forming any part of a flue other than a flue pipe, while a flue can be described as a passage for conveying the discharge of an appliance to the external air and includes any part of the passage in an appliance ventilation duct which serves the purpose of a flue. A flue pipe is a pipe forming a flue, but does not usually include a pipe built as a lining into either a chimney or an appliance ventilation duct.

Flue Installation

Approved Document J (1989) prescribes that a chimney serving a solid fuel burning appliance with a rated output up to 45 kW (Section 2 appliance) may be lined with rebated or socketed linings of clay to BS 1181[11] or kiln-burnt aggregate and high alumina cement or imperforate clay pipes and fittings to BS 65[12], or be constructed of a refractory material without a lining. A flue when measured in cross-section shall contain a circle with a diameter of not less than 175 mm, in a chimney for an appliance with an output rating of 30 to 45 kW. Flues shall be separated by solid materials not less than 100 mm thick.

As far as practicable bends in a flue should be kept to a minimum. Where bends are necessary the angle of travel should be kept as steep as possible, preferably not less than 60°.[13] It is good practice to take the flue as high as possible above a fireplace opening before introducing a bend. The inner surface of the flue should be smooth to prevent the accumulation of soot, which in its turn reduces the size of the flue and retards the flow of combustion gases. Flue liners provide a smooth inner surface and also protect the surrounding brickwork from the harmful effect of acid gases and condensation, which is particularly prevalent with boiler flues. The majority of flues in houses built before 1965 are 'parged' internally with lime mortar and, where this has cracked, smoke may penetrate the building and the fire risk is increased.

No timbers shall be built into a wall within 200 mm of the inside of a flue or fireplace recess. No timber, other than a floorboard, skirting board, dado rail, picture rail, mantelshelf or architrave, can be placed within 40 mm of the outer surface of the chimney, unless the flue surround of non-combustible material is at least 200 mm thick. Flue surrounds in domestic buildings rarely exceed half-a-brick in thickness (102.5 mm), and care is needed in the placing of trimming timbers in floors and roofs. No metal which is in contact with timber or other combustible material shall be placed within 50 mm of the inside of the flue or fireplace recess (Approved Document J: 1989).

Approved Document J1/2/3 (1989) also permits the use of flue pipes to connect a section 2 appliance to a chimney provided they do not pass through a roof space. Acceptable materials for flue pipes are listed as

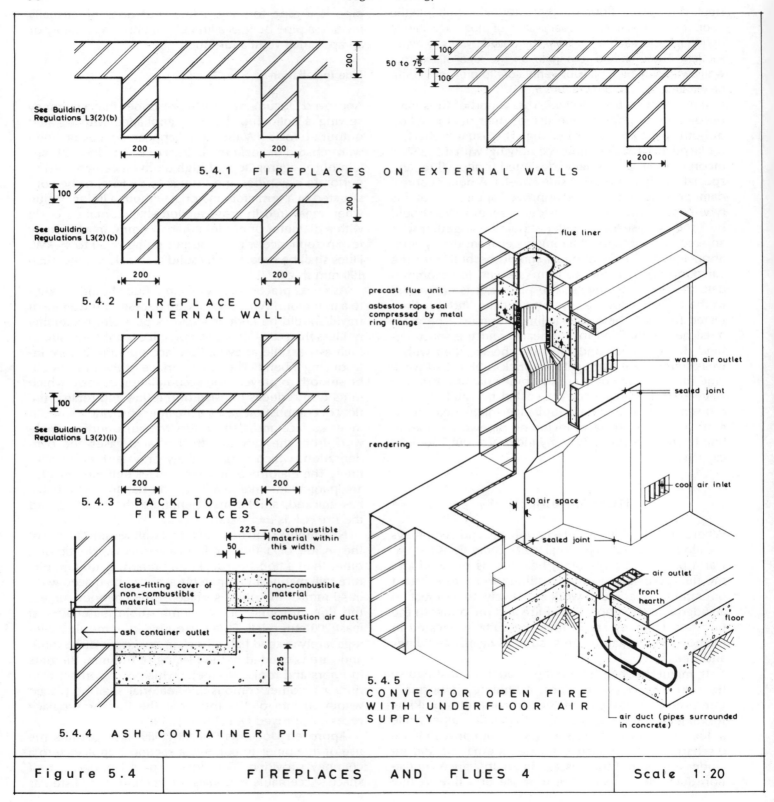

See Building
Regulations L3(2)(b)

200   200

200

5.4.1 FIREPLACES ON EXTERNAL WALLS

100

50 to 75

100

200

See Building
Regulations L3(2)(b)

100

200

200   200

5.4.2 FIREPLACE ON
         INTERNAL WALL

See Building
Regulations L3(2)(ii)

100

200   200

5.4.3 BACK TO BACK
         FIREPLACES

225 no combustible
      material within
50   this width

close-fitting cover of
non-combustible
material

non-combustible
material

combustion air duct

ash container outlet

225

5.4.4 ASH CONTAINER PIT

flue liner

precast flue unit

asbestos rope seal
compressed by metal
ring flange

warm air outlet

sealed joint

rendering

cool air inlet

50 air space

sealed joint

air outlet

front
hearth

floor

5.4.5 CONVECTOR OPEN FIRE
        WITH UNDERFLOOR AIR
        SUPPLY

air duct (pipes surrounded
in concrete)

| Figure 5.4 | FIREPLACES AND FLUES 4 | Scale 1:20 |

cast iron with sockets uppermost to BS 41,[14] mild steel at least 3 mm thick, stainless steel at least 1 mm thick, and vitreous enamelled steel to BS 6999.[15] There is also provision for the use of factory-made insulated chimneys to BS 4543[16] and BS 6461.[17] The provision of flue pipes and brick and blockwork chimneys for use with section 4 (oil burning) appliances must be able to withstand the temperature of the flue gases under the worst operating conditions.

The same Approved Document also incorporates acceptable arrangements for the placing and shielding of flue pipes where they pass through roofs or external walls. The deemed-to-satisfy provisions are

(1) a distance of not less than three times the external diameter of the flue pipe from any combustible material; or

(2) separated from any combustible material by solid non-combustible material not less than 200 mm thick; or

(3) enclosed in a sleeve of metal or asbestos cement with a space of not less than 25 mm packed with non-combustible thermal insulating material between the sleeve and the pipe, and subject to various other specific requirements, illustrated in Approved Document J (1989).

Approved Document J (1989) also prescribes requirements for flues serving section 3 gas burning appliances with an input rating up to 60 kW. These chimneys can be constructed of prescribed flue linings. Alternatively, they may be constructed of precast concrete or clay blocks complying with BS 1289.[18] The concrete blocks are sized to bond with brickwork and can be obtained within the thickness of a wall.

A common size of flue block is 425 × 102 × 215 mm containing an aperture 300 × 50 mm. Some blocks have spigot and socket joints which minimise the risk of jointing mortar entering the flue during construction in addition to ensuring true alignment of the blocks and a smooth flueway. The concrete flue blocks can terminate through an approved lining in a brick chimney stack finished with a clay pot, through a flexible steel tube terminating in a steel terminal or may be connected to a ridge terminal with asbestos cement or steel ducting (figure 5.5.6).[19]

No internal flue dimension for section 3 appliances shall be less then 90 mm and the cross-sectional area of a flue serving one gas-fire shall be not less than 12 000 mm² for a round flue or 16 500 mm² for a rec-

tangular flue, and that serving an appliance other than a gas-fire shall be not less than the area of the outlet of the appliance.

No part of the flue pipe serving a section 3 gas burning appliance shall be less than 25 mm from any combustible material, and where it passes through a wall, floor or roof, constructed of combustible materials, the flue pipe shall be enclosed in a sleeve of non-combustible material and be separated from the sleeve by an air space of not less than 25 mm.

Approved Document J (1989) permits room-sealed section 3 gas appliances to discharge directly into the external air through a balanced flue outlet (figure 5.5.1), subject to the following conditions

(1) The inlet and outlet of the appliance are incorporated in a terminal which is designed to allow free intake of combustion air and discharge of the products of combustion and to prevent the entry of any matter which may restrict the inlet or outlet.

(2) Where the outlet is located near any opening in the building, no part of the outlet is within 300 mm of the opening.

(3) Where the outlet could come into contact with people near the building or be subject to damage it shall be protected by a terminal guard of durable material.

CHIMNEY STACKS AND FLUE OUTLETS

Section 2 Appliances

Approved Document J (1989) prescribes that the outlet of any flue serving a section 2 appliance which is solid fuel burning with a rated output up to 45 kW shall be so positioned that the top of the chimney or flue pipe is not less than

(1) One metre above the highest point of contact between the chimney or flue pipe and the roof; provided that where a roof has a pitch on both sides of the ridge of 10° or greater with the horizontal, and the chimney or flue pipe passes through the roof at or within 600 mm of the ridge, the top of the chimney or flue pipe shall be not less than 600 mm above the ridge (figure 5.5.5).

(2) One metre above the top of an openable part of window or skylight, or of any ventilator or air inlet to a ventilation system, which is situated in any roof or

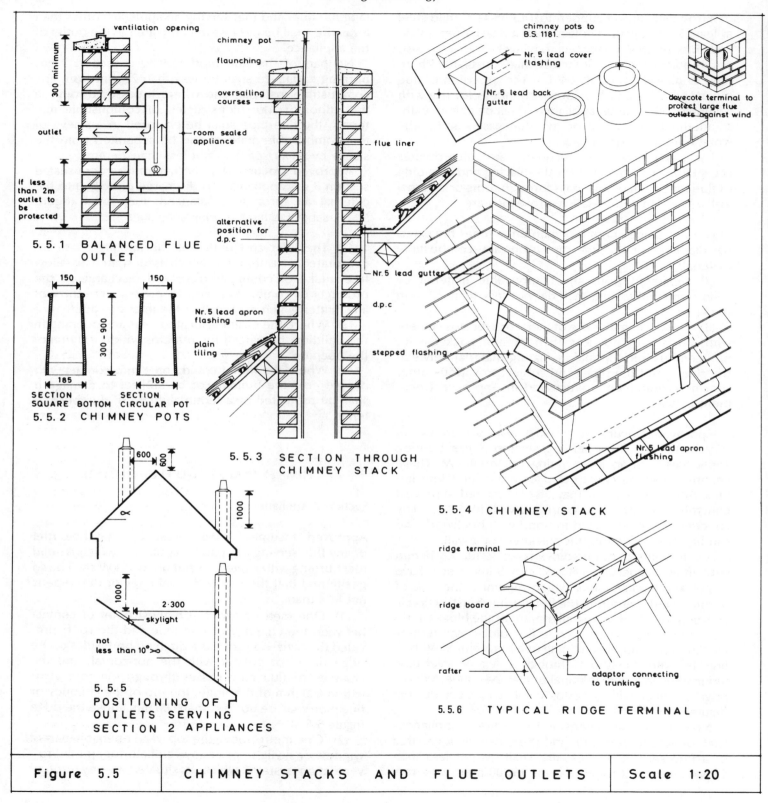

ventilation opening

300 minimum

outlet

if less than 2m outlet to be protected

5.5.1 BALANCED FLUE OUTLET

room sealed appliance

150   150

300 - 900

185   185

SECTION   SECTION
SQUARE BOTTOM CIRCULAR POT

5.5.2 CHIMNEY POTS

chimney pot

flaunching

oversailing courses

flue liner

alternative position for d.p.c

Nr.5 lead apron flashing

plain tiling

d.p.c

5.5.3 SECTION THROUGH CHIMNEY STACK

600   600

1000

1000   2·300

skylight

not less than 10°

5.5.5 POSITIONING OF OUTLETS SERVING SECTION 2 APPLIANCES

chimney pots to B.S. 1181

Nr.5 lead cover flashing

Nr.5 lead back gutter

dovecote terminal to protect large flue outlets against wind

Nr.5 lead gutter

stepped flashing

Nr.5 lead apron flashing

5.5.4 CHIMNEY STACK

ridge terminal

ridge board

rafter

adaptor connecting to trunking

5.5.6 TYPICAL RIDGE TERMINAL

| Figure 5.5 | CHIMNEY STACKS AND FLUE OUTLETS | Scale 1:20 |

external wall of a building and is not more than 2.30 m, measured horizontally, from the top of the chimney or flue pipe (figure 5.5.5).

(3) One metre above the top of any part of a building (other than a roof, parapet wall or another chimney or flue pipe) which is not more than 2.30 m, measured horizontally, from the top of the chimney or flue pipe.

Section 3 Appliances

With section *3* appliances (gas burning with a rated input up to 60 kW), the flue outlet shall be

(1) fitted with a flue terminal, designed to allow free discharge, to minimise downdraught and to prevent the entry of any matter which might restrict the flue, where any dimension across the axis of the flue outlet is less than 175 mm (figure 5.5.6);

(2) so situated externally that a current of air may pass freely across it at all times;

(3) so situated in relation to any opening (as previously described) that no part of the outlet is less than 600 mm from the opening.

Hence with flues serving section *3* appliances, the outlets do not have to project above the ridge.

Section 4 Appliances

With section *4* oil burning appliances with a rated output up to 45 kW, the flue size for a flue pipe shall be the same as for the flue outlet from the appliance, and for chimneys the diameters are 100, 125 and 150 mm according to the rated output (up to 20 kW, 20–32 kW and 32–45 kW). Flue outlets shall be at least 600 mm from any opening in the building and the outlet from a flue serving a pressure jet may be terminated anywhere above the roof line.

Chimneys and Chimney Stacks

When determining the height of a chimney or flue pipe, the pot or other flue terminal is disregarded. Furthermore, the height of a chimney stack should not exceed 4.5 times the least width to ensure adequate stability (Approved Document A1/2: 1991). Chimney pots as illustrated in figure 5.5.2 provide a neat and effective finish to the outlet of a flue. The taper of a pot reduces the entry of rain and improves draught and flow of gases. It is best to use square-based pots with square flues. Pots may be 300, 450, 600, 750 or 900 mm high.[11]

Chimney pots should be built into the stack not less than 150 mm, or one-quarter the length of the pot, whichever is the greater, to ensure ample support. The pots are flaunched around in cement and sand (1:3) to throw water off the stack and give additional support to the pots (figures 5.5.3 and 5.5.4). Pots should extend above the highest point of the flaunching for at least 50 mm.

The head of a chimney stack may be finished in different ways. One method is to build brick oversailing courses (each projecting about 28 mm) and to flaunch around the pots as shown in figures 5.5.3 and 5.5.4. Another method is to use a precast concrete capping which is weathered and throated on all sides, overhangs the stack by about 38 to 50 mm in all directions, and is generally bedded on a damp-proof course. Another damp-proof course is inserted lower down the stack where it passes through the roof (figure 5.5.3).

Rafters and ceiling joists are trimmed around the chimney and these timbers must be kept at least 38 mm clear of the outside face of the brick stack. A watertight joint has to be provided around the chimney where the roof covering of tiles or slates adjoins the brickwork. A gutter and flashing are normally used at the top edge, an apron flashing at the bottom edge and soakers and stepped flashings at the sides, as illustrated in figures 5.5.3 and 5.5.4. Lead is often used in these positions but various other materials are available, including zinc, aluminium, copper and nuralite. The stepped flashings overhang the right-angled soakers, and the top edges of the flashings are wedged into brick joints and pointed. At the top edge the gutter sheeting is dressed 150 mm up the brickwork, across the gutter board, and up the roof slope over a tilting fillet, whilst a cover flashing with its top edge built into the brickwork is dressed down over the gutter upstand. The constructional aspects will be considered in greater detail in chapter 7.

CURING SMOKY CHIMNEYS

As described in the early part of this chapter smoky chimneys, whereby the chimney does not draw satisfactorily causing smoke to be blown back back into the room through a fireplace opening, can result from a number of factors. These include a blocked flue, insufficiency of air supply, poor design of throat, unsatisfactory gathering of flue, and overshadowing of flue outlet by tall trees or buildings, setting up pressure and suction zones.

A normal open fire needs between 110 and 170 m^3 of air per hour and, in the absence of open windows or doors, this must find its way into the room through cracks. Some occupants aggravate the problem by sealing many of the cracks with draught excluders. In extreme cases an underfloor air supply to the hearth will overcome the difficulty, whilst a throat restrictor may be introduced to reduce the amount of air passing up the chimney.

When inspecting a smoky chimney, the following procedures should be adopted.

(1) Ensure that the chimney is swept; excessive quantities of soot may be the result of using unsuitable fuel. Check on the size of throat, shape and position of lintel, and similar matters.

(2) Open the window(s) and door(s) of the room in which the fire is located; if smoking ceases then the fire is starved of air. There is probably a need for ventilators or underfloor ducts.

(3) Determine the effect of reducing the height of the fireplace opening with a strip of sheet metal. Where an improvement results from a restriction of the opening by up to 100 mm, then the fitting of a permanent canopy will probably be the answer.

(4) On occasions the streamlining of entrances and restriction of the throat with a piece of bent sheet metal may produce a considerable improvement. In these circumstances a variable throat restrictor could be built into the fireplace.

(5) Determine the effect of increasing the height of the chimney stack with pieces of sheet metal pipe of varying lengths, with a throat restrictor in position in the fireplace. Where good results are secured with a certain length of pipe, it will be advisable to fit a longer pot, increase the height of the stack, or possibly to fit a cowl or louvred pot.

REFERENCES

1. *The Building Regulations 1991 and Approved Documents A (1991) and J (1989)*. HMSO
2. *BS 476, Part 4: 1970/1984* Non-combustibility tests for materials
3. *BS 1251: 1970/1987* Open fireplace components
4. *BS 3376: 1982* Solid mineral fuel open fires with convection, with or without boilers
5. *BS 3377: 1985* Boilers for use with domestic solid mineral fuel appliances
6. *BS 4834: 1990* Inset open fires without convection with or without boilers, burning solid mineral fuels
7. *BS 8303: 1986* Installation of domestic heating and cooking appliances burning solid mineral fuels
8. *BS 3378: 1982* Room heaters burning solid mineral fuels
9. *BS 5258* Domestic gas appliances, *Part 5: 1989* gas fires; *Part 8: 1980* combined appliances: gas fire/back boiler; *Part 12: 1980* decorative gas log and other fuel effect appliances
10. *BS 5871: 1980* Gas fires, convectors and fire/back boilers
11. *BS 1181: 1971* Clay flue linings and flue terminals
12. *BS 65: 1988* Vitrified clay pipes, fittings, joints and ducts
13. *BS 6461* Installation of chimneys and flues for domestic appliances burning solid fuel (including wood and peat), *Part 1: 1984* Code of practice for masonry chimneys and flue pipes
14. *BS 41: 1973* (1981) Cast iron spigot and socket flue or smoke pipes and fittings
15. *BS 6999: 1989* Specification for vitreous enamel low carbon steel flue pipes, other components and accessories for solid burning appliances with a maximum rated output of 45 kW
16. *BS 4543* Factory-made insulated chimneys, *Part 1: 1990* Methods of test for factory-made chimneys; *Part 2: 1990* Specification for chimneys for solid fuel fired appliances; *Part 3: 1976* (1981) Specification for chimneys for oil fired appliances
17. *BS 6461* Installation of chimneys and flues for domestic appliances burning solid fuel (including wood and peat), *Part 2: 1984* Code of practice for factory-made insulated chimneys for internal appliances
18. *BS 1289* Flue blocks and masonry terminals for gas appliances, *Part 1: 1986* Specification for precast concrete flue blocks and terminals; *Part 2: 1989* Specification for clay flue blocks and terminals
19. *BS 5440* Installation of flues and ventilation for gas appliances of rated input not exceeding 60 kW, *Part 1: 1990* Specification for installation of flues; *Part 2: 1989* Specification for installation of ventilation for gas

6 FLOORS: STRUCTURES AND FINISHINGS; TIMBER CHARACTERISTICS AND DEFECTS

This chapter examines the various types of floor and the constructional techniques and materials used in their provision. The techniques employed must be considered in relation to the functional requirements, in order to achieve a satisfactory form of construction. The form of construction selected may also influence the provision of other elements. For instance, the thickness of floors will affect the height of external cladding, and the choice between a solid or suspended ground floor can affect the location of services, the size of the heating installation, provision of thermal insulation, susceptibility to rot in timbers and the choice of floor finishes.

FUNCTIONS OF FLOORS

Floors need to satisfy a number of functional requirements which are now outlined.

Ground Floors

(1) To withstand the loads that will be imposed upon them; with domestic buildings they are normally confined to persons and furniture, but in other classes of building, such as factories, warehouses and libraries, the floors may be subjected to much heavier loads and must be of sufficient strength to carry them.

(2) To prevent the growth of vegetable matter inside the building, by the provision of the concrete oversite.

(3) To prevent damp penetrating the building by inserting a damp-proof membrane in or below the floor; suspended ground floors also require under-floor ventilation to prevent stagnant, moist air accumulating below them.

(4) To meet certain prescribed thermal insulation standards by incorporating a layer of insulating material to reduce the heat loss into the ground below, which will be further examined in chapter 15.

(5) To be reasonably durable and so reduce the amount of maintenance or replacement work to a minimum.

(6) To provide an acceptable surface finish which will meet the needs of the users with regard to appearance, comfort, safety, cleanliness and associated matters.

Upper Floors

The functional requirements of upper floors differ considerably from those of ground floors. Their requirements are as follows.

(1) To support their own weight, ceilings and superimposed loads.

(2) To restrict the passage of fire; this is particularly important in high-rise buildings, and in buildings where parts are in different ownerships, where there are many occupants or where large quantities of combustible goods are stored.

(3) To restrict the transmission of sound from one floor to another, particularly where this may seriously interfere with the activities undertaken.

(4) To possess an adequate standard of durability.

(5) To bridge the specific span economically and be capable of fairly quick erection.

(6) To accommodate services readily.

(7) To provide an acceptable surface finish in the manner described for ground floors.

SOLID GROUND FLOORS

With domestic buildings a choice often has to be made between solid (ground supported) and suspended ground floors. Solid floors are likely to prove cheaper on fairly level sites, may reduce the quantity of walling, eliminate the need for underfloor ventilation and hence reduce heat loss through the floor, can avoid the risk of dry rot and offer a greater selection of floor finishes. For these reasons the majority of ground floors in modern dwellings are of solid construction, comprising a concrete slab laid on a bed of hardcore. The concrete slab usually supports a cement and sand screed on which the floor finish is bedded. There is also a need to provide a suitable thermal insulant as shown in figures 6.1.1 to 6.1.6 and 6.2.1 and 6.2.2 and described in chapter 15, where all aspects of thermal insulation are considered.

Hardcore

A bed of hardcore may be required on a building site to fill hollows and to raise the finished level of an oversite concrete slab after removal of turf and other vegetation.[1] On wet sites it may be used to provide a firm working surface and to prevent contamination of the lower part of the wet concrete during placing and compaction. It can reduce the amount of rising ground moisture but it cannot eliminate the need for a waterproof membrane.[2]

The best filling materials are those with particles that are hard and durable and chemically inert and will not attack concrete or brickwork mortar, and that can readily be placed in a compact and dense condition, requiring only limited consolidation. Materials with fairly large particles, such as hard rock wastes, gravels and coarse sands, drain easily and consolidate quickly. If the material is well graded, with a mixture of coarser and finer grains, it will form a dense fill. The concrete oversite slab can be protected against excess sulphates by a layer of bitumen felt or plastics sheeting placed on the hardcore; alternatively, or in addition, the concrete can contain a more resistant type of cement.[3]

Some of the more commonly used materials for hardcore beds are now considered.

Brick or *tile rubble*. This is clean, hard and chemically inert but it may contain little or no fine material and so may not consolidate readily.

Clinker. Ideally this should be hard-burnt with the material fused and sintered into hard lumps.

Colliery shale. Ideally this should be well-burnt brick-red shale, but some may contain sulphates.

Gravel. If well graded, it meets most requirements but is expensive. A cheaper alternative is to use coarse screened gravel in excess of 40 mm for the bottom layer with a consolidated upper layer of well-graded material.

Quarry waste. This is clean, hard and safe to use, but it may be unevenly graded and consequently difficult to consolidate.

Pulverised-fuel ash (PFA). This is recovered from flue gases produced by boilers with pulverised coal, as at power stations. It is lightweight, and when conditioned with the correct amount of water and compacted, has self-hardening properties.

Concrete Oversite

The oversite bed of concrete should not be less than 100 mm thick, although it is often 150 mm thick. The mix of concrete should be at least 1:3:6 with a maximum size of coarse aggregate of 38 mm, but a mix of 1:2:4 is to be preferred incorporating coarse aggregate with a maximum size of 19 mm. Approved Document C4 (1991) of the Building Regulations 1991[4] in section 3 recommends concrete of a mix of 50 kg of cement to not more than 0.11 m^3 of fine aggregate and 0.16 m^3 of coarse aggregate or BS 5328[5] mix ST2. Figures 6.1.1, 6.1.2, 6.2.1 and 6.2.2 show typical concrete ground floor slabs. It should be noted that the edges of the slab are not built into the surrounding walls to allow the two elements with their differing loads to move independently of one another.

The Building Regulations 1991[4] (paragraph C4 of Schedule 1) prescribe that 'the floors of the building shall resist the passage of moisture to the inside of the building.' Oversite concrete and floor screeds as commonly laid cannot by themselves keep back all ground moisture. In the absence of a satisfactory waterproof membrane, the moisture rises to the surface where it evaporates or accumulates under a less pervious material. Some floor finishes are particularly susceptible to damp, such as magnesium oxychloride, PVA emulsion/cement, rubber, flexible PVC, linoleum, cork-carpet or tiles, and wood flooring.

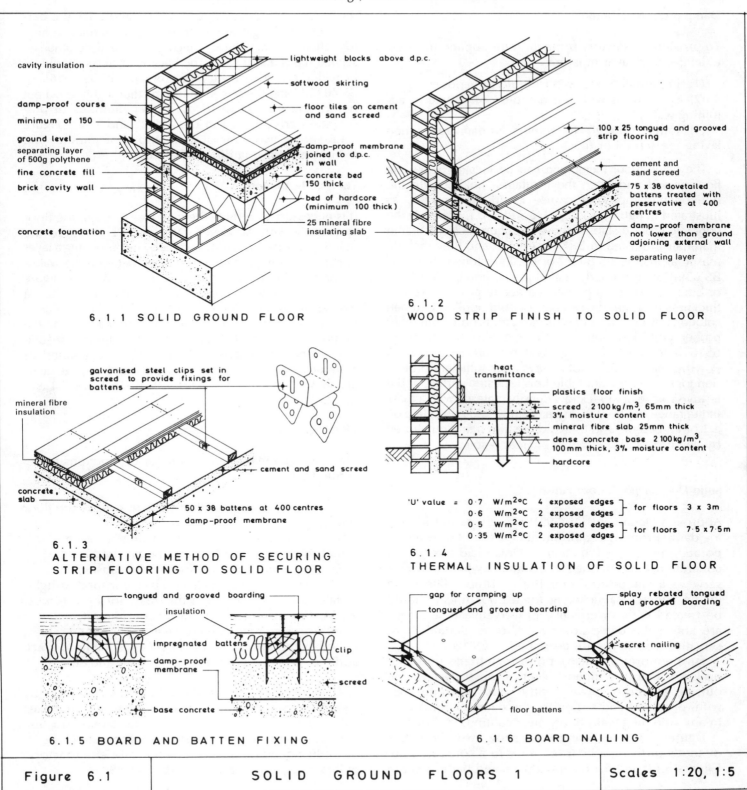

cavity insulation

lightweight blocks above d.p.c.

softwood skirting

floor tiles on cement and sand screed

damp-proof course

minimum of 150

ground level

separating layer of 500g polythene

fine concrete fill

brick cavity wall

damp-proof membrane joined to d.p.c. in wall

concrete bed 150 thick

bed of hardcore (minimum 100 thick)

concrete foundation

25 mineral fibre insulating slab

6.1.1 SOLID GROUND FLOOR

100 x 25 tongued and grooved strip flooring

cement and sand screed

75 x 38 dovetailed battens treated with preservative at 400 centres

damp-proof membrane not lower than ground adjoining external wall

separating layer

6.1.2 WOOD STRIP FINISH TO SOLID FLOOR

galvanised steel clips set in screed to provide fixings for battens

mineral fibre insulation

cement and sand screed

concrete slab

50 x 38 battens at 400 centres

damp-proof membrane

6.1.3 ALTERNATIVE METHOD OF SECURING STRIP FLOORING TO SOLID FLOOR

heat transmittance

plastics floor finish

screed 2 100 kg/m³, 65mm thick 3% moisture content

mineral fibre slab 25mm thick

dense concrete base 2 100 kg/m³, 100mm thick, 3% moisture content

hardcore

'U' value =   0·7   W/m²°C   4 exposed edges ⎤ for floors 3 x 3m
           0·6   W/m²°C   2 exposed edges ⎦
           0·5   W/m²°C   4 exposed edges ⎤ for floors 7·5 x 7·5m
          0·35 W/m²°C   2 exposed edges ⎦

6.1.4 THERMAL INSULATION OF SOLID FLOOR

tongued and grooved boarding

insulation

impregnated battens

clip

damp-proof membrane

screed

base concrete

6.1.5 BOARD AND BATTEN FIXING

gap for cramping up

tongued and grooved boarding

splay rebated tongued and grooved boarding

secret nailing

floor battens

6.1.6 BOARD NAILING

| Figure 6.1 | SOLID GROUND FLOORS 1 | Scales 1:20, 1:5 |

Damp-proof Membrane

To provide satisfactory protection for applied finishes, a damp-proof layer must be

(1) impermeable to water (liquid or vapour);

(2) continuous with the damp-proof course in adjoining walls;[2]

(3) tough enough to remain undamaged when laying the screed or finish.

The damp-proof membrane can be located below the floor screed and often the insulation linking with the horizontal damp-proof course in adjoining walls, as illustrated in figures 6.1.1, 6.1.2 and 6.1.3, or below the concrete ground slab as shown in figure 6.2.2. Suitable materials are listed in BRE Digest 54[2] and they include mastic asphalt to BS 6925[6] or 6577;[7] bitumen sheets to BS 6398[8] with properly sealed joints; hot-applied pitch or bitumen laid on a primed surface to give an average thickness of 3 mm; three coats of cold-applied bitumen solutions or coal tar pitch/rubber emulsion or bitumen/rubber emulsion; and polythylene film sheets to BS 6515[9] at least 0.46 mm thick with properly sealed joints. Polythylene film may not always offer sufficient protection for the more vulnerable finishes listed earlier, and it is advisable to use black low density material with a mass in the range 0.425–0.60 kg/m^2 with sealed joints, laid on a 12 mm bed of sand to avoid the film being punctured by any irregularities in the concrete below.

Solid Floors with Timber Finish

Approved Document C4 (1991) prescribes the following damp-proofing requirements for solid floors incorporating timber. A timber floor finish laid directly on concrete may be bedded in a material which may also serve as a damp-proof membrane. Timber fillets laid in the concrete as a fixing for a floor finish should be treated with an effective preservative unless they are above the damp-proof membrane. Suitable preservative treatments are described in BS 1282.[10]

Furthermore, the damp-proof membrane must not be lower than the highest level of the surface of the outside ground or paving, and it must be continuous with, or joined and sealed to, the damp-proof course in any adjoining wall, floor, pier, column or chimney.

Figures 6.1.2, 6.1.3 and 6.1.5 show two alternative methods of fixing a boarded floor to a concrete slab. Figures 6.1.2 and 6.1.6 show dovetailed battens set in

the screed, whereas in figures 6.1.3 and 6.1.5 the battens are held by galvanised steel clips set in mineral fibre insulation 38 mm thick, which can also accommodate service pipes and cables. The battens or fillets should be treated in accordance with the provisions of BS 5056[11] or BS 4072,[12] and the screed should be thoroughly dried out before the boards are fixed. Other methods of preservation are detailed in BSs 1282[10] and 144.[13] Figure 6.1.6 shows two alternative methods of jointing and fixing the floor boarding to battens.

Thermal Insulation

It is necessary to reduce the heat loss through the floor to comply with Approved Document L (1995) to the Building Regulations 1991, such as by incorporating a layer of material of high thermal resistance between the waterproof membrane and the screed as shown in figure 6.1.4[14]. A mineral fibre slab, 25 mm thick, has a thermal resistance of 0.66 to 0.71 m^2 °C/W. The U-value of such a floor increases as the area of floor reduces and the number of exposed edges increases. Thermal insulation aspects of exposed floors will be examined in more detail in chapter 15. Other insulating arrangements are shown in figures 6.1.1, 6.1.2, 6.1.3, 6.1.5, 6.2.1 and 6.2.2.

Floor Screeds

Floor screeds may serve a number of functions as described below.

(1) to provide a smooth surface to receive the floor finish;

(2) to bring a number of floor finishes each of different thicknesses up to the same finished level;

(3) to provide falls for drainage purposes;

(4) to give thermal insulation by incorporating lightweight concrete, although this may not be the most efficient or most economical way of doing it;

(5) to accommodate service pipes and cables, although the thin screed over pipes is liable to crack and access to defective services is costly and difficult.[15]

Cement and sand screeds in the ratios of 1:3–4½ (by weight) are suitable for thicknesses up to 40 mm. Mixes with a lower cement content will be subject to less shrinkage. For thicker screeds, fine concrete (1:1½:3) using 10 mm maximum coarse aggregate is satisfactory. Suitable aggregates are prescribed in BS 882.[16]

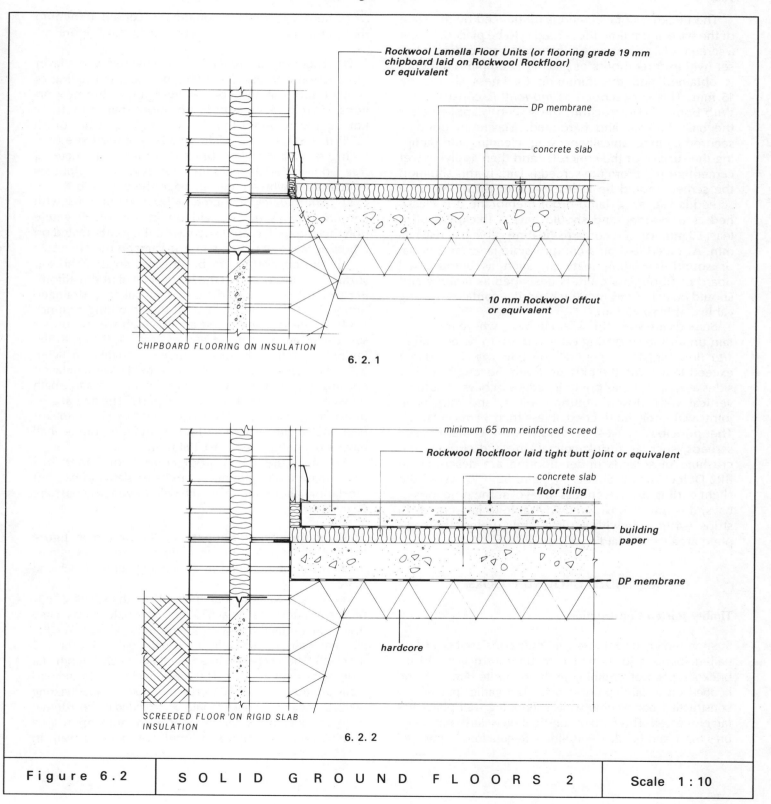

Rockwool Lamella Floor Units (or flooring grade 19 mm chipboard laid on Rockwool Rockfloor) or equivalent

DP membrane

concrete slab

10 mm Rockwool offcut or equivalent

CHIPBOARD FLOORING ON INSULATION

6. 2. 1

minimum 65 mm reinforced screed

Rockwool Rockfloor laid tight butt joint or equivalent

concrete slab

floor tiling

building paper

DP membrane

hardcore

SCREEDED FLOOR ON RIGID SLAB INSULATION

6. 2. 2

| Figure 6.2 | S O L I D   G R O U N D   F L O O R S   2 | Scale 1 : 10 |

The thickness of a screed is influenced by the state of the base at the time the screed is to be placed. When a screed is laid on an *in situ* concrete base before it has set (within three hours of placing), complete bonding is obtained and the minimum thickness should be 15 mm. This is described as *monolithic* construction. With bonded construction, the screed is applied after the base has set and hardened. Maximum bond is secured by mechanically hacking, cleaning and damping the surface of the concrete, and then applying wet cement grout before the screed is laid. In this situation the screed should be at least 40 mm thick. Where a screed is laid on a damp-proof membrane it is classified as *unbonded* and should have a thickness of at least 50 mm, or, if it contains heating cables, at least 60 mm. A screed laid on a compressible layer of thermal or sound insulating material, such as compressed board or fibre glass quilt, is described as *floating* and should be at least 65 mm thick, or, if it contains heating cables, at least 75 mm.[15]

Screeds are sometimes laid in bays, where they contain underfloor warming cables or are to receive an *in situ* floor finish. In general the bay size should not exceed 15 m² with the ratio between the lengths of bay sides as near 1:1½ as possible. Edges of bays should be vertical with closely abutting joints, and expansion joints will rarely be needed at less than 30 m centres.[15] The principal causes of failure with cement-based screeds, such as shrinkage-cracking and curling, and crushing of screeds under flooring are described in BRE Defect Action Sheets 51[17] and 52.[18] To avoid the slight curling or unevenness in level at the junctions of bays, it is often considered advisable to lay screeds in strips 3–4 m wide with vertical joints, unless the complete area can be laid in a day.[19, 20]

SUSPENDED GROUND FLOORS

Timber Joisted Construction

Suspended ground floors generally consist of boarding nailed to floor joists which in their turn are laid on brick sleeper walls running in the opposite direction and bedded on a damp-proof course. The earlier practice of constructing honeycomb walls supporting wall plates has largely ceased. This form of construction is illustrated in figures 6.3.1 and 6.3.2. The Building Regulations[4] Approved

Document C4 (1991) recommends certain minimum requirements for suspended timber ground floors.

(1) The ground surface is to be covered with a layer of concrete not less than 100 mm thick, consisting of cement and fine and coarse aggregate in the proportions of 50 kg of cement to not more than 0.13 m³ of fine aggregate and 0.18 m³ of coarse aggregate, or BS 5328[5] mix STI, properly laid on a bed of hardcore consisting of clean broken brick or other inert material free from materials including water-soluble sulphates in quantities which could damage the concrete.

(2) Alternatively, concrete at least 50 mm thick with composition as in (1), laid on at least 1200 gauge polythylene (polythene) sheet with the joints sealed on a bed of material which will not damage the sheet.

(3) The concrete is to be finished so that the top surface is entirely above the highest level of the adjoining ground or it should be laid to fall to a drainage outlet above the lowest level of the adjoining ground.

(4) There is a ventilated air space above the upper surface of the concrete of not less than 150 mm to the underside of any suspended timbers as shown in figure 6.3.2. Two opposing external walls should have ventilation openings placed so that ventilating air will have a free path between opposite sides and to all parts. The actual area of opening shall be at least equivalent to 1500 mm² for each metre run of wall and any ventilating pipes shall have a diameter of at least 100 mm.

(5) There are damp-proof courses of impervious sheet material, engineering bricks or slates in cement mortar or other material which will prevent the passage of moisture.

All of these requirements are illustrated in figure 6.3.2. A good understanding of timber joisted floor construction can be obtained by examining figures 6.3.1 and 6.3.2.

The sleeper walls are built half-a-brick thick with 215 × 65 mm airbricks spaced at 675 mm centres, to permit a free flow of air under the floors from the airbricks built into the external walls. These walls are usually either 150 or 225 mm high. The sleeper walls are built across each room at a spacing of from 1200 to 1800 mm centres, with the end walls positioned about 50 to 100 mm from the loadbearing walls. Sufficient airbricks should be provided in the external walls to give the recommended minimum open area of 1500 mm² per metre of length of external wall, in

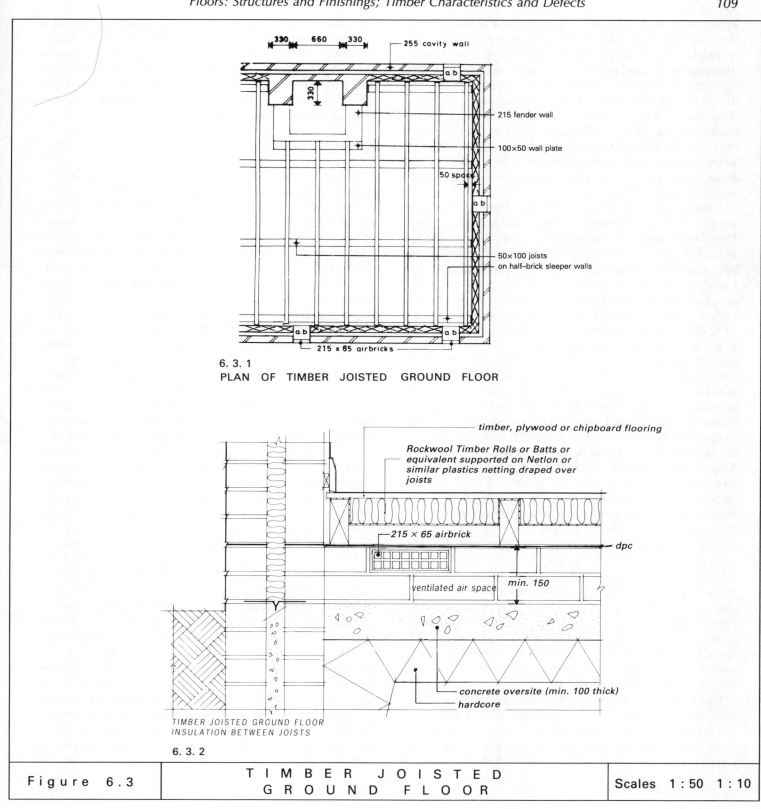

330 660 330 — 255 cavity wall

a.b

330

— 215 fender wall

— 100×50 wall plate

50 space

a.b

— 50×100 joists
on half–brick sleeper walls

a.b   a.b

— 215 × 65 airbricks

6. 3. 1
PLAN OF TIMBER JOISTED GROUND FLOOR

— timber, plywood or chipboard flooring

*Rockwool Timber Rolls or Batts or
equivalent supported on Netlon or
similar plastics netting draped over
joists*

—215 × 65 airbrick

— dpc

ventilated air space   *min. 150*

— concrete oversite (min. 100 thick)

— hardcore

*TIMBER JOISTED GROUND FLOOR
INSULATION BETWEEN JOISTS*

6. 3. 2

| Figure 6.3 | TIMBER JOISTED GROUND FLOOR | Scales 1 : 50 1 : 10 |

accordance with BS 5250,[21] whereas openings need to be provided with grills to prevent entry of rodents.[22] Airbricks should be positioned as high as possible on opposite walls, and are normally provided at about 3 m centres, and care should be taken to avoid unventilated airpockets such as may occur near corners, projections or bay windows. A typical clay airbrick conforming to BS 493[23] is illustrated in figure 6.3.2, and a duct may be formed through the cavity wall behind the airbrick with slates. Another method is to use hollow tile ducts, while yet another alternative is to install purpose-made asbestos cement ducts.

A damp-proof course must be provided on top of all sleeper walls to prevent moisture reaching any timber. The damp-proof course should ideally be flexible to allow for any movement without fracturing and must be lapped at any joints. A timber wall plate, often 100 × 50 mm, may be bedded in mortar above the damp-proof course and this assists in spreading the load from each joist over the wall below. The floor joists are often 50 × 100 mm (actual sizes – 47 or 50 × 97 mm), spaced at about 400 mm centres and skew-nailed to the wall plate or laid on the dpc at the top of the sleeper walls. The actual spacing of the joists is influenced by the thickness of the floorboarding and the load on it, whereas the size of joists is affected by the strength of timber used and the loading and spacing of joists and sleeper walls. The sizes of joists can be determined quite easily from tables A1 and A2 in the Approved Document A1/2 (1991) of the Building Regulations.[4] The ends of joists are sometimes cut on the splay and are best kept back a short distance from the walls to prevent any moisture being absorbed by the vulnerable open grain at the ends of joists. A suitable distance would be 20 mm. In like manner, the end joists should be kept 50 mm from the loadbearing walls as shown in figure 6.3.1.

Approved Document L (1989) to the Building Regulations 1991 require all exposed floors to be insulated, and this entails the provision of insulation between the joists, normally comprising insulation rolls or batts supported by plastics netting draped over the joists as shown in figure 6.3.2 or rigid insulation boards on battens fixed with corrosion resisting nails or slips. Further information on this aspect is provided in chapter 15.

The normal arrangement at a ground floor hearth is illustrated in figure 6.3.1. A one-brick thick fender wall encloses and supports the filling material under the front hearth and can also carry a wall plate which supports the ends of joists. It is important to provide adequate ventilation where a suspended floor joins a solid floor, and particularly to avoid any stagnant corners. Ventilating pipes or ducts should be laid below the solid floor, connected to airbricks in the external wall. The damp-proof membrane under the solid floor should be linked with the damp-proof course in the division wall by a vertical damp-course.

Timber Floorboarding

A timber joisted floor may be covered with timber boards or strips or sheets of chipboard or plywood. The majority of timber floors are finished with softwood boards between 100 and 150 mm wide, with a nominal thickness of 25 mm. The boards are laid at right angles to the joists with each board nailed to each joist with two floor brads, with the brads being 40 mm longer than the thickness of the boards and the heads of brads well punched down below the top surface of the boards. Floorboards are normally joined together with tongued and grooved joints of the type shown in figure 6.1.6, although in cheap work, plain or square-edged boards may be used. The use of plain-edge boards is not recommended, as with ventilated suspended floors draughts can penetrate through the joints, and with timber finished solid floors spilt liquid or washing-down water could penetrate the joints, would only dry out very slowly and might result in an outbreak of dry rot. With timber-covered solid floors it is important to pressure-impregnate the battens and to apply a brush coat of preservative to the underside of the boards as an additional safeguard against decay. Where it is considered desirable to avoid punch holes for brads, boards may be secret nailed as shown in figure 6.1.6, using splay rebated, tongued and grooved joints, to reduce the risk of splitting the tongue when nailing. This method of jointing and nailing is particularly suited for hardwood floorboarding and narrow strip boarding where appearance is especially important.

Floorboards should preferably be rift sawn, when they are cut as near radially from the log as practicable on conversion. This reduces later warping of the boards to a minimum. Heading or end joints of floor boards are usually splayed and should be made over joists. They should be staggered in adjoining boards for greater strength. The edges of floorboards should be kept 13 mm away from surrounding walls to allow for movement and reduce the risk of damp penetration. The gap is closed by a skirting at the base of the wall or partition, which also masks the gap between the bottom edge of the wall plaster and the floor and provides adequate resistance to kicks. Skirtings are considered in more detail in chapter 10.

Chipboard Flooring

BRE Defect Action Sheets 31[24] and 32[25] describe how chipboard can provide a satisfactory floor deck, provided the basic requirements for grade, thickness in relation to span, fixings and edge support are all met satisfactorily. If boards are not of moisture resistant grade or not protected from moisture in potentially wet areas, such as kitchens and bathrooms, significant permanent loss of strength may occur.

Chipboard should conform to BS 5669[26] type C2 (red stripe) or, for improved moisture resistance, type C3 (red/green stripe), and be 18 mm thick for joist spacings up to 450 mm and 22 mm thick for joist spacings up to 600 mm. All edges of boards should be supported on joists or noggings at least 38 mm wide. A gap for movement is required at the perimeter of each floor deck. The gap should be 2 mm per metre run and never less than 10 mm. Boards should be fixed with 10 gauge (3.35 mm) ring-shank nails of a length of at least 2½ × the board thickness, spaced at not more than 300 mm centres at board perimeters and not more than 500 mm elsewhere.

Concrete Construction

Building Regulations Approved Document C4 (1991) lists the following ways in which a suspended concrete ground floor may be constructed to conform to the requirements of Schedule 1 to the Regulations with regard to the prevention of moisture penetration.

(1) *in situ* concrete at least 100 mm thick and containing at least 300 kg of cement for each cubic metre of concrete; or

(2) precast concrete construction with or without infilling slabs; and

(3) reinforcing steel protected by concrete cover of at least 40 mm if the concrete is *in situ*, and at least the thickness required for moderate exposure if the concrete is precast.

A suspended concrete floor should incorporate

(a) a damp-proof membrane if the ground below the floor has been excavated below the lowest level of the surrounding ground and will not be effectively drained.

(b) a ventilated air space at least 150 mm deep where there is a risk of an accumulation of gas which might lead to an explosion.

An example of an efficient type of suspended concrete ground floor is Jetfloor which consists of lightweight prestressed reinforced concrete inverted tee joists with a clear span of up to 4.50 m, supporting standard precast concrete blocks and a live load of 1.50 kN/m². They are easily and economically laid to provide a sound working platform which is normally covered with 20 mm of expanded polystyrene board, a 1000 g polythylene vapour check and 18 mm chipboard, giving an all-dry, rot and damp-proof and durable floor, with a high thermal insulation value.

UPPER FLOORS

Timber Floor Construction

Most upper floors in domestic buildings are constructed of timber floorboarding supported by timber joists which, in their turn, bear upon the loadbearing walls below. Hence the main difference between suspended floors at ground level and those on upper floors is the longer spans of joists in the latter case, with resultant deeper joists. The sizes of joists vary with the grade of timber, the dead load (including ceiling below), the span of the joists and their spacing. Table B1 of the Building Regulations Approved Document A1/2 (1991) categorises the grades of different species of timber into strength classes SC3 and SC4. Tables A1 and A2 of the same Approved Document show joist sizes varying from 38 × 97 mm up to 75 × 220 mm for dead loads ranging from 0.25 to 1.25 kN/m² and an imposed load not exceeding 1.5 kN/m², joist spacings of 400 to 600 mm and clear spans ranging from 1.04 to 5.42 m. A useful rule-of-thumb method for calculating the depth of floor joists is $\frac{1}{25}$ span in mm + 50 = depth in mm. For example if the span of the joists is 3 m, then their depth = 3000/25 + 50 = 170 mm (170 mm is a standard size). In fact, 50 × 170 mm floor joists at 400 m centres are quite common in normal domestic construction. The strength of joists can be significantly reduced by large or badly located holes or notches for services.

The ends of upper floor joists are usually supported by the inner leaves of external cavity walls or loadbearing internal walls or partitions, as illustrated in figure 6.4.6. Joists should run the shortest span wherever practicable to economise in timber. The ends of the joists may be supported in a number of different ways.

(1) Ends of joists are treated with preservative and built into walls, preferably with a space all round. The joists may require packing and cutting of brickwork to keep their tops level. Care must be taken to prevent joists penetrating into the cavity and thus providing a bridge for moisture and probable deterioration of the timber. It is common practice to taper the ends of joists or to keep them clear of the wall face.

(2) Ends of joists may rest on a wall plate bedded on top of an internal partition (figure 6.4.4), although in practice the wall plate is often omitted. This provides a sound, uniform bearing with maximum distribution of loads. It is not suitable for external walls, where brickwork will be superimposed upon the plate and damp from the cavity might cause decay of the non-ventilated timber. Another alternative with external walls is to support the joists on wall plates located clear of the wall on wrought iron or mild steel corbels built into the wall at about 750 mm centres (figure 6.4.3). In the latter situation a cornice is needed to mask the projecting plate in the rooms below.

(3) Ends of joists can rest on a galvanised mild steel or wrought iron bearing bar built into an external wall to assist in spreading the load, although this is not usually vital in domestic construction.

(4) Ends of joists are supported on galvanised steel hangers which are built into the wall, and the joists rest on the bottoms of hangers and are nailed from the sides (figure 6.4.2). Another form of steel joist hanger is shown in figure 6.4.10; this is welded. This is the best method of providing support to joists from external walls as it avoids building timber into the walls with its resultant risks and the hangers can usually be adjusted on the site to suit the particular conditions. BRE Defect Action Sheet 58[27] describes the following defects relating to joist hangers in masonry walls: coursing of masonry not level; packing under hangers; wrong grade of hanger used (hangers to BS 6178[28] are marked for minimum masonry crushing strengths of 2.8, 3.5. and 7.0 N/mm^2); hangers not tight to wall; gaps too great between joist ends and back plates; brick course used on internal block wall; and absence of perimeter noggings to give support to edges of plasterboard.

Where joists from either side meet on a loadbearing internal wall, they are usually placed side by side and nailed to each other and possibly a wall plate (figure 6.4.4), with passings of about 150 mm. Where the span of joists exceeds 3 m, they should be strutted at about 2 m centres to stiffen the joists and prevent them twisting, and also to strengthen the floor. The most common form is herringbone strutting made up of approximately 40 × 40 mm or 40 × 50 mm timbers with their ends cut on the splay and nailed to each other where they intersect and to the upper part of one joist and the lower part of the next (figure 6.4.5). The space between the first and last joists and the adjoining wall is wedged to take the thrust of the strutting. Another alternative is to use solid strutting, usually about 32 mm thick, which may be staggered to allow nailing of the ends to the joists, or laid in a continuous line and skew-nailed to joists. This method is not so effective as herringbone strutting, as the strutting may become loose due to rounding of joist sides or shrinkage in timbers. The floor joists should also provide lateral support to the walls (see Approved Document A1/2: 1991).

The internal walls on the ground and first floors of a dwelling do not always coincide, and lightweight block partitions may be required on the first floor where there are no loadbearing walls beneath. Where the partition on the upper floor is parallel to the floor joists, two joists can be bolted together to form a base on which the partition can be built. Another alternative is to fix solid bridging pieces 50 mm thick at about 400 mm centres between a pair of joists, and the bridging pieces support floorboarding and a wall plate on which the partition is built. Where the partition runs at right angles to the floor joists, solid bridging pieces can be fixed between the joists on the line of the partition, with boarding and a wall plate above them.

Construction around Openings

Figure 6.4.6 shows the method of trimming floor joists around a fireplace opening. A trimming joist is placed about 50 mm from each outside edge of the chimney breast, and a trimmer is provided adjacent to the front edge of the hearth and is supported at each end by the trimming joists. Both trimmers and trimming joists are 25 mm thicker than the normal floor or bridging joists as they have to withstand heavier loads and it is not possible to increase their depth. Strong and soundly constructed joints are required between the trimming

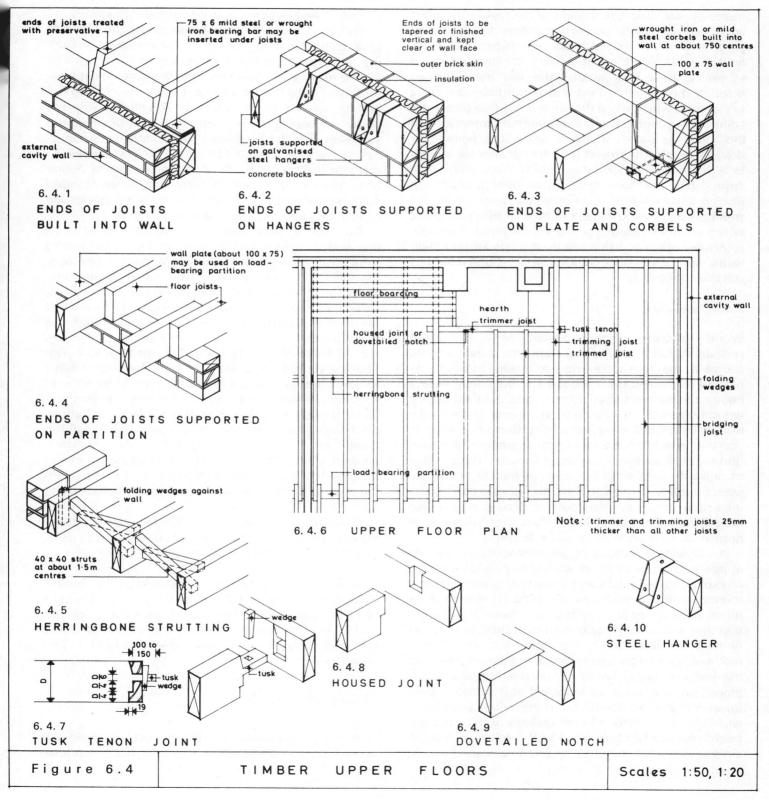

6. 4. 1
ENDS OF JOISTS BUILT INTO WALL

ends of joists treated with preservative
external cavity wall

6. 4. 2
ENDS OF JOISTS SUPPORTED ON HANGERS

75 x 6 mild steel or wrought iron bearing bar may be inserted under joists
Ends of joists to be tapered or finished vertical and kept clear of wall face
outer brick skin
insulation
joists supported on galvanised steel hangers
concrete blocks

6. 4. 3
ENDS OF JOISTS SUPPORTED ON PLATE AND CORBELS

wrought iron or mild steel corbels built into wall at about 750 centres
100 x 75 wall plate

6. 4. 4
ENDS OF JOISTS SUPPORTED ON PARTITION

wall plate (about 100 x 75) may be used on load-bearing partition
floor joists

6. 4. 6   UPPER FLOOR PLAN

floor boarding
hearth
trimmer joist
housed joint or dovetailed notch
tusk tenon
trimming joist
trimmed joist
herringbone strutting
load-bearing partition
external cavity wall
folding wedges
bridging joist

Note: trimmer and trimming joists 25mm thicker than all other joists

6. 4. 5
HERRINGBONE STRUTTING

folding wedges against wall
40 x 40 struts at about 1·5 m centres

6. 4. 7
TUSK TENON JOINT

100 to 150
tusk wedge
19

6. 4. 8
HOUSED JOINT

wedge
tusk

6. 4. 9
DOVETAILED NOTCH

6. 4. 10
STEEL HANGER

Figure 6.4	TIMBER UPPER FLOORS	Scales 1:50, 1:20

members, and tusk tenon joints (figure 6.4.7) are often advocated for joints between trimmers and trimming joists and either dovetail notches (figure 6.4.9) or housed joints (figure 6.4.8) between trimmed joists and trimmers. In a tusk tenon joint the tenon passes through the trimming joist and is morticed out for a key or tusk, which when driven into position brings the shoulders up tight. The recommended dimensions of this joint are shown in figure 6.4.7. The housing or dovetail notch will project into the trimmer for at least one-half the depth of the trimmer. An alternative approach is to hang galvanised steel brackets or hangers (figure 6.4.10) over trimming joists to receive trimmers and over trimmers to receive trimmed joists, which are both nailed to prevent movement. The same principles apply in trimming floor joists around stairwells and rafters and ceiling joists around chimney stacks and rooflights.

Sound Insulation

Wood-joist floors are not sufficiently heavy and stiff to restrain vibration of the walls and the maximum net sound insulation is accordingly controlled by the thickness of the walls, even though the floor may have a higher sound insulation. Hence insulation values for wood floors can only be given in conjunction with the wall system, and the recommended forms of construction aim at achieving a satisfactory standard of insulation for both airborne and impact sound. The passage of noise through walls is known as *flanking transmission*. Sound transmission through floors is particularly important in the construction of flats, but may also have significance in ordinary dwellings where first-floor rooms are used as study bedrooms.

A satisfactory method of soundproofing wood-joist upper floors adjoining thin walls is to provide a ceiling of expanded metal lath and three-coat plaster, loaded directly with a *pugging* of dry sand 50 mm thick or other loose material weighing not less than 80 kg/m^2, together with a properly constructed floating floor, as illustrated in figure 6.5.1. The sand must not be omitted from the narrow spaces between the end joists and the wall; it should be as dry as possible when it is placed in the floor and should not contain deliquescent salts. An alternative to metal lath and plaster would be to use strong board ceilings of thick plasterboard that are firmly bonded to the walls, with solid, airtight joints between boards and a combined weight

of ceiling and pugging of not less than 120 kg/m^2.

With thick walls, it is possible to use a lighter form of floor construction (figure 6.5.2), incorporating a floating floor but reducing the pugging to 15 kg/m^2. The ceiling may be of plasterboard with a single-coat plaster finish, but a heavier ceiling is to be preferred. Wire netting stapled to the joists is sometimes inserted above the ceiling to keep the pugging in position and so ensure that the floor has a half-hour fire resistance. Another way of achieving the same fire resistance is to increase the plaster finish on the plasterboard ceiling to a thickness of 13 mm. A commonly recommended pugging material is high density slag wool (about 200 kg/m^3), laid to a thickness of about 75 mm.

The floating floor consists of woodboard or strip, plywood, or chipboard flooring, nailed to battens to form a *raft*, which rests on resilient quilt draped over the joists. The raft should not be nailed to the joists and should be isolated from the surrounding walls, either by turning up the quilt at the edges or by the less satisfactory alternative of leaving a gap round the edges, which can be covered by a skirting. Skirtings should be fixed only to walls and not to floors. The flooring should be at least 20 mm thick, preferably tongued and grooved, and battens should be 50 × 40 to 50 mm in size, laid parallel to the joists. A common method of constructing the raft is to place the battens on the quilt along the top of each joist and to nail the boards to the battens (figure 6.5.1). An alternative method is to prefabricate the raft in separate panels for the length of the room and up to one metre wide, with the battens across the panels positioned between the joists (figure 6.5.2).

Other forms of sound insulation for use with floors are examined in chapter 15, which covers sound insulation in a wider context. These include:

(1) flooring of hardboard over plain edge floorboards, mineral fibre between joists and a ceiling of a double layer of 13 mm plasterboard; (2) flooring of 18 mm chipboard on 19 mm plasterboard, mineral fibre on 16 mm plywood, with mineral fibre between joists and ceiling of 19 + 12 mm plasterboard.

Double Timber Floors

The longest span which can be effectively bridged with a normal joist floor is 5.42 m, using 75 × 220 mm joists of timber class SC4 at 400 mm centres with a dead load not exceeding 0.25 kN/m^2 (table A2 of Approved Document A1/2: 1991). Spans in excess of this are broken down into convenient lengths by the use of cross-beams forming what is commonly known as a *double floor*. The majority of double floors consist of timber joists supported on steel beams in the manner shown in figure 6.5.3. The steel beams bridge the shortest span of the room and are generally placed at about 3 m centres with 50 × 175 mm timber floor joists running between them. The ends of steel beams are often supported on precast concrete padstones built into the walls, and should be protected against corrosion where they adjoin a cavity in an external wall. The ends of the timber joists are notched around the steel beam and supported on a plate, often about 50 × 75 or 75 × 75 mm, bolted to the beam at about 750 to 900 mm centres. Bracketing or cradling of 50 × 38 mm timber fillets around the bottom of the steel beam provides a framework to receive the plasterboard, finished with a setting coat of plaster. Timber bearers or fillets, often about 50 × 50 mm, are nailed to the upper part of each joist and bridge the gap over the steel beam to provide a bearing for the floorboards. The size of the steel beam will depend on the span, the load to be carried and the shape and thickness of the flange and web of the beam. In normal domestic work the depth of the beam is often about $\frac{1}{20}$ of the span and some idea of the range of sizes can be obtained from steelwork tables or from BS 4.[29]

An alternative to the steel beam is an exposed timber beam, which might be of oak or other suitable hardwood, probably with chamfered bottom edges (figure 6.5.4). Timbers beams can provide an attractive feature to the ceiling of the room below.

Fire-resisting Floors

The selection of the most suitable form of fire-resisting floor in a particular situation will be influenced by such factors as cost, availability, rigidity, ease and speed of erection, sound and thermal insulating properties, permissible span and load, and ease of variation if required. A selection of floor types is now considered.

Filler joist floors. These represent the earliest method of combining steel and concrete in floor construction. Concrete panels are formed between the main steel beams and are supported by light steel beams or filler joists, often at about 750 mm centres, encased in the concrete. The light steel beams may rest on the tops of the supporting beams or on a shelf angle riveted to the webs of the main beams. The concrete in the floor requires shuttering, but this can be supported from the beams and filler joists. Details of a typical filler joist floor are illustrated in figure 6.6.1.

To overcome objections to the standard filler joist floor arising from its weight, need for shuttering and likelihood of sound transmission, hollow clay tiles or tubes are sometimes placed between filler joists in place of concrete. The tiles vary in depth from 100 to 175 mm and the filler joists are usually at about 900 mm centres. Another alternative is to use sheets of expanded metal instead of shuttering to retain the concrete.

Reinforced concrete floors. Concrete is strong in compression but relatively weak in tension, whereas steel is strong in tension. In reinforced concrete the steel makes good the inadequacies of the concrete and the concrete protects the steel from the effects of fire. The steel reinforcement is generally in the form of round mild steel bars ranging from 6 to 40 mm in diameter. These bars are positioned where the tensile stresses are greatest. For instance the bars will be placed near the bottom of a suspended slab, but will be fixed near the top where the slab passes over or rests on supports. All bars must have adequate cover of concrete, normally not less than 19 mm in slabs and 25 mm in beams, or the diameter of the bar in either case, whichever is the greater.

Larger floors are split into bays which are supported at their edges on beams or intermediate walls. Reinforcing bars lap each other where extended and hooked ends are introduced where the bond stress (friction between concrete and steel) is high. Shear bars are incorporated to strengthen concrete which is subject to shear stresses.

For short spans of up to 5 m, a reinforced concrete floor can span between walls as in figure 6.6.2, but for larger spans it is advisable to introduce secondary beams. Reinforced concrete floors carry heavier loads than timber floors of equal thickness, provide good lateral rigidity, and offer good fire resistance and insulation against airborne sound. On the other hand, they need support from formwork and props, as illustrated in figure 6.6.2, for about seven days.

Hollow pots. These form a popular variety of patent floor as they are economical, of lightweight material,

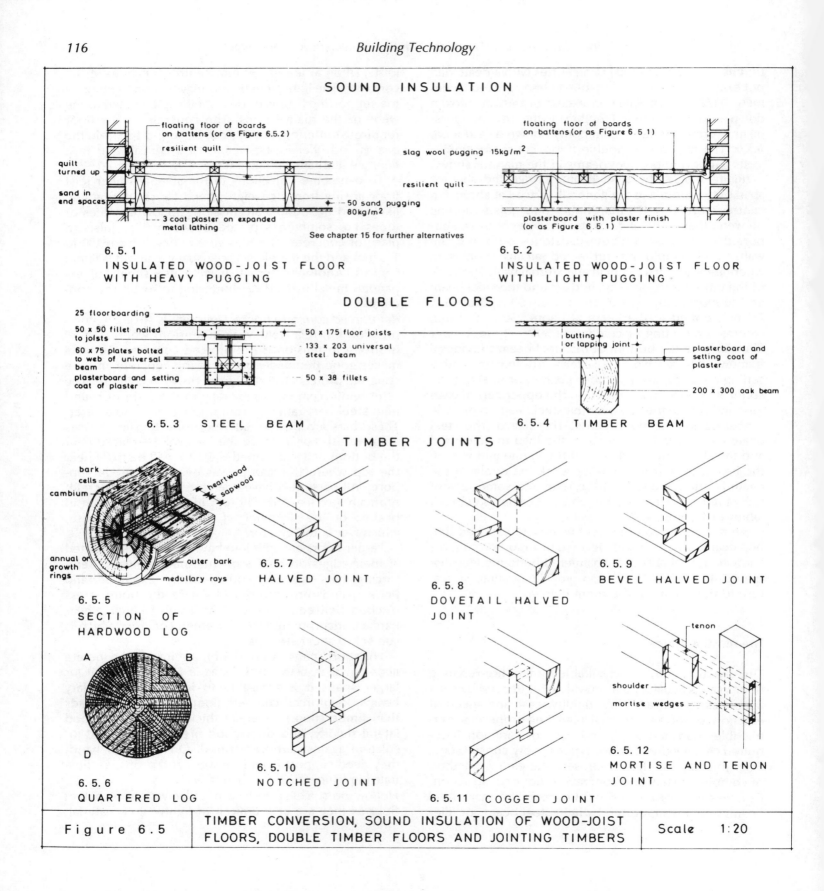

SOUND INSULATION

floating floor of boards
on boards (or as Figure 6.5.2)

resilient quilt

quilt
turned up

sand in
end spaces

50 sand pugging
80kg/m²

3 coat plaster on expanded
metal lathing

See chapter 15 for further alternatives

floating floor of boards
on battens (or as Figure 6.5.1)

slag wool pugging 15kg/m²

resilient quilt

plasterboard with plaster finish
(or as Figure 6.5.1)

6.5.1
INSULATED WOOD-JOIST FLOOR
WITH HEAVY PUGGING

6.5.2
INSULATED WOOD-JOIST FLOOR
WITH LIGHT PUGGING

DOUBLE FLOORS

25 floorboarding

50 x 50 fillet nailed
to joists

60 x 75 plates bolted
to web of universal
beam

plasterboard and setting
coat of plaster

50 x 175 floor joists

133 x 203 universal
steel beam

50 x 38 fillets

butting
or lapping joint

plasterboard and
setting coat of
plaster

200 x 300 oak beam

6.5.3   STEEL   BEAM

6.5.4   TIMBER   BEAM

TIMBER JOINTS

bark
cells

cambium

heartwood
sapwood

annual or
growth
rings

outer bark

medullary rays

6.5.5
SECTION OF
HARDWOOD LOG

6.5.7
HALVED JOINT

6.5.8
DOVETAIL HALVED
JOINT

6.5.9
BEVEL HALVED JOINT

A   B

D   C

6.5.6
QUARTERED LOG

6.5.10
NOTCHED JOINT

6.5.11   COGGED JOINT

tenon

shoulder

mortise wedges

6.5.12
MORTISE AND TENON
JOINT

| Figure 6.5 | TIMBER CONVERSION, SOUND INSULATION OF WOOD-JOIST FLOORS, DOUBLE TIMBER FLOORS AND JOINTING TIMBERS | Scale   1:20 |

offer good thermal and sound insulation, economise in concrete and have good fire resistance. The floor consists of *in situ* reinforced concrete tee beams with hollow clay or terra cotta blocks cast in between the beams or ribs, as illustrated in figure 6.6.3, to provide a flat ceiling. The blocks are laid on shuttering and the reinforcing bars placed in position. The concrete filling is then poured from the top, together with the topping which forms part of the concrete beams. The thickness of the floor and quantity and size of reinforcement varies with the span and superimposed load. In some cases the concrete beams or ribs extend the full depth of the floor, while in others clay slip tiles are inserted at the bottom of ribs and this reduces the possibility of pattern staining, gives better fire resistance and even suction for plaster.

Precast concrete beams. These also provide a useful form of floor. Figure 6.6.4 shows hollow precast beams laid side by side. Each beam is reinforced at the bottom corners and the gaps between beams are filled with jointing material. When the floor is supported by a main beam, continuity bars are inserted in the top of the slab or between the precast beams. The hollow beams have rather thin walls and a constructional concrete topping is laid over the tops of beams to spread superimposed loads. Figure 6.6.5 shows another type of floor using precast concrete arched beams. Precast concrete beam floors require no formwork and can be quickly erected.

Another alternative type of floor using precast concrete beams suitable for upper floors is Jetfloor, as described under suspended ground floors.

Methods of increasing the fire resistance of existing timber floors are described in BRE Digest 208,[30] and include the fixing of plasterboard to the existing ceiling, hardboard or plywood sheeting to the existing floor, and mineral fibre insulation between the joists.

TIMBER

This section is concerned with the properties, types, conversion processes, diseases and methods of jointing timber. Timber is probably the most important flooring material.

Properties of Timber

Timber is composed of cells which conduct sap (water fluid), give strength and provide storage space for food materials such as salts, starches, sugars and resins. A layer or band of cellular tissue is deposited round the trunk or stem and branches each year, and these are termed *annual rings*. In many timbers the wood formed in the early part of the growing season (spring or early wood) is more open, porous, and lighter in colour than the timber produced in the summer (summer or late wood). There is also a series of horizontal tubes or open cells radiating from the pith at the centre of the stem to the outside of the tree, known as *medullary rays*. Some down-flowing sap passes through medullary rays to interior layers which become durable *heartwood*, while the other layers remain as lighter *sapwood*. Bark consists of cells and woody fibres and forms the outside protective layer of the tree. The wood from trees of rapid growth will have wide annual rings and is described as *coarse-grained*, while slow-growing trees produce narrow growth rings and the wood is termed *close-grained*. Many of these terms are illustrated in figure 6.5.5.

Timber Types

Botanically trees are grouped into two classifications: *Broad-leaved trees* or *hardwoods*. These are generally hard, tough and dark-coloured with acrid, aromatic or even poisonous secretions, although not all hardwoods are hard. Typical examples of hardwoods are oak, teak, mahogany, walnut and elm.

Needle-leaved trees or *softwoods*. These are from coniferous trees which have cone-shaped seed vessels and narrow, needle-shaped leaves. They are usually elastic and easy to work, with resinous or sweet secretions. Some softwoods, like pitch pine, are quite hard. Typical examples of softwoods are European redwood, yellow pine, Douglas fir, whitewood and Canadian spruce.

Some of the more commonly used timbers are detailed in table 6.1, but readers requiring more detailed information are referred to BRE publications references 31 and 32 and BRE Digest 296.[33] There is also strong pressure to avoid the use of timbers from tropical rain forests.

Conversion of Timber

Felling of trees is best undertaken when there is minimum moisture in the tree, usually between mid-

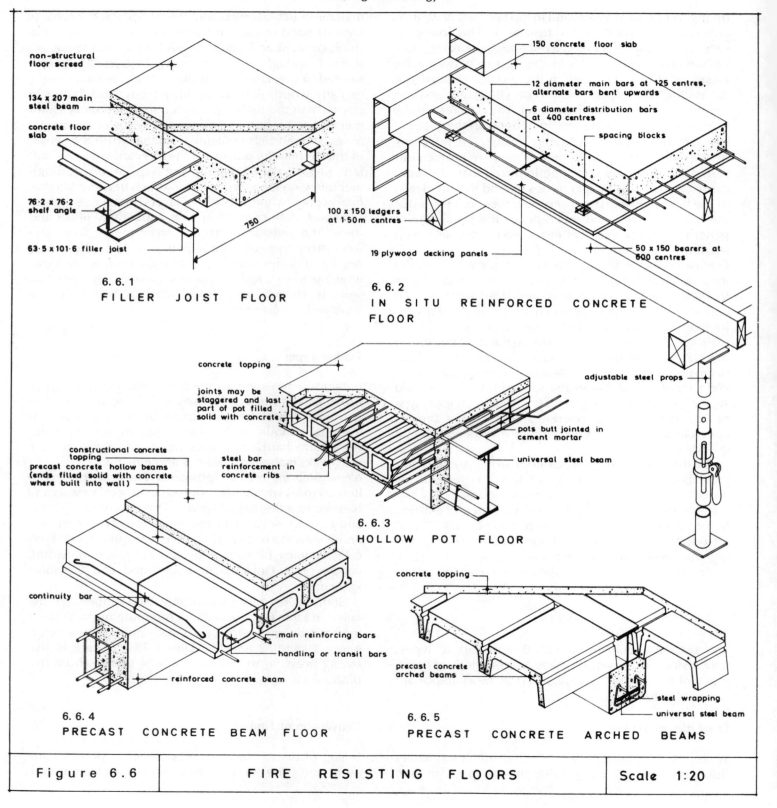

non-structural
floor screed

134 x 207 main
steel beam

concrete floor
slab

76·2 x 76·2
shelf angle

63·5 x 101·6 filler joist

750

6. 6. 1
FILLER JOIST FLOOR

150 concrete floor slab

12 diameter main bars at 125 centres,
alternate bars bent upwards

6 diameter distribution bars
at 400 centres

spacing blocks

100 x 150 ledgers
at 1·50m centres

19 plywood decking panels

50 x 150 bearers at
600 centres

6. 6. 2
IN SITU REINFORCED CONCRETE
FLOOR

concrete topping

joints may be
staggered and last
part of pot filled
solid with concrete

steel bar
reinforcement in
concrete ribs

pots butt jointed in
cement mortar

universal steel beam

6. 6. 3
HOLLOW POT FLOOR

adjustable steel props

constructional concrete
topping
precast concrete hollow beams
(ends filled solid with concrete
where built into wall)

continuity bar

main reinforcing bars

handling or transit bars

reinforced concrete beam

6. 6. 4
PRECAST CONCRETE BEAM FLOOR

concrete topping

precast concrete
arched beams

steel wrapping

universal steel beam

6. 6. 5
PRECAST CONCRETE ARCHED BEAMS

| Figure 6.6 | FIRE RESISTING FLOORS | Scale 1:20 |

Table 6.1   Timbers in common use

Variety	Hardwood or softwood	Average weight (kg/m^3)	Origin	Main properties	Principal uses
Beech	hardwood	720	Europe and Japan	Usually straight-grained with fine texture, stronger than oak, works easily and stains and polishes well	Veneers, kitchen equipment, cabinet-making and block and strip floors
Cedar (Western red)	softwood	360	British Columbia and Western United States	Straight-grained, coarse and soft, corrodes iron	Flush doors, weatherboarding and ceiling finishes
Elm	hardwood	560	English and Continental varieties	Twisty grain and warps unless carefully seasoned, tough and durable under wet conditions	Coffins, weatherboarding, piling, furniture and panelling
Fir (Douglas)	softwood	530	Pacific coast of United States and Canada	Straight-grained with pronounced figure, readily worked and stains well	Carpentry timbers, doors, flooring and panelling
Greenheart	hardwood	990 to 1090	Guyana	Very hard, heavy and strong, difficult to work, resists fungus attack and marine borers	Heavy construction, flooring, piling, dock and harbour work
Iroko	hardwood	660	W. Africa	Very resistant to decay, high strength, yellow to brown in colour	Exterior joinery and interior joinery, fittings and furniture
Jarrah	hardwood	900	Western Australia	Hard, heavy, coarse timber, generally straight-grained, dull red in colour and resists decay and fire	Heavy construction, flooring and stair treads
Mahogany (African)	hardwood	560	West Africa, Nigeria and Ghana	Variable quality; not very easy to work, cheapest variety of mahogany	Veneers, joinery, and interior decorative work
Mahogany (American)	hardwood	540	Central and South America	Dark red with good figuring, strong but works well	High-class joinery, veneers and panelling
Oak (European)	hardwood	740 to 770	European countries	Strong and durable with colour ranging from light to dark brown and has pronounced silver grain when quarter sawn	Furniture, internal and external joinery, gates and posts
Pitch pine	softwood	660	United States	Very hard, tough and resinous with many hardwood characteristics, works and finishes well	Heavy construction, piling and flooring

Table 6.1   (cont.)

Variety	Hardwood or softwood	Average weight (kg/m^3)	Origin	Main properties	Principal uses
Pine (Scots)	softwood	530	Europe	Straight-grained, resinous and strong, may be excessively knotty, finishes well, yellow in colour	Carpentry timbers, flooring and joinery
Sapele	hardwood	640	W. Africa	Pronounced stripe when quarter sawn, reddish brown in colour	Exterior joinery and interior joinery, fittings and furniture
Sycamore	hardwood	610	Great Britain and temperate Europe	Straight-grained, fine texture, good figure but not very durable, white in colour	Joinery, cabinet making and ladders
Teak	hardwood	640 to 720	Burma, India and Thailand	Very strong, durable and resistant to decay, moisture and fire, brown in colour	Garden furniture, doors and windows, flooring and panelling
Walnut (European)	hardwood	610 to 770	Europe	Good figure, hard, easy to work and polishes well	Veneers, cabinet work and panelling
Whitewood (European)	softwood	430	Northern Europe	White to light yellow, not very strong or durable, works easily but obstructed by knots, finishes well	Interior joinery, flooring and scaffold poles

October and mid-January. An axe or cross-cut saw is generally used for felling, although sometimes they are uprooted with winches. The felled tree trunks are converted into logs by sawing of all branches and the logs are converted to timbers of marketable sizes (scantlings) using an axe or saw. Four methods of quartering a log are illustrated in figure 6.5.6. The quartering method is valuable when cutting oak and similar boards parallel to the medullary rays to show the figuring or grain to best advantage. Method A is considered the best but is also the most expensive, whereas B and C are less wasteful of timber and hence more economical. Method D is the most economical and is sometimes called the *slash* method. Methods A and B are sometimes described as *rift* or *radial* sawing. To reduce waste still further, the log can be sawn into boards with all cuts parallel ('through and through') but some boards will twist on shrinking. Another method of conversion is to cut the timbers parallel to the annual rings, leaving a 'boxed heart' in the centre of the log. Veneers can be obtained by mounting the log horizontally along its longitudinal axis and cutting off a fine circumferential shaving with a long knife edge.

Seasoning of Timber

In 'green' timber large quantities of free water are present in the cell cavities and the cell walls are also saturated. Seasoning consists of drying out the free water and some of the water from the cell walls, which on withdrawal causes the timber to shrink, with the object of reducing the moisture content to a level consistent with the humidity of the air in which the timber will be placed. There are two principal methods of seasoning timber.

Air seasoning. This is where the 'green' timber is stacked with laths or 'stickers' between the timbers to allow the passage of air and assist in the evaporation of moisture from the timber. A suitable roof is needed to protect the timber from sun and rain. Air seasoning is unlikely to reduce the moisture content below 17 per cent even under ideal conditions and may take up to two years. Timber used internally in centrally heated buildings should not have a moisture content exceeding ten per cent to prevent shrinkage and warping.

Kiln seasoning. This is normally carried out in a forced draught compartment kiln, in which the air is heated by steam pipes and humidified by water sprays or steam jets. The temperature, degree of humidity and rate of air flow are all controlled from outside the kiln.

A maximum moisture content of 22 per cent for green timber should be specified. For structural timbers the moisture content in service is likely to vary between 12 per cent in continuously heated buildings and 20 per cent in unheated buildings.[34]

Defects in Timber

The strength and usefulness of timber can be affected by a wide variety of defects, some of which can occur during natural growth, others during seasoning or manufacture, while others result from attack by fungi or insects. The principal defects are listed and briefly described, with some definitions extracted from BSI: Glossary of Building and Civil Engineering Terms.[35]

A. Defects arising from natural causes

(1) *Knots*. These are portions of branches enclosed in the wood by the natural growth of the tree; they affect the strength of the timber as they cause a deviation of the grain and may leave a hole. There are several types of knot. A *sound* knot is one free from decay, solid across its face and at least as hard as the surrounding wood. A *dead* knot has its fibres intergrown with those of the surrounding wood to an extent of up to one-quarter of the cross-sectional perimeter; a *loose* knot is a dead knot not held firmly in place. *Rind gall* is a surface wound that has been enclosed by the growth of the tree.

(2) *Shakes*. These consist of a separation of fibres along the grain as a result of stresses developing in the standing tree, or during felling or seasoning. A *cross* shake occurs in cross-grained timber following the grain; a *heart* shake is a radial shake originating at the heart; a *ring* shake follows the line of a growth ring; and a *star* shake consists of a number of heart shakes resembling a star.

(3) *Bark* pocket. Bark in a pocket associated with a knot which has been partially or wholly enclosed by the growth of the tree.

(4) *Deadwood*. Timber produced from dead standing trees.

(5) *Resin pocket*. Lens-shaped cavities in timber containing or that has contained a resinous substance.

B. Defects due mainly to seasoning

(1) *Check*. A separation of fibres along the grain forming a crack or fissure in the timber, not extending through the piece from one surface to another.

(2) *Ribbing*. A more or less regular corrugation of the surface of the timber caused by differential shrinkage of spring wood and summer wood (crimping).

(3) *Split*. A separation of fibres along the grain forming a crack or fissure that extends through the piece from one surface to another.

(4) *Warp*. A distortion in converted timber causing departure from its original plane. *Cupping* is a curvature occurring in the cross-section of a piece and a *bow* is a curvature of a piece of timber in the direction of its length.

C. Defects due to manufacture
(including conversion)

(1) *Chipped grain*. The breaking away of the wood below the finished surface by the action of a cutter or other tool.

(2) *Imperfect manufacture*. Any defect, blemish or imperfection incidental to the conversion or machining of timber, such as variation in sawing, torn grain, chipped grain or cutter or chip marks.

(3) *Torn grain*. Tearing of the wood below the finished surface by the action of a cutter or other tool.

(4) *Waney edge*. This is the original rounded surface of a tree remaining on a piece of converted timber.

D. Fungal attack

(1) *Fungal decay*. This is decomposition of timber caused by fungi and other micro-organisms, resulting in softening, progressive loss of strength and weight and often a change of texture or colour. Fungi are living plants and require food supply, moisture, oxygen and a suitable temperature. A fungus is made up of cells called *hyphae* and a mass of hyphae is termed *mycelium*. It also contains fruit bodies within which very fine spores are formed. When timber is infected by spores being blown on to it, hyphae are then formed which penetrate the timber and break down the wood as food by means of *enzymes*. The Building Research Establishment[36] distinguishes two main forms of wood-rotting fungi according to their effect on the wood. (1) *Brown rots* cause the wood to become darker in colour and to crack along and across the grain; when dry, very decayed wood will crumble to dust. Many common wet rots and dry rot are in this group. (2) *White rots* cause the wood to become lighter in colour and lint-like in texture, without cross cracks. All white rots are wet rots.

(2) *Dote*. The early stages of decay characterised by bleached or discoloured streaks or patches in wood, the general texture remaining more or less unchanged. This defect is also known as doaty, dosy, dozy and foxy.

(3) *Dry rot*. This is a serious form of decay of timber in contact with wet brickwork caused by a fungus, *Serpula lacrymans*. The fungus develops from rust-coloured spores, which can be carried by wind, animals or insects, and throw out minute hollow white silky threads (hyphae). The fungus also produces grey or white strands 2 to 8 mm thick which can travel considerable distances and penetrate brick walls through mortar joints. The strands throw off hyphae forming white to grey silky sheets of mycelium, often with lemon or lilac patches, whenever they meet timber and carry the water supply for digestion of the timber, which becomes friable, powdery and dull brown in colour, accompanied by a distinctive mushroom-like smell. Often the timber shrinks and splits into brick-shaped pieces formed by deep longitudinal and cross cracks.[37]

The fungus needs a source of damp timber with a moisture content above 20 per cent. Damp, still air will encourage the establishment and spread of the fungus, particularly if these conditions are maintained for long periods. To eradicate an outbreak of dry rot, all affected timber and timber for 300 to 450 mm beyond must be cut away and burnt, preferably on the site. Surrounding masonry should be sterilised with a suitable fungicide such as sodium pentachlorophenoxide or sodium 2-phenylphenoxide. All apparently sound timber which is at risk should receive three brush coats of a suitable preservative, such as type F in BS 5707.[38] Where it is likely that damp conditions will persist, as in a damp cellar, replacement timber should be pressure impregnated with a copper/chromium/arsenic preservative conforming to BS 4072.[12] Where dampness is not expected to persist, an organic solvent type of preservative can be applied by immersion or the double vacuum process in accordance with BS 5707.[38] Levels of treatment are detailed in BS 5268[39] and BS 5589.[40]

The cause of the outbreak must be established and rectified, usually by preventing damp penetration and improving ventilation. The most vulnerable locations are cellars, inadequately ventilated floors, ends of timbers built into walls, backs of joinery fixed to walls and beneath sanitary appliances. The design of timber floors to prevent dry rot is covered in BRE Digest 364.[22]

(4) *Coniophora puteana (cellar fungus) and Coniophora marmorata*. These are the commonest cause of rots in timbers which have become soaked by water leakage, and are often found in badly ventilated basements and bathrooms. The fruit bodies are in sheets, olive green to olive brown with a cream margin with C. puteana and pinkish brown with C. marmorata.

(5) *Wet rot*. Originally this was diagnosed as chemical decomposition arising from alternate wet and dry conditions, as occurs with timber fencing posts at ground level. The decay may be accelerated by fungus attack. More recently[36] the term has been applied to fungus attacks other than dry rot, for instance Coniophora puteana, Poria placenta and Fibroporia vaillantii (white strands) in areas of higher temperature, and Paxillus panuoides (yellow to amber strands) in very wet situations; all being brown rots. Other fungi associated with wet rot, all in the white rot category, are Asterostroma spp (cream to beige to light tan, flat fruit bodies) on skirting boards and Phellinus contiguus (ochre to dark brown fruit bodies covered in minute pores) on exterior joinery. BS 345[36] gives a

detailed summary of the identifying characteristics of all the common wood-rotting fungi.

Dealing with an outbreak of wet rot is a relatively straightforward process, encompassing identifying the extent of the outbreak, removal of the source of dampness and subsequent drying out, and removal of all affected timber and its replacement with preservative treated timber. Timbers which have become structurally unsound require replacement or strengthening.

E. Insect attack

The life cycle of an insect is in four stages: the egg, the larva, the pupa and the adult. The egg is laid on the surface of the timber in a crack or crevice and the larva when hatched bores into the wood which provides its food. The larva makes a special chamber near the surface and then changes to a pupa. Finally it develops into the adult insect and bores its way out through the surface. The principal wood-boring insects are listed. BRE Digest 307[41] lists the damage features used in the identification of wood-boring insects ranging from type and condition of wood to sizes and shape of surface holes; colour, shape and texture of bore dust; and size and shape of tunnels in wood.

(1) *Common furniture beetle or wood worm.* This is about 2.5 to 5 mm long and dark brown in colour. Eggs are laid during June to August and larvae subsequently bore into the timber for a year or two, before the beetle emerges, chiefly in June and July leaving an exit hole about 1 to 2 mm diameter. They prefer seasoned hardwoods and softwoods, both sound and decayed; plywood made from birch or alder and bonded with animal glues is particularly susceptible.

(2) *Death-watch beetle* is about 6 to 9 mm long, dark brown in colour and mainly attacks hardwood, especially in large structural timbers (3 mm diameter exit holes).

(3) *Powder-post beetle* is about 5 mm long and reddish-brown in colour. Sapwood of new hardwood is particularly vulnerable (exit holes about 1.5 mm diameter).

(4) *House longhorn beetle* is about 15 mm long and is generally brown or black in colour. It is a most destructive pest on the Continent but, except in some parts of Surrey, is comparatively rare in this country.

This beetle favours sapwood of new softwood, especially in roof spaces. Exit holes are oval shaped and approximately 10 × 5 mm in size.

There are many proprietary preparations available for treating infested timber, most of which contain chemicals such as chlorinated naphthalenes, metallic naphthenates and pentachlorophenol. Insecticides should be brushed or sprayed over the surface during spring and early summer and injected into exit holes. The treatment should be repeated at least once each year during the summer months when beetles are active until there is no sign of continued activity. For further information on insecticidal treatments against wood-boring insects, the reader is referred to BRE Digest 327.[42]

Preservation of Timber

Few timbers are resistant to decay or insect attack for long periods of time, and in many cases the length of life can be much increased by preservative treatment.[39] The need for preservative treatment is largely dependent on the severity of the service environment.[34] The principal protective liquids are toxic oils, such as coal tar creosote; water-borne inorganic salts such as copper/chrome/arsenic,[12] which are suitable for exterior and interior use; and organic solvent solutions, such as copper and zinc naphthenate,[11] which are also suitable for exterior and interior use.[43]

Preservatives can be applied by non-pressure methods such as brush application, spraying, immersion and steeping.[44] The double vacuum method uses a closed cylinder. A vacuum is created, the organic solvent preservative flows in, the vacuum is released and the preservative is taken into the wood. Sometimes the cylinder is pressurised to increase penetration, then a second vacuum is created to drain surplus preservative. For lasting preservation, a pressure method is preferable. In the 'full-cell' process, the timber is placed in a closed cylinder and a partial vacuum applied to draw out air from the cells, hot preservative admitted, air pressure applied for one to six hours and a partial vacuum reapplied to remove excess liquid. The 'empty-cell' process is cleaner and more economical and a deeper penetration can be obtained with only limited excess preservative. The timber is subjected to air pressure, the preservative admitted and a higher pressure applied causing the liquid to penetrate the timber and compress air in the cells. When the timber is extracted, the air trapped in the cells forces out excess liquid leaving the cells empty but impregnating the cell walls.[43]

Strength of Timber

There are thousands of different species of timber but relatively few produce timber for structural use. For structural use, where appearance is generally not important, strength properties and durability are usually the prime considerations when making the choice of species.[34] Tables of reliable strength properties for each timber species and stress grade combination are contained in BS 5268.[45]

Stress grading is the process by which individual pieces of sawn structural timber are sorted into grades to which strength values are assigned for each species. It can be carried out by a visual inspection described for softwoods in BS 4978,[46] taking account of the strength-reducing characteristics, such as straightness of grain and size and position of knots, and the Standard prescribes two visual grades: GS and SS. Stress grading can also be carried out by machine to assess the strength of each piece of timber by measuring its stiffness. BS 4978[46] also requires that stress graded timber is marked with its stress grade.

BS 5268[45] categorises the strength requirements for structural timber into nine strength classes. The softwood species fall into the first five classes. SC1 (lowest) to SC5, and strength classes SC3 and SC4 are used when determining acceptable levels of performance for floor and roof timbers in Building Regulations Approved Document A1/2 (1991).

Jointing Timbers

Timber joints can be classified in a number of ways. One method would be to distinguish between carpentry, joinery and general-purpose joints, although some overlapping is bound to occur. BS 1186[47] distinguishes between fixed joints and joints permitting movement in joinery work. Another approach would be to classify the joints according to function: *width* joints for joining up boards such as tongued and grooved; *angle* joints for forming angles; *framing* joints such as mortise and tenon; *lengthening* joints such as scarfed and lapped joints; and *shutting* and *hanging* joints used in hanging doors and windows.

The simplest form of joint is the *butt* joint when adjoining members merely butt one another as with floorboards on occasions. Other methods of connecting floorboards are tongued and grooved (figure 6.1.6), splay rebated tongued and grooved (figure 6.1.6), and ploughed and tongued often incorporating a strip of three-ply as a continuous tongue.

There is a wide range of framing joints from the very complicated tusk tenon joint (figure 6.4.7) sometimes used in trimming work, to the much more widely used mortise and tenon (figure 6.5.12), in which one member is reduced to about one-third of its thickness (the tenon) which is wedged into a recess in the other member (the mortise). Where the penetrating member is not reduced in thickness it is termed a housed joint (figure 6.4.8). A stronger joint is obtained with a dovetail joint (figure 6.4.9). Where one member crosses over another the intersecting joint may be notched (figure 6.5.10) or cogged (figure 6.5.11). The jointing of wall plates in their length or at angles is normally undertaken with halved joints (figure 6.5.7), where each member is halved in depth for the length of the overlap. A stronger joint can be obtained by dovetail-halved (figure 6.5.8) or bevel-halved (figure 6.5.9) joints. Lengthening and birdsmouth joints are described in chapter 7 and other joinery joints in chapters 8, 9 and 10.

FLOOR FINISHINGS

Many factors deserve consideration when selecting a floor finish but not all the factors are of equal importance. Furthermore, requirements vary in different parts of the building. For instance, resistance to oil, grease and moisture is relevant in a kitchen but not a bedroom, and appearance could be important in a lounge but is of little consequence in a store. Some of the principal matters to be considered are

(1) *Durability*. The material must have a reasonable life to avoid premature replacement with resultant extra cost and inconvenience.

(2) *Resistance to wear*. This includes resistance to indentation where the floor has to withstand heavy furniture, fittings or equipment, and resistance to abrasion in buildings subject to heavy pedestrian traffic and movable equipment.

(3) *Economy*. Reasonable initial and maintenance costs, having regard to the class of building and the particular location within the building.

(4) *Resistance to oil, grease and chemicals*. This is particularly important in domestic kitchens, laboratories and some factories.

(5) *Resistance to moisture*. This is important in domestic kitchens, bathrooms, entrance passages and halls, and in some industrial buildings.

(6) *Ease of cleaning*. This is of increasing importance in many classes of building as the labour-intensive cleaning costs continue to rise at a disproportionate rate.

(7) *Warmth*. Some finishes are much warmer than others and this may be an important consideration.

(8) *Non-slip qualities*. These are particularly important in bathrooms and kitchens where floors may become damp.

(9) *Sound absorption*. Some buildings such as hospitals and libraries need floor finishes with a high degree of sound absorption.

(10) *Appearance*. This is an important consideration in many rooms of domestic buildings, although the current tendency to fully carpet rooms may not justify the provision of the more expensive but attractive floor finishes such as wood blocks and strip flooring.

(11) *Resilience*. Some flexibility or 'give' is often desirable.

Some of the more common floor finishes are now described.

Asphalt. The basic aggregates of mastic asphalt flooring are natural rock asphalt[7] or limestone.[6] Asphalt is available in dark colours, principally red, brown and black. It provides a jointless floor which is dustless and impervious to moisture. It is extremely durable but concentrated loads may cause indentations. A thickness of 15 to 20 mm is suitable for light use, increasing to 25 or even 38 mm for heavy duty. Floors over 20 mm thick may be laid directly on new concrete, but an isolating layer of black sheathing felt to BS 747[48] should be provided for thinner floors.

Pitch mastic. This can be black or coloured, is similar to mastic asphalt but is less affected by oils and fats. It is suitable for a wide range of situations, although little used nowadays. The thickness normally varies between 16 and 25 mm. It is reasonably quiet, warm, resilient, non-slip, dustless, hardwearing and resistant to moisture.

Rubber latex cement. This is a mixture of Portland cement, aggregate, fillers and pigment, gauged on the site with a stabilised aqueous emulsion of rubber latex. It is usually laid to a thickness of 6 mm, resists damp, is non-slip, reasonably hardwearing, quiet and warm and can be laid in a variety of colours, but has largely fallen into disuse.

Granolithic concrete. This is primarily used for factory flooring because of its hardwearing qualities. A typical mix is 1:1:2 of Portland cement, sand and granite or whinstone. The best finish is obtained by laying a 20 mm granolithic topping in bays of about 30 m^2 within three hours of laying the base, to obtain monolithic construction. Otherwise the topping should be at least 40 mm thick in bays of about 15 m^2 and adequately bonded to the base. After compacting the granolithic should be trowelled to a level surface; two hours later it should be re-trowelled to close surface pores and finally trowelled when the surface is hard. The tendency to produce dust can be reduced by applying sodium silicate or magnesium or zinc silicofluoride or by using a sealer such as polyurethane. A non-slip finish can be obtained by sprinkling carborundum over the surface (1.35 kg/m^2) and trowelling it in before the granolithic has set. A coloured finish can be obtained by using pigmented Portland cement. A minimum of seven days is required for curing. It provides a cold, unattractive finish, but at a moderate price.

Terrazzo. This is a decorative form of concrete usually made from white Portland cement and crushed marble aggregate to a mix of 1:2$\frac{1}{2}$ and may be *in situ* or precast. Terrazzo tiles have largely superseded *in situ* terrazzo in recent times. *In situ* terrazzo can either be laid with all the aggregate incorporated in the mix, or a mix wherein the fine aggregate is spread in position and the larger aggregate is then sprinkled on the surface and beaten and trowelled in. Both monolithic and separate construction from the base methods are used. The following precautions should be taken to prevent cracking

(1) floor divided into panels not exceeding 1 m^2 in area, with sides in ratio of 3:1, and separated by dividing strips of metal, ebonite or plastics, with a topping thickness of 20 mm;

(2) water/cement ratio kept as low as possible;

(3) no smaller aggregate than 3 mm used;

(4) terrazzo allowed to dry out slowly, and if possible the building should not be heated for six to eight weeks after the floor finish has been laid.

The material must be thoroughly compacted by tamping, trowelling and rolling, with further trowelling at intervals to produce a dense, smooth surface. After about four days, the surface is ground by machine and finally polished using a fine abrasive stone.[50] It is attractive in appearance, hardwearing, easily cleaned, but is noisy, cold and expensive. Further advice on the design and installation of this type of flooring is given in BS 5385, Part 5.[51]

Magnesium oxychloride or *magnesite*. This is a mixture of calcined magnesite, fillers such as sawdust, wood flour, ground silica, talc or powdered asbestos, to which a solution of magnesium chloride is added. It needs to be laid by specialists and should comply with BS 776.[52] Magnesite flooring is available in various colours to mottled or grained finishes. It is suitable for industrial, commercial or domestic use, but should be laid on a damp-proof membrane as the material tends to soften and disintegrate when wet. The thickness may vary from 10 mm for single-coat work to 50 mm for two or three-coat work. It is reasonably warm, hardwearing, easily cleaned and of good appearance, but noisy and is little used nowadays.

Composition blocks. These are generally about 150 × 50 × 9 mm and are usually made from cement, sawdust, fillers and binders. The blocks are bedded on a 13 mm bed of cement and sand, jointed in a composition material, and sanded and sealed. They are obtainable in a limited range of colours and can be laid in similar patterns to wood block flooring, are moderately priced, warm and quiet, hardwearing, resistant to heat, oils and greases, soon wear to a non-slip surface and are used extensively in general-purpose sports halls. Guidance on design and installation can be obtained from BS 5385, Part 5.[51]

Cork tiles. These are made from granulated cork, compressed and bonded with synthetic resins. Tiles are made to various sizes but 300 × 300 mm are probably the most popular, in thicknesses varying from 5 mm upwards depending on the amount of wear they will receive. They provide a warm, quiet and resilient floor finish at a reasonable cost. They should be laid in accordance with BS 8203.[53]

Cork carpet. This is a type of sheet linoleum with an attractive appearance and natural finish in a limited range of colours. It has a softer finish than linoleum, is quiet to the tread, non-slippery and resistant to water and weak acids, but requires sealing. It contains a high percentage of cork granules and is made in thicknesses of 3–6 mm and in rolls 1.8 m wide and in lengths of over 15 m and should comply with BS 6826.[54]

Carpet. This is now one of the most important finishes and is widely used in domestic properties and public areas of commercial buildings. A wide range of qualities exists and often expert advice is required on selection. The quality is dependent on the yarn, construction, backing, density, face pile weight and pile height. It needs to be laid in dry situations on an underlay of felt or latex and secured by adhesives, nailed around the perimeter or stretched over and attached to special fixing strips. Carpet is supplied in rolls of varying widths and carpet tiles are often 600 × 600 × 25 mm thick.

Linoleum. This is manufactured from linseed oil, cork, resin gums, fibres, wood flour and pigments applied to a canvas or glass fibre backing. It is best bonded to the base with a waterproof adhesive and for heavy use should be at least 4.5 mm thick. Linoleum forms a resilient, quiet and comfortable floor, to a wide range of colours and patterns, is easily cleaned, wears well and is resistant to oils and fats.[53, 54]

Rubber tiles and sheets. These are made of natural or synthetic rubber, the latter being most widely used. Tiles are normally 2–4 mm thick for housing work and 4–6 mm in public buildings. Rubber flooring must be fixed with epoxy adhesive and is available in a variety of colours, patterns and textures. This flooring is quiet, warm, resilient, non-slip, resistant to moisture, dilute acids, alkalis, oils and fats, but it is expensive.[53] A variety of adhesives for use with flooring materials are listed in BS 5442.[55]

Thermoplastic tiles. These are made from asbestos fibres, mineral fillers, pigments and a thermoplastic binder and can be obtained in a variety of colours and patterns. The most common size is 250 × 250 × 2.5 or 3 mm thick. The tiles become flexible when heated and are fixed with a bituminous adhesive to a cement and sand screed. They are moderately warm, quiet and resistant to water, oil and grease.[56]

Vinyl asbestos tiles. These were a later development of thermoplastic tiles and are made of PVC resin, asbestos fibre, powdered mineral fillers and pigments.[57] They have fair resistance to grease and hence are suited for use in kitchens and will withstand underfloor heating at moderate temperatures.

Flexible PVC tiles and sheet. These contain a substantial proportion of PVC resin, which gives added flexibility and resilience and provides good wearing qualities. A common size of tile is 305 × 305 × 2.4 mm thick and some possess realistic imitations of expensive traditional finishes, such as terrazzo and marble. They should comply with BS 3261.[58] BS 5085[59] specifies the requirements for flexible PVC flooring with a wear layer, an underlayer and a backing of needle-loom felt or cellular PVC.

Clay tiles. There are two main types of clay tile: floor quarries (20 to 32 mm thick) made from unrefined clays in variations of red, brown, beige and blue, and ceramic floor tiles (9.5 to 19 mm thick). When bedding tiles a separate layer of bitumen or polythene sheet should be laid over the base to permit variations in movement between the tiles and the base.[60] The tiles are soaked in clean water and laid on a bed of cement, lime and sand (1:½:4 or 5) 13 mm thick. Alternatively a thicker bed (40 mm) can be used without a separate layer. Adhesives should be those specified in BS 5385, Part 3.[61] Tiles should be laid on a bitumen bed where subjected to high temperatures and an expansion joint should be provided around the perimeter of tiled floors. Ceramic tiles are made from refined clays and obtainable in a variety of colours, patterns and sizes (75 × 75 mm to 150 × 150 mm and often about 12.7 mm thick). They are useful for kitchens and toilets, being hardwearing, easily cleaned and resistant to water, oils and most acids and alkalis; but they are also cold, noisy and relatively expensive.[62]

Concrete tiles. These are made from coloured cement and hardwearing aggregate set on a normal concrete backing, varying in size from 150 × 150 mm to 500 × 500 mm in thicknesses ranging from 15 to 40 mm.[63] They are usually laid on a bed of cement mortar on a separating layer, as for clay tiles. They have attractive finishes in a wide range of colours, are hard, durable, dustless, easily cleaned and resistant to damp, but are also cold, noisy and attacked by oils.

Wood blocks. These are made in lengths of 150 to 380 mm, widths of up to 90 mm and in thicknesses of 19 to 38 mm, with 25 mm as the most common nominal thickness.[64] Various interlocking systems are used, the commonest being the tongue and groove and the dowel. The blocks are usually laid on a screeded concrete floor and dipped in a cold bitumen latex emulsion adhesive.[65] They are often laid to herringbone or basket-weave patterns with the surface of the blocks sealed and waxed. Wood blocks provide a warm,

quiet, resilient, attractive, hardwearing floor, but are expensive.

Other timber floor finishes. In addition to wood blocks there are a number of other types of timber floor finish, and the principal ones are now described.

(1) *Strip and board*, where strip includes widths of 100 mm and under and board refers to anything wider, with common thicknesses of 20, 25, 32 and 38 mm. BS 1297[66] covers tongued and grooved softwood flooring, and this was described earlier in the chapter and illustrated in figure 6.1.6 and strip flooring in figure 6.1.2.

(2) *Parquet flooring* is one of the most attractive floor coverings embracing a variety of patterns and different species of wood but is, unfortunately, not often specified nowadays as skilled craft operatives are scarce and suitable hardwoods are expensive. In the traditional approach, flooring sections are laid on to a suitable wood sub-floor or pressed on to a backing and laid as panels. The parquet battens vary in thickness from 6 to 9 mm and the panels including the base are 25 to 32 mm thick, and the panels are normally about 305 to 610 m² and fixed by glueing and secret nailing.

(3) *Wood mosaic* provides a warm, decorative, resilient and durable flooring and consists of fingers of hardwood approximately 115 × 25 × 7.5 mm thick, arranged in groups to form 115 mm squares laid to form a basket weave pattern. They are normally prefabricated to form panels about 475 mm square and are either mounted on a sheet of permanent backing material or attached to a sheet of suitable paper on the upper surface which is removed after laying. The panels are usually laid with a bituminous adhesive.[64]

REFERENCES

1. *BRE Digests 274* and *275*: Fill (1990)
2. *BRE Digest 54*: Damp-proofing solid floors (1971)
3. *BRE Digest 363*: Sulphates and acid resistance of concrete in the ground (1991)
4. *The Building Regulations 1991 and Approved Documents A(1991), C(1991) and L(1995)*. HMSO
5. *BS 5328: 1981 (1986)* Specifying concrete, including ready-mixed concrete

6. *BS 6925: 1988* Specification for mastic asphalt for building and engineering (limestone aggregate)
7. *BS 6577: 1985* Specification for mastic asphalt for building (natural rock asphalt aggregate)
8. *BS 6398: 1983* Bitumen damp-proof courses for masonry
9. *BS 6515: 1984 (1987)* Polyethylene damp-proof courses for masonry
10. *BS 1282: 1975* Guide to the choice, use and application of wood preservatives
11. *BS 5056: 1974* Copper naphthenate wood preservatives
12. *BS 4072* Wood preservation by means of copper/chrome/arsenic compositions, *Part 1: 1987* Specification for preservatives; *Part 2: 1987* Method for timber treatment
13. *BS 144* Wood preservatives using coal tar creosote, *Part 1: 1990* Specification for preservative; *Part 2: 1990* Methods for timber treatment
14. *BRE Digest 145*: Heat losses through ground floors (1984)
15. *BRE Digest 104*: Floor screeds (1979)
16. *BS 882: 1983 (1986)* Specification for aggregates from natural sources for concrete
17. *BRE Defect Action Sheet 51*: Floors: cement-based screeds – specification (1984)
18. *BRE Defect Action Sheet 52*: Floors: cement-based screeds – mixing and laying (1984)
19. *BS 8204* In-situ floorings, *Part 1: 1987* Code of practice for concrete bases and screeds to receive in-situ floorings
20. *BS 8000* Workmanship on building sites, *Part 9: 1989* Code of practice for cement sand floor screeds and concrete floor toppings
21. *BS 5250: 1989* Code of practice for control of condensation in buildings
22. *BRE Digest 364*: Design of timber floors to prevent decay (1991)
23. *BS 493: 1970* Air bricks and gratings for wall ventilation
24. *BRE Defect Action Sheet 31*: Suspended timber floors: chipboard flooring – specification (1983)
25. *BRE Defect Action Sheet 32*: Suspended timber floors: chipboard flooring – storage and installation (1983)
26. *BS 5669* Particleboard, *Part 2: 1989* Specification for wood chipboard
27. *BRE Defect Action Sheet 58*: Suspended timber floors: joist hangers in masonry walls – installation (1984)
28. *BS 6178, Part 1: 1982* Specification for joist hangers for building into masonry walls of domestic dwellings
29. *BS 4* Structural steel sections, *Part 1: 1980* Specification for hot-rolled sections
30. *BRE Digest 208*: Increasing the fire resistance of timber floors (1988)
31. BRE. *Handbook of hardwoods* (1975)
32. BRE. *Handbook of softwoods* (1983)
33. *BRE Digest 296*: Timbers: their natural durability and resistance to preservative treatment (1985)
34. *BRE Digest 287*: Specifying structural timbers (1984)
35. *BSI* Glossary of building and civil engineering terms. *Sec. 410*. BSP (1993)
36. *BRE Digest 345*: Wet rots: recognition and control (1989)
37. *BRE Digest 299*: Dry rot: its recognition and control (1993)
38. *BS 5707* Solutions of wood preservatives in organic solvents, *Part 1: 1979* Specification for solutions for general purpose applications, including timber that is to be painted; *Part 3; 1980* Methods of treatment
39. *BS 5268* Structural use of timber, *Part 5: 1988* Code of practice for the preservative treatment of structural timber
40. *BS 5589: 1989* Code of practice for preservation of timber
41. *BRE Digest 307*: Identifying damage by wood-boring insects (1992)
42. *BRE Digest 327*: Insecticidal treatments against wood-boring insects (1993)
43. *BRE Digest 378*: Wood preservatives: application methods (1993)
44. BRE *Methods of applying wood preservatives* (1974)
45. *BS 5268* Structural use of timber, *Part 2: 1988* Code of practice for permissible strength design, materials and workmanship
46. *BS 4978: 1988* Timber grades for structural use
47. *BS 1186* Timber for and workmanship in joinery, *Part 1: 1986* Specification for timber; *Part 2: 1988* Specification for workmanship
48. *BS 747:1977* Roofing felts
49. *BS 4131: 1973* Terrazzo tiles
50. *CP 204* In-situ floor finishes, *Part 2: 1970* Metric units
51. *BS 5385* Wall and floor tiling, *Part 5: 1990* Code of

practice for the design and installation of terrazzo tile and slab, natural stone and composition block floorings

52. *BS 776* Materials for magnesium oxychloride (magnesite) flooring, *Part 2: 1972* Metric units

53. *BS 8203: 1987* Code of practice for installation of sheet and tile flooring

54. *BS 6826: 1987* Linoleum and cork carpet sheet and tiles

55. *BS 5442* Adhesives for construction, *Part 1: 1989* Classification of adhesives for use with flooring materials

56. *BS 2592: 1973* Thermoplastic flooring tiles

57. *BS 3260: 1969 (1984)* Specification for semi-flexible PVC floor tiles

58. *BS 3261* Unbacked flexible PVC flooring, *Part 1: 1973* Homogeneous flooring

59. *BS 5085* Backed flexible PVC flooring, *Part 1: 1974* Needle-loom felt backed flooring; *Part 2: 1976* Cellular PVC backing

60. *BRE Digest 79*: Clay tile flooring (1976)

61. *BS 5385* Wall and floor tiling, *Part 3: 1989* Code of practice for the design and installation of ceramic floor tiles and mosaics

62. *BS 6431: 1983/86* Ceramic floor and wall tiles

63. *BS 1197* Concrete flooring tiles and fittings, *Part 2: 1973* Metric units

64. *BS 1187: 1959 (1968)* Wood blocks for floors

65. *BS 8201: 1987* Code of practice for flooring of timber, timber products and wood based panel products

66. *BS 1297: 1987* Tongued and grooved softwood flooring

7 ROOFS: STRUCTURES AND COVERINGS

This chapter examines the general principles involved in the design of roofs; the various forms of construction and coverings; the techniques used in roof drainage; and the materials employed.

GENERAL PRINCIPLES OF DESIGN

Roofs have to perform a number of functions and some of the more important are listed and described.

Weather resistance

A roof which is not weatherproof is of little value and will not comply with paragraph C4 of Schedule 1 to the Building Regulations 1991,[1] which requires the roof of a building to resist the passage of moisture to the inside of the building. The slope of the roof and lap of the roof coverings must be considered, as well as the degree of exposure. For example, plain tiles are not suitable for use on slopes of less than 40°.

Strength

The roof structure or framework must be of adequate strength to carry its own weight together with the superimposed loads of snow, wind and foot traffic. In addition to the loading conditions specified by BS 6399,[2] checks of structural designs of roofs for adequacy against high local snow loads are necessary where proposed roof slopes and associated architectural features are likely to cause accumulations of drifted snow.[3] The roof structure should also provide lateral strength to walls (see Approved Document A: 1991 to the Building Regulations).

Wind pressure requires special consideration where a light roof covering is laid to a low slope. The greatest suction occurs at slopes below 15°, and this diminishes to around zero at 30°, with maximum suction at the edges of roofs. In conditions of high wind pressure the uplift can exceed the dead weight of the covering and so adequate fixing is necessary to prevent stripping of the coverings. CP 3[4] gives guidance on the determination and effect of wind loads and further guidance on the assessment of wind loads is provided in BRE Digest 346;[5] Part 3 contains a map of basic hourly mean wind speeds (based on annual probability), which are then modified by various factors to take account of topography, terrain and building factors, and gust peak factors in order to obtain a design wind speed, and loading coefficients for typical buildings, as described and illustrated in parts 4 to 6 of the digest.[5]

Durability

The coverings should be able to withstand atmospheric pollution, frost and other harmful conditions. With large concrete roofs and sheet-metal coverings provision must be made to accommodate thermal expansion. There should also be effective means for the speedy removal of rainwater from the roof, which might otherwise cause deterioration of the roof covering.

Fire resistance

The roof of a building is required under paragraph B4 of Schedule 1 to the Building Regulations 1991[1] to resist the spread of fire over the roof and from one building to another, having regard to the use and position of the building. Performance for the external fire exposure of roofs is determined by reference to the methods specified in BS 476[6] under which constructions were designated by two letters in the range A to D in the 1958 edition of the standard, with an AA designation being the best. The notional designations of roof coverings are listed in table A5 of Approved Document B2/3/4(1991). Natural slates; fibre reinforced cement slates; clay and concrete tiles; galvanised steel, aluminium, copper, lead and zinc sheets; mastic asphalt; vitreous enamelled steel; lead/tin alloy coated steel sheet; zinc/aluminium alloy coated steel sheet; and pre-painted (coil coated) steel sheet including liquid-applied PVC coatings, each with a suitable supporting structure, all have an AA designation. While bitumen felt has lower designations varying from AB to CC according to the type of felt and supporting structure.

Insulation

Thermal insulation of roofs is necessary to reduce heat losses to an acceptable level and to prevent excessive solar heat gains in hot weather, thus ensuring a reasonable standard of comfort within the building. The Building Regulations 1991, incorporating Approved Document L1 (1989) require the elemental U-Value of roofs of dwellings to be 0.25 W/m^2 K and for other buildings: 0.45 W/m^2K. Detailed information on the thermal insulation of roofs is given in chapter 15. Sound insulation on the other hand is rarely an important consideration in roof design.

Condensation

Schedule 1 of the Building Regulations 1991 (paragraph F2) requires that adequate provision shall be made to prevent excessive condensation in a roof; or in a roof void above an insulated ceiling, and this aspect will be examined in greater depth later in this chapter, and in chapter 15 and illustrated in figure 15.3.

Appearance

Roof design can have an important influence on the appearance of a building, both in regard to the form and shape of the roof, and as to the colour and texture of the covering material. Plain tiles are often better suited in scale to small domestic buildings than the larger single lap tiles, whilst clay tiles generally hold their colours better than concrete tiles. Improved appearance often results from the use of roof coverings which are of darker colour than the wall cladding.

CHOICE OF ROOF TYPES

There is a wide range of roof types available and the choice may be influenced by a number of factors. Roofs vary from flat roofs, not exceeding 10° slope, of timber or concrete, covered with one of a number of materials, to pitched roofs of various structural forms which are usually covered with slates or tiles. Factors influencing the choice of roof type follow.

Size and shape of buildings. Buildings of simple shape are readily covered with pitched roofs, whereas flat roofs are better suited for irregularly shaped buildings. The clear spans required will also influence the choice of roof.

Appearance. Aesthetic considerations might well dictate a pitched roof for a small building and a flat roof for a large building.

Economics. Both capital and maintenance costs should be considered in selecting a roof type. Some coverings such as zinc will not last the life of the building and replacement costs must be included in the calculations, using a discounting method as illustrated in *Building Economics*.[7]

Other considerations. Ease with which services can be accommodated in roof space; weatherproofing; condensation; maintenance and similar matters.

PITCHED ROOF CONSTRUCTION

Some of the more common roofing terms are explained by reference to figure 7.1.1. *Hipped* end is where the roof slope is continued around the end of a building, whereas the wall is carried up to the underside of the roof at a *gable* end. *Hip rafters* frame the external angles at the intersection of roof slopes, while *valley rafters* are used at internal angles. The shortened rafters running from hip rafters to plate and from ridge to valley rafters are termed *jack rafters*, while full-length rafters are often called *common rafters*. The bottom portion of the roof overhanging the wall is known as the *eaves*. Where the roof covering overhangs the gable end, it is termed the *verge*. *Purlins* are horizontal roof members which give intermediate support to rafters. Rafters are splay cut or bevelled and nailed to the ridge board at the upper end and birdsmouthed and nailed to the wall plate at the lower end (figure 7.1.9). Where purlins or rafters need to

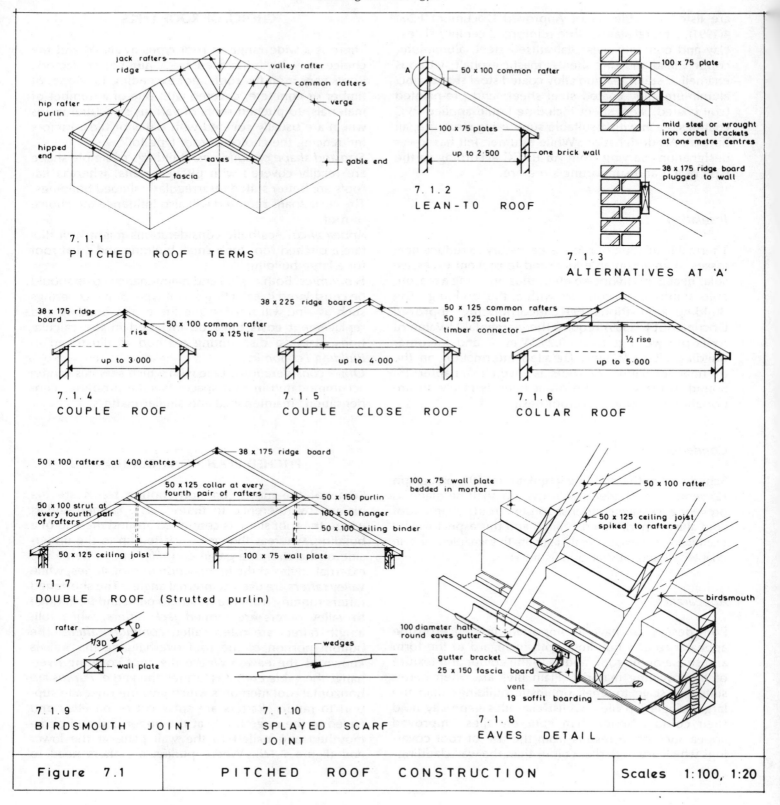

7.1.1 PITCHED ROOF TERMS

jack rafters
ridge
valley rafter
common rafters
hip rafter
purlin
verge
hipped end
eaves
fascia
gable end

7.1.2 LEAN-TO ROOF

50 x 100 common rafter
A
100 x 75 plates
up to 2·500
one brick wall

7.1.3 ALTERNATIVES AT 'A'

100 x 75 plate
mild steel or wrought iron corbel brackets at one metre centres
38 x 175 ridge board plugged to wall

7.1.4 COUPLE ROOF

38 x 175 ridge board
50 x 100 common rafter
rise
up to 3·000

7.1.5 COUPLE CLOSE ROOF

38 x 225 ridge board
50 x 125 tie
up to 4·000

7.1.6 COLLAR ROOF

50 x 125 common rafters
50 x 125 collar
timber connector
½ rise
up to 5·000

7.1.7 DOUBLE ROOF (Strutted purlin)

38 x 175 ridge board
50 x 100 rafters at 400 centres
50 x 125 collar at every fourth pair of rafters
50 x 150 purlin
50 x 100 strut at every fourth pair of rafters
100 x 50 hanger
50 x 100 ceiling binder
50 x 125 ceiling joist
100 x 75 wall plate

7.1.9 BIRDSMOUTH JOINT

rafter
1/3D
D
wall plate

7.1.10 SPLAYED SCARF JOINT

wedges

7.1.8 EAVES DETAIL

100 x 75 wall plate bedded in mortar
50 x 100 rafter
50 x 125 ceiling joist spiked to rafters
birdsmouth
100 diameter half round eaves gutter
gutter bracket
25 x 150 fascia board
vent
19 soffit boarding

| Figure 7.1 | PITCHED ROOF CONSTRUCTION | Scales 1:100, 1:20 |

be lengthened, a scarfed joint should be used (figure 7.1.10).

The slope of a roof is usually given in degrees, whereas the pitch is the ratio of rise to span. The rise is the vertical distance between the ridge and the wall plate, while the span is the clear distance between walls. In a half pitch or 'square pitched' roof, the span is twice the rise, for example 3.5 m rise with a 7 m span. The minimum pitch or slope is determined by the roof covering material, as described later in the chapter.

Pitched roofs can be broadly classified into three main categories as follows.

Single Roofs

This is where rafters are supported at the ends only. The simplest form is the *lean-to* or *pent* roof (figure 7.1.2) where one wall is carried up to a higher level than the other and the rafters bridge the space between. It is especially suitable for outbuildings and domestic garages with a span not exceeding 2.5 m. The upper ends of rafters in lean-to roofs are normally supported by a ridge board plugged to the wall or a plate resting on corbel brackets (figure 7.1.3).
Couple roofs (figure 7.1.4). These consist of rafters supported by a ridge board and wall plates. In the absence of a tie, the rafters exert an outward thrust on the walls, and this type of roof should not be used for spans exceeding 3 m. An improvement of this design is to nail ceiling joists to each pair of rafters when the roof is termed a *couple close* roof (figure 7.1.5), and the span may be increased to 4 m.
Collar roofs (figure 7.1.6). These may be used for spans up to 5 m. Each pair of rafters is framed up with a horizontal collar or collar tie, which should not be placed higher than one-half of the vertical height from wall plate to ridge. Collars should preferably be jointed to rafters with a suitable timber connector to give increased strength. This form of roof is valuable when additional headroom is needed.

Double Roofs

When the span exceeds 5 m, single roofs are no longer suitable as their use would entail excessively large and uneconomical timbers. A typical double or purlin roof is shown in figure 7.1.7, in which purlins are used at about 3 m centres to give intermediate support to rafters. The purlins are in turn supported by struts which bear on loadbearing partitions or ceiling beams. To further strengthen and stiffen the roof, collars may be provided at each fourth pair of rafters. For larger spans, the introduction of hangers at every fourth rafter with connections to the purlin and ceiling binder, which is in turn nailed to the ceiling joists, provides additional strengthening of the roof.

Figure 7.1.8 illustrates a typical form of construction at the eaves. The feet of the rafters are splay cut to receive the fascia and soffit boarding. The soffit boarding may be tongued to the fascia and both soffit and fascia are nailed to the rafters.

The sizes of ceiling joists, binders, common or jack rafters and purlins to pitched roofs are dependent upon the dead load to be carried, the strength class of the timber (SC3 or SC4), and the span and spacing of the members. Suitable sizes can be obtained from tables A3 to A16 of Building Regulations Approved Document A1/2 (1991).

Trussed or Framed Roofs

These are mainly used over large spans (often in excess of 7.5 m), where further support to purlins is provided in the form of roof trusses.
King and queen post timber trusses
The original type of timber roof truss consisted of large timber members bolted and strapped together at joints. Although these have been superseded by TRADA (Timber Research and Development Association) timber trusses and steel strusses, the student should be aware of the earlier trusses and the principles on which they were based. The king post truss (figure 7.3.1) was used for spans of 6 to 9 m, whilst the queen post truss (figure 7.3.2) was used for spans of from 9 to 12.75 m. The mansard truss (figure 7.3.3) incorporated both types of truss and enabled accommodation to be provided in the roof space. The members of the roof trusses were arranged so that their centre lines intersected at joints.

The king post truss (figure 7.3.1) consisted of a triangular frame which supported the ridge and purlins; these spanned from truss to truss and carried the common rafters and roof covering. The roof load was primarily transmitted through principal rafters to the walls below, and the feet of the principal rafters were strapped and bolted to a tie beam to prevent them spreading under load. A strut was introduced under the midpoint of the principal rafter to prevent it sagging, and the central post or king post prevented the tie beam from sagging and effectively stiffened and strengthened the truss.

Figure 7.3.4 shows the detailed arrangements for a queen post truss which is similar to a king post but has two vertical posts (queen posts), which are strutted apart at their heads by a straining beam. Principal rafters are supported by two purlins and the feet of queen posts are held in position by tenons, straps and a straining sill. Metal straps are also provided at the other intersections of the principal members.

TRADA roof trusses

The Timber Research and Development Association has designed standard timber roof trusses which are strong and effective and dispense with the need for support from internal loadbearing partitions, allowing greater flexibility in internal planning. Figure 7.2.1 shows a TRADA *timber truss* with a slope of 30° and a span of 11 m, capable of supporting slates or tiles weighing not more than 73 kg/m² on slope, together with a superimposed load including snow of 73.2 kg/m² measured on plan, and a ceiling load of 48.8 kg/m². Trusses will be spaced at 1.8 m centres with common rafters at 450 mm centres. Timber is to be S2-50 grade to BS 5268, Part 2.[8] The trusses can be made up in sections for transporting and then be assembled on site.

Joints between truss members are mainly formed with timber connectors complying with BS 1579,[9] as they increase considerably the strength of a bolted joint. The connectors must be adequately sheradised or galvanised to give protection against corrosion. The most common form of connector is the *toothed plate* (figure 7.2.2) consisting of round or possibly square metal plates with projecting teeth around the edge. Another type of connector is the *split ring* (figure 7.2.3) comprising a steel ring, 63 to 100 mm diameter, cut at one point, and the split so formed permits the ring to adapt itself to any movement of the timber and thus maintain a tight connection. Split rings possess a higher bearing resistance than toothed plates. In both cases the connectors are fitted into precut grooves formed with a special tool. Joints can also be formed with *truss plates* (figures 7.2.5 and 7.2.6). One variety has projecting teeth, while another contains predrilled holes to take 32 mm sheradised nails. The plates are generally made from galvanised mild steel sheet about 1 to 2 mm thick and are fitted to each side of the joint (see figure 7.2.6).

TRADA *trussed rafters* have been designed for the shallower slopes (10°, 15°, 20° and 25°), serving spans of 5.1, 6.0, 6.9 and 8.1 m. Figure 7.2.4 shows a truss of this type suitable for a 6 m span and a lightweight roof covering, such as woodwool slabs and felt. This form of roof covering is supported directly on the trusses fabricated prior to erection spaced at 600 mm centres, without the need for purlins, common rafters and ceiling joists. The trusses at slopes of 20° and 25° can carry tiles or slates, but if the load exceeds 50 kg/m² on slope the trusses must be spaced closer together. These have largely superseded the TRADA roof trusses. Trussed rafters should comply with BS 5268, Part 3.[10]

A TRADA trussed rafter suitable for a *monopitch* roof with a slope of 10° and a span of 6 m is illustrated in figure 7.2.7. It is designed to carry a roof covering of the type previously described (woodwool slabs and felt). All the designs are based on the use of S2-50 grade timber complying with BS 5268, Part 2.[8] Suitable timbers include western hemlock, parana pine, European redwood, whitewood and Canadian spruce or Scots pine of appropriate quality.

Trussed rafters which are badly stacked on site may become permanently distorted, and if permitted to become very wet during storage may result in rotting timber and corrosion of the connector plates.[11] Care is also needed in their installation as inadequately braced roofs may move because of the trussed rafters leaning sideways. If raking braces, ceiling and ridge level binders are wrongly positioned, badly fixed or cut to make room for items such as pipes and flues, the roof will not be structurally sound.[12] The roof structure must also be adequately supported to counteract wind uplift. BRE Defect Action Sheet 43[13] has shown the importance of supporting cold water storage tanks on three or four trusses, according to the tank capacity, located close to centres of trusses, with adequately sized bearers on a sound platform, not using chipboard, to prevent distortion of trussed rafter members, deterioration of tank platform and disturbance of plumbing joints.

Steel roof trusses are very popular for use in industrial, commercial and agricultural buildings. They are made up of members of small section, can be used for spans which would be beyond the economical capacity of timber, and steel is more dependable than timber in character and quality. The general principles of design are similar to those adopted for timber trusses, with members giving direct support at purlin points and

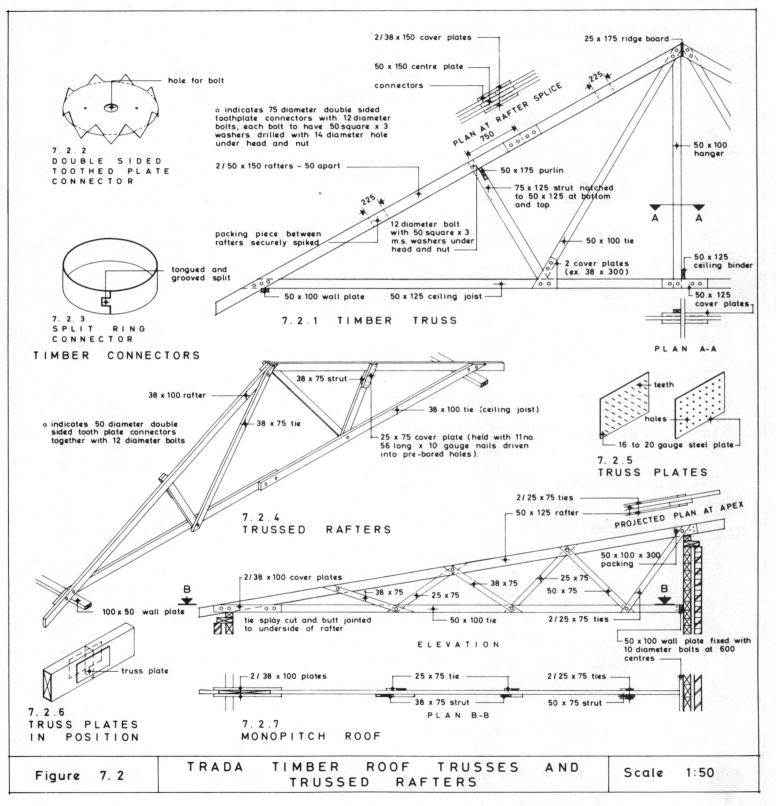

hole for bolt

**7.2.2
DOUBLE SIDED
TOOTHED PLATE
CONNECTOR**

tongued and
grooved split

**7.2.3 SPLIT RING
CONNECTOR**

TIMBER CONNECTORS

2/ 38 x 150 cover plates

50 x 150 centre plate

connectors

25 x 175 ridge board

PLAN AT RAFTER SPLICE

225

750

o indicates 75 diameter double sided
toothplate connectors with 12 diameter
bolts, each bolt to have 50 square x 3
washers drilled with 14 diameter hole
under head and nut

2/ 50 x 150 rafters – 50 apart

50 x 175 purlin

50 x 100
hanger

75 x 125 strut notched
to 50 x 125 at bottom
and top

A   A

225

50 x 100 tie

packing piece between
rafters securely spiked

12 diameter bolt
with 50 square x 3
m.s. washers under
head and nut

2 cover plates
(ex. 38 x 300)

50 x 125
ceiling binder

50 x 100 wall plate

50 x 125 ceiling joist

50 x 125
cover plates

7.2.1 TIMBER TRUSS

PLAN A-A

38 x 75 strut

38 x 100 rafter

38 x 100 tie (ceiling joist)

o indicates 50 diameter double
sided tooth plate connectors
together with 12 diameter bolts

38 x 75 tie

25 x 75 cover plate (held with 11no.
56 long x 10 gauge nails driven
into pre-bored holes).

**7.2.4
TRUSSED RAFTERS**

teeth

holes

16 to 20 gauge steel plate

**7.2.5
TRUSS PLATES**

2/ 25 x 75 ties

50 x 125 rafter

PROJECTED PLAN AT APEX

50 x 100 x 300
packing

100 x 50 wall plate

2/ 38 x 100 cover plates

B

38 x 75

25 x 75

38 x 75

50 x 75

25 x 75

B

tie splay cut and butt jointed
to underside of rafter

50 x 100 tie

2/ 25 x 75 ties

50 x 100 wall plate fixed with
10 diameter bolts at 600
centres

ELEVATION

truss plate

**7.2.6
TRUSS PLATES
IN POSITION**

2/ 38 x 100 plates

25 x 75 tie

2/ 25 x 75 ties

38 x 75 strut

50 x 75 strut

PLAN B-B

**7.2.7
MONOPITCH ROOF**

| Figure 7.2 | TRADA TIMBER ROOF TRUSSES AND
TRUSSED RAFTERS | Scale 1:50 |

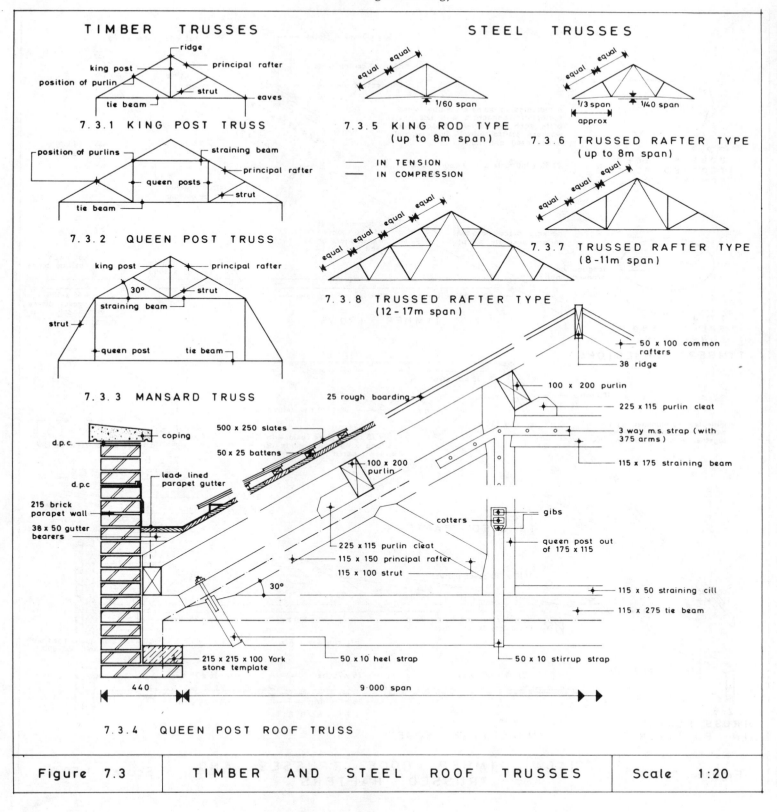

TIMBER TRUSSES

7.3.1 KING POST TRUSS
- ridge
- king post
- principal rafter
- position of purlin
- strut
- tie beam
- eaves

7.3.2 QUEEN POST TRUSS
- position of purlins
- straining beam
- principal rafter
- queen posts
- strut
- tie beam

7.3.3 MANSARD TRUSS
- king post
- principal rafter
- 30°
- strut
- straining beam
- strut
- queen post
- tie beam

STEEL TRUSSES

7.3.5 KING ROD TYPE
(up to 8m span)
- equal
- equal
- 1/60 span

IN TENSION
IN COMPRESSION

7.3.6 TRUSSED RAFTER TYPE
(up to 8m span)
- equal
- equal
- 1/3 span
- 1/40 span
- approx

7.3.7 TRUSSED RAFTER TYPE
(8–11m span)
- equal
- equal
- equal

7.3.8 TRUSSED RAFTER TYPE
(12–17m span)
- equal
- equal
- equal
- equal

7.3.4 QUEEN POST ROOF TRUSS
- coping
- d.p.c.
- d.p.c
- 215 brick parapet wall
- 38 x 50 gutter bearers
- 500 x 250 slates
- 50 x 25 battens
- lead lined parapet gutter
- 100 x 200 purlin
- 25 rough boarding
- 215 x 215 x 100 York stone template
- 50 x 10 heel strap
- 225 x 115 purlin cleat
- 115 x 150 principal rafter
- 115 x 100 strut
- 30°
- 50 x 100 common rafters
- 38 ridge
- 100 x 200 purlin
- 225 x 115 purlin cleat
- 3 way m.s. strap (with 375 arms)
- 115 x 175 straining beam
- cotters
- gibs
- queen post out of 175 x 115
- 115 x 50 straining cill
- 115 x 275 tie beam
- 50 x 10 stirrup strap
- 440
- 9·000 span

| Figure 7.3 | TIMBER AND STEEL ROOF TRUSSES | Scale 1:20 |

those in compression being kept as short as possible. Purlins normally consist of steel angles spaced to suit the roof covering material and bolted to short angle cleats, which may be riveted, bolted or welded to the principal rafters. Members are connected by gusset plates to which they are usually riveted. A selection of steel roof trusses for use with varying spans is shown by line diagrams in figures 7.3.5 to 7.3.8.

TRIMMING PITCHED ROOF AROUND CHIMNEY STACK

Where a chimney stack passes through a pitched roof, it is necessary to trim the rafters and ceiling joints around the stack, allowing a minimum space of 40 mm between the timbers and the outside surface of the stack in accordance with Approved Document J1/2/3. The trimmers and trimming rafters are 25 mm thicker than the common rafters (figure 7.4.2). Special care must be taken to obtain a watertight joint between the stack and the roof covering on all sides. On the sloping sides the best approach is to use stepped flashings in conjunction with soakers (figures 7.4.1, 7.4.3, 7.4.4 and 7.4.7), whereby rainwater runs down the stepped flashing on to a soaker, fixed between tiles or slates, and then runs down the roof slope from underneath one tile or slate on to the upper surface of the one below. The upper horizontal edges of the stepped flashings are turned about 25 mm into brick bed joints, secured by lead wedges and pointed. A gutter is formed on the top edge of the stack with one edge turned upwards against the stack and covered by a flashing, and the other edge dressed over a tilting fillet under the roof covering (figure 7.4.6). At the bottom side of the stack, an apron flashing is dressed over the adjacent roof covering (figures 7.4.6 and 7.4.8). A horizontal damp-proof course or tray should be provided where the stack emerges from the roof, preferably combined with the apron flashing as recommended in BRE Digest 380.[14] Good protection is also needed at the top of the stack (figure 7.4.5).

PITCHED ROOF COVERINGS

Roof Slopes

Table 7.1 shows the minimum slopes to which various coverings should be laid.

Table 7.1 *Minimum slopes for pitched roof coverings*

Material	Slope
Plain tiles	40°
Pantiles (double clay interlocking)	22½°
Concrete single lap tiles (interlocking), clip fixing, 75 mm lap*	17½°
Natural slates, green and blue (minimum 300 mm wide)	20°
Natural slates in diminishing courses	30°
Fibre-reinforced cement slates (larger sizes)	17½°
Thatch	45°
Profiled materials (protected metal)	15°
Wood shingles (with 125 mm gauge)	30°

*Certain types can be laid as flat as 12½°

In very exposed positions it would be advisable to lay roofing materials at slopes greater than the minimum. Some materials can be laid to flatter slopes by increasing the lap and hence the number of roofing units, although with wood shingles this could lead to greater susceptibility to decay. When using sprockets at eaves to improve appearance by introducing a bell-cast, care must be taken not to come below the minimum acceptable slopes at the most vulnerable point in the roof receiving maximum rainwater runoff.

Plain Tiles

Plain or double lap tiles can be made of clay or concrete in a wide range of colours, although the clay tiles generally retain their colour better than concrete tiles. Plain tiles measure 265 × 165 × 12 mm, while under-eaves and top-course tiles are each 215 mm long and tile-and-a-half tiles for use at verges are 248 mm wide.[15] They are slightly cambered in their length so that the tails bed tightly and they are sometimes cross-cambered in addition to prevent the entry of water by capillary action and to ventilate the undersides of tiles to accelerate drying out after rain. Each tile has two nibs for hanging over battens and has two holes for nails near its head. Plain tiles should be nailed with 38 mm nails of aluminium alloy, copper, stainless steel or silicon bronze at every fourth or fifth course and at eaves, top courses and verges.[16,17] Sarking felt should be provided under the tiling battens to prevent driving rain or snow from penetrating the roof. The felt should be of the untearable variety on a woven base and improved thermal insulation will result from the use of aluminium foil faced felt. BRE Defect Action Sheet 10[18] describes how dampness and staining of ceilings, wetting of loft insulation and dampness in eaves and barge boards have been caused by sarking felt not being

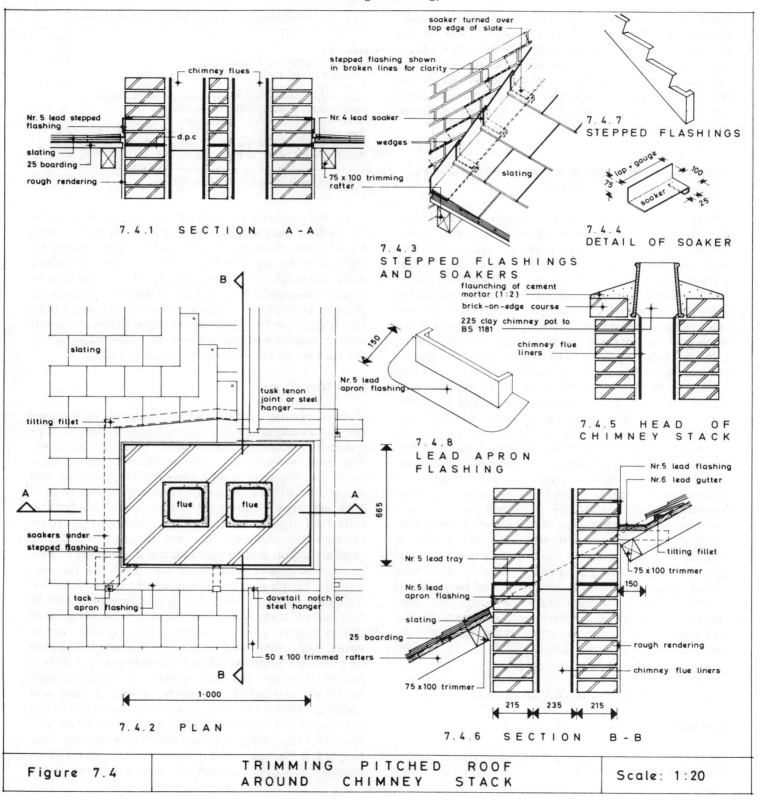

chimney flues

Nr. 5 lead stepped flashing

d.p.c

slating
25 boarding
rough rendering

7.4.1 S E C T I O N   A - A

soaker turned over top edge of slate

stepped flashing shown in broken lines for clarity

Nr. 4 lead soaker

wedges

75 x 100 trimming rafter

slating

7.4.3 S T E P P E D   F L A S H I N G S   A N D   S O A K E R S

7.4.7 S T E P P E D   F L A S H I N G S

lap + gauge
75
100
soaker
25

7.4.4 D E T A I L   O F   S O A K E R

B

slating

tusk tenon joint or steel hanger

tilting fillet

Nr. 5 lead apron flashing

150

7.4.8 L E A D   A P R O N   F L A S H I N G

flaunching of cement mortar (1:2)
brick-on-edge course
225 clay chimney pot to BS 1181
chimney flue liners

7.4.5 H E A D   O F   C H I M N E Y   S T A C K

A

flue

flue

A

665

soakers under stepped flashing

tack apron flashing

dovetail notch or steel hanger

50 x 100 trimmed rafters

B

1·000

7.4.2 P L A N

Nr.5 lead flashing
Nr.6 lead gutter

tilting fillet

75 x 100 trimmer

Nr. 5 lead tray

Nr.5 lead apron flashing

slating

25 boarding

75 x 100 trimmer

150

rough rendering

chimney flue liners

215   235   215

7.4.6 S E C T I O N   B - B

| Figure 7.4 | TRIMMING   PITCHED   ROOF AROUND   CHIMNEY   STACK | Scale: 1:20 |

properly lapped, not dressed out over eaves gutters and barge boards, or not fitting closely around soil and vent pipes. Various ventilating tiles have been produced which are watertight, blend in with the adjoining tiling and provide ventilation of the roof space to assist in preventing the occurrence of condensation.

A typical eaves detail is shown in figure 7.5.1 to which some of the more commonly used roofing terms have been added. The *lap* is the amount by which the tails of tiles in one course overlap the heads of tiles in the next course but one below, and for plain tiles should not be less than 65 mm or 75 mm in exposed positions. Shorter under-eaves tiles are provided at the eaves to maintain the lap. *Gauge* is the distance between centres of battens and is calculated by the formula: gauge = length of tile minus lap/2, hence the gauge of plain tiles to a 65 mm lap = $(265 - 65)/2 = 100$ mm. The *margin* is the exposed area of each tile on the roof and the length of the margin is the same as the gauge.

A typical *verge* detail is shown in figure 7.5.5 using tile-and-a-half tiles to maintain the bond bedded on and pointed in cement mortar on an undercloak of plain tiles. The tiles should overhang the wall by 50 to 75 mm to give protection against the weather. Half-round tiles are commonly used to cover *ridges* (figure 7.5.13) and these are bedded in cement mortar. Exposed ends are filled solid with mortar and often incorporate horizontal tile slips. Alternatively, segmental or angular ridge tiles may be used.

Hips of plain tiled roofs may be covered in a variety of ways. Half-round tiles, similar to those used at ridges, are popular and they require a galvanised or wrought-iron hip iron or bracket fixed to the bottom of the hip rafter to give support (figure 7.5.9). Other alternatives include bonnet hip tiles (figure 7.5.7) which are bedded at the tails, and angular hip tiles (figure 7.5.8) which are bedded at the heads of tiles.

Valleys are a particularly vulnerable part of a roof as the slope is several degrees less than that of the general roof surface and they have to provide a channel for water converging on it from two slopes. One approach is to form an open gutter with a timber sole covered with metal which is dressed over a tilting fillet and under the tiles on each side of the gutter. Purpose-made valley tiles (figure 7.5.10) provide a sound and attractive finish to a valley and do not require nailing or bedding. Swept valleys (figure 7.5.11) consist of tiles cut to the required sweep on a board and strip of felt not less than 600 mm wide, laid the full length of the valley and turned into the gutter at the bottom end. Laced valleys (figure 7.5.12) have a board not less than 225 mm wide laid in the valley and the plain tiling radiates from a tile-and-a-half tile placed obliquely in the middle of the valley in each course.

Single Lap Tiles

In single lap tiling each tile overlaps the head of the tile in the course below and there is also a side lap, the dimensions of which are usually fixed by the design of the tile. Thus there is only one thickness of tile on the greater part of the roof with two thicknesses at the ends and sides of each tile. Single lap tiles can be laid at a flatter slope than double lap or plain tiles, as they are obtainable in larger sizes and the actual inclination of the tiles for any given roof slope is greater. Some of the more common types of single lap tile are illustrated: Italian tiles (figure 7.5.2); Spanish tiles (figure 7.5.3); double Roman tiles (figure 7.5.4); and pantiles (figure 7.5.14). Pantiles are the oldest form of single lap tile and are used extensively in East Anglia. There are many forms of interlocking tiles manufactured in concrete complying with BS 473 and 550,[19] a common size being 418×330 mm – a typical tile is shown in figure 7.5.6. The principal advantages of single lap tiles over plain tiles is that they give a lighter roof covering and permit a flatter slope of roof. They are, however, more difficult to replace, loss of a tile results in water penetration and they are not so adaptable to complicated roof designs. A typical eaves detail using single lap tiles is illustrated in figure 7.6.7 and shows suitable thermal insulation and roof ventilation arrangements.

Vertical Tile Hanging

Vertical tiling may be fixed either direct to a wall, to battens or to battens and counter battens, treated with preservative and fixed to the wall face, to give added interest to a façade of a building. If vertical tiling is used as a facing for timber-framed construction, a suitable underlay must be provided in the form of a breather membrane which permits limited transfer of moisture vapour, or a vapour barrier, whichever is appropriate. The lap of tiles must be not less than 32 mm, tiles must be nailed with two nails to each tile, and the bottom edge of vertical tiling must finish with an undercourse as described for eaves of pitched roofs.[16]

Slates

Although slates have been superseded in many parts of the country by clay or concrete tiles and other forms of

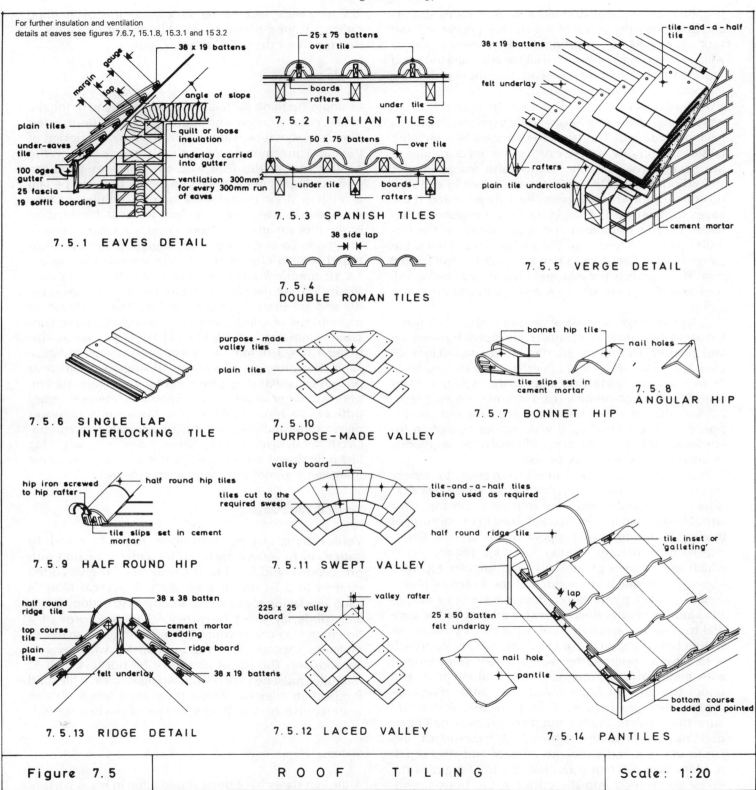

For further insulation and ventilation
details at eaves see figures 7.6.7, 15.1.8, 15.3.1 and 15 3.2

38 x 19 battens

gauge

margin

lap

plain tiles

under-eaves tile

100 ogee gutter

25 fascia

19 soffit boarding

angle of slope

quilt or loose insulation

underlay carried into gutter

ventilation 300mm² for every 300mm run of eaves

7.5.1 EAVES DETAIL

25 x 75 battens over tile

boards

rafters

under tile

7.5.2 ITALIAN TILES

50 x 75 battens

over tile

under tile

boards

rafters

7.5.3 SPANISH TILES

38 side lap

7.5.4 DOUBLE ROMAN TILES

tile-and-a-half tile

38 x 19 battens

felt underlay

rafters

plain tile undercloak

cement mortar

7.5.5 VERGE DETAIL

7.5.6 SINGLE LAP INTERLOCKING TILE

purpose-made valley tiles

plain tiles

7.5.10 PURPOSE-MADE VALLEY

bonnet hip tile

nail holes

tile slips set in cement mortar

7.5.7 BONNET HIP

7.5.8 ANGULAR HIP

hip iron screwed to hip rafter

half round hip tiles

tile slips set in cement mortar

7.5.9 HALF ROUND HIP

valley board

tiles cut to the required sweep

tile-and-a-half tiles being used as required

7.5.11 SWEPT VALLEY

half round ridge tile

tile inset or 'galleting'

lap

25 x 50 batten
felt underlay

nail hole

pantile

bottom course bedded and pointed

7.5.14 PANTILES

half round ridge tile

top course tile

plain tile

felt underlay

38 x 38 batten

cement mortar bedding

ridge board

38 x 19 battens

7.5.13 RIDGE DETAIL

225 x 25 valley board

valley rafter

7.5.12 LACED VALLEY

| Figure 7.5 | ROOF   TILING | Scale: 1:20 |

roofing material, they are still used predominantly in slate producing districts. They vary from 255 × 150 mm to 610 × 355 mm in 27 different sizes.[20] Each slate is secured by two nails, at the head or centre of the slate, and the nails may be composition, copper, aluminium alloy or silicon bronze and they vary in length from 32 to 63 mm according to the weight of slate. In the early 1990s, slates satisfying the British Standard requirements, were being imported into the UK from other EC countries at competitive prices, and were fixed with hooks instead of nails, which is an advantage in high winds. A typical eaves detail is shown in figure 7.6.1, from which it will be apparent that the lap is the amount by which the tails of slates in one course overlap the heads of slates in the next course but one below, as for plain tiles. It is customary to centre nail all but the smallest slates as there is a tendency for the larger head nailed slates to lift in high winds. The main advantage claimed for head nailed slates is that there are two thicknesses of slate covering the nails, but this involves the use of a larger number of slates and they are not so easily repaired. Nails should be not less than 30 mm from the edges and 25 mm from the heads of slates. The gauge is the distance between the nail holes in one slate from those in the adjoining slate, and for centre nailed slates = (length − lap)/2, whereas for head nailed slates it is [length − (lap + 25 mm)]/2. Thus, taking 460 mm long slates head nailed, the gauge becomes [460 − (75 + 25)]/2 = 180 mm.

Figure 7.6.4 illustrates centre nailed slates abutting a verge incorporating slate-and-a-half slates. The slates are arranged to bond so that side joints in one course are over the centre of slates in the course below. The slates may be nailed to battens fixed direct to rafters, to boarding fixed to rafters, to battens on boarding, or to battens on counter battens. The main advantage of counter battens is that any rain driven between the slates can run down the sarking felt between the sloping battens. The short slates at eaves are head nailed and they should overhang the gutter by 50 mm.

Ridges and hips may be covered by half-round, segmental or angular tiles (figure 7.6.6), preferably Staffordshire blue or tinted terracotta or be formed with a 50 mm diameter wood roll covered with code nr 5 or code nr 6 lead sheet, 450 to 500 mm wide, lapped 75 mm at joints and secured by screws and lead tacks (figure 7.6.3). The most common form of valley for slated roofs is the open metal valley illustrated in figure 7.6.2. The metal (lead, copper or zinc) should extend for at least 38 mm beyond the tilting fillets and should be nailed at both edges. Alternative valley treatments include swept or laced valleys which require highly skilled and experienced slaters, and mitred and secret valleys, neither of which are entirely satisfactory.[16] The best form of treatment at abutments is the use of soakers and cover flashings (figure 7.4.3). Perforations for pipes and similar fittings should be made weathertight by dressing over and under the slating or tiling a lead slate to which a lead sleeve is burned or soldered (figure 7.6.5). The sleeve should be bossed up around the pipe and suitably sealed at the top. Alternatively copper, 'Naturalite' or 'Zincon' can be used for this purpose.

Asbestos Cement and Fibre-reinforced Cement Slates

Asbestos cement slates are supplied in sizes the width of which is half their length, and they are suitable for use at a slope of not less than 35° if laid with a 75 mm lap. They are sometimes laid to a diagonal pattern. Their life depends on the degree of acid pollution of the atmosphere to which they are exposed. Asbestos cement slates are centre nailed with two copper wire nails to each slate, and the tails are prevented from lifting by a copper rivet passing through the tail and between the edges of the two slates of the course below. These slates are so light that rafters can be spaced up to 750 mm apart.[16]

Asbestos poses serious hazards to health and the Advisory Committee on Asbestos[21] recommended in 1979 that the use of asbestos should be phased out. As a result fibre-reinforced cement slates using other fibres have been produced and were used extensively in the late 1980s. They were covered by Agrément certificates giving them a minimum life of 30 years, although BRE Information Paper IP 1/91[22] shows that tests undertaken by BRE indicate that these materials are inferior to asbestos cement products in relation to mechanical properties.

Wood Shingles

Wood shingles are normally obtained in red cedar as it is very durable and resistant to insect attack, although BRE has recommended that they should be pressure impregnated with a copper-chrome-arsenic preservative. Cedar shingles weather rapidly to silver grey, or they may be dipped in oil to retain the natural reddish-

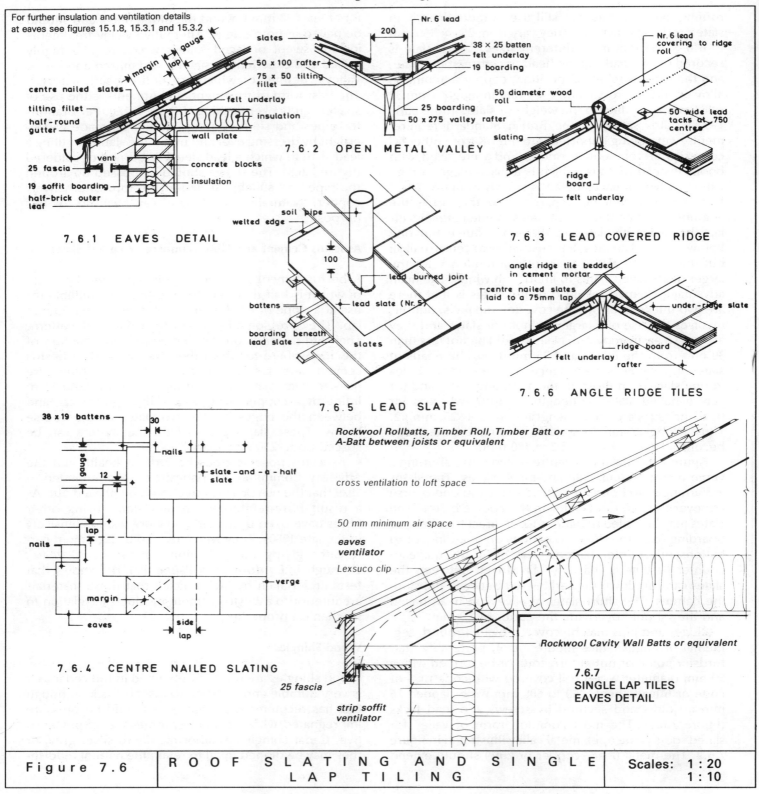

For further insulation and ventilation details at eaves see figures 15.1.8, 15.3.1 and 15.3.2

7.6.1   EAVES   DETAIL

margin   gauge   lap
slates
50 x 100 rafter
75 x 50 tilting fillet
centre nailed slates
felt underlay
tilting fillet
insulation
half-round gutter
wall plate
25 fascia
vent
19 soffit boarding
concrete blocks
half-brick outer leaf
insulation

7.6.2   OPEN METAL VALLEY

Nr. 6 lead
200
38 x 25 batten
felt underlay
19 boarding
25 boarding
50 x 275 valley rafter

7.6.3   LEAD COVERED RIDGE

Nr. 6 lead covering to ridge roll
50 diameter wood roll
50 wide lead tacks at 750 centres
slating
ridge board
felt underlay

7.6.5   LEAD SLATE

soil pipe
welted edge
100
lead burned joint
battens
lead slate (Nr. 5)
boarding beneath lead slate
slates

7.6.6   ANGLE RIDGE TILES

angle ridge tile bedded in cement mortar
centre nailed slates laid to a 75mm lap
under-ridge slate
ridge board
felt underlay
rafter

7.6.4   CENTRE NAILED SLATING

38 x 19 battens
30
nails
gauge
12
slate - and - a - half slate
nails
lap
nails
verge
margin
eaves
side lap

7.6.7 SINGLE LAP TILES EAVES DETAIL

Rockwool Rollbatts, Timber Roll, Timber Batt or A-Batt between joists or equivalent
cross ventilation to loft space
50 mm minimum air space
eaves ventilator
Lexsuco clip
Rockwool Cavity Wall Batts or equivalent
25 fascia
strip soffit ventilator

| Figure 7.6 | ROOF   SLATING   AND   SINGLE LAP   TILING | Scales: 1 : 20
1 : 10 |

brown colour. They are imported in lengths varying from 390 to 410 mm and widths of from 75 to 300 mm, and they are normally laid in random widths, to a minimum slope of 30° with a 125 mm gauge. Each shingle is nailed twice in its length with a minimum side lap of 40 mm and a gap of 3 mm between shingles in the same course. As shingles are light in weight rafters are often placed at 600 mm centres or more. One or two undercourses are used at eaves, hips may be cut and mitred with sheet metal soakers, ridges may be covered with narrow width shingles or a lead roll may be used, and valleys are best formed with open sheet-metal gutters and cut shingles. Shingles form an attractive finish but are more expensive than clay tiles and some fire risk is involved. Some species of Western Red Cedar used for wood shingles have over a period of years been subject to fungal attack in areas of high rainfall.[23]

Thatch

Thatching with reed from the Norfolk Broads is traditional in East Anglia. It provides a most attractive finish to steep roofs of at least 45° and preferably 55°. Reed is the most durable thatching material with a life of 50 to 75 years, but ridges are made of softer sedge with a life of about 20 years. Reed is laid about 300 mm thick with high thermal insulation properties, but it is more prone to fire damage than other coverings, is liable to attack by birds and vermin, is costly and is best confined to simple roofs. Rafters spaced at about 400 mm centres support battens (25 × 19 mm at 225 mm centres) which carry the reed. Eaves and verges have tilting or tilter boards to tighten the reed. The reed is used in bundles as cut, laid butt downwards, and tapering upwards in courses in a similar manner to tiling. The first course is tied in position but each subsequent course is held in place with a hazel rod or sway (about 19 mm diameter and 2.00 to 2.25 m long) laid across and fixed with hooks driven through the reed into the rafters. The ridge is formed of sedge bundles bent over a roll and down both roof slopes, laid alternate ways to correct the taper of the bundles, to a thickness of about 75 mm.

Sheet Coverings

Sheet coverings are available in a number of different materials and are particularly well suited for garages, stores, agricultural and industrial buildings. Probably the most commonly used material in years past has been asbestos cement which is made of fibreised asbestos and Portland cement; the natural colour is grey but it is also obtainable in other colours. Asbestos cement sheeting is unattractive in appearance, and although it is incombustible and light in weight, the surface softens with weathering and the material becomes increasingly brittle with age and is unlikely to have a life exceeding 30 years. As described earlier, there has been an increasing tendency in recent years to replace asbestos cement with fibre-resisting cement, but tests by BRE indicate some reduction in its mechanical properties compared with asbestos cement.[22]

To overcome the brittleness of asbestos cement sheets other materials have been produced, such as sheets incorporating a core of corrugated steel covered with layers of asbestos and bitumen. This combines the strength of steel with the corrosion-resisting properties of asbestos. Corrugated galvanised steel is inclined to be fairly shortlived, noisy, subject to condensation and rusting at bolt-holes, and is not well suited for most purposes. Aluminium sheets, of alloy NS3-H (international grading 3103) complying with BS 1470[24] are useful for roofing purposes. They are corrosion resistant, of light weight and reasonably good appearance, and their reflective value has some thermal insulation properties. The minimum recommended slope is 15° and the sheets are fixed in a similar manner to other corrugated sheet materials.[25]

Profiled metal sheeting is used extensively for cladding and roofing a wide variety of modern buildings from factories to office blocks. Coated or galvanised steel is the most common on grounds of cost as compared with aluminium and stainless steel. The range of colours in coloured coated steel and aluminium is considerable. Profiled metal roofs are vulnerable to water leaks at fixing points often because the fixings have loosened, or at lap joints resulting from inadequate overlaps or sealing of overlaps, or corrosion of protective coating to steel sheeting, particularly at cut edges.[23]

Skylights

Modern designs of skylight are now used extensively to light accommodation in the roof space. An attractive and effective double glazed top hung skylight or roof window is available in PVC-U for use with tiled or slated roofs and with pretensioned support arms and trickle ventilation.

Skylights/roof windows manufactured by Klober Plastics vary in glass area from 680 × 390 to 1130 × 780 mm.

FLAT ROOF CONSTRUCTION

A flat roof may be the only practicable form of roof for many large buildings or those of complicated shape and can be a more economical proposition than a pitched roof. It does unfortunately constitute a common area for premature failure in modern building. Most flat roof failures could have been avoided if the design principles now outlined had been adhered to. The technical options are described in BRE Digest 312[26] and Flat Roofing Design and Good Practice.[57]

Movement. Continuous coverings on flat roofs are much more susceptible to the effect of movement than the small units on pitched roofs. All forms of roof construction are liable to thermal movement and deflection under load, and in the case of a timber roof can subject the roof finish to considerable strain. Timber construction is subject to moisture movement, concrete to drying shrinkage and new brickwork to expansion. Hence only those roof finishes which are able to withstand some movement should be used on the more flexible types of roof, and asphalt, which is ill suited to accept movement, should not be used on timber roofs. Upstands should not be less than 150 mm high and if movement between the vertical and horizontal sections is possible, a separate metal or lead flashing should be provided to cap the top of the turned-up edge of the covering. Adequate expansion joints should also be provided, consisting of an upstand not less than 150 mm high with a metal capping.

Falls. The retention of water on most forms of roof covering is undesirable and constitutes a common cause of failure. This often results from ponding caused by inadequate falls. In the past the normally accepted minimum finished fall has been 1 in 80, but after making allowance for building inaccuracies and structural deflection, BS 6229[27] recommends a fall of 1 in 40.

Insulation. The standards of insulation prescribed in Part L of The Building Regulations 1991[1] should be regarded as minima. Care must be taken to keep all insulating materials dry as they cease to be effective and may deteriorate when wet.

Solar protection. With higher standards of insulation, solar reflective treatment is a necessity for asphalt and bitumen felt roofs. White spar chippings are useful, while concrete tiles may be justified if the roof is accessible and likely to receive regular use. Upstands which cannot be treated with chippings should receive an applied reflective coating, such as metal foil on a felt backing. Moreover, solar reflective finishes prolong the life of the roof covering.

Condensation. The humidity and vapour pressure are normally higher inside an occupied building than outside. Water vapour will usually penetrate most of the internal surfaces of a building and when the outside temperature is lower, it may condense at some point in the roof structure, often on the surface of insulation, and this is known as interstitial condensation.[28] Where the insulation is immediately under the roof covering it must be placed on a vapour barrier, often consisting of bitumen felt laid and lapped in hot bitumen and turned past the edges of the insulation to meet the roof covering. With timber roofs it is necessary to cross ventilate the spaces between joists[29] and where the roof extends over a cavity wall the cavity should be sealed at the top to prevent moist air entering the roof. Where cavity barriers restrict the crossflow of air in a flat roof structure, cowl type ventilators penetrating through the roof surface are needed. Insulation of pitched roofs is examined in chapter 15 and illustrated in figure 15.3.

Flat Roof Design

The Bituminous Roofing Council[30] has defined the following three principal types of flat roof.

(1) *Cold deck roof.* In this type of construction the thermal insulation material is placed below the roof deck, normally at ceiling level. Heat loss through the ceiling is thus restricted, keeping the cavity, roof deck and covering at low temperature during winter. If condensation problems are to be avoided in cold roofs, adequate provision must be made for efficient ventilation of the roof space. It is important to provide a sufficient open area on each side of the roof cavity to ensure a free, unobstructed path for ventilation purposes. It is also necessary to provide a vapour control membrane above the ceiling as shown in figures 7.9.1, 7.9.2 and 7.9.3.

(2) *Warm deck roof.* In the construction of warm roofs the thermal insulant is placed immediately below the

waterproof covering and on top of the roof deck and vapour barrier. The deck is thus maintained at warm temperatures during the winter. The thermal insulant is secured to the deck by bonding or mechanical fasteners, while the waterproof covering is bonded to the top surface of the insulation. With this type of construction high levels of thermal insulation are more easily achievable and more positive condensation control is possible by the selective use of high efficiency insulation material and vapour barriers. See figures 7.7.1, 7.7.3, 7.8.1 and 7.8.3. for roofing details.

(3) *Inverted warm deck roof.* In this type of popular construction, the thermal insulation material is placed on top of the waterproof covering so that the complete roof construction, including roof covering, is kept at warm temperatures during winter and at moderate temperatures during summer. With inverted roofs, the important requirement is that the thermal insulation material has low moisture absorption and water vapour transmission characteristics. An important advantage is that the insulation protects the waterproof covering from extremes of temperature and differential movements within the roof structure are thus reduced to a minimum. Furthermore, the waterproof covering is protected from various forms of damage, but it must be able to withstand continually wet conditions and is not immediately accessible for inspection and repair. See figures 7.7.2, 7.7.4, 7.8.2 and 7.8.4 for roofing details.

Vapour control layers. The Property Services Agency[31] (DOE) has emphasised that all insulants must be laid on and protected by an efficient and properly laid vapour control layer, although with inverted warm deck roofs no separate vapour control layer is required.[57] Failure to do this in warm deck roof construction is likely to result in moisture vapour from the building affecting the insulation, reducing its thermal efficiency and starting up cyclic interstitial condensation. Feedback reports to PSA have indicated that far too many roof defects are due to ineffective vapour control layers. High performance felts are most likely to achieve the required level of performance, and these include polyester bitumen, bitumen polymer and pitch polymer. Vapour control layers should be fully bonded to the substrate (deck, topping or screed) with hot bitumen, and be provided with end and side laps with a minimum width of 100 mm.

In cold deck roof construction, it is more difficult to ensure that the vapour barrier will remain effective by always being above dewpoint temperature because: (1) it is very difficult to construct an efficient vapour

barrier at ceiling level; and (2) the waterproof covering, which is the first real vapour barrier, is during the winter at a temperature well below dewpoint for the internal conditions, and good roof cavity insulation is essential, but may be interrupted or reduced by structural support with the risk of cold bridging.[57]

Design Details for Warm Deck and Inverted Warm Deck Roofs

The following design details are taken from *Flat Roofing: Design and Good Practice* by kind permission of the joint copyright holders and publishers: British Flat Roofing Council and CIRIA.

Eaves with gutter: warm deck roof (built up roofing) (figure 7.7.1)

Eaves should be formed to ensure that water is discharged into the gutter over an adequate welted drip, not less than 50 mm deep and with a 25 mm minimum projection beyond the fascia, which also has a drip and stands clear of the wall. The vapour control layer in a warm deck roof can be either turned back at least 75 mm over the insulation or taken over the fascia/batten. The insulation should be protected at the edge by an upstand or timber bearer.

Eaves with gutter: inverted warm deck roof (built up roofing) (figure 7.7.2)

On eaves to inverted roofs, insulation and protection need physical restraint to prevent movement resulting from wind uplift and thermal expansion, and figure 7.7.2 shows stainless steel strapping through the concrete paving.

Verge: warm deck roof (built up roofing) (figure 7.7.3)

Verges must provide an adequate check to prevent the discharge of water over the side of the roof, and a minimum depth of 75 mm is recommended. In a warm deck roof the vapour control layer can be either turned back at least 75 mm over the insulation or taken over the fascia/batten.

Verge: inverted warm deck roof (built up roofing) (figure 7.7.4)

This detail shows a substantial concrete upstand with insulation attached to the outer vertical face and projecting soffit to avoid cold bridges occurring. The roof surface shows concrete paving with surrounding pebbles on a filter membrane above the insulation.

Parapet abutment: warm deck roof (built up roofing) (figure 7.8.1)

The damp-proof course under the coping is supported by a slate or other suitable form of cavity closer. Cavity trays must protrude over flashings and be sealed at laps which should have a minimum height of 100 mm. Weep holes should be provided at 900 mm centres. Copings are weathered and have adequate overhangs and drips. Cavity trays should not be formed of lead unless provision is made for movement.

Parapet abutment: inverted warm deck roof (built up roofing) (figure 7.8.2)

In an inverted warm deck roof, insulation to the upstand membrane is strongly recommended as it reduces heat ageing of the membrane, protects the membrane from damage and eliminates the risk of cold bridging of the wall/ceiling abutment.

Eaves with gutter: warm deck roof (mastic asphalt) (figure 7.8.3)

Eaves should be formed to ensure that water is discharged into the gutter with a suitable drip. The drip, if it is formed in mastic asphalt, will require a key of expanded metal lath. The use of lead drips is no longer recommended because of differential thermal movement and low bond strength between the two materials. Insulation should be protected at the edge by an upstand fascia and timber bearer.

Eaves with gutter: inverted warm deck roof (mastic asphalt) (figure 7.8.4)

Eaves drainage to inverted roofs should be avoided as far as possible, as effective *U*-values may be reduced by water flow below the insulation and restraint to paving may be difficult to achieve. On eaves to inverted roofs the insulation and protection require physical restraint to prevent wind uplift and subsequent dislodgement, as provided by the stainless steel strapping shown in figure 7.8.4. An effective solar reflective finish must be provided to exposed mastic asphalt.

Timber Roofs

Timber is reasonably adaptable, easily worked and hence quite popular for small buildings, such as domestic garages and stores. Timber flat roofs generally consist of roof joists, 50 mm thick and spaced at 400 to 450 mm centres, carrying tapering firring pieces to give the necessary fall and support to the boarding which carries the roof covering (figures 7.9.1 and 7.9.3). It is good practice to provide galvanised steel straps at intervals, screwed to the roof joists and built into the brickwork to prevent the possibility of the roof being lifted in a high wind. The fall should ideally be 1 in 40, with the surface water usually drained to an eaves gutter (figure 7.9.3) or a wall gutter. The roof joists generally bridge the shortest span and the boarding is nailed at right angles to them, although the boarding or its grain should preferably follow the fall to avoid warping boards retarding the flow of water. Square-edged boarding is sometimes used but tongued and grooved boarding or exterior quality plywood give a much sounder job. Each board should be nailed with two nails to each joist with the nailheads well punched down below the surface of the boarding. The moisture content of the boarding should not exceed 15 per cent and the roof space should be ventilated (at least 300 mm^2 free opening for every 300 mm run of eaves) as shown in figure 7.9.3, to prevent fungal growth and condensation. The timber may be treated with preservative but creosote must not be used with built-up felt roofs as it is injurious to bitumen.

Built-up roofing (BUR). This is now a very popular covering to this type of roof owing to the high cost of metal coverings. It normally consists of three layers of felt laid breaking joint and bonded together with a hot bitumen adhesive, with each length of felt overlapping the adjoining sheet by at least 50 mm. It is possible to reduce the number of layers to two by using high performance polyester-cored felts and polymer modified bitumen for the coating, offering improved flexibility and performance. To secure a good standard of thermal insulation, mineral wool roll or batt or other suitable insulant preferably at least 50 mm thick and often up to 150 mm thick may be used, supported on roof joists at about 1.80 m centres (figure 7.9.1). The insulation board should fit tightly between the roof joists. The second and top or cap sheets of felt are fixed with bitumen adhesive, staggering the joints in successive sheets. The roof surface

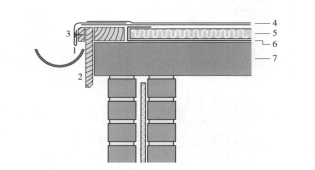

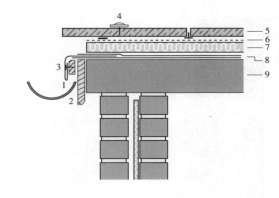

Warm deck roof.

1. Mineral faced elastomeric layer with welted drip
2. Fascia
3. **Pre-treated timber batten (minimum 25 mm thick × 50 mm high) secured to deck or fascia as appropriate.**
4. Built up roofing
5. Insulation
6. Vapour control layer
7. Deck

7. 7. 1

EAVES WITH GUTTER (built up roofing)

Inverted warm deck roof.

1. Mineral faced elastomeric layer with welted drip.
2. Timber fascia
3. **Pre-treated timber batten (minimum 25 mm thick × 50 mm high) secured to deck or fascia as appropriate**
4. Stainless steel restraint strapping
5. Concrete paving on pads
6. Filter layer
7. Insulation
8. Built up roofing
9. Deck

7. 7. 2

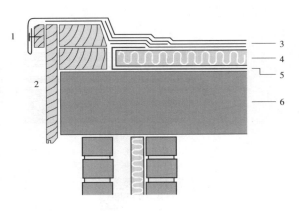

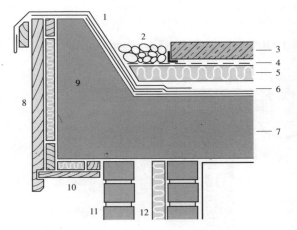

Warm deck roof.

1. Mineral elastomeric layer with welted drips
2. Timber fascia
3. Built up roofing
4. Insulation
5. Vapour control layer
6. Deck

7. 7. 3

VERGE (built up roofing)

Inverted warm deck roof.

1. Mineral faced elastomeric layer
2. **Pebbles 20 mm–40 mm**
3. Concrete paving on pads
4. Filter membrane
5. Insulation
6. Built up roofing
7. Deck
8. Timber fascia
9. Concrete upstand
10. Timber soffit
11. Brickwork
12. Insulation

7. 7. 4

| Figure 7.7 | WARM AND INVERTED WARM DECK ROOFING DETAILS 1 |

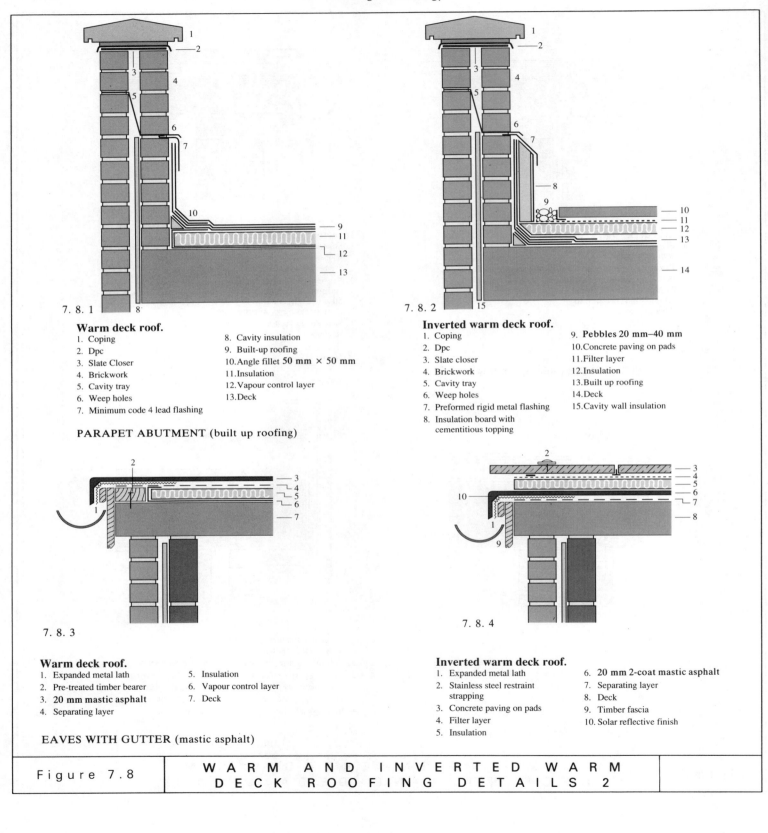

7.8.1

Warm deck roof.

1. Coping
2. Dpc
3. Slate Closer
4. Brickwork
5. Cavity tray
6. Weep holes
7. Minimum code 4 lead flashing
8. Cavity insulation
9. Built-up roofing
10. Angle fillet **50 mm × 50 mm**
11. Insulation
12. Vapour control layer
13. Deck

PARAPET ABUTMENT (built up roofing)

7.8.2

Inverted warm deck roof.

1. Coping
2. Dpc
3. Slate closer
4. Brickwork
5. Cavity tray
6. Weep holes
7. Preformed rigid metal flashing
8. Insulation board with cementitious topping
9. **Pebbles 20 mm–40 mm**
10. Concrete paving on pads
11. Filter layer
12. Insulation
13. Built up roofing
14. Deck
15. Cavity wall insulation

7.8.3

Warm deck roof.

1. Expanded metal lath
2. Pre-treated timber bearer
3. **20 mm mastic asphalt**
4. Separating layer
5. Insulation
6. Vapour control layer
7. Deck

EAVES WITH GUTTER (mastic asphalt)

7.8.4

Inverted warm deck roof.

1. Expanded metal lath
2. Stainless steel restraint strapping
3. Concrete paving on pads
4. Filter layer
5. Insulation
6. **20 mm 2-coat mastic asphalt**
7. Separating layer
8. Deck
9. Timber fascia
10. Solar reflective finish

Figure 7.8 | **WARM AND INVERTED WARM DECK ROOFING DETAILS 2**

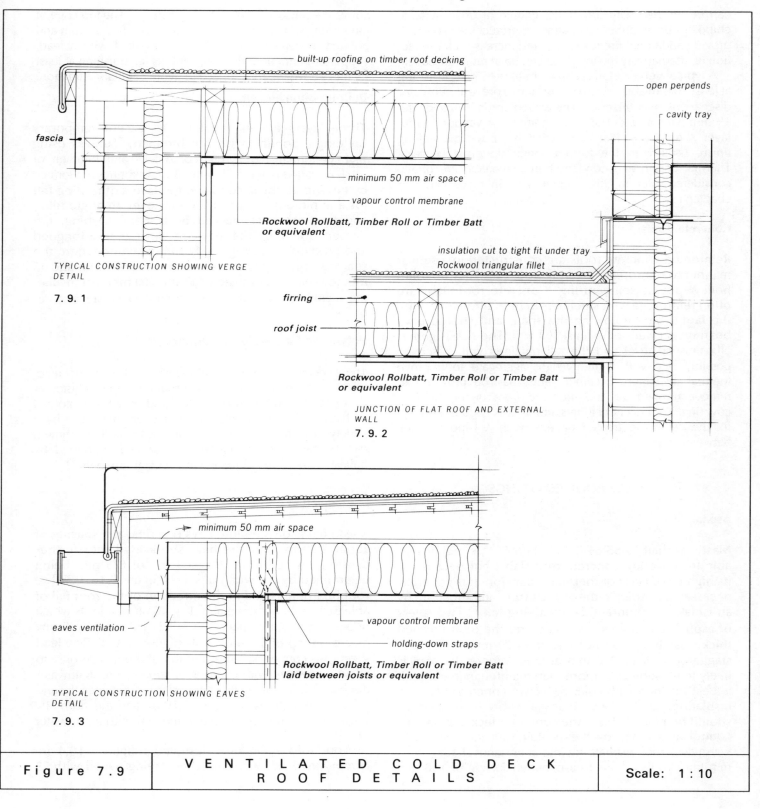

built-up roofing on timber roof decking

open perpends

cavity tray

fascia

minimum 50 mm air space

vapour control membrane

Rockwool Rollbatt, Timber Roll or Timber Batt or equivalent

TYPICAL CONSTRUCTION SHOWING VERGE DETAIL

7. 9. 1

insulation cut to tight fit under tray

Rockwool triangular fillet

firring

roof joist

Rockwool Rollbatt, Timber Roll or Timber Batt or equivalent

JUNCTION OF FLAT ROOF AND EXTERNAL WALL

7. 9. 2

minimum 50 mm air space

eaves ventilation

vapour control membrane

holding-down straps

Rockwool Rollbatt, Timber Roll or Timber Batt laid between joists or equivalent

TYPICAL CONSTRUCTION SHOWING EAVES DETAIL

7. 9. 3

| Figure 7.9 | VENTILATED COLD DECK ROOF DETAILS | Scale: 1 : 10 |

can be finished with limestone, granite or other suitable chippings on an adhesive coating, to protect the cap sheet, provide additional fire resistance and increase solar reflection or alternatively the top sheet can be of mineralised felt.

A typical verge detail is shown in figure 7.9.1, finishing at least 30 mm above the roof surface to prevent rainwater discharging over the edge. The welted apron is nailed with 19 mm nails at about 50 mm centres.[32] A vapour control layer should be placed immediately above the ceiling finish, usually in the form of high performance felts. Thermal design and condensation prevention are briefly considered later in this chapter, and in more detail in chapter 15.

Concrete Roofs

Reinforced concrete roofs are constructed in a similar manner to reinforced concrete floors and may be solid, hollow pot or self-centring. Concrete roof slabs are often reinforced with steel bars in both directions, with the larger bars following the span, and should have bearings on walls of at least 100 mm. The slab is generally finished level and the fall obtained with a screed, possibly one with a lightweight aggregate to improve thermal insulation. Breather vents may be provided to remove trapped air and moisture from under the roof covering. To prevent ceiling staining, it is good practice to fix the ceiling to treated battens, with a vapour check between them.

FLAT ROOF COVERINGS

Asphalt

Mastic asphalt to BS 6577[33] or BS 6925[34] is highly suitable for covering concrete roof slabs, but because of its high coefficient of thermal expansion, it is generally necessary to separate the asphalt from any substrate by an isolating membrane of sheathing felt.[35] Two layers of asphalt are always necessary and the total finished thickness should be not less than 20 mm with joints staggered at least 150 mm at laps. Where the roof is likely to be subject to more than maintenance traffic, it is best finished with solar reflective concrete tiles. An insulating membrane such as glass fibre or cork board should be placed above the concrete deck and vapour control layer and below the sheathing felt and asphalt in a warm deck roof. Asphalt skirtings at upstands should be 13 mm thick in two coats to a minimum height of 150 mm

above the finished level of the asphalt flat. The top edge of the skirting should be tucked into a chase 25 × 25 mm and pointed in cement mortar, and be masked with a lead, copper or aluminium flashing, with its tail finishing at least 75 mm above the roof surface to avoid capillary attraction.

Built-up Bitumen Felt

Three layers are essential for all except temporary buildings, bonded with hot bitumen,[35] unless using high performance polyester-cored felts and bitumen or pitch modified polymer. Upstands and skirtings are formed by turning up the second and top layers of roofing felt for a minimum height of 150 mm over an angle fillet, and they are best masked by a metal flashing. On timber roofs the base may consist of 25 mm tongued and grooved boarding, or 12 mm plywood where the spacing of roof joists does not exceed 400 mm and 15 mm for spacings between 400 and 600 mm. An insulating membrane and a vapour control layer are also required.

Polymeric Single-ply Membranes

These membranes are normally preformed sheet manufactured from synthetic polymers or rubbers, can be elastomeric or thermoplastic in character, and may be reinforced with a fabric core. Joints are formed on or off site by a variety of gluing or welding techniques,[57] and attachment can be by adhesive bonding, loose laid supported by ballast or mechanical fixing with special fasteners.[58]

Lead Sheeting

A lead flat is divided into bays by drips, at a spacing of about 2.50 m, and the bays subdivided into further divisions by rolls with the spacing of 675 mm being determined by the width of sheets (figure 7.10.3).[36] The roof boarding is laid in the direction of the roof fall of at least 1 in 80 but preferably 1 in 40 on roof joists which bridge the shortest span. The fall is obtained by firring pieces on top of the roof joists (figure 7.10.3). The lead sheets are copper nailed at the top and one side only to allow movement with changes of temperature. Rainwater discharges into a parallel gutter positioned between the end joist and parapet wall (figure 7.10.6), and this normally has an outlet through a hole in the wall into a rainwater head.

Wood rolls of the form illustrated in figure 7.10.4 are commonly used, with one sheet being nailed with 25 mm copper nails to the roll and the adjoining sheet passing over it with a 38 mm splash lap. A junction of a roll and drip are shown in figure 7.10.1. Drips are normally 50 mm deep and may incorporate an anti-capillary groove and a rebate into which the lower sheet is nailed (figure 7.10.1 and 7.10.5). At abutments the lead sheet is turned up the wall face 150 mm as an upstand and a 150 mm wide cover flashing passes over the top of the upstand to form a watertight joint (figure 7.10.2). The top edge of the cover flashing is tucked and wedged into a brick joint and lead tacks at 750 mm centres prevent the bottom edge of the flashing from curling.

Lead, which should conform to BS 1178,[37] is ductile and flexible, easily cut and shaped, highly resistant to corrosion and has a long life. It is however of low strength, high cost, very heavy in weight, creeps on all but the flattest slopes, can be attacked by damp cement mortar and is supplied in smallish sheets entailing many expensive joints.

Zinc Sheeting

Zinc may be used as a cheaper alternative to lead but has a shorter life and is not so well suited for use in heavy industrialised areas. Zinc is now alloyed with titanium or lead to give a better quality material for roofing purposes, complying with BS 6561.[38] It does not creep, is light in weight and reasonably ductile. The sheets are normally about 2500 × 1000 mm and should have a minimum thickness of 0.80 mm. Wood or batten rolls (figure 7.11.1) are used to provide joints running with the fall of the roof, at about 985 mm centres, where a zinc capping covers the turn-up of the sheets on either side of the roll, and these are secured by zinc clips at 750 mm centres.[39] Drips may be beaded (figure 7.11.3) or welted (figure 7.11.2) and are provided at about 2.35 m centres. Cover flashings (figure 7.11.4) are used at abutments as with lead roofs.

Copper Sheeting

Copper sheet and strip for flat roofs should conform to BS 2870[40] and a common thickness is 0.60 mm. A thin, stable insoluble film (patina) forms on the copper on exposure to air, consisting of a combination of copper oxide, sulphate, carbonate and chloride. The coating is green in colour and improves the appearance in addition to giving protection. Copper sheet is very tough and durable, readily cut and bent, is light in weight, does not creep, and resists corrosion reasonably well. On the other hand it is fairly costly, the smallish sheets involve many joints and electrolytic corrosion may take place if it comes into contact with metals other than lead.

The minimum fall for a copper roof is 1 in 60. Drips should be used on roofs of 5° slope or less, should be spaced at not more than 3 m centres and be 65 mm deep. A copper roof covering consists of a number of sheets joined along the edges, and held by clips inserted in the folds (figure 7.11.6). In the direction of the fall the joints are raised either as a standing seam (a double-welted joint formed between sides of adjacent bays and left standing approximately 20 mm) as shown in figure 7.11.7, or dressed to a wood roll with a separate welted cap (figure 7.11.5), at about 525 mm centres, the actual spacing varying with the thickness of the sheet. The choice of standing seams or rolls is influenced by the architectural treatment required, but where the roofs will be subject to foot traffic wood rolls are preferable. The joints across the fall are formed by means of flattened welted seams. Where the pitch is greater than 45°, a single lock welt seam is used (figure 7.11.9), and for flatter pitches, a double-lock cross welt is essential (figure 7.11.10). Welts on the side of a drip can be single lock (figure 7.11.8). Where sheets are jointed by double-lock cross welts between standing seams, the cross joints should be staggered to avoid welting too many thicknesses of copper into the standing seam (figure 7.11.6).[41]

Aluminium and Other Sheetings

The most readily available material is 1050 A grade (international nomenclature) and the thickness is often 0.80 mm. The method of fixing and jointing is similar to copper, using batten rolls or standing seams.[42] The strength of aluminium is increased by the addition of alloys, although they reduce its ductility. Aluminium roof flashings are normally laid in natural 'mill finish' condition, requiring no protective treatment and weathering to a matt silver grey. Another alternative is welded stainless steel flat roofing to BS 1449, Part 2[43], normally type 304S16 containing 18 per cent chromium and 10 per cent nickel alloy, and using roll cap or standing seam joints, giving a light, ductile finish with good resistance to chemicals. There is also a wide range of proprietary roof decks and coverings on the market.

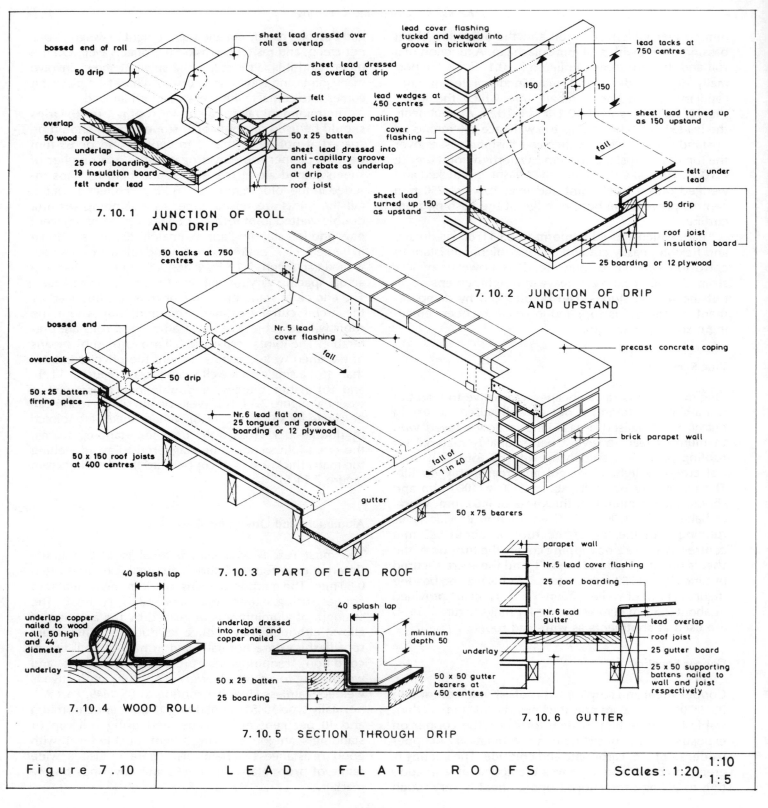

7.10.1 JUNCTION OF ROLL AND DRIP

bossed end of roll
50 drip
overlap
50 wood roll
underlap
25 roof boarding
19 insulation board
felt under lead
sheet lead dressed over roll as overlap
sheet lead dressed as overlap at drip
felt
close copper nailing
50 x 25 batten
sheet lead dressed into anti-capillary groove and rebate as underlap at drip
roof joist

7.10.2 JUNCTION OF DRIP AND UPSTAND

lead cover flashing tucked and wedged into groove in brickwork
lead tacks at 750 centres
lead wedges at 450 centres
150
150
sheet lead turned up as 150 upstand
cover flashing
fall
felt under lead
sheet lead turned up 150 as upstand
50 drip
roof joist
insulation board
25 boarding or 12 plywood

7.10.3 PART OF LEAD ROOF

50 tacks at 750 centres
bossed end
overcloak
50 x 25 batten firring piece
50 x 150 roof joists at 400 centres
50 drip
Nr. 5 lead cover flashing
fall
Nr.6 lead flat on 25 tongued and grooved boarding or 12 plywood
fall of 1 in 40
gutter
50 x 75 bearers
precast concrete coping
brick parapet wall

7.10.4 WOOD ROLL

40 splash lap
underlap copper nailed to wood roll, 50 high and 44 diameter
underlay

7.10.5 SECTION THROUGH DRIP

40 splash lap
underlap dressed into rebate and copper nailed
minimum depth 50
50 x 25 batten
25 boarding

7.10.6 GUTTER

parapet wall
Nr. 5 lead cover flashing
25 roof boarding
Nr.6 lead gutter
lead overlap
roof joist
underlay
50 x 50 gutter bearers at 450 centres
25 gutter board
25 x 50 supporting battens nailed to wall and joist respectively

| Figure 7.10 | LEAD FLAT ROOFS | Scales: 1:20, 1:10 1:5 |

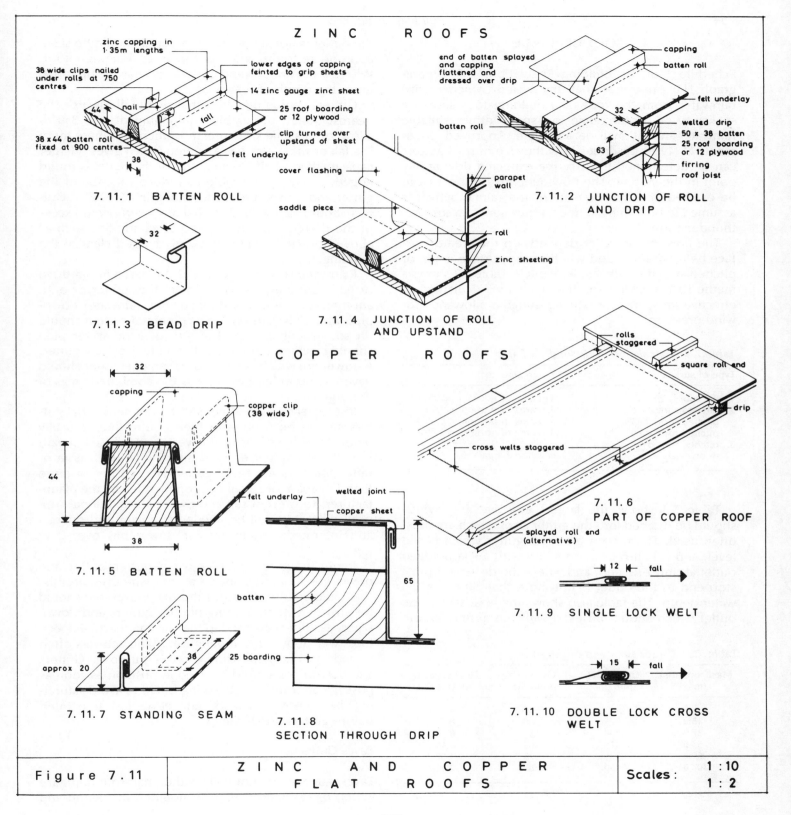

ZINC ROOFS

7.11.1 BATTEN ROLL

zinc capping in 1·35m lengths
38 wide clips nailed under rolls at 750 centres
lower edges of capping feinted to grip sheets
14 zinc gauge zinc sheet
25 roof boarding or 12 plywood
nail
fall
clip turned over upstand of sheet
38 x 44 batten roll fixed at 900 centres
felt underlay
44
38

7.11.2 JUNCTION OF ROLL AND DRIP

end of batten splayed and capping flattened and dressed over drip
capping
batten roll
batten roll
felt underlay
welted drip
50 x 38 batten
25 roof boarding or 12 plywood
firring
roof joist
32
63

7.11.3 BEAD DRIP

32

7.11.4 JUNCTION OF ROLL AND UPSTAND

cover flashing
saddle piece
parapet wall
roll
zinc sheeting

COPPER ROOFS

7.11.5 BATTEN ROLL

32
capping
copper clip (38 wide)
felt underlay
44
38

7.11.6 PART OF COPPER ROOF

rolls staggered
square roll end
drip
cross welts staggered
splayed roll end (alternative)

7.11.7 STANDING SEAM

approx 20
36

7.11.8 SECTION THROUGH DRIP

welted joint
copper sheet
batten
25 boarding
65

7.11.9 SINGLE LOCK WELT

12
fall

7.11.10 DOUBLE LOCK CROSS WELT

15
fall

Figure 7.11	ZINC AND COPPER FLAT ROOFS	Scales: 1:10 1:2

ROOF DRAINAGE

Schedule 1 of the Building Regulations 1991[1] (paragraph H3) prescribes that any system which carries rainwater from the roof of a building to a sewer, a soakaway, a watercourse or some other suitable rainwater outfall shall be adequate. Approved Document H3 (1989) describes how the size of roof gutters can be determined to carry the expected flow at any point in the system. The flow depends on the area to be drained and the intensity of the rainfall which is assumed to be 75 mm an hour; equivalent to a summer thunderstorm.[44]

The flow into a gutter depends on the area of surface being drained and whether the surface is flat or pitched and, if so, the angle of pitch. Table 7.2 shows a method of allowing for the pitch by calculating an effective area, incorporating a suitable allowance for wind pressure.

Table 7.2   Calculation of drained roof area

Type of surface	Design area (m^2)
1. flat roof	plan area of relevant portion
2. pitched roof at 30°	plan area of portion × 1.15
pitched roof at 45°	plan area of portion × 1.40
pitched roof at 60°	plan area of portion × 2.00
3. pitched roof over 70° or any wall	elevational area × 0.5

Table 7.3 shows the largest effective area which should be drained into the gutter sizes which are most often used. These sizes are for a gutter which is laid level, and is half-round in section with a sharp-edged outlet at only one end, and where the distance from a stop end to the outlet is not more than 50 times the water depth. The table also shows the least size of the outlet which should be used with the gutter. Where

Table 7.3   Gutter sizes and outlet sizes

Max. roof area (m^2)	Gutter size (mm dia.)	Outlet size (mm dia.)	Flow capacity (litres/s)
6.0	–	–	–
18.0	75	50	0.38
37.0	100	63	0.78
53.0	115	63	1.11
65.0	125	75	1.37
103.0	150	89	2.16

the outlet is not at the end, the gutter should be sized appropriate to the larger of the areas draining into it. Where there are two end outlets they may be up to 100 times the depth of flow apart.

Gutters should be laid with a slight fall towards the nearest outlet, preferably a minimum of 1 in 350 to allow for possible structural movement.[45] Where there is a fall or the gutter has a section which gives it larger capacity than a half-round gutter or the outlet is round edged, it may be possible to reduce the size of the gutter and downpipe, as described in BS 6367.[44] Gutters should also be laid so that any overflow in excess of the design capacity, caused by conditions such as above normal rainfall, will be discharged clear of the building.

Rainwater downpipes should discharge into a drain or gully but may discharge to another gutter or onto another surface if it is drained. Any rainwater downpipe which discharges into a combined system should do so through a trap. The size of a rainwater pipe should be at least the size of the outlet from the gutter. A downpipe which serves more than one gutter should have an area at least as large as the combined areas of the outlets.

BRE Defect Action Sheet 56[46] recommends that gutters should be fixed with their centre line vertically below the edge of the roof covering and close beneath it, with sarking felt dressed into the gutter. Fascia or rafter brackets for eaves gutters should be no more than 1 m apart, or closer if recommended by the manufacturer, to prevent sag and overspill. Additional support for gutters will be needed at angles and outlets, and intermediate support for downpipes over 2 m long.

Joints in gutters and between gutter outlet and downpipe should be sealed as and when required by the manufacturer. Gaps for thermal movement should be provided when jointing plastics gutters and downpipes, as gutters exposed to the sun can reach temperatures well above air temperature. Manufacturers often provide pre-formed strips for this purpose. Gutters and downpipes should be fixed in position as soon as possible after the roof covering is laid to avoid saturating the constructed work, and eaves fascias painted before gutters are installed.

Eaves Gutters

Eaves gutters are normally either half-round (figure 7.1.8), ogee (figure 7.5.1) or moulded in section and are made in a variety of materials – cast iron, asbestos cement, PVC-U, pressed steel, aluminium, and wrought copper and wrought zinc, together with the necessary fittings (stop ends, angles and outlets).

Cast iron gutters. These are supplied in 1830 mm lengths to BS 460[47] and in sizes varying from 75 to 150 mm, with a shallow socket at one end to receive the adjoining length. The joint is made with red lead and putty or mastic jointing compound and is secured by a small bolt. Half-round gutters are supported by brackets at 915 mm centres and these are screwed to the fascia or to the tops or sides of rafter feet, while ogee gutters are screwed through the back at 610 mm centres to the fascia.

Asbestos cement gutters. These are supplied in 1800 mm lengths to BS 569[48], and in sizes from 75 to 200 mm, screwed together and jointed either with a special jointing compound, or synthetic rubber pads and joint often fixed by galvanised mild steel brackets to feet of rafters or to fascia boards. They are however little used nowadays.

PVC-U gutters. These are made in unplasticised PVC (polyvinyl chloride) to BS 4576[49], are black, light grey and, in some cases, white in colour and are subject to movement and slight colour change in use. Their expected life is 20 years or more. They are made in a variety of sections, including square, in lengths varying from 2 to 4 m and in widths varying from 75 to 150 mm, and are jointed with a factory fitting sealing pad and usually supported by vinyl brackets. They are, however, vulnerable to damage with heavy snow falls.

Pressed steel galvanised gutters. These are made in light gauge to BS 1091[50] in sizes ranging from 75 to 150 mm in both half-round and ogee sections, in lengths of 910, 1220 and 1830 mm. Gutters are bolted and jointed with either red lead and putty or mastic jointing compound. Heavy pressed valley, box and half-round gutters are made for use in industrial buildings.

Aluminium gutters. These are made by casting, extrusion or pressing from sheet, in a variety of grades, sections and sizes to BS 2997.[51] They give good performance with most roof coverings, except copper, with very little maintenance. Aluminium gutters are jointed with bitumen butyl rubber or other plastic compound and supported every 1830 mm by aluminium or galvanised mild steel brackets.

Wrought copper and wrought zinc gutters. These are made to BS 1431[52] and require little maintenance and are light in weight. They are made in various sizes up to 125 mm and are supported by stays at 380 mm centres and brackets at 760 mm centres.

Precast concrete eaves gutters. These have a minimum internal diameter of 125 mm. The units are jointed with a mastic or a suitable mortar and they should be lined with a non-ferrous metal or other suitable protective coating. Other types of gutter include valley gutters and parapet gutters, which may be tapering (figure 7.3.4) or parallel (figure 7.10.6).

Downpipes

Downpipes convey rainwater from roof gutters to underground drains, often through a back entry rainwater gully at ground level. When used with projecting eaves they generally require a swan-neck consisting of a fitting with two bends to negotiate the soffit. Flat roof parapet gutters may discharge into rainwater heads at the top of downpipes.

Cast iron spigot and socket downpipes. These are made to BS 460[47] in 1830 mm lengths and diameters from 50 to 150 mm. Pipes are generally unjointed, but joints may be filled with red and white lead putty or an approved mastic. Fixing of pipes is normally performed by nails through ears and distance pieces into hardwood plugs built into walls. Cast iron pipes are usually painted or coated with composition.

Asbestos cement spigot and socket downpipes. These are made to BS 569[48] in diameters of 50 to 150 mm and lengths from 1.80 to 3.00 m. Joints are usually left unfilled but are otherwise jointed with cement and sand (1:2) or a jointing mastic. Pipes are fixed with galvanised mild steel ring clips to keep them 38 mm clear of the wall. These pipes are now little used.

PVC-U downpipes. Made in unplasticised PVC to BS 4576[49] in lengths ranging from 2 to 5.5 m, diameters of 50 to 100 mm and rectangular sections of 65 × 50 mm. No jointing material is necessary.

Pressed steel galvanised light gauge downpipes. These are made to BS 1091[50] are of limited durability and best confined to temporary work. They are made in lengths of 900, 1120 and 1830 mm and in nominal diameters from 50 to 100 mm. Joints are left loose and pipes are fixed by galvanised pipe nails through flat ears welded on the pipes and hardwood plugs, keeping them 38 mm clear of the wall.

Aluminium downpipes. Made to BS 2997[51] in a variety of shapes, thicknesses and sizes. Joints may be loose or caulked and fixing is by pipe nails through ears and driven into hardwood plugs.

Wrought copper and wrought zinc downpipes. These are made to BS 1431[52] in lengths of 1.8, 2.13 and 2.44 m and in diameters of 50 to 100 mm. Joints are made in telescoped form without any jointing compound. Pipes are fixed through ears and hardwood plugs, keeping them at least 38 mm from the wall.

Wire balloons of galvanised steel, aluminium or copper should be inserted in gutter outlets to prevent blockages occurring in downpipes.

THERMAL INSULATION

The Building Regulations 1991 and Approved Document L1 (1995)[1] require the thermal transmittance coefficient of the roof of a dwelling to be not more than 0.25 W/m^2 K, and further information on the requirements and application of Approved Document LI is given in chapter 15. Insulating materials used in flat roofs include the following

(1) Boards or slabs: rock fibre, glass fibre, fibre board/expanded polystyrene, cellular glass, Perlite, expanded or extruded (beaded) polystyrene, polyisocyanurate and polyurethane.

(2) Mats and quilts: glass fibre, rock fibre or other fibrous materials enclosed between sheets of waterproof paper.

(3) Loose fills: glass fibre, rock fibre, cellulose fibre, Perlite, polystyrene granules and beads.

(4) Aluminium foil: single or double-sided paper reinforced or combined corrugated and flat aluminium foil.

(5) Insulating screeds: made with lightweight aggregates such as vermiculite, expanded clay, foamed slag or sintered pulverised fuel ash or aerated or cellular concrete.

(6) Spray-applied polyurethane or polyisocyanurate foams. BS 5803[53] covers the requirements and application of the various types of insulant and guidance on the choice and use of insulation for flat roofs is given in BRE Digest 324.[54]

With pitched roofs, it is more economical to position the insulating membrane across the ceiling rather than in the plane of the rafters. Insulation fixed to rafters is generally in the form of boards or slabs, whilst that at ceiling level is usually a quilt laid over the ceiling joists or loose fill between them. Any tanks or pipes above the ceiling insulation will need insulating treatment, but leaving the bottom of the tank free of insulation.

With timber flat roofs, board or slab insulation may be provided between, under or over the roof joists. In the case of concrete flat roofs, insulation can take the form of a lightweight insulating screed or as permanent formwork of wood-wool slabs or fibre board under the concrete slab. In many cases it is necessary to provide a vapour control layer, preferably of high performance felt, on the warm side of the insulation to prevent water vapour condensing on the underside of the flat roof covering and causing saturation inside the roof structure.

CONDENSATION

In Approved Document F (1995) to the Building Regulations,[1] the mandatory requirement F2 was extended to all building types and to the roof as a whole. The guidance contained in Approved Document F2 aimed at limiting the amount of condensation in a roof or in the spaces above an insulated ceiling, so that the performance of the roof structure and of the thermal insulation will not be impaired by adequate ventilation of the roof space. This aspect is considered in more detail in chapter 15, which considers condensation in its wider context.

PRINCIPAL MATERIALS USED IN ROOFS

Clay Roofing Tiles

These are principally manufactured in districts producing clay bricks from well-weathered or well-prepared clay or marl. The process of manufacture is similar to that for bricks and they may be machine-made or hand-made, rough or smooth in texture, and even or mottled in colour. Tiles should be free from particles of lime or fire cracks, true in shape, dense, tough, show a clean fracture when broken and be well burnt throughout.[15]

Concrete Roofing Tiles

These are made from fine concrete with the addition of a colouring pigment and may have a textured surface. They have a dense structure which is highly resistant to lamination and frost damage. Concrete tiles are required by BS 473[19] to be true to shape and show a uniform structure on fracture.

Slates

Slates are a type of rock which can be split into very thin layers. They form an excellent roof covering material of good durability, impermeability and lightweight. BS 680[20] prescribes that they shall be of reasonably straight cleavage, ring true when struck, and the grain shall run longitudinally. Uniform length slates shall be to one of the sizes listed in the standard. The principal sources and colours of roofing slates in this country are as follows

Location	*Colours*
North Wales	Mainly blue, blue-purple, blue-grey and grey
South Wales	Green, silver grey and rustic
Cornwall	Grey and grey-green
Cumbria	Various shades of green; rough texture
North Lancashire	Soft blue-grey
Scotland (Argyllshire)	Blue

Asphalt

This is a mixture of bitumen and inert mineral matter, and may be lake asphalt from the West Indies or rock asphalt from central Europe. Most asphalt used in roofing is *mastic asphalt*, which is a synthetic substitute for natural asphalt, and is made from bitumen and fillers.

Bitumen Felts

These are covered by BS 747[55] which recognises four main categories of felt used in built-up roofing: class 1 – fibre base; class 2 – asbestos base; class 3 – glass fibre base; and class 5 felts with a polyester base. There are various types of felt in each category; for instance, fibre base felt is subdivided into

(3B) fine granule surfaced bitumen – coated on both sides and suitable for use as lower layers of built-up roofing;

(3E) mineral surfaced bitumen – finished with mineral granules on upper side and surfacing material on the other and suitable as external layer on sloping roofs; and

(3G) venting base layer bitumen – specially perforated, covered on lower side with mineral granules and on the other side with surfacing material and is suitable for use as a first layer when partial bonding and/or venting of the first layer of built-up felt roofing is required.

Readers requiring more detailed information on mastic asphalt and bitumen felts and their application to flat roofs are referred to *Flat Roofing*,[56] and Flat Roofing Design and Good Practice.[57]

Metals used in Roofing

Lead. A grey, soft, heavy, malleable non-ferrous metal, smelted from lead ores, impurities removed, heated and cast into sheets or pigs. Most lead used in building work is 'milled' by rolling into sheets of the required thickness. It is particularly useful for flashings, but its wider use is restricted by its high cost.

Zinc. A light grey non-ferrous metal, made by smelting zinc ores. It is relatively strong and ductile, but is not very resistant to polluted atmospheres. A more durable material can be obtained by using zinc alloyed with titanium or lead. Its use in roofing is restricted by its limited durability but it is used widely as a protective coating to other metals.

Copper. A reddish-brown, non-ferrous metal which is made by smelting copper ores. It soon obtains a greenish protective coating on exposure to the atmosphere, and is durable, tough and ductile, and does not creep on slopes.

Aluminium. A silvery white non-ferrous metal obtained from bauxite by electrical processes, used principally as an alloy. It is light in weight, resistant to corrosion, fairly soft and reasonably ductile. It is used for structural members, windows and doors, roof coverings and wall claddings, rainwater goods and for thermal insulation.

REFERENCES

1. *The Building Regulations 1991 and Approved Documents A(1991), B(1991), C(1991), F(1995), H(1989), J(1989) and L(1995)*. HMSO

2. *BS 6399* Loading for buildings, *Part 1: 1984* Code of practice for dead and imposed loads; *Part 3: 1988* Code of practice for imposed roof loads

3. *BRE Digest 332*: Loads on roofs from snow drifting against vertical obstructions and in valleys (1988)

4. *CP 3*: Code of basic data for the design of buildings; chapter V Loading: *Part 2: 1972 (1989)* Wind loads

5. *BRE Digest 346*: The assessment of roof loads (1989/1992); *Part 1*: Background and method; *Part 2*: Classification of structures; *Part 3*: Wind climate in the UK; *Part 4*: Terrain, climate and building factors and gust peak factors; *Part 5*: Assessment of wind speed over topography; *Part 6*: Loading coefficients for typical buildings; *Part 7*: Wind speeds for serviceability and fatigue assessments

6. *BS 476* Fire tests on building materials and structures, *Part 3: 1958/75* External fire exposure roof test

7. I.H. Seeley. *Building Economics*. Macmillan (1995)

8. *BS 5268* Structural use of timber, *Part 2: 1988* Code of practice for permissible stress design, materials and workmanship

9. *BS 1579: 1960* Connectors for timber

10. *BS 5268* Structural use of timber, *Part 3: 1985* Code of practice for trussed rafter roofs

11. *BRE Defect Action Sheet 5*: Pitched roofs; truss rafters – site storage (1982)

12. *BRE Defect Action Sheet 24*: Pitched roofs: trussed rafters bracings and binders – installation (1983)

13. *BRE Defect Action Sheet 43*: Trussed rafter roofs: tank supports – specification (1982)

14. *BRE Digest 380*: Damp-proof courses (1993)

15. *BS 402* Clay roofing tiles and fittings, *Part 1: 1990* Specification for plain tiles and fittings

16. *BS 5534* Slating and tiling, *Part 1: 1990* Design

17. *BS 8000* Workmanship on building sites, *Part 6: 1990* Code of practice for slating and tiling of roofs and claddings

18. *BRE Defect Action Sheet 10*: Pitched roofs: sarking felt underlay – watertightness (1982)

19. *BS 473* and *550: 1990* Concrete roofing tiles and fittings

20. *BS 680* Roofing slates, *Part 2: 1971* Metric units

21. W. Simpson. *Final Report of the Advisory Committee on Asbestos, Vols 1 and 2*, HMSO (1979)

22. *BRE Information Paper IP 1/91*. Durability of non-asbestos fibre-reinforced cement (1991)

23. PSA. *Defects in buildings*. HMSO (1989)

24. *BS 1470: 1987* Wrought aluminium and aluminium alloys for general engineering purposes: plate, sheet and strip

25. *CP 143* Sheet roof and wall coverings, *Part 1: 1958* Aluminium, corrugated and troughed

26. *BRE Digest 312*: Flat roof design: the technical options (1986)

27. *BS 6229: 1982* Code of practice for flat roofs with continously supported coverings

28. *BRE Digest 180*: Condensation in roofs (1986)

29. *BRE Information Paper 35/79*: Moisture in a timber-based flat roof of cold deck construction (1979)

30. Bituminous Roofing Council. *Information Sheet 1*: Flat roof design and construction: types of flat roof (1983)

31. DOE, Property Services Agency. *Flat roofs technical guide: Vol 1*. Design. HMSO (1981)

32. *CP 144* Roof coverings, *Part 3: 1970* Built-up bitumen felt. Metric units

33. *BS 6577: 1985* Specification for mastic asphalt for building (natural rock asphalt aggregate)

34. *BS 6925: 1988* Specification for mastic asphalt for building and engineering (limestone aggregate)

35. *BRE Digest 144*: Asphalt and built-up felt roofings: durability (1972)

36. *CP 143*: Sheet roof and wall coverings, *Part 11: 1970* Lead. Metric units

37. *BS 1178: 1982* Milled lead sheet for building purposes

38. *BS 6561: 1985* Zinc alloy sheet and strip for building

39. *CP 143*: Sheet roof and wall coverings, *Part 5: 1964* Zinc

40. *BS 2870: 1980* Specification for rolled copper and copper alloys: sheet, strip and foil

41. *CP 143*: Sheet roof and wall coverings, *Part 12: 1970* Copper. Metric units

42. *CP 143*: Sheet roof and wall coverings, *Part 15: 1973* Aluminium. Metric units

43. *BS 1449* Steel plate, sheet and strip, *Part 2: 1983* Specification for stainless and heat resisting steel plate, sheet and strip

44. *BS 6367: 1983* Code of practice for drainage of roofs and paved areas

45. *BRE Defect Action Sheet 55*: Roofs: eaves gutters and downpipes – specification (1984)

46. *BRE Defect Action Sheet 56*: Roofs: eaves gutters and downpipes – installation (1984)

47. *BS 460: 1964 (1981)* Cast iron rainwater goods

48. *BS 569: 1973 (1983)* Asbestos cement rainwater goods

49. *BS 4576* Unplasticised polyvinyl chloride (PVC-U) rainwater goods and accessories, *Part 1: 1989* Half-round gutters and pipes of circular cross section
50. *BS 1091: 1963* Pressed steel gutters, rainwater pipes, fittings and accessories
51. *BS 2997: 1958 (1971)* Aluminium rainwater goods
52. *BS 1431: 1960* Wrought copper and wrought zinc rainwater goods
53. *BS 5803*: Thermal insulation for use in pitched roof spaces in dwellings, *Part 1: 1985* Specification for man-made mineral fibre thermal insulation mats; *Part 2: 1985* Specification for man-made mineral fibre thermal insulation in pelleted or granular form for application by blowing; *Part 3: 1985* Specification for cellulose fibre thermal insulation for application by blowing; *Part 4: 1985* Methods for determining flammability and resistance to smouldering; *Part 5: 1985* Specification for installation of man-made mineral fibre and cellulose fibre insulation
54. *BRE Digest 324*: Flat roof design: thermal insulation (1987)
55. *BS 747: 1977* Roofing felts
56. Tarmac. *Flat roofing: a guide to good practice* (1982)
57. CIRIA and British Flat Roofing Council. *Flat Roofing Design and Good Practice* (1993)
58. *BRE Digest 372*: Flat roof design: waterproof membranes (1992)

8 WINDOWS AND GLAZING

This chapter examines the general principles of design of windows; the construction, detailing, fixing and uses of different types of window; glazing techniques; and the forms and uses of double glazing and double windows.

GENERAL PRINCIPLES OF WINDOW DESIGN

Functions of Windows

The primary functions of a window are to admit light and air into a room in a building. They frequently also provide occupants of the building with an outside view. A number of other factors deserve consideration when designing windows, such as thermal and sound insulation, avoidance of excessive sunglare and solar heat, security and safety.

Ventilation Requirements

Schedule 1 of the Building Regulations 1991[1] in paragraph F1 prescribes that there shall be adequate means of ventilation provided for people in the building.

Approved Document F1 (1995) to the Building Regulations shows how ventilation by natural means will meet the performance standards where habitable rooms, and sanitary accommodation each have one or more ventilation openings with a minimum total area of one-twentieth of the floor area of the room, and some part of the ventilation opening is at least 1.75 m above floor level. Habitable rooms also require minimum background ventilation of 8000 mm². An alternative for sanitary accommodation is mechanical extract at 6 litres per second in lieu of a ventilation opening.

There are two main types of proprietary ventilation to windows made from aluminium or galvanised steel, fitted into the head member or directly into a glazed area. Trickle ventilation is provided by fine-tuned ventilators to limit condensation risk and to provide adequate ventilation. The aim is to provide a 10 to 20 per cent increase in room ventilation without resultant excessive heat losses. All openable windows should desirably provide low trickle ventilation.

The wider aspects of ventilation of dwellings is covered in chapter 15, and this includes reference to figure 8.1.1.

Solar Heat and Daylight Admittance

Windows which admit sunlight also admit solar heat, and although heat from the sun is welcome in cool buildings, in excess it can make buildings uncomfortably hot in summer. There is thus an upper limit to the size of windows that can be used without thermal discomfort in sunny spells. Window design is therefore a compromise; if the window size is increased, the daylight illumination and the view through the window are improved but, beyond a certain size, overheating problems arise. The only visual effects limiting window size are glare from sun and sky and loss of privacy, but these can be overcome by fitting internal blinds usually of the venetian type. Although internal blinds can protect occupants from the direct heating effect of sunshine, they do not lower internal temperatures appreciably. External blinds are more effective but high cost and maintenance difficulties usually prevent their provision. In addition, large windows will result in excessive heat loss in cold weather.

BRE Digests 309 and 310[2] explain how the amount of daylight received in buildings is most conveniently expressed in terms of the percentage ratio of indoor-to-outdoor illumination, the ratio being called the *daylight factor*; the method of computation is detailed in the digests. For any given situation, the value of the factor varies with the sky conditions, the size, shape and position of the windows, the effect of any obstructions outside the windows and reflectivity of the external and internal surfaces. Its value can be determined at the design stage by measurement from a model of the building or, more frequently, by calculation from drawings and other data. The quality and intensity of daylight varies with latitude, season, time of day and local weather conditions. The value is generally based on a heavily overcast sky, although improved conditions are likely to exist for about 85 per cent of normal working time throughout the year.

Positioning and Subdivision of Windows

The glass line or sill level is often about 675 or 750 mm above floor level in living rooms to give maximum vision, possibly increasing to about 900 mm in bedrooms, where rather more privacy is usually required, and 1050 mm in the bathrooms and kitchens, where fittings are often located under the windows. The tops of windows should be fairly close to the ceiling to obtain good ventilation and lighting. Horizontal framing members (transoms) and glazing bars should not be positioned at heights where they will restrict the vision of occupants (eye level of persons standing; 1500 to 1600 mm and persons sitting in dining chairs: 1100 to 1150 mm above floor level).

It is now less common to subdivide windows into small panes with glazing bars, as they interfere with the vision of the occupants, make cleaning more difficult and increase painting costs. Some however favour their inclusion on aesthetic grounds, arguing that they give 'scale' to a small dwelling. Where window panes are used, the ratio of width to height should desirably be 2 : 3 to ensure good proportions. Leaded lights give character to a small dwelling but they are costly, break up vision, restrict light and make cleaning more difficult. Diamond-shaped leaded lights increase the risk of leakage.

Window Types

Windows may be classified in three different ways or a combination of them.

(1) The method of opening

(a) casements which are side hung, top hung, bottom hung, or tilt and turn;

(b) pivot hung either horizontal (reversible), vertical or coupled sash (dual sash);
(c) sliding, either vertical (double hung sash) or horizontal;
(d) miscellaneous: louvre or roof (basically a horizontal pivot).

(2) The materials from which they are made: steel; aluminium; timber; PVC-U.
(3) Size of window.

Each of the different types of window are now examined.

WOOD CASEMENT WINDOWS

With wood casement windows, a solid frame is fixed to the edges of the opening and this receives the glazed casements which may be side hung, top hung or bottom hung. Small top or bottom hung casements are used for controlled ventilation, whereas the larger side hung casements can be opened for greater ventilation in warmer weather. Side hung casements can create a safety hazard with young children and where located above ground floor level are more difficult to clean unless provided with 'easy clean' hinges. This type of window is best provided with weather stripping to resist wind and rain penetration. The window frame is sometimes subdivided; the vertical divisions are known as mullions and the horizontal members as transoms (figure 8.1.2). Casements may be subdivided into smaller areas by glazing bars. BS 644, Part 1[3] is no longer divided into different parts for the various window types, but adopts a performance-based format to include casements, projecting, pivoting, reversible tilt and turn and sliding sashes. Window sizes and other details are left to the manufacturer to provide in accordance with the performance requirements, which are to be assessed in accordance with BS 6375, Parts 1 and 2[4], which draws on the testing methods in BS 5368.[5] CP 153[6] details timber species and protective treatments. Side hung wood casements are generally made in widths of 450 to 650 mm and heights of 600 to 1400 mm, while fixed lights are usually made in widths varying from 450 to 1850 mm and heights from 200 to 1500 mm.

Figure 8.1.2 illustrates a typical two light wood casement window (the two refers to the number of lights in the width of the window). The window opening is

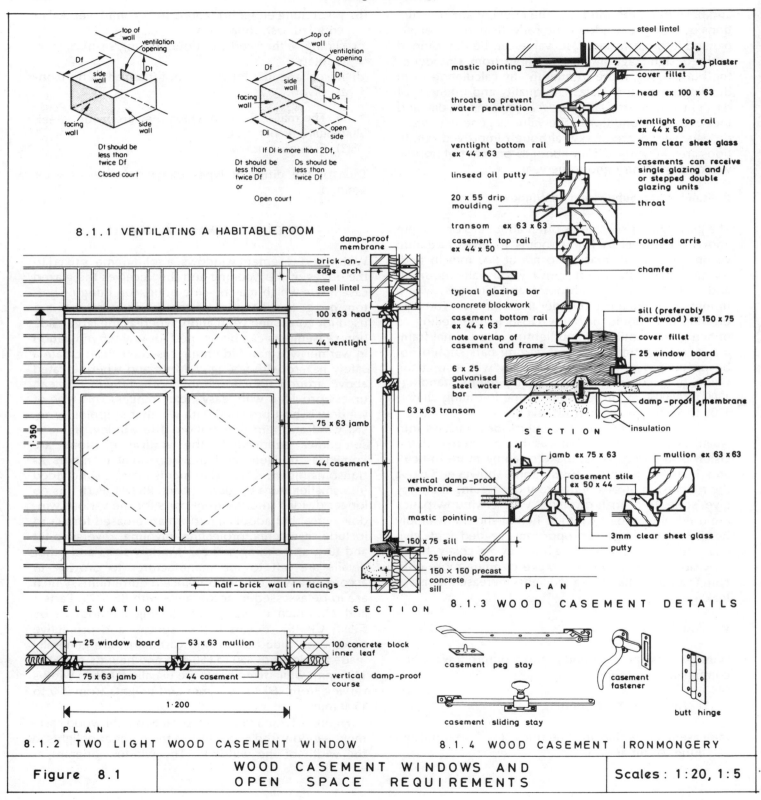

8.1.1 VENTILATING A HABITABLE ROOM

Dt should be less than twice Df

Closed court

If Dl is more than 2Df,

Dt should be less than twice Df

or

Ds should be less than twice Df

Open court

ELEVATION

SECTION

8.1.2 TWO LIGHT WOOD CASEMENT WINDOW

PLAN

- 25 window board
- 63 x 63 mullion
- 75 x 63 jamb
- 44 casement
- 100 concrete block inner leaf
- vertical damp-proof course
- 1·200

damp-proof membrane
brick-on-edge arch
steel lintel
100 x 63 head
44 ventlight
75 x 63 jamb
44 casement
63 x 63 transom
150 x 75 sill
25 window board
150 x 150 precast concrete sill
half-brick wall in facings
1·350

8.1.3 WOOD CASEMENT DETAILS

SECTION

steel lintel
mastic pointing
throats to prevent water penetration
ventlight bottom rail ex 44 x 63
linseed oil putty
20 x 55 drip moulding
transom   ex 63 x 63
casement top rail ex 44 x 50
typical glazing bar
casement bottom rail ex 44 x 63
note overlap of casement and frame
6 x 25 galvanised steel water bar
plaster
cover fillet
head ex 100 x 63
ventlight top rail ex 44 x 50
3mm clear sheet glass
casements can receive single glazing and/or stepped double glazing units
throat
rounded arris
chamfer
sill (preferably hardwood) ex 150 x 75
cover fillet
25 window board
damp-proof membrane
insulation

PLAN

vertical damp-proof membrane
mastic pointing
jamb ex 75 x 63
casement stile ex 50 x 44
mullion ex 63 x 63
3mm clear sheet glass
putty

8.1.4 WOOD CASEMENT IRONMONGERY

casement peg stay
casement sliding stay
casement fastener
butt hinge

| Figure 8.1 | WOOD CASEMENT WINDOWS AND OPEN SPACE REQUIREMENTS | Scales : 1:20, 1:5 |

spanned externally by a brick-on-edge arch backed by a reinforced concrete lintel. As the flat arch has little strength it is supported by a mild steel angle, often 75 × 75 × 6 mm, with ends built into the brickwork and the exposed edge painted for protection. Other variations would be to use a precast reinforced concrete boot lintel, a proprietary insulated steel lintel (figure 8.2.2) or a galvanised steel lintel (figure 8.2.1). The window frame may be fixed to the sides of brick jambs as in figures 8.1.2 and 8.1.3 or be set behind recessed jambs to give additional protection from the weather. With cavity walls, the frame is often set about 38 mm back from the outer wall face and so receives little protection from the weather. The joint between the frame and brick jambs should be sealed effectively with a suitable sealant. The wood frame or jamb may be fixed to the brickwork by screwing or nailing to hardwood plugs let into mortar joints or by right-angled galvanised steel cramps, with one leg screwed to the back of the frame and the other built into a mortar joint. The external sill in figure 8.2.1 consists of a wood sill, preferably of hardwood to improve its weathering qualities, which overhangs the face of the brickwork. Where the window is set well back from the outer wall face, it is necessary to incorporate a subsill below the main sill. The subsill could be made of timber, precast concrete, stone, bricks or a double course of roofing tiles.[7]

Large scale details of the various window members with common sizes are shown in figure 8.1.3. Casements are often 44 mm thick, and consist of top and bottom rails with stiles at each side. The corner joints of casements should be scribed and framed together with close-fitting combed joints having two tongues on each member, which are glued with durable glues and pinned together with at least one non-ferrous metal or sherardised or galvanised steel pin or 6 mm wood peg. Where casements occur below ventlights, drip mouldings should be housed, glued and pinned to the bottom rails of the ventlights (figure 8.1.3). Glazing bars are generally about 44 × 22 mm in size, rebated on both sides to receive glass, and intersecting joints are scribed, mortised and tenoned (figure 8.1.3). Anticapillary grooves are formed in the outer edges of casement members opposite corresponding grooves in the frame members, to prevent water penetration by capillary action. The top rails and stiles are usually 50 mm wide, but bottom rails are deeper for added strength and are invariably stepped over the sill to increase the weatherproofing qualities. The glazing rebate is often about 19 × 19 mm, and the glass is secured by putty or glazing beads.

Heads and jambs of frames vary in size between 115 × 75 mm and 75 × 63 mm, while sills are larger with a common range between 175 × 75 mm and 100 × 63 mm. The dimensions of the sill are influenced by the position of the window in relation to the external wall face and the existence or otherwise of some form of subsill. The upper exposed surface of a sill should be suitably weathered for speedy removal of rainwater and the underside of the sill should be throated (figure 8.1.3) so that water drips clear of the wall face to prevent staining. A second groove is usually provided to the underside of the sill to accommodate a metal water bar, which is usually bedded in red lead and will form a barrier to water penetration by capillary action. The external faces of heads and jambs are provided with bedding grooves to give a good key for mortar, and the head may project over the casement to give added protection (figure 8.1.3). All corner joints of these members should be scribed and framed together either with close-fitting combed joints or glued mortise and tenon joints. In the latter case heads and sills should project a minimum of 38 mm into the brickwork at either side of the opening. These projecting pieces are termed *horns*. Mullions should be through-tenoned into heads and sills, and transoms stub-tenoned into jambs and mullions. Alternatively, where transoms extend from jamb to jamb of multi-light windows, the transoms can be through-tenoned into the jambs and the mullions stub-tenoned into the transoms.

Opening casements not exceeding 1200 mm in height and all ventlights are each hung on a pair of butts, but 1½ pairs (three butts) are required for taller casements. The butts (figure 8.1.4) are usually of sherardised steel 50 or 63 mm long and each butt is screwed to the frame and casement with three 50 mm sherardised steel screws to each flap. Alternatively, easy-clean hinges of steel or aluminium alloy may be used to assist in the cleaning of external glass surfaces from inside the building. Opening casements are held in the closed position by casement fasteners (figure 8.1.4), while both casement and ventlights can be fixed in a number of open positions by means of casement stays, which may be of either the peg or sliding varieties (figure 8.1.4).

The arrangement of the component parts of the drawing in figure 8.1.2. deserves attention. The plan and section are drawn first in suitable positions whereby the elevation can be produced by extending up-

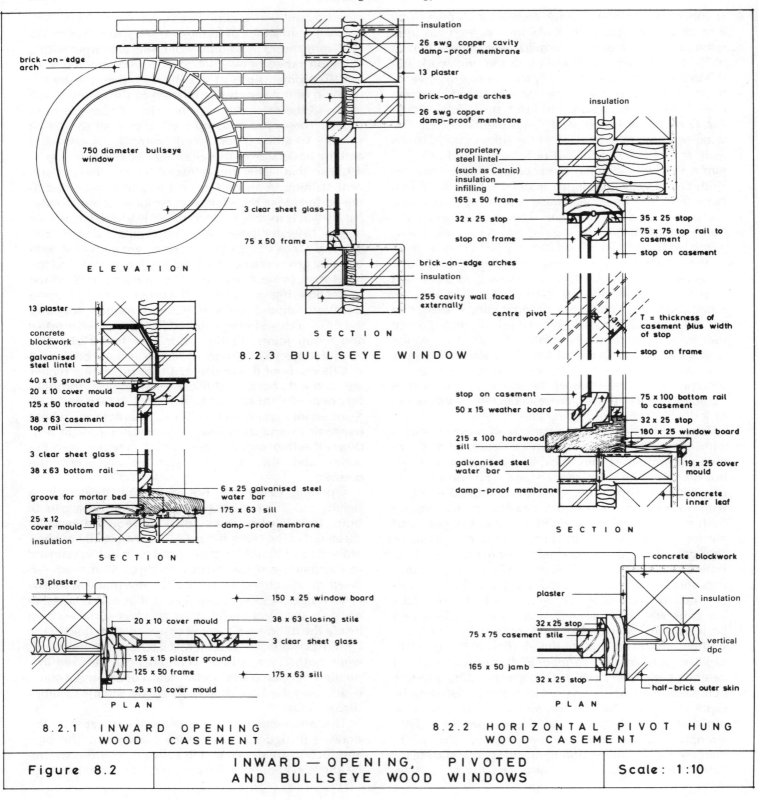

brick-on-edge arch

750 diameter bullseye window

E L E V A T I O N

insulation

26 swg copper cavity damp-proof membrane

13 plaster

brick-on-edge arches

26 swg copper damp-proof membrane

3 clear sheet glass

75 x 50 frame

brick-on-edge arches

insulation

255 cavity wall faced externally

S E C T I O N

8.2.3 BULLSEYE WINDOW

insulation

proprietary steel lintel (such as Catnic) insulation infilling

165 x 50 frame

32 x 25 stop

stop on frame

35 x 25 stop

75 x 75 top rail to casement

stop on casement

centre pivot

T = thickness of casement plus width of stop

stop on frame

stop on casement

50 x 15 weather board

215 x 100 hardwood sill

galvanised steel water bar

damp-proof membrane

75 x 100 bottom rail to casement

32 x 25 stop

180 x 25 window board

19 x 25 cover mould

concrete inner leaf

S E C T I O N

13 plaster

concrete blockwork

galvanised steel lintel

40 x 15 ground

20 x 10 cover mould

125 x 50 throated head

38 x 63 casement top rail

3 clear sheet glass

38 x 63 bottom rail

groove for mortar bed

25 x 12 cover mould

insulation

6 x 25 galvanised steel water bar

175 x 63 sill

damp-proof membrane

S E C T I O N

13 plaster

20 x 10 cover mould

125 x 15 plaster ground

125 x 50 frame

25 x 10 cover mould

150 x 25 window board

38 x 63 closing stile

3 clear sheet glass

175 x 63 sill

P L A N

8.2.1 INWARD OPENING
WOOD CASEMENT

concrete blockwork

plaster

insulation

32 x 25 stop

75 x 75 casement stile

165 x 50 jamb

32 x 25 stop

vertical dpc

half-brick outer skin

P L A N

8.2.2 HORIZONTAL PIVOT HUNG
WOOD CASEMENT

| Figure 8.2 | INWARD—OPENING, PIVOTED
AND BULLSEYE WOOD WINDOWS | Scale: 1:10 |

wards from the plan and across from the section, to reduce the amount of scaling to a minimum. All the component parts on plan and section should be fully described and dimensioned to provide well-detailed working drawings. All materials shown in section, whether horizontal or vertical, should be suitably hatched for ease of identification.

Timber windows are usually made of softwood, most commonly European Redwood. Hardwoods are used in many instances for high performance windows, although the use of tropical hardwoods is causing concern on environmental grounds. It is advisable to use hardwood sills with softwood casements.

Inward-opening Casements

Inward-opening casements are occasionally provided where windows abut verandahs, narrow passageways or public thoroughfares, where outward-opening casements could be dangerous. They have been used for dormers and similar situations to overcome the difficulty of cleaning the windows but a simpler solution is to use easy-clean hinges. It is difficult to make inward-opening casements watertight and they cause problems with curtains. Details of a typical inward-opening wood casement are shown in figure 8.2.1, incorporating closing or meeting stiles and the bottom rails are throated and rebated over a galvanised steel water bar let into the wood sill below. Alternatively, a wood weather fillet can be screwed to the sill.

Tilt and Turn Windows

These are windows introduced from Germany and are inward opening with a double action sash – bottom hung, inward opening (tilt) for draught free ventilation and side hung opening-in (turn) for cleaning. They are gaining popularity in the UK.

Pivot-hung Casements

Centre pivot-hung windows are quite popular for use with small windows to toilets, larders and the like, in positions high up from the floor and in upper storeys of buildings to make window cleaning easier. This type of window has a solid frame, without rebates, and the casement is pivoted to allow it to open with the top rail swinging inwards (figure 8.2.2). The pivots are fixed about 25 mm above the horizontal centre line of the

casement, so that it will be self-closing. Beads or stops are fixed to certain sections of frame to replace the rebate and to form a stop for the casement. As shown in figure 8.2.2 the external stop to the upper part is fixed to the frame and the lower part to the casement; this is reversed for internal stops where the internal stop to the upper part is fixed to the casement and the lower part to the frame. The stops must be cut in the correct positions to enable the casement to open and close freely. A convenient way to obtain the cut lines on the stops is to draw a circle around the pivot point having a radius equal to the thickness of the casement plus the width of one stop plus 3 mm. A weather board about 50 × 15 mm in size may be fitted to the bottom rail of the casement (figure 8.2.2) to direct rainwater away from the foot of the window and to help stiffen the stops fixed to the outside of the casement. Pivot-hung casements are generally hung with adjustable friction pivot hinges, with the initial opening controlled by a cranked roller armstay, and there is usually a four-point locking system. When fully reversed the casement can be locked by a catch and held for cleaning purposes. It is also possible for casements to be pivoted vertically. They are weatherstripped for higher thermal properties but draughts and water penetration can arise as the seals and draught stripping become worn.

Horizontal Sliding Casements

Casements can be fitted to slide horizontally in a similar manner to sliding doors. They can be fitted with rollers to the bottom rails running on a brass track fixed to the sill and with the top rail sliding in a guide channel incorporated in the frame. Cheaper methods incorporate fibre or plastic gliders and tracks. Recently developed sliding windows have made them suitable for use in conditions of severe exposure but they are relatively expensive.

French Casements

When casements extend to the floor so that they can be used as doors to give access to balconies, gardens and the like, they are termed *French casements*. French casements may be provided singly or in pairs and frequently open inwards. They may have sidelights added to give additional light to rooms and entrance halls. The bottom or kicking rail is normally not less

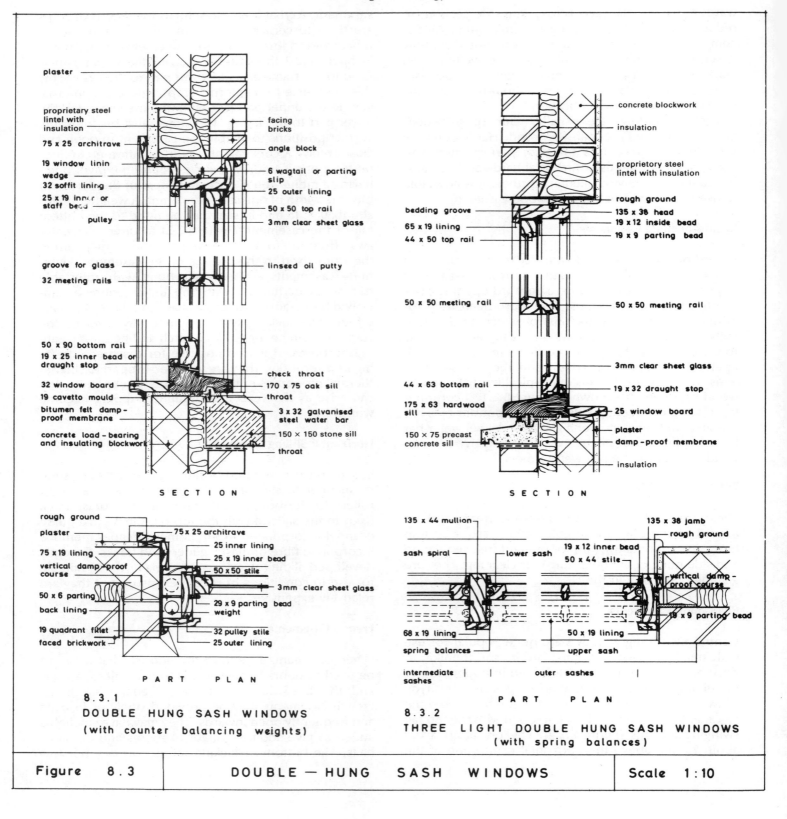

plaster

proprietary steel
lintel with
insulation

75 x 25 architrave

19 window linin
wedge
32 soffit lining
25 x 19 inner or
staff bead
pulley

facing
bricks

angle block

6 wagtail or parting
slip
25 outer lining
50 x 50 top rail
3 mm clear sheet glass

groove for glass
32 meeting rails

linseed oil putty

50 x 90 bottom rail
19 x 25 inner bead or
draught stop
32 window board
19 cavetto mould
bitumen felt damp-
proof membrane
concrete load-bearing
and insulating blockwork

check throat
170 x 75 oak sill
throat
3 x 32 galvanised
steel water bar
150 × 150 stone sill
throat

SECTION

rough ground
plaster
75 x 19 lining
vertical damp-proof
course
50 x 6 parting
back lining
19 quadrant fillet
faced brickwork

75 x 25 architrave
25 inner lining
25 x 19 inner bead
50 x 50 stile
3mm clear sheet glass
29 x 9 parting bead
weight
32 pulley stile
25 outer lining

PART   PLAN

8.3.1
DOUBLE HUNG SASH WINDOWS
(with counter balancing weights)

concrete blockwork
insulation
proprietory steel
lintel with insulation
rough ground
135 x 38 head
19 x 12 inside bead
19 x 9 parting bead

bedding groove
65 x 19 lining
44 x 50 top rail

50 x 50 meeting rail
50 x 50 meeting rail

3mm clear sheet glass
44 x 63 bottom rail
19 x 32 draught stop
175 x 63 hardwood
sill
25 window board
plaster
150 × 75 precast
concrete sill
damp-proof membrane
insulation

SECTION

135 x 44 mullion
sash spiral
lower sash
68 x 19 lining
spring balances
intermediate
sashes

19 x 12 inner bead
50 x 44 stile
135 x 38 jamb
rough ground
vertical damp-
proof course
50 x 19 lining
upper sash
19 x 9 parting bead
outer sashes

PART   PLAN

8.3.2
THREE LIGHT DOUBLE HUNG SASH WINDOWS
(with spring balances)

| Figure   8.3 | DOUBLE — HUNG   SASH   WINDOWS | Scale   1:10 |

than 200 mm wide and the glass should be toughened glass fixed with beads to reduce the risk of breakage and danger to young children. PVC-U patio doors are described in chapter 9.

Bullseye Windows

Circular or bullseye windows are sometimes used as features to give added interest to an elevation. A typical bullseye window is illustrated in figure 8.2.3 in the form of a fixed light with the glass bedded into the frame. The opening is formed with brick-on-edge arches with a copper vertical damp-proof course sealing the cavity. An additional damp-proof membrane is inserted above the window to disperse water in the cavity at this point. The internal reveals to bullseye windows are usually plastered and decorated.

Louvre Windows

These consist of a number of horizontal glazed units fixed at their ends so as to move from a vertical plane when closed to a diagonal or horizontal plane when open. They can provide up to 90 per cent of open area with little projection from the vertical plane, are safe for children, relatively easy to clean from the inside and recent developments have resulted in improved security and rain and air resistance.

DOUBLE-HUNG SASH WINDOWS

Sash windows are those in which the sashes slide up and down and they normally consist of two sashes, placed one above the other. Where both sashes open they are termed *double-hung sash windows* and if the top sash only opens it is known as a *single-hung sash window*. BS 644, Part 2[3] describes how the sashes may either be hung from cased frames with counterbalancing weights, usually about 50 mm diameter of iron (figure 8.3.1), or from solid frames with spring devices (figure 8.3.2). With cased frames, two-light windows have boxed centre mullions and three-light windows have fixed side lights and solid mullions, with the cords passing over the sidelights. With solid frames, three-light windows normally have fixed sidelights, only the centre sashes being made to open.

The cased frame to accommodate the weights consists of an inner and outer lining, pulley stile and back lining. The parting bead between the two sashes is housed into the pulley stile. The construction of the head is often similar to that of the sides, with the omission of the back lining. Figure 8.3.1 shows how the sashes can slide past one another, the lower sash being placed on the inside. The outer lining, parting bead and inner or staff bead form the recesses in which the sashes slide. The sashes are hung to cords or chains which pass over a pulley, often 44 mm diameter in cast iron with steel axles and brass bushes, in the pulley stiles and to which the weights are attached.

To allow the removal of the sashes for adjustment or the renewal of broken cords, the inner lining is stopped and a removable bead nailed to it. When the inside bead is removed the sash can be swung inwards. Similarly, by removing the parting bead the top sash can be taken out. A pocket is formed in the pulley stile to provide access to the weights. To prevent the weights colliding or the cords becoming entangled, a parting slip is inserted in the cased frame, and is suspended from the soffit lining by a wedge passing through it. An inner and removable bead is fixed all round the frame, and a deeper bead (draught stop) is often fixed to the sill (figure 8.3.2). This permits some ventilation to be obtained between the meeting rails without having an opening at the bottom of the window. The meeting rails comprise the top rail of the lower sash and the bottom rail of the upper sash and they are kept as shallow as possible to cause the least possible obstruction. They are thicker than the other sash members to accommodate the space occupied by the parting bead, and are splayed and rebated on their adjoining faces to fit tightly together when the window is closed. A rebated splay drawn in the wrong direction would keep the sashes permanently closed. The lower sash is normally provided with sash lifts on the bottom rail for raising, and a sash fastener on the meeting rails provides a means of fastening the sashes in the closed position.

A number of balancing devices are now available to dispense with cords and expensive cased or boxed frames. Often spring balances with a metal case receive a rustless steel tape which winds onto a revolving drum leaving its free end to be fixed to a sash. The required degree of balance is obtained by adjusting the spring drum. Another form of spring balance consists of a torsion spring and helical rod enclosed in a metal tube, with the rod passing through a nylon bush which causes the spring to wind or unwind as the sash moves. The barrel of the balance is housed in either the sash or frame.

Sash windows vary between 350 and 1100 mm in width and 900 and 1800 mm in height and, for stability, the width of each sash should not exceed twice its height. Panes in sash windows are rarely less than 225 mm wide × 300 mm high. The dimensions and rather formal character of sash windows generally make them unsuitable for small dwellings, although they are often considered to be superior to casements both on aesthetic and functional grounds, through better control of ventilation. One major disadvantage of sash windows in the past has been high maintenance costs stemming from the replacement of broken sash cords, but the use of chains, albeit noisy, and the more recently introduced spring balances have overcome this problem. The construction is rather complex making them more expensive than other types of wood window.

DEFECTS IN WOOD WINDOWS

Building Research Establishment Digest 304[8] describes how window joinery, particularly in houses built in the nineteen-sixties and early nineteen-seventies, often gave cause for complaint and problems are still extensive. Decay is particularly marked in ground floor windows, especially in kitchens and bathrooms, and the lower parts of the windows (sills, bottoms of jambs and mullions, and lower rails of opening lights) are most vulnerable. Decay is generally of the 'wet rot' variety resulting from the use of timber of low natural resistance. Hence it is advisable to either use well seasoned heartwood from timber, which is naturally resistant to decay, or timber treated with preservative, followed in both cases by the application of a suitable primer under favourable conditions. Surfaces should be designed to shed rainwater satisfactorily on the outside and condensation internally. The use of flimsy sections in window joinery, particularly where weather conditions are severe, should be avoided. The sealing of end grain at joints is especially important.

Suitable timbers for joinery are specified in BS 1186[9] which lists six softwoods and 27 hardwoods. However, joinery for domestic buildings is made mainly of softwood, especially European redwood (*Pinus sylvestris*) as it machines to a good finish, has heartwood with moderate fungal resistance and is easily treated with preservative, but unfortunately has low decay resistance. Furthermore, specifying hardwood does not necessarily imply the use of a timber having an accept-

able level of durability for outdoor use. Ideally all timbers should be specified by name, such as Douglas fir (softwood) or Iroko (hardwood), and sapwood to be excluded if it is a durable species to be used without preservative treatment. Wood windows can be painted, stained or finished with a plastics skin, 1 to 1.5 mm thick.

Fortunately much of the external joinery used in the nineteen-eighties and nineties is treated with preservative. Suitable methods are pressure impregnation with waterborne preservative to BS 4072[10] or double vacuum treatment with an organic solvent to BS 5707,[11] and the latter method is particularly suitable for the treatment of European redwood. Excellent guidance is given on preservative treatment in BS 5589.[12] The decay of softwood windows made of fast grown redwood, often comprising mainly sapwood and having a high moisture content, has caused serious problems in numerous modern dwellings, particularly in the bottom rails and bottom ends of stiles to wood casements. All decayed timber must be removed and the adjoining timber effectively treated, such as by using boron-based rods or tablets as described in BRE Information Paper IP 14/91,[13] prior to filling and painting. In cases where the decay has progressed so far that repair of the window is no longer feasible, then the window will have to be replaced, following the guidance notes on installation given in BS 8213, Part 4.[14]

Joinery is frequently left unprotected on building sites prior to installation, creating an additional hazard as damp timber can support the growth of mould fungi capable of the later disruption of paint films. All timber should be stored internally and stacked clear of the ground. Advice on good practice for the protection of windows on site is given in CP 153[6] and BRE Defect Action Sheet 11.[15] The backs of window frames should be liberally treated with priming paint before fixing against damp masonry, and any cut surfaces should be given two good brush applications of preservative. High rates of condensation in bathrooms and kitchens can also cause decay of timber, often because of defective back putty.

METAL WINDOWS

Steel Windows

Steel windows are fabricated from hot-rolled steel sections, mitred and welded at the corners, while subdividing bars are hot-tenon riveted to the frames and each other. Steel windows are hot-dip galvanised to BS 729[16] to resist corrosion, and some manufacturers supply windows with decorative coatings such as polyester applied over the galvanising. Stainless steel windows should be made from chromium nickel–molybdenum stainless steels. Windows may be coupled together by the use of mullions (vertical coupling bars) and transoms (horizontal coupling bars) as shown in figures 8.4.4 and 8.4.5. Steel windows for domestic and similar buildings normally meet the requirements of BS 6510[17] where windows can be fixed, side, top and bottom hung or be horizontally or vertically pivoted, including reversible pivoted casements, and can vary in width from 600 to 1800 mm and in height from 1300 to 2400 mm. All steel windows are supplied complete with fittings (hinges, handles and stays), but the design and quality of fittings may vary between manufacturers. Handles and stays are usually of brass, zinc-based or aluminium alloy to various finishes. Optional accessories include safety devices, ventilators, remote controls, insect screens and pressed metal sills. Glazing should be carried out in accordance with BS 6262[18] using metal casement putty, and spring wire glazing clips or beads.

Steel windows are fixed to brick walls with metal or wire lugs (figure 8.4.3), to concrete with fibre plugs or screwed to timber surrounds (figure 8.4.2). The use of timber surrounds (figure 8.4.1) as described in BS 1285[19] improves the appearance of steel windows considerably and reduces condensation staining, although they do increase the cost. It is necessary to point with a suitable sealant the gap between the metal window and the adjoining masonry or wood surround, and between wood surround and masonry to ensure a watertight joint in these vulnerable positions. Figure 8.4.1 also illustrates the use of an external tile subsill and an internal quarry tile sill. Steel windows with their lighter and thinner sections cause less obstruction than wood windows and they are not subject to warping, rot or insect attack. They do however need careful handling on the site, ample protection from corrosion by a good paint film and must not be subject to loads. Steel sections are accurately rolled to provide a close fit between members, but additional draught-proofing or weatherstripping can be obtained by inserting a neoprene weatherstrip in a groove inside the opening frame.

Aluminium Windows

Aluminium windows are supplied in a variety of forms: fixed; bottom, side and top-hung casements; horizontally and vertically pivoted, including reversible pivoted windows; horizontal and vertical sliding windows; and horizontally adjustable louvred ventilating windows – all should comply with BS 4873.[20] The windows are fabricated from aluminium alloy extrusions,[20] and are provided with a standard range of hardware. The structural members can be single web or tubular sections which are mechanically jointed, cleated or welded. Glass may be inserted in glazing gaskets or non-setting compound with clip-on or screw-fixed glazing beads. The glass louvres in ventilating windows fit into blade holders of polypropylene or aluminium. Finishes to aluminium windows include mill finish, anodising often to a bronze finish, organic (stoved acrylic and polyester) and stainless steel clad. Top and bottom-hung casements vary from 400 to 2400 mm in width and 400 to 1800 mm in height, whereas horizontal sliding windows have widths varying from 900 to 3000 mm and heights from 500 to 1500 mm. Non-standard sizes are produced readily. Casement, sliding and pivoted windows can be coupled with mullion and transom bars to form co-ordinated assemblies. Aluminium windows need careful handling to avoid scratching surface coatings or bending members. Aluminium windows have slender sections, are attractive and durable but require regular cleaning; they may be double the initial cost of similar type wood and steel windows. Maintenance treatments are described in CP 153.[6] Double aluminium windows separated by 75 to 200 mm can give good sound reduction up to 40 dB.

Figure 8.5 illustrates a sophisticated aluminium clad timber framed, triple glazed Modul coupled sash window supplied by Sampson Windows of Needham Market, Suffolk. It probably offers the ultimate in window design, providing the warmth, visual appeal and structural rigidity of high grade Swedish redwood, slow grown and matured for over 100 years, on the inside, and the strength, durability and clean appearance of aluminium on the outside. The external face of the aluminium can either be coated with polyester powder or be colour anodised to give an attractive finish. The triple glazing provides a very high standard of thermal and acoustic insulation, which can be further supplemented by solar control blinds located in the 55 to 85 mm wide ventilated air gap.

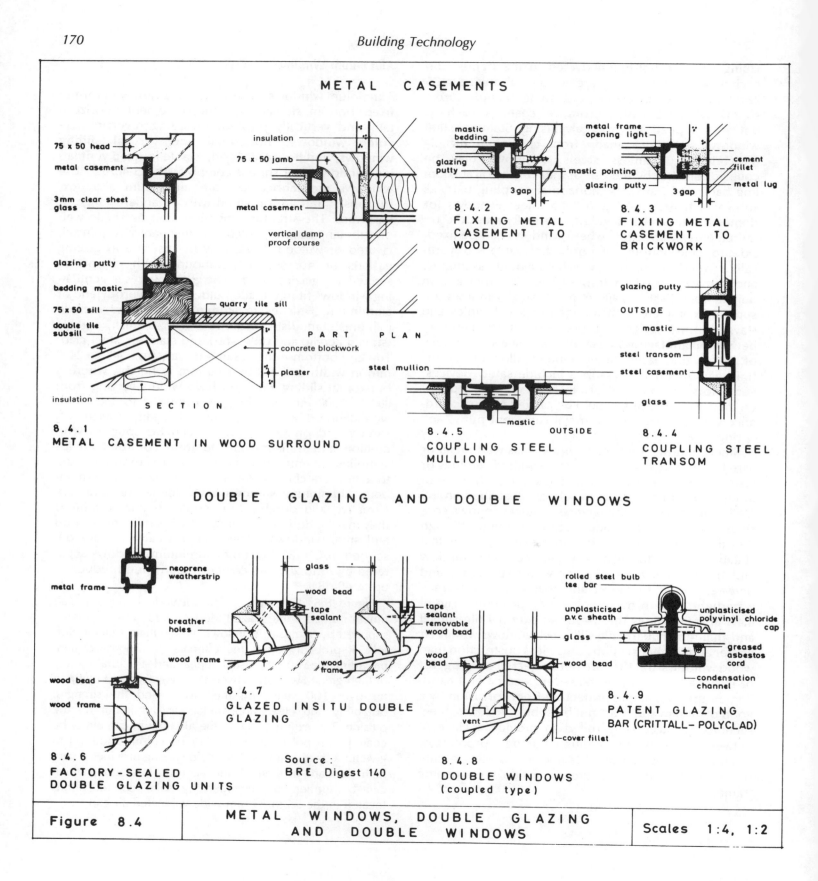

METAL CASEMENTS

75 x 50 head
metal casement
3mm clear sheet glass
glazing putty
bedding mastic
75 x 50 sill
double tile subsill
insulation
SECTION

8.4.1
METAL CASEMENT IN WOOD SURROUND

insulation
75 x 50 jamb
metal casement
vertical damp proof course
quarry tile sill
concrete blockwork
plaster
PART PLAN

mastic bedding
glazing putty
3 gap

8.4.2
FIXING METAL CASEMENT TO WOOD

metal frame opening light
mastic pointing
glazing putty
3 gap
cement fillet
metal lug

8.4.3
FIXING METAL CASEMENT TO BRICKWORK

glazing putty
OUTSIDE
mastic
steel transom
steel casement
glass

8.4.4
COUPLING STEEL TRANSOM

steel mullion
mastic
OUTSIDE

8.4.5
COUPLING STEEL MULLION

DOUBLE GLAZING AND DOUBLE WINDOWS

metal frame
neoprene weatherstrip
wood bead
wood frame

8.4.6
FACTORY-SEALED DOUBLE GLAZING UNITS

glass
wood bead
tape sealant
breather holes
wood frame
tape sealant
removable wood bead
wood frame

8.4.7
GLAZED INSITU DOUBLE GLAZING

Source:
BRE Digest 140

glass
wood bead
wood bead
vent
cover fillet

8.4.8
DOUBLE WINDOWS (coupled type)

rolled steel bulb tee bar
unplasticised p.v.c sheath
glass
unplasticised polyvinyl chloride cap
greased asbestos cord
condensation channel

8.4.9
PATENT GLAZING BAR (CRITTALL— POLYCLAD)

| Figure 8.4 | METAL WINDOWS, DOUBLE GLAZING AND DOUBLE WINDOWS | Scales 1:4, 1:2 |

This method of construction originated in Scandinavia to provide the high standard of thermal performance required in conditions of extreme cold. The inner sash is fitted with a sealed glazing unit and the outer sash with a single pane. Both sashes open as one but can be uncoupled if required. Concealed multi-point espagnolette locking permits the window to be locked at a number of points from a single operating handle, providing improved airtightness, weathertightness, sound insulation, safety and security, which is becoming so important nowadays with the increasing crime rate.

Replaceable composite EPDM weatherstrips, dustproof gaskets, brush seals and silicone capping are all used in various combinations and then tested against water penetration and air infiltration. Modul windows can be supplied to meet *U*-values down to as low as 1.2 W/m²K. Concealed slot ventilators as illustrated in figure 8.5 can be fitted to window frame heads to provide a controlled airflow when the windows are in the closed position.

Plastics Windows

Plastics (PVC-U) windows to BS 7413 are self-finished and available in various colours, of which white is the most common and of proven performance, and they can be readily fabricated to any required size and, where necessary, are reinforced with a galvanised mild steel core. They should not require painting for 15 to 20 years, but need regular cleaning and have an anticipated life expectancy of 30 to 40 years. Painting may impair the impact resistance of PVC-U.[21] Side hung plastics casements vary in width from 575 to 1200 mm and in height from 800 to 1900 mm. They are normally made to order and non-standard sizes are produced readily. In the long term PVC-U windows are proving to be the most economical proposition showing a probable 30 per cent saving in cost compared with wood windows over a 25 year period, and they are becoming increasingly popular.

Figure 8.6 shows the main details of a Schüco casement window made from high impact modified PVC-U to DIN 7748, with a wall thickness of up to 3.8 mm and multi-chamber construction for strength and insulation. Glazing can be double glazing up to 45 mm thick or single glazing with a condensation channel, with the glass secured by snap-in glazing beads and long life EPDM gaskets. Drainage is provided on either face or concealed through chambers which are isolated from the reinforcing. The reinforcing is normally galvanised steel as shown in the sill in figure 8.6.3, as it is three times as strong as aluminium. Joints are mainly welded and weatherseals are of EPDM.

Colours are white, brown, mahogany, and light or dark oak. PVC-U windows can incorporate various forms of ventilation, including trickle ventilators fitted into the outer frame. For outward opening windows there is a choice between cockspur or espagnolette locking, with or without locking handles. A selection of different window styles is available.

A cross-sectional detail of an opening Sheerframe high impact PVC-U window supplied by L B Plastics Ltd of Nether Heague, Derby is illustrated in figure 8.7. It shows an outward opening casement with double glazing, although inward opening casements and tilt and turn windows are also available, and glazing can be either single, double or triple. The triple chamber system provides large insulating cavities for improved thermal insulation. Glazing beads can be fitted internally or externally with adequate regard to security and aesthetics. All casement profiles are designed with Eurogrooves for the ease of location of espagnolettes and friction hinges. Both aluminium and steel are available as optional materials for reinforcement and these metals are sealed from water and air in the centre profile chambers as shown in figure 8.7. The colour options are white, brown and woodgrain.

GLASS AND GLAZING

Glass

Glass is one of our oldest materials but through modern research work it has been possible to alter the properties of glass to make it a more versatile material. The constituent materials are normally sand, soda ash, limestone, dolomite, felspar, sodium sulphate and cullet (broken glass), which are mixed, melted and refined; the glass is then either drawn, cast or rolled, annealed, possibly polished, and cut to the required sizes. BS 952[22] recognises and describes a number of different types of glass, and should be read in conjunction with BS 6262.[18]

Transparent glasses. These transmit light and permit clear vision through them; they include sheet glass and clear float sheet or polished plate glass. *Sheet glass* has natural fire-finished surfaces and as the two surfaces are never perfectly flat and parallel, there is always some distortion of vision and reflection. It varies in thickness from 2 to 6 mm, can be coloured and is supplied in four qualities:

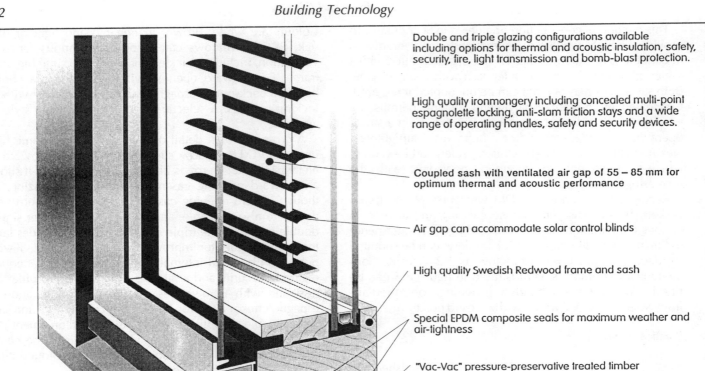

Double and triple glazing configurations available including options for thermal and acoustic insulation, safety, security, fire, light transmission and bomb-blast protection.

High quality ironmongery including concealed multi-point espagnolette locking, anti-slam friction stays and a wide range of operating handles, safety and security devices.

Coupled sash with ventilated air gap of 55 – 85 mm for optimum thermal and acoustic performance

Air gap can accommodate solar control blinds

High quality Swedish Redwood frame and sash

Special EPDM composite seals for maximum weather and air-tightness

"Vac-Vac" pressure-preservative treated timber

Frame depth of 100 mm, 105 mm or 118 mm giving excellent thermal properties with good cavity closing characteristics and easy fixing conditions in refurbishment situations

Heavy duty timber frame with pinned and bonded mortice and tenon joints

Outer profiles polyester powder coated or colour anodised

Aluminium exterior cladding for long-term durability and appearance

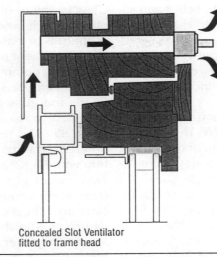

Concealed Slot Ventilator fitted to frame head

| Figure 8.5 | M O D U L   A L U M I N I U M / T I M B E R
C O U P L E D   S A S H   W I N D O W | |

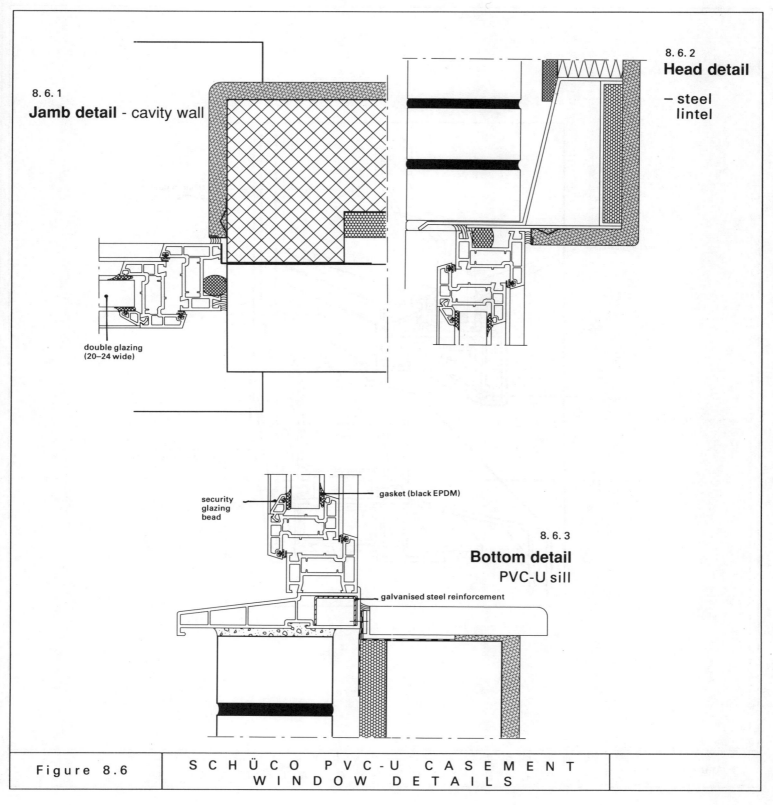

8. 6. 1
Jamb detail - cavity wall

double glazing
(20–24 wide)

8. 6. 2
Head detail

– steel
lintel

security
glazing
bead

gasket (black EPDM)

8. 6. 3
Bottom detail
PVC-U sill

galvanised steel reinforcement

Figure 8.6   S C H Ü C O   P V C - U   C A S E M E N T
W I N D O W   D E T A I L S

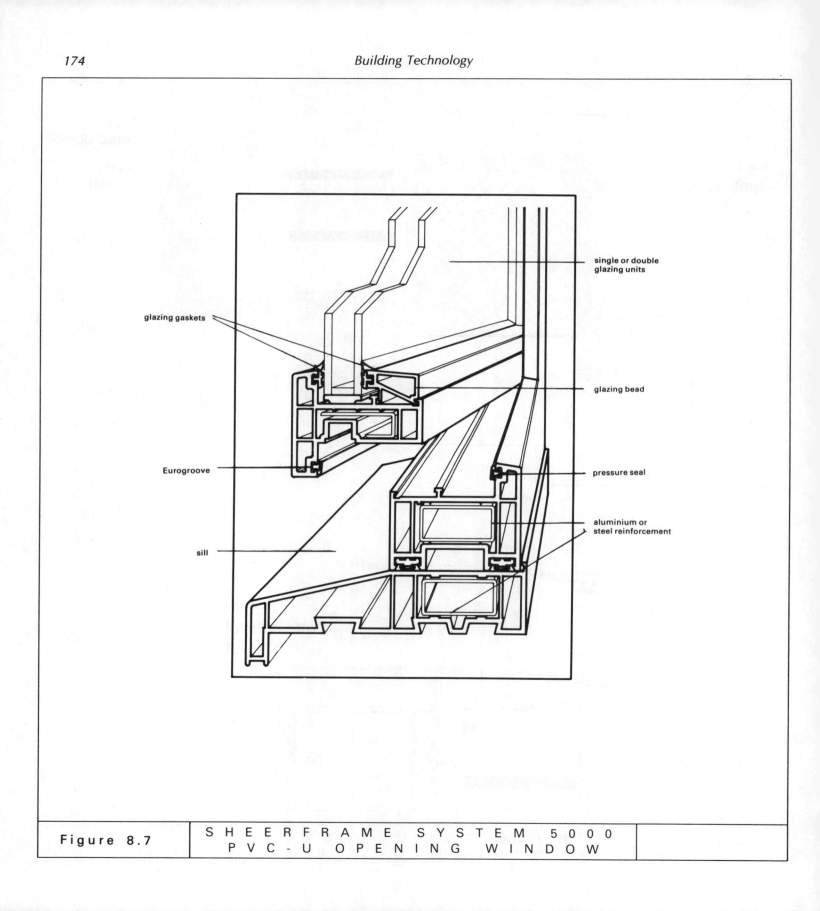

single or double
glazing units

glazing gaskets

glazing bead

Eurogroove

pressure seal

sill

aluminium or
steel reinforcement

Figure 8.7   SHEERFRAME SYSTEM 5000
PVC-U OPENING WINDOW

(1) ordinary glazing quality (OQ) for general glazing purposes;

(2) selected glazing quality (SQ);

(3) special selected quality (SSQ) for high-grade work such as pictures and cabinets;

(4) horticultural: an inferior quality.

For panes exceeding 1 m² clear sheet glass should be at least 4 mm thick.

Float or *polished plate glass* has flat and parallel surfaces providing clear undistorted vision and reflection, produced either by grinding and polishing or by the float process. Generally, clear float glass has superseded polished plate glass in thicknesses up to 25 mm, and it can also be supplied as body tinted, surface coated or surface modified tinted. It is normally 6 mm thick and is supplied in three qualities:

(1) glazing quality (GG);

(2) selected glazing quality (SG) and is also suitable for mirrors and bevelling;

(3) silvering quality (SQ) used for high class mirrors and wherever a superfine glass is required.

Translucent glasses. These cast or patterned glasses transmit light with varying degrees of diffusion so that vision is not clear. They include rough cast glass (textured on one surface), rolled glass (narrow parallel ribs on one surface), fluted and ribbed glass (wider flutes or ribs), reeded glass (various patterns of ribs or flutes), and cathedral and figured rolled glass (one surface textured and the other patterned). Shallow patterns give a partial degree of diffusion while the deeper patterns are almost completely obscure.

Opal glasses. These may be white or coloured and have light-scattering properties because of the inclusion of small particles in the glass.

Glasses for special purposes. Probably the most important is *wired glass* with a wire mesh embedded in it, which holds the glass together on fracture, and is well suited for rooflights and similar situations. Georgian wired cast glass is a translucent glass with rough cast finish and contains electrically welded 12 mm square wire mesh; Georgian polished wired glass is a transparent glass similarly wired but with two ground and polished surfaces; and hexagonal wired cast glass is a translucent glass with rough cast finish containing 22

mm hexagonal wire mesh. *Tinted glass* and *surface coated* float glass are used extensively in office buildings to reduce the transmission of the sun's radiation.

Prismatic glass is a translucent rolled glass, one surface of which consists of parallel prisms, while *heat-absorbing glass* is almost opaque to infra-red radiation and usually has a bluish-green tint. *Heat-resisting sheet glass* has a low coefficient of expansion and greater resistance to changes of temperature; *toughened glass* has increased resistance to external forces; *laminated safety glass* is less likely to cause severe cuts on fracture and other types of laminated glass include laminated security glasses comprising anti-bandit glass, bullet-resistant glass, blast-resistant glass and solar control laminated glasses; and *mirror glass* is clear plate glass silvered on one face. *Leaded lights* are panels consisting of small pieces of glass held together with lead cames; while copper sections are used in *copper lights*, and these are sometimes used in the windows of small houses to give 'scale'.

Safety films with a polyester base can be applied to the inner surfaces of glass to enhance safety and security, eliminate most ultra-violet light and reduce solar heating. Low emissivity glass can also be used effectively to reduce heat loss.

Hollow glass blocks can be used in non-loadbearing partitions to permit light transmittance. The blocks are available in various colours and four standard sizes, 240 × 240 × 80 mm, 240 × 115 × 80 mm, 190 × 190 × 80 mm and 115 × 115 × 80 mm. They are hollow, translucent glass units with various patterns moulded on their interior faces; they are normally jointed in a weak-gauged mortar such as 1:1:8, although sometimes in cement mortar and are often pointed both sides in mastic. They diffuse illumination, reduce heat loss and give a higher level of sound insulation than single glazing. The blocks are non-loadbearing but are normally self-supporting for heights up to 6 m.

Glazing

Prior to glazing, timber rebates should be cleaned, primed and painted with one coat of oil paint, and metal rebates cleaned and primed in accordance with the recommendations of BS 6262.[18] Glass should be cut to allow a small clearance at all edges and then be back-puttied, by laying putty along the entire rebates and bedding the glass solidly, sprigged for timber rebates (using small square nails without heads) and pegged for metal rebates, and neatly front puttied, taking care to ensure that the putty does not appear above the sight lines. Putty used for timber rebates should be linseed oil putty conforming to BS 544,[23] while that for metal rebates should be an approved metal casement putty. Putty should be adequately protected with paint as soon as the putty has hardened (7 to 21 days) as its continuing effectiveness depends on the paint-to-glass seal. Glass to doors, screens and borrowed lights is best bedded in wash leather or plastics and held in place by wood beads fixed with brass screws in cups. For large areas of glass and in severely exposed situations, additional protection can be provided by inserting a mastic strip around the edges of the glass and finishing externally with a sealant capping, and possibly an internal trim of PVC or synthetic rubber wedged between the glass and the bead.

DOUBLE GLAZING AND DOUBLE WINDOWS

Various double glazing and double window systems are available stemming from the increasing demand for improved heat insulation in buildings and the more searching requirements of the Building Regulations 1991 and Approved Document L (1995)[1]. They range from simple 'do it yourself' *in situ* systems (figure 8.4.7) to the more sophisticated factory-produced hermetically sealed double-glazing units (figure 8.4.6) and from double-rebated frames (figure 8.4.7) to openable coupled casements and sashes (figure 8.4.8) or separate secondary windows. The optimum width of air space for vertical double glazing is usually taken as 20 mm, although widths down to 12 mm are almost equally effective.[24] An air space 12 to 20 mm wide halves the thermal transmittance, thus with normal exposure the thermal transmittance (U) of single glazing is 5.7 and that of double glazing 2.8 W/m^2 K in sheltered situations. For sound insulation, however, a minimum air space of 150 mm is required, although 200 to 300 mm is desirable.

Double glazing can reduce the risk of condensation on the glass because the surface exposed to the room is warmer than single glazing and is more likely to be above the prevailing dewpoint temperature, but it is less effective with high humidities, as in kitchens, and low standards of heating. Double glazing results in some reduction in light transmission as it could be about 70 per cent compared with 85 to 90 per cent for single glazing with clear glass up to 6 mm thick.

Insulating Glass Units

Factory made, hermetically sealed, flat double glazing units with fused all-glass edges are being increasingly used as well as those with a welded glass-to-metal seal (figure 8.4.6), but where use is made of flat sheets of glass bonded to spacing strips and sealed, the edges must be kept dry. Some manufacturers produce double-glazing units to fit into the rebates of standard wood or metal sections, without the use of beads. Stepped units are used where the frames are too small or unsuitable for enlargement. Airspace widths may vary from 5 to 20 mm.[22] These units should comply with BS 5713,[25] whereby if the edge protection takes the form of a channel or tape as a permanent component, this shall be of corrosion resistant material, such as specified aluminium alloy or specified stainless steel. It also gives dimensional tolerances and performance requirements. Further examples of double glazing are illustrated in figures 8.5, 8.6 and 8.7.

Single-frame Double-glazing Systems Sealed *in situ*

Various arrangements are available including glazing to double rebated frames or fixing a second line of glazing with wood or plastic face beads, generally to existing frames (figure 8.4.7). No matter how well the glazing seals are made, the cavities cannot be expected to remain airtight indefinitely. The seals may disintegrate under movement and shrinkage, allowing water vapour to enter the air space and to condense on the inside of the outer glazing. The wood exposed to the air space should be painted or varnished to reduce the evaporation of moisture from the timber into the air space and breather holes should be provided at the rate of one 6 mm diameter hole per 0.5 m^2 of window (figure 8.4.7). The holes should be plugged with glass fibre or nylon to exclude dust and insects.[24]

Coupled or Sliding Double Sashes

These may be of wood or metal pivoted with openable coupled sashes (figure 8.4.8), or sliding with pairs of metal sliders in the same frame. The inner sashes should be well sealed when closed together, and in the coupled pivoted type, a ventilating slot is often left around the periphery of the outer sash, to ventilate the air space externally when the sashes are closed (figure 8.4.8), and so reduce the risk of condensation.

PATENT GLAZING

Patent glazing, incorporating a loose bar system of construction providing support for glass on two long edges and normally omitting glazing compounds or gaskets, is used extensively for roof and vertical glazing, particularly in industrial and commercial buildings, on account of its high durability and light transmittance properties. The recommended minimum slope when used in roofs is 15°. Bars are available in a range of designs and thicknesses for differing loads, with the stalk of the bar projecting either outwards or inwards. The load bearing element of the bar is normally aluminium or steel. A typical patent glazing bar is illustrated in figure 8.4.9, incorporating a rolled steel bulb tee bar, dipped in calcium plumbate paint stoved on, sheathed in extruded white PVC-U, hermetically sealed at ends, to form internal condensation channels and an independent PVC-U snap-on capping available in various colours to form a watertight joint with the glass. The glass is seated on greased asbestos cord. Another form of patent glazing bar is of aluminium alloy with aluminium wings or cap to provide a watertight joint. Requirements for patent glazing are specified in BS 5516.[26]

GLAZING MATERIALS AND PROTECTION

Part N of Schedule 1 to the Building Regulations 1991 prescribes in paragraph N1 that glazing with which people are likely to come into contact while in passage in or about the building, shall –

(a) if broken on impact, break in a way which is unlikely to cause injury; or

(b) resist impact without breaking; or

(c) be shielded or protected from impact.

Requirement N2, which does not apply to dwellings, prescribes that transparent glazing, with which people are likely to collide, while in passage in or about the building, shall incorporate features which make it apparent.

Critical locations in terms of safety are defined in Approved Document N1 (1991) as (a) between finished floor level and 800 mm above that level in internal walls and partitions and (b) between finished floor level and 1500 mm above that level in a door or in a side panel close to either edge of the door, as illustrated in the Approved Document.

Safety Measures to reduce the risks of glazing in critical locations are listed in Approved Document N1 (1991) as either (a) break safely, if it breaks, (b) be robust or in small panes, or (c) be permanently protected.

(a) *Safe breakage* is concerned with the performance of laminated and toughened glass, as defined in BS 6206,[27] and is based on an impact test.

(b) *Robustness* Some glazing materials, such as annealed glass, gain strength through thickness, while others such as polycarbonates or glass blocks are inherently strong. *Small panes* should have a maximum width of 250 mm and an area not exceeding 0.5 m², each measured between glazing beads or similar fixings. Annealed glass in a small pane should not be less than 6 mm in thickness.

(c) *Permanent screen protection* should prevent a sphere of 75 mm from coming into contact with the glazing, be robust and, if it is intended to protect glazing that forms part of protection from falling, be difficult to climb.

Manifestation of Glazing is covered in Approved Document N2 (1991) and is only necessary in critical locations in which people in passage in or about the building might not be aware of the presence of the glazing and may collide with it, and includes large uninterrupted areas of clear glazing in non-domestic buildings. Measures used include the provision of mullions, transoms, door framing or large pull or push handles, or broken or solid lines, patterns or company logos at appropriate heights and intervals.

REFERENCES

1. *The Building Regulations 1991 and Approved Documents F (1995), L (1995) and N (1991).* HMSO
2. *BRE Digests 309* and *310*: Estimating daylight in buildings (1986)
3. *BS 644* Wood windows, *Part 1: 1989* Specification for factory assembled windows of various types; *Part 2: 1958* Wood double hung sash windows
4. *BS 6375* Performance of windows, *Part 1: 1989* Classification for weathertightness (including guidance on selection and specification); *Part 2: 1987* Specification for operation and strength characteristics
5. *BS 5368* Method of testing windows, *Part 1: 1976* Air permeability test; *Part 2: 1980* Water tightness test under static pressure; *Part 3: 1978* Wind resistance tests; *Part 4: 1978* Form of test report
6. *CP 153* Windows and rooflights, *Part 2: 1970* Durability and maintenance
7. *BS 5642* Sills and copings, *Part 1: 1978* Specification for window sills of precast concrete, cast stone, clayware, slate and natural stone
8. *BRE Digest 304*: Preventing decay in external joinery (1985)
9. *BS 1186* Timber for and workmanship in joinery, *Part 1: 1986* Specification for timber; *Part 2: 1988* Specification for workmanship
10. *BS 4072* Wood preservation by means of copper/chrome/arsenic compositions, *Part 1: 1987* Specification for preservation; *Part 2: 1987* Method for timber treatment
11. *BS 5707* Solutions of wood preservatives in organic solvents, *Parts 1 to 3: 1979/1980*
12. *BS 5589: 1989* Code of practice for preservation of timber
13. *BRE Information Paper IP 14/91* In-situ treatment of exterior joinery using boron-based implants (1991)
14. *BS 8213* Windows, doors and rooflights, *Part 4: 1990* Code of practice for installation of replacement windows and door sets in dwellings
15. *BRE Defect Action Sheet 11*: Wood windows and door frames: care on site during storage and installation (1982)
16. *BS 729: 1971 (1986)* Hot dip galvanised coatings on iron and steel articles
17. *BS 6510: 1984* Specification for steel windows, sills, window boards and doors
18. *BS 6262: 1982* Code of practice for glazing for buildings
19. *BS 1285: 1980* Specification for wood surrounds for steel windows and doors
20. *BS 4873: 1986* Specification for aluminium alloy windows
21. *BRE Digest 377*: Selecting windows by performance (1993)
22. *BS 952* Glass for glazing, *Part 1: 1978* Classification
23. *BS 544: 1969 (1987)* Linseed oil putty for use in wooden frames
24. *BRE Digest 379*: Double glazing for heat and sound insulation (1993)
25. *BS 5713: 1991* Specification for hermetically sealed flat double glazing units
26. *BS 5516: 1977* Code of practice for patent glazing
27. *BS 6206: 1981* Specification for impact performance requirements for flat safety glass and safety plastics for use in buildings

9 DOORS, DOORSETS AND IRONMONGERY

Consideration of general design principles is followed by an examination of the various types of door and their uses and the constructional techniques employed, together with the main characteristics of frames and linings. Finally, metal and plastics doors and ironmongery are investigated.

GENERAL PRINCIPLES OF DESIGN OF DOORS

Doors form an important part of joinery work, which can be defined as 'the art of preparing and fixing the wood finishings of buildings'. Carpentry work primarily makes use of sawn or unwrought timber, whilst joinery work embraces almost entirely planed or wrought timber, including hardwoods as well as softwoods. Most joinery work is exposed to view and is usually painted or polished.

The main principles to be observed in the construction of doors and framing of joiner's work generally, may be summarised as follows

(1) Timber should be dry and well seasoned, preferably with a moisture content within the following limits

Type of joinery	per cent
External joinery (floor level hardwood sills and thresholds)	19 ± 3
External joinery (all other)	16 ± 3
Internal joinery (for buildings with intermittent heating)	15 ± 2
Internal joinery (for buildings with continuous heating: 12 to 19° C)	12 ± 2
Internal joinery (for buildings with continuous heating: 20 to 24° C)	10 ± 2
(Source: BS 1186[1])	

(2) Timber should be free from serious defects as listed in BS 1186,[1] such as excessive deviation from straightness of grain; large checks, splits and shakes; knots except sound, tight, small knots; pitch pockets; decay and insect attack.

(3) The work should consist of timber which is suitable for the particular situation; for example, Parana pine and sycamore are suitable for internal doors but unsuitable for external doors. BS 1186[1] gives guidance on the uses of a wide range of different species of timber.

(4) The joints between timbers should permit movement due to variations in temperature or humidity without exposing open joints.

(5) The faces of members joined shall be flush with one another unless the design requires otherwise.

(6) The haunch in a tenon joint or the tongue in a dowel joint shall be a push fit in its groove.

Finished Sizes

It is common to specify the sizes out of which a joinery member is to be worked, and 3 mm should then be allowed for each wrought face. Thus a door frame specified as 100 × 75 mm (nominal) will have a finished size of 94 × 69 mm and a door with a nominal thickness of 38 mm will actually be 32 mm. Specification clauses for joinery work should distinguish between nominal and actual sizes. Full-size and 1:5 joinery details should be drawn to finished sizes, whereas nominal sizes can be used for smaller scale drawings.

DOOR TYPES

A variety of matters need consideration when deciding on the type of door and the associated constructional work around the door opening. This is now illustrated by reference to the front entrance door of a good quality dwelling house.

(1) Size of door to be adequate for all needs including passage of perambulators and furniture, probably 806 × 1994 mm.

(2) Adequate strength and durability and dimensional stability, panelled or flush, normally 40 or 44 mm thick, well constructed and hung on adequate butt hinges.

(3) Attractive appearance: careful design of door including mouldings and door furniture.

(4) Weatherproofing qualities: consider provision of water bar in threshold, weatherboard to bottom rail of door, throats to frame and effective pointing with suitable sealant of joint between frame and reveal.

(5) Type of timber and finish: painted softwood or polished hardwood and whether glazing is required to give natural light in house.

(6) Adequate frame to support door and nature and extent of mouldings to frame.

(7) Treatment of reveals: for example, internal, whether plain or panelled linings or plastered and decorated; external, brick, stone or rendered.

(8) Head of opening: consider various alternatives; provision of arch, boot lintel, steel lintel, among other possibilities, and whether a fanlight is desirable.

(9) Threshold and steps: consider alternatives, such as brick, concrete, stone, terrazzo and clay tiles.

There is a wide range of door types available each with their own particular uses, and they can be broadly classified as panelled, flush and matchboarded. The majority of doors used in domestic work are standard doors made of timber and plywood conforming to BS 1186[1] or BS 6566[2] respectively. A common range of sizes of doors and doorsets (doors and frames), as detailed in BS 4787,[3] is as follows

	Overall size of door opening (mm)	Door size (mm)
(1)	900 × 2100	826 × 2040 × 40 or 44 thick
(2)	800 × 2100	726 × 2040 × 40 or 44 thick
(3)	700 × 2100	626 × 2040 × 40 or 44 thick
(4)	600 × 2100	526 × 2040 × 40 or 44 thick

Type (1) doors are especially suitable for external doors, type (2) doors for most internal doors, type (3) doors may be useful for cloakrooms and cupboards and type (4) doors for small cupboards.

Panelled Doors

Panelled doors are usually described by the number of panels which they contain and which may vary from one to six as shown in figure 9.1, and the thickness and finish to the edges of the framing (stiles and rails). Panelled doors are framed by joining the members where they intersect by dowels (figure 9.2.4) or mortises and tenons (figure 9.2.3). Common finished sizes for framing are 94 × 40 or 44 mm stiles, muntins (intermediate vertical members) and top and intermediate rails, and 194 × 40 or 44 mm bottom and lock rails, with a minimum plywood panel thickness of 6 mm for doors with more than one panel and 9 mm for single panel doors. Panels are framed into grooves in the rails and stiles and panels should be about 2 mm smaller in height and width than the overall distance between grooves. Mouldings to the edges of panel openings may take one of several forms; typical solid mouldings are shown in figures 9.1.10 and 9.2.1, while figure 9.2.2 shows a separate mould which is planted or nailed to the frame, but this constitutes a less satisfactory finish. In high-class work, doors may be provided with solid moulded and raised panels where the centre portion of the panel is thicker than the edges or margins (figure 9.1.10), or bolection mouldings (figure 9.1.11) where the moulding projects beyond the face of the framing and covers any shrinkage in the panels. A flush panel with a recessed moulding incorporated in it is described as bead and butt (figure 9.1.9), and this gives a strong door and conceals the joint between the panel and frame. Openings for glazing are often rebated and moulded out of the solid (figures 9.1.2 and 9.1.4), whereas fully glazed doors (figure 9.1.5) are usually provided with separate mitred glazing beads. Some exterior doors (figure 9.1.4) are prepared to receive letter plates to BS 2911[4] with apertures 250 × 38 mm. It is advisable to provide weatherboards and water bars (figure 9.2.1) to external doors to ensure adequate protection against driving rain. Doors either hung singly or in pairs with substantial glazed areas (figures 9.1.5 and 9.1.6) are referred to as *wood casement doors* or *French casements*, and are useful for increasing light transmittance within a dwelling and providing access to balconies, patios and the like. A typical contemporary hardwood domestic front entrance door is illustrated in figure 9.1.12.

Mortise and tenon joints between framed members. These are illustrated in figure 9.2.3. It is advisable that top and bottom rails and at least one other rail should be through mortised and tenoned. Other intermediate rails, muntins and glazing bars should be stub-tenoned to the maximum depth possible, often about 25 mm. Haunchings, to prevent rails twisting, should be not less than 10 mm deep and no tenon should be within 38 mm of the top or bottom of the door. Through tenons are wedged, and glue is applied to the tenon and shoulders before being inserted in the mortise. The thickness of the tenon should equal one-third of the stile and its width should not exceed five times this thickness or a maximum of 125 mm, whichever is the least.

Dowelled joints. These (figure 9.2.4) are becoming increasingly common in machine-made doors as they are cheaper than mortise and tenon joints and are usually equally satisfactory. The dowels are usually of hardwood but may be of the same material as the framing members. They should have a minimum size of 16 × 122 mm, be slightly grooved to give a key for the glue and be equally spaced at distances not exceeding 56 mm centre to centre. There should be at least three dowels for lock and bottom rails, two for top rails and one for intermediate rails. The lowest dowel in a bottom rail should not be less than 44 mm from the bottom of the door.

Panelled doors have to some extent been superseded by flush doors, which with their large smooth surfaces are devoid of dust-catching edges. Panelled doors are, however, used extensively as wood casement doors and, in high-class work, designs similar to those illustrated in figures 9.1.10 and 9.1.11 may be popular, possibly with matching wall panelling.

Flush Doors

The majority of flush doors are made by specialist manufacturers with normal finished thicknesses of 40 or 44 mm. Three of the more commonly used constructional methods are illustrated in figures 9.3.1, 9.3.2 and 9.3.3, and in every case the framework is covered with plywood on both faces and a hardwood edging strip, 6 or 9 mm thick, on both long edges to protect the plywood (figure 9.3.6). The strongest form of flush door is the *solid core* (figure 9.3.1) often made up of longitudinal laminations of precision-planed timber, butt-jointed with resin-based adhesive under hydraulic pressure, and it has excellent fire-check and sound-reducing qualities. A cheaper and lighter alternative is the *half-solid* door (figure 9.3.2) made up of a timber frame incorporating horizontal rails not more than 63 mm apart, and the whole forming a 50 per cent solid timber core. Typical widths of the framed members are shown in figure 9.3.2. This is a strong door which can

satisfactorily accommodate standard ironmongery. An even lighter flush door is the *timber-railed* door (figure 9.3.3) consisting of horizontal rails not more than 125 mm apart. This is used extensively in local authority and private housing developments where stringent cost limits prevail, possibly faced with hardboard, although this is not recommended. Other light duty doors contain a core of expanded cellular paper or spiral paper coils and are often classified as skeleton or cellular doors. Both half-solid and timber-railed doors have a ventilation channel in each rail for air circulation. Suitable timbers are Canadian western red cedar for the framing and beech for the plywood. The resin used in the plywood for external doors should be WBP quality (weather and boil-proof), complying with BS 1203.[5]

The normal square or plain edging strip, to protect the edges of plywood facing, is illustrated in figure 9.3.6. With doors hung folding, a rounded strip (figure 9.3.4) or rebated meeting stiles (figure 9.3.5) may be used. Flush doors can be provided with square, rectangular or circular apertures to receive glass. The glass is best held by glazing beads of which two varieties are illustrated – flush (figure 9.3.7) and bolection (figure 9.3.8). Figure 9.3.9 shows a one-hour fire-check door containing a 5 mm asbestos wallboard bonded to each side of the solid timber core and faced with plywood, with a finished thickness of 54 mm, and complying with BS 476.[6] Further information on timber fire doors can be found in BRE Digest 320,[7] which lists other door constructions which provide similar performances, such as blockboard, compressed straw and cork cores with asbestos facings and veneer. A fire door should prevent excessive transmission of products of combustion which can interfere with the safe use of escape routes, and it should maintain the effectiveness as a fire-barrier of the wall in which it is located.

The appearance of flush doors can be enhanced by the use of polished hardwood veneers selected from a wide range of timbers including African walnut, afromosia, Honduras mahogany, iroko, oak, rosewood and teak. In addition to plywoods and veneers, there are many other facing materials which combine decoration with protection and utility. These include laminated plastics, aluminium plymax and composite veneer panels. Leaderflush market a 'velvet superfine' finish whereby the doors are machine coated with a melamine-based clear wood lacquer, heat impregnated into the veneer and burnished between coats, to enhance

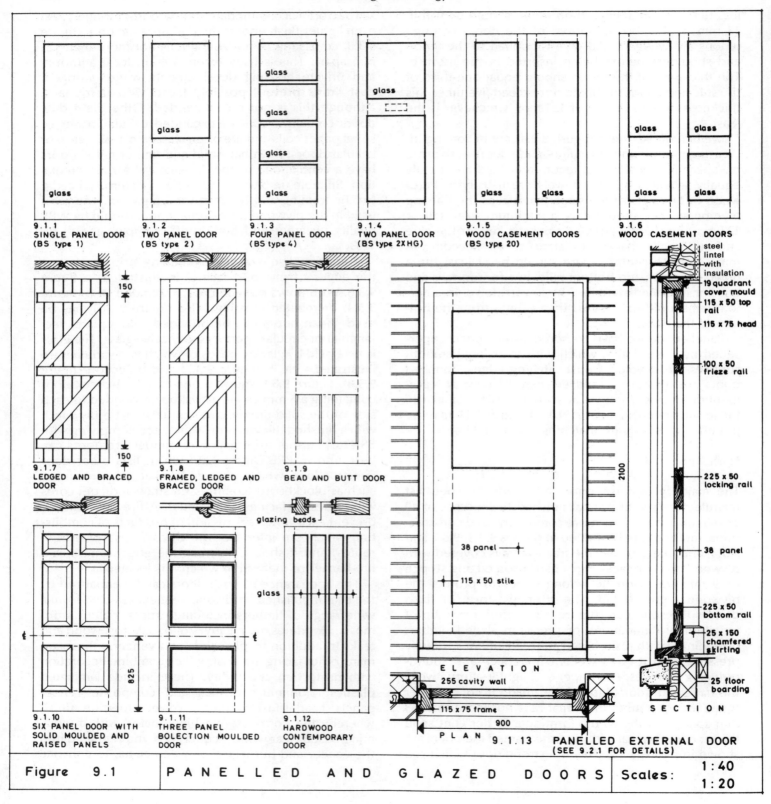

9.1.1 SINGLE PANEL DOOR (BS type 1)

9.1.2 TWO PANEL DOOR (BS type 2)

9.1.3 FOUR PANEL DOOR (BS type 4)

9.1.4 TWO PANEL DOOR (BS type 2XHG)

9.1.5 WOOD CASEMENT DOORS (BS type 20)

9.1.6 WOOD CASEMENT DOORS

9.1.7 LEDGED AND BRACED DOOR

9.1.8 FRAMED, LEDGED AND BRACED DOOR

9.1.9 BEAD AND BUTT DOOR

9.1.10 SIX PANEL DOOR WITH SOLID MOULDED AND RAISED PANELS

9.1.11 THREE PANEL BOLECTION MOULDED DOOR

9.1.12 HARDWOOD CONTEMPORARY DOOR

9.1.13 PANELLED EXTERNAL DOOR (SEE 9.2.1 FOR DETAILS)

ELEVATION

PLAN

SECTION

steel lintel with insulation
19 quadrant cover mould
115 x 50 top rail
115 x 75 head
100 x 50 frieze rail
225 x 50 locking rail
38 panel
225 x 50 bottom rail
25 x 150 chamfered skirting
25 floor boarding

2100
38 panel
115 x 50 stile

255 cavity wall
115 x 75 frame
900

| Figure 9.1 | PANELLED AND GLAZED DOORS | Scales: 1:40 1:20 |

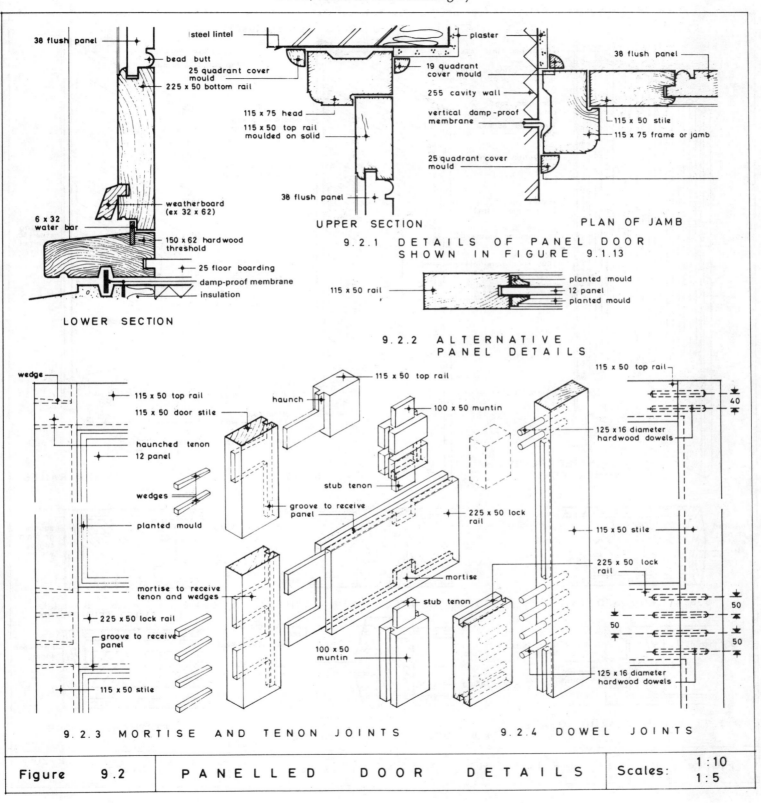

38 flush panel

bead butt

25 quadrant cover mould

225 x 50 bottom rail

steel lintel

plaster

19 quadrant cover mould

255 cavity wall

vertical damp-proof membrane

25 quadrant cover mould

38 flush panel

115 x 75 head

115 x 50 top rail moulded on solid

38 flush panel

115 x 50 stile

115 x 75 frame or jamb

weatherboard (ex 32 x 62)

6 x 32 water bar

150 x 62 hardwood threshold

25 floor boarding

damp-proof membrane

insulation

UPPER SECTION

PLAN OF JAMB

LOWER SECTION

9.2.1 DETAILS OF PANEL DOOR SHOWN IN FIGURE 9.1.13

115 x 50 rail

planted mould

12 panel

planted mould

9.2.2 ALTERNATIVE PANEL DETAILS

wedge

115 x 50 top rail

115 x 50 door stile

haunched tenon

12 panel

wedges

planted mould

mortise to receive tenon and wedges

225 x 50 lock rail

groove to receive panel

115 x 50 stile

haunch

115 x 50 top rail

100 x 50 muntin

stub tenon

groove to receive panel

225 x 50 lock rail

mortise

stub tenon

100 x 50 muntin

115 x 50 top rail

125 x 16 diameter hardwood dowels

40

115 x 50 stile

225 x 50 lock rail

50

50

50

125 x 16 diameter hardwood dowels

9.2.3 MORTISE AND TENON JOINTS

9.2.4 DOWEL JOINTS

| Figure 9.2 | PANELLED DOOR DETAILS | Scales: 1:10 1:5 |

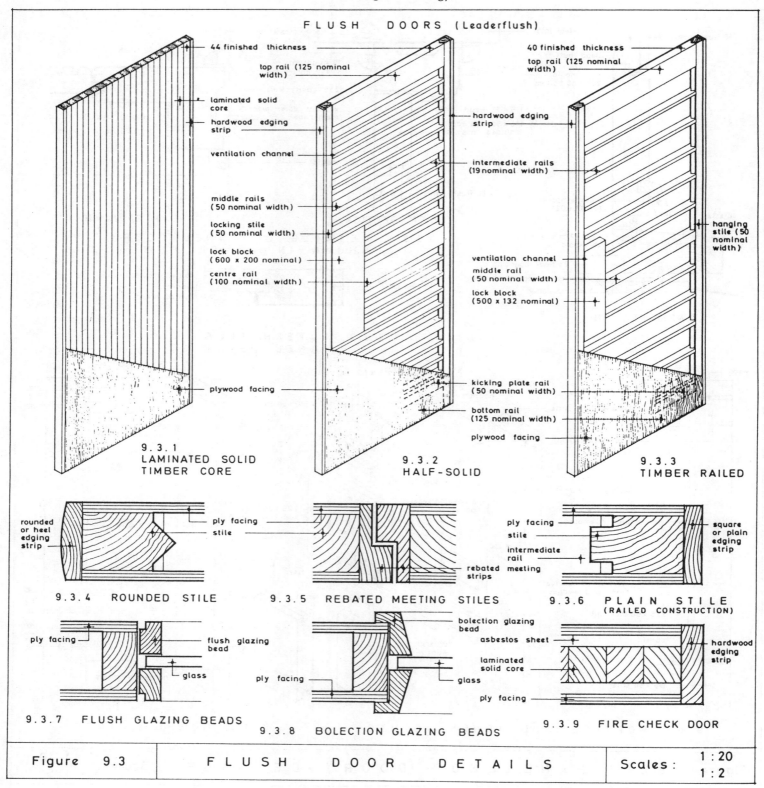

FLUSH DOORS (Leaderflush)

44 finished thickness

top rail (125 nominal width)

laminated solid core

hardwood edging strip

ventilation channel

middle rails (50 nominal width)

locking stile (50 nominal width)

lock block (600 x 200 nominal)

centre rail (100 nominal width)

plywood facing

9.3.1 LAMINATED SOLID TIMBER CORE

40 finished thickness

top rail (125 nominal width)

hardwood edging strip

intermediate rails (19 nominal width)

ventilation channel

middle rail (50 nominal width)

lock block (500 x 132 nominal)

kicking plate rail (50 nominal width)

bottom rail (125 nominal width)

plywood facing

9.3.2 HALF-SOLID

hanging stile (50 nominal width)

plywood facing

9.3.3 TIMBER RAILED

rounded or heel edging strip

ply facing

stile

9.3.4 ROUNDED STILE

ply facing

stile

rebated strips

rebated meeting

9.3.5 REBATED MEETING STILES

ply facing

stile

intermediate rail

square or plain edging strip

9.3.6 PLAIN STILE (RAILED CONSTRUCTION)

ply facing

flush glazing bead

glass

9.3.7 FLUSH GLAZING BEADS

bolection glazing bead

ply facing

glass

9.3.8 BOLECTION GLAZING BEADS

bolection glazing bead

asbestos sheet

laminated solid core

ply facing

hardwood edging strip

9.3.9 FIRE CHECK DOOR

| Figure 9.3 | FLUSH DOOR DETAILS | Scales: 1:20 1:2 |

the natural beauty of the wood and to produce a hard-wearing, long-lasting finish to the door.

Matchboarded Doors

These doors are relatively inexpensive and are mainly used as external doors in inconspicuous positions, such as for outbuildings and fuel stores. They consist of a matchboarded face fixed to a timber framework and are not particularly attractive in appearance. The simplest form of matchboarded door is the *ledged and braced door* (figure 9.1.7) which consists of three horizontal ledges and two parallel braces, with their lower ends abutting the hanging edge of the door, all square edged with a maximum size of 1075 × 2135 mm.[8] The ledged and battened door without braces, sometimes found in old cottages, cannot be recommended because of its liability to twist and drop at the opening edge. With ledged and braced doors, the minimum finished sizes of ledges and braces are 91 × 20 mm, and the matchboarding is not less than 15 mm thick, tongued and grooved, and V-jointed on both faces. The width of the boards, excluding the tongues, should be between 70 and 115 mm, with all other than edge boards of the same width. The braces should fit closely against the ledges and the ends of the ledges should be set back 15 mm from the edges of the door, to clear the door stop or rebate, and the top and bottom ledges set 150 mm from the ends of the door. Each board should be nailed to the ledges and braces using not less than two steel nails, 50 mm long, placed diagonally at each ledge and one at each brace. Alternatively, boards may be stapled to ledges and braces with 18 gauge clenching staples 30 mm long, driven by mechanically operated tools. Some advocate screwing the ledges to the end boards for increased strength.

Framed, Ledged and Braced Doors

These (figure 9.1.8) consist of stiles and top rail (probably 91 × 44 mm, middle and bottom rails (142 × 27 mm) and two parallel braces (91 × 27 mm)). The lower edge of the bottom rail should be between 20 and 40 mm above the bottom of the door, and the middle rail positioned centrally between the top and bottom rails. The rails should be through-tenoned into the stiles and the tenons to top and bottom rails should be haunched. Each tenon should be secured by two wedges and either pinned with a 10 mm diameter hard-wood pin or 6 mm non-ferrous metal star dowel, or bedded in weather-resistant adhesive. Alternatively, framing members may be dowel jointed using two 15 × 110 mm wooden dowels at each joint. Matchboarding is either tongued or rebated into the top rail and stiles, and nailed or stapled to the rails and braces in the manner described for ledged and braced doors. Framed, ledged and braced doors are more expensive but much stronger and of better appearance than unframed doors.

Folding Doors

Folding doors are those which close an opening with two or more leaves. Where there are two leaves they are usually each hinged to opposite jambs of the opening, and the meeting stiles are generally rebated to close (figure 9.3.5). Double doors may be hung to swing both ways, when the edges of the meeting stiles must be rounded (figure 9.3.4) to allow them to pass. The hanging stiles of the doors will also be rounded to fit into a corresponding hollow on the frame. These doors swing on centres, the lower being connected to a spring contained in a box set in the floor (floor spring) and which acts as a check to the swing of the door.

Garage Doors

Doors to domestic garages provide a measure of security for the contents of the garage and a barrier against the weather. The doors may be hung to swing, slide, fold or move 'up-and-over'. The majority of garage doors currently being installed are of the 'up-and-over' variety largely because of their simple and effective means of operation, often enhanced by electronic remote control systems. Methods of suspension vary widely, but all are counterbalanced by weight or spring. Traditional double doors hung to swing, although much less popular are still available. Horizontal sliding leaves are usually sectional and slide 'round-the-corner'. Bottom roller tracks for sliding doors are made of nylon for light use and brass or steel for heavier applications. Overhead tracks are usually rolled galvanised mild steel, although aluminium is suitable for light use. Wheels for heavy-duty gear are generally made of brass or steel although nylon is often suitable for domestic use. Bottom tracks are liable to clog with dirt while overhead gear entails the weight of the doors being carried by the lintel. Folding and roller shutter doors are little used for domestic garages because of their high cost. The usual door height is 2100 mm.

Wood leaves can be of panelled or matchboarded construction, and may include glazed panels. Many garage doors are now manufactured from steel or aluminium or PVC-U and these will be described later in the chapter.

DOOR FRAMES AND LININGS

Doors may be hung to solid frames or to linings. Frames are mainly used with external doors (figure 9.4.7) and to support internal doors in partitions not exceeding 75 mm thick (figure 9.5.4).

Door Frames

Door frames are often 100 × 75 mm in size with a 12 mm rebate to receive the door. Where the thickness of an internal partition exceeds the width of the frame, the additional thickness may be made up with wrought grounds as in figure 9.4.3. An internal door frame may be rebated to receive a thin partition (figure 9.4.4). With door openings in 50 and 62 mm partitions, it is advisable to incorporate a storey-height frame (figure 9.4.2), which can be fixed at both top and bottom to increase stability. Door frames are usually fixed with three metal anchors each side built into brick joints (figure 9.4.7), but on occasions fixing bricks or timber pallets may be used.

Door frames consist of two uprights (jambs) and a head, which should be scribed and framed together with either mortise and tenon or combed joints, in accordance with BS 1567.[9] Heads and sills should be provided with projections or 'horns' not less than 38 mm long. The tenons should be close-fitting in the mortises and be pinned with corrosion-resisting metal pins, star-shaped and not less than 8 mm diameter, or with wood dowels not less than 10 mm diameter. Alternatively, the tenons may fit into tapered mortises and be wedged. In either case the joints should be made either with an adhesive or with lead base paint.

Sills should preferably be made of hardwood with a weathered upper surface and may be grooved for and fitted with a 25 × 6 mm sherardised or galvanised mild steel water bar bedded in mastic (figure 9.5.1). When door frames are used without sills, the feet of the jambs should be fixed with 12 mm diameter corrosion-

resisting metal dowels. Fanlight openings should be provided with glazing fillets or beads. The heads of door frames must be relieved of the weight of brick and block partitions by providing reinforced concrete lintels or by filling hollow blocks with concrete and inserting steel reinforcing bars in the bottom cavities (figure 9.4.1).

An alternative to wood frames is to use *metal door frames* (figure 9.4.6). These are made of good commercial mild steel not less than 1.2 mm nominal thickness to the profiles shown in BS 1245.[10] Metal door frames are fitted with fixing lugs, hinges, lock strike plate and rubber buffers. The finish may take various forms including hot dip galvanising, metal spray and various priming treatments.

It is conventional for front and rear entrance doors to a dwelling to open inwards, but doors leading to a garden, terrace or balcony, as well as doors to outbuildings, often open outwards. The chief problem with external door frames is to prevent the entry of rain, particularly driving rain, at the bottom of the door. The best solution is to provide a galvanised or sherardised mild steel water bar set into the top of the threshold (figure 9.5.1). This should be set at a point halfway in the thickness of the door and preferably housed into the jambs about 6 mm, or at least provide a tight fit between them. The bottom of the door is rebated over the water bar. In particularly exposed situations it is advisable to provide, in addition, a weatherboard housed or tongued into the bottom rail and finishing flush with the edges of the door, and it may be throated to assist in shedding water. This positioning of the water bar ensures that any rainwater passing down the joint between the jamb of the frame and the door will finish on the outside of the water bar.

Door Linings

Many internal doors are hung from linings which have a width equal to the thickness of the wall or partition plus the plaster or other finish on either side. Hence the width of a lining in a one-brick wall could be 245 to 255 mm. The rebate for the door may be formed in the lining, where the thickness is adequate, or by means of a planted (nailed) stop. Linings can be broadly classified into three types

Plain. These are usually prepared from boards 38 to 40 mm thick which are either single rebated (figure 9.5.3) or double rebated (figures 9.5.5 and 9.5.6). A

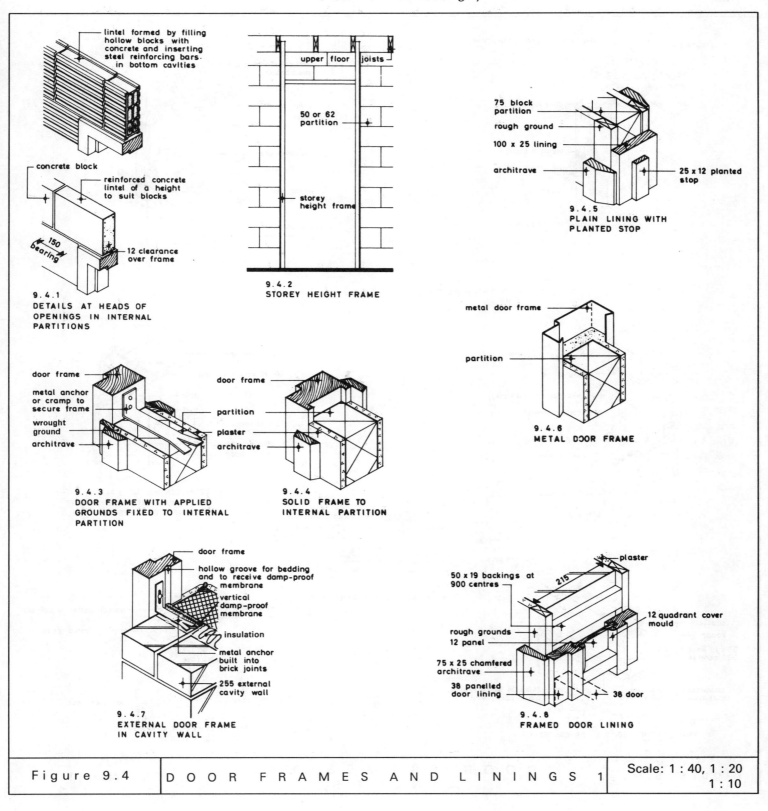

lintel formed by filling hollow blocks with concrete and inserting steel reinforcing bars in bottom cavities

concrete block

reinforced concrete lintel of a height to suit blocks

150 bearing

12 clearance over frame

9.4.1
DETAILS AT HEADS OF OPENINGS IN INTERNAL PARTITIONS

upper floor joists

50 or 62 partition

storey height frame

9.4.2
STOREY HEIGHT FRAME

75 block partition

rough ground

100 x 25 lining

architrave

25 x 12 planted stop

9.4.5
PLAIN LINING WITH PLANTED STOP

door frame

metal anchor or cramp to secure frame

wrought ground

architrave

9.4.3
DOOR FRAME WITH APPLIED GROUNDS FIXED TO INTERNAL PARTITION

door frame

partition

plaster

architrave

9.4.4
SOLID FRAME TO INTERNAL PARTITION

metal door frame

partition

9.4.6
METAL DOOR FRAME

door frame

hollow groove for bedding and to receive damp-proof membrane

vertical damp-proof membrane

insulation

metal anchor built into brick joints

255 external cavity wall

9.4.7
EXTERNAL DOOR FRAME IN CAVITY WALL

plaster

50 x 19 backings at 900 centres

215

12 quadrant cover mould

rough grounds

12 panel

75 x 25 chamfered architrave

38 panelled door lining

38 door

9.4.8
FRAMED DOOR LINING

| Figure 9.4 | D O O R   F R A M E S   A N D   L I N I N G S   1 | Scale: 1 : 40, 1 : 20 1 : 10 |

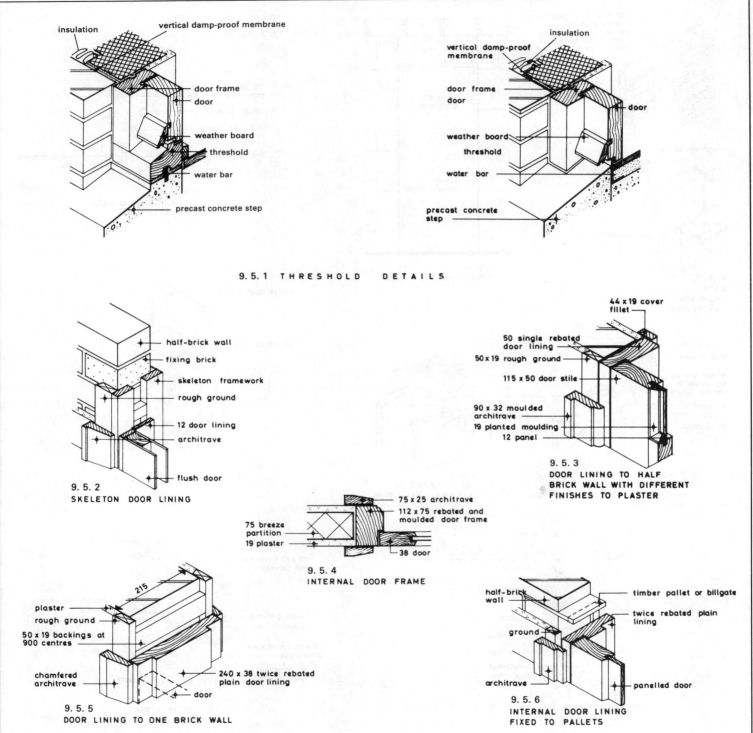

9.5.1 THRESHOLD DETAILS

insulation
vertical damp-proof membrane
door frame
door
weather board
threshold
water bar
precast concrete step

vertical damp-proof membrane
insulation
door frame
door
door
weather board
threshold
water bar
precast concrete step

9.5.2 SKELETON DOOR LINING

half-brick wall
fixing brick
skeleton framework
rough ground
12 door lining
architrave
flush door

9.5.3 DOOR LINING TO HALF BRICK WALL WITH DIFFERENT FINISHES TO PLASTER

44 x 19 cover fillet
50 single rebated door lining
50 x 19 rough ground
115 x 50 door stile
90 x 32 moulded architrave
19 planted moulding
12 panel

9.5.4 INTERNAL DOOR FRAME

75 x 25 architrave
112 x 75 rebated and moulded door frame
75 breeze partition
19 plaster
38 door

9.5.5 DOOR LINING TO ONE BRICK WALL

215
plaster
rough ground
50 x 19 backings at 900 centres
chamfered architrave
240 x 38 twice rebated plain door lining
door

9.5.6 INTERNAL DOOR LINING FIXED TO PALLETS

half-brick wall
timber pallet or billgate
twice rebated plain lining
ground
architrave
panelled door

| Figure 9.5 | DOOR FRAMES AND LININGS 2 | Scale: 1:10 |

cheaper but less satisfactory alternative is to use a 25 mm thick lining with a 12 mm stop planted on (figure 9.4.5); this, however, economises in timber and can be adjusted to take up irregularities in the door.

Skeleton. This type of lining (figure 9.5.2) is particularly well suited for wider jambs and consists of a skeleton timber framework made up of two jambs or uprights, often 75 × 32 mm, with cross-rails tenoned to them at intervals. A board 10 or 12 mm thick is planted on to the framework to form a door stop and to give the appearance of a double rebated lining. The framework is fixed to rough grounds which are plugged to the brickwork.

Framed. This is the best form of lining for an opening in a thick wall (figure 9.4.8). It consists of framed panels to the jambs and soffit similar to a panelled door, and the mouldings should match those on the adjoining door.

Linings can be fixed in a variety of ways and one of the more common is to use rough grounds, often about 50 × 19 mm, plugged at about 900 mm intervals in the height of the opening (figures 9.4.8 and 9.5.5). The architrave, which is an ornamental timber member masking the gap between the plaster and the lining, is attached to a continuous vertical ground adjoining the lining. For walls of 215 mm thickness or more, the grounds are normally framed together (figure 9.4.8). A cheaper form of fixing is to nail the lining to wood pads or pallets inserted in brick joints and projecting from the brickwork to give a 'plumbed' face for the lining (figure 9.5.6). Another alternative is to nail the lining to three fixing bricks on each side of the opening. Figure 9.5.3 shows the use of a cover fillet to replace the architrave and rough ground; the architrave is mitred at the top angles of the opening and is continued down to floor level, where the skirting stops against it.

METAL DOORS

Metal doors are used for two principal purposes in domestic construction, namely patio and garage doors. Aluminium sliding patio doors are becoming increasingly popular, being of very attractive appearance with a minimum obstruction of vision and with excellent weatherproofing qualities and little maintenance. One popular variety[10] consists of horizontal sliding doors with single or double glass sealed units of 5 mm safety float glass, adjustable rollers, continuous stainless steel track and an integral weathering of neoprene and siliconised wool pile. Flush glazing is obtained through neoprene gaskets fitted into the frame which is extruded from aluminium alloy with a natural anodised finish. The aluminium unit is often set in a rebated hardwood frame. Acceptable standards for single and double glazed aluminium framed sliding doors for general purposes are prescribed in BS 5286,[12] with preferred co-ordinating sizes ranging from 1800 to 4800 × 2100 mm in five widths. More recently, sliding patio doors of white PVC-U extrusion, often comprising multi chamber profiles, reinforced with galvanised mild steel where necessary, have become very popular.

'Up-and-over' garage doors are often manufactured from steel or aluminium and consist of pressed panels, frequently ribbed to provide extra strength, mounted on a suitably braced framework, and supplied complete with suspension and locking systems. Suspension systems may incorporate torsion springs or counterbalancing weights and are very efficient in operation. Most steel doors are hot-dip galvanised but should still be painted, using a suitable metal primer, to prevent corrosion and improve appearance. Alternatively, doors can be supplied in 'colorcoated' steel sheet. Aluminium doors may be left unpainted provided they are cleaned regularly, but a painted finish may be preferred on grounds of appearance.

BS 8213[13] gives recommendations for the installation, into existing structural openings, of windows and doorsets (with hinged and sliding leaves), manufactured from metal, plastics or wood, and supplied with or without surrounds.

PLASTICS DOORS

Plastics doors can take various forms ranging from up and over garage doors to residential or entrance doors, French doors, and sliding or patio doors. The sliding or patio doors in PVC-U have become very popular as they are waterproof, draughtproof, attractive and secure, and have become a good selling point in many dwellings. They are normally constructed in high impact PVC-U to DIN 7748 in a similar manner to the PVC-U windows described and illustrated in chapter 8, but are fully reinforced.

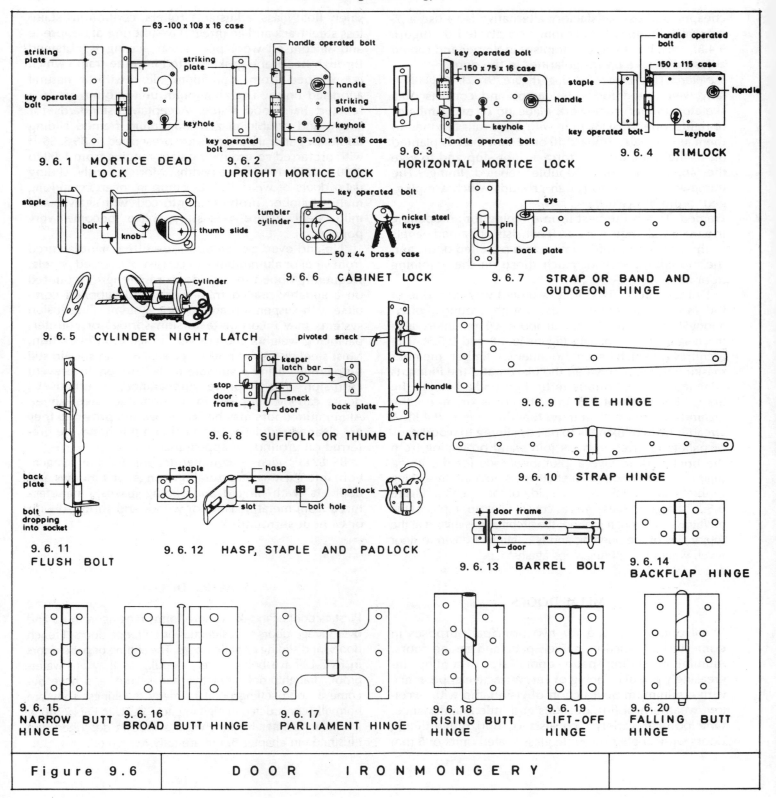

9.6.1　MORTICE DEAD LOCK

9.6.2　UPRIGHT MORTICE LOCK

9.6.3　HORIZONTAL MORTICE LOCK

9.6.4　RIMLOCK

9.6.5　CYLINDER NIGHT LATCH

9.6.6　CABINET LOCK

9.6.7　STRAP OR BAND AND GUDGEON HINGE

9.6.8　SUFFOLK OR THUMB LATCH

9.6.9　TEE HINGE

9.6.10　STRAP HINGE

9.6.11　FLUSH BOLT

9.6.12　HASP, STAPLE AND PADLOCK

9.6.13　BARREL BOLT

9.6.14　BACKFLAP HINGE

9.6.15　NARROW BUTT HINGE

9.6.16　BROAD BUTT HINGE

9.6.17　PARLIAMENT HINGE

9.6.18　RISING BUTT HINGE

9.6.19　LIFT-OFF HINGE

9.6.20　FALLING BUTT HINGE

Figure 9.6　　　DOOR　　IRONMONGERY

An attractive and soundly constructed PVC-U patio door system is that produced by Rehau with ample steel reinforcement, varying in cross-section from 40 × 20 to 50 × 20 mm. Glazing units up to 28 mm thick can be accommodated. An aluminium running strip is provided for the lower member of the main frame. Adjustable tandem rollers with ball-bearing movement are provided for silent and smooth running of the sliding leaf, with the rollers fixed to the lower member of the sliding leaf to ensure that the leaf is securely borne. Cylinder locks with upper and lower locking points can be operated from both inside and outside the door. Sealing between the main frame, fixed leaf and sliding leaf is by means of brush seals. The system is so designed that wind pressure increases the tightness of the seal of the sliding leaf, which has three continuous seals.

IRONMONGERY

The term *ironmongery* or *hardware* includes locks, latches, bolts, furniture (handles and coverplates), suspension gear and closing and check gear. A glossary of terms relating to builder's hardware is contained in BS 3827.[14] BS 5872[15] covers locks and latches for doors and gives the range of materials and finishes and essential minimum dimensions and weights. The requirements as to materials, workmanship, construction, dimensions and weight of a wide range of hinges are given in BS 1227.[16]

Locks, Latches, Bolts and Furniture

When selecting locks it should be borne in mind that larger locks can house stronger working parts, bolt 'shoots' should be of adequate length (16 mm for a locking or key-operated bolt and 13 mm for latch or handle-operated bolt) and the latch bolt mechanism should be of 'easy action'.

The *mortice lock* has two bolts, one a dead or locking bolt operated by a key and the other a latch bolt operated by a handle. The lock case, usually of wrought iron or pressed steel, is fitted into a mortice cut in the door and, in consequence, the lock cannot be removed when the door is locked. The door must be of sufficient thickness to receive it without being weakened unduly. The *horizontal mortice lock* (figure 9.6.3) can be inserted in the middle or lock rail of a panelled door, whereas the *upright mortice lock* (fig-

ure 9.6.2) is used with single panel, glazed and flush doors. The *mortice dead lock* (figure 9.6.1) has only one bolt, a locking bolt operated by a key, and is especially suitable for hotel bedrooms. The *rimlock* (figure 9.6.4), often of japanned steel, is fitted on the door face and is used mainly with matchboarded doors and the thinner panelled doors. It is less attractive and less efficient than the mortice lock. *Rim latches* with a finger-operated sliding bolt or snib in addition to the spring bolt are useful for WCs and bathrooms and are called *pulpit latches*.

The *cylinder night latch* (figure 9.6.5), often referred to as a 'Yale lock', is frequently fitted on the main external doors to dwellings. The spring-loaded bolt is opened by a key from the outside or a round knob on the inside, and the bolt can be held in or out by a slide. A *cabinet lock* (figure 9.6.6) generally consists of a small mortice dead lock with cylinder action. In the interests of safety a second lock, preferably a mortice dead lock, should be provided about 600 mm up from the bottom of the door.

Door knobs are available in non-ferrous, alloy and plastics in a variety of shapes, and are generally supplied in pairs with a steel spindle. *Lever door handles* in non-ferrous, plastics or aluminium, usually have a minimum length of 90 mm. A *lockset* consists of a pair of lever handles attached to plates for screwing to the face of the door. *Finger plates*, made in the same range of materials and normally 300 × 63 mm in size, are fitted to prevent the door finish from being dirtied.

The *Suffolk* or *thumb latch* (figure 9.6.8) is used on garden gates and matchboarded doors to outbuildings, with a latch bar on one side operated by a pivoted sneck or catch on the other. Store doors are sometimes fastened with a *hasp and staple* (figure 9.6.12), with a hasp bolted to the door, a staple screwed to the frame and locked together with a padlock. The most common form of bolt is the *barrel bolt* (figure 9.6.13) of mild steel or brass varying in length from 75 to 300 mm. The *tower bolt* is less enclosed than the barrel bolt and passes through rings. A more attractive bolt is the *flush bolt* (figure 9.6.11), where the back plate finishes flush with the face of the door and the bolt is recessed into the door.

Hinges

There are various types and sizes (63 to 150 mm long) of hinges available for hanging doors. In general one pair of butts is adequate for hollow flush internal door leaves, while wider and heavier leaves, such as solid flush or external doors, are best hung on one-and-a-half pairs. *Butt hinges* (figure 9.6.15) in pressed steel, cast iron or brass are widely used on domestic doors. They have narrow flaps that are concealed within the thickness of the door and they require letting-in flush on both meeting faces. *Rising butts* (figure 9.6.18) have knuckles so shaped that when the door is opened the leaf rises to clear carpets, and this action also causes the door to close itself. *Falling butts* (figure 9.6.20) have the reverse effect and are used in public WC apartments. *Pin hinges* (figure 9.6.16) have a loose pin to permit separation of leaf and frame. *Lift-off hinges* (figure 9.6.19) have a pin attached to one flap which engages in the knuckle of the other so that the two may be readily disengaged. *Parliament hinges* (figure 9.6.17) have flaps extended so that the knuckle projects clear of the leaf and frame, thus enabling the door to swing through 180°, clearing obstructions such as architraves and skirtings. *Back-flap hinges* (figure 9.6.14) have square or rectangular flaps and are usually screwed to the face of leaf and frame where the leaf is too thin to take a butt hinge.

Tee hinges or 'cross garnets' (figure 9.6.9) have one elongated flap, 150 to 450 mm long, and are used where the weight must be carried over a large area, such as with ledged and braced leaves. *Strap, band and gudgeon*, or *hook and eye hinges* (figure 9.6.7) are used with heavy gates and garage doors, and are usually about 50 × 6 mm in section and are supplied in lengths varying from 250 to 1050 mm. Another form of *strap hinge* (figure 9.6.10) has two elongated flaps for use between folding leaves.

Knuckle type hinges are made of steel, cast iron, solid drawn (extruded) brass and aluminium, and band type hinges are made of steel.[16]

REFERENCES

1. *BS 1186*: Timber for and workmanship in joinery; *Part 1: 1986* Specification for timber; *Part 2: 1988* Specification for workmanship
2. *BS 6566*: Plywood, *Part 1: 1985* Specification for construction of panels and characteristics of plies including marking
3. *BS 4787*: Internal and external wood doorsets, door leaves and frames; *Part 1: 1980* Specification for dimensional requirements
4. *BS 2911: 1974* Letter plates
5. *BS 1203: 1979* Specification for synthetic resin adhesives (phenolic and aminoplastic) for plywood
6. *BS 476*: Fire tests on building materials and structures; *Part 7: 1971* Surface spread of flame test for materials; *Part 8: 1972 (1985)* Test methods and criteria for the fire resistance of elements of building construction
7. *BRE Digest 320*: Fire doors (1988)
8. *BS 459: 1988* Matchboarded wooden door leaves for external use
9. *BS 1567: 1953 (1971)* Wood door frames and linings
10. *BS 1245: 1975* Metal door frames (steel)
11. Hillaldam Coburn. *Solair aluminium sliding patio doors*
12. *BS 5286: 1978 (1984)* Specification for aluminium framed sliding glass doors
13. *BS 8213*: Windows, doors and rooflights, *Part 4: 1990* Code of practice for the installation of replacement windows and doorsets in dwellings
14. *BS 3827*: Glossary of terms relating to builder's hardware, *Part 1: 1964* Locks (including locks and latches in one case); *Part 2: 1967* Latches; *Part 3: 1967* Catches; *Part 4: 1967* Door, drawer, cupboard and gate furniture
15. *BS 5872: 1980* Specification for locks and latches for doors in buildings
16. *BS 1227*: Hinges; *Part 1A: 1967 (1974)* Hinges for general building purposes

10 STAIRS AND FITTINGS

This chapter completes the study of joinery work by examining the design and construction of staircases and associated constructional works and simple joinery fitments.

STAIRWAY, RAMP AND PEDESTRIAN GUARDING DESIGN

Stair Types

Stairs consist of a succession of steps and landings that make it possible to pass on foot to other levels, whereas the term *staircase* or *stairway* is often applied to the complete system of treads, risers, strings, landings, balustrades and other component parts, in one or more successive flights of stairs. The space occupied by a staircase is termed a *stairwell*, with the vertical distance between the floors served by a staircase described as the *lift*.

The simplest form of stair is the *straight flight* stair (figure 10.1.1) consisting of a straight continuous flight or run of parallel steps or a continuous slope. A *quarter-turn* stair is one containing a flight with a landing and a right-angle turn to left or right; the landing is termed a quarter-space landing (figure 10.1.3). A *dogleg* or *half-turn* stair (figure 10.1.2) has one flight rising to an intermediate half-space landing, with the second flight travelling in the opposite direction to the first flight. An *open well* or *open newel* stair (figure 10.1.3) contains a central well, with newels at each change of direction and two or more flights of steps around the outside of the well. A *geometrical stair* (figure 10.1.4) takes the form of a spiral, with the face of steps radiating from the centre of a circle which forms the plan of the outer string, and incorporates an open well. A *spiral* stair is a form of geometrical stair without a well (a helix round a central column). A *helical* stair contains a helix round a central void.

Requirements for Domestic Staircases

Schedule 1 to the Building Regulations 1991[1] in paragraph K1, provides that stairs, ladders and ramps shall offer safety to users moving between levels of the building. Approved Document K (1991) prescribes certain minimum requirements for stairs serving single dwellings (private stairs) and those stairs serving a place where a substantial number of people will gather (institutional and assembly) and other stairs used in all other buildings. In this chapter we are mainly concerned with the first category of stairs.

A *parallel* tread is one having a uniform width throughout that part of its length within the width of the stairway (figure 10.1.1), whereas a *tapered* tread (figure 10.1.4) is one which has a greater width at one side than at the other and a going which changes at a constant rate throughout its length. *Pitch line* (figure 10.1.6) is a notional line which connects the nosings of all treads in a flight with the nosing of the landing at the top of the flight down to the ramp or landing at the bottom of the flight. The *going* of a tread (figure 10.1.6) is measured on plan between the nosing of the tread and the nosing of the tread, ramp or landing next above it, or, alternatively, it can be expressed as the horizontal dimension from the front to back of a tread less any overlap with the next tread above. The rise of a step means the height between consecutive treads. For practical purposes the width of a stairway is its unobstructed width, clear of handrails and other obstructions (figure 10.1.5). The minimum width for a private stair should normally be 800 mm and there are no recommendations in Approved Document K (1991).

A stairway must be designed to provide a safe, serviceable and commodious means of access from one floor to another of a building and Approved Document K to the Building Regulations 1991[1] has been framed to secure this objective. The Approved Document K1 (1991) prescribes that in a flight the steps shall have the

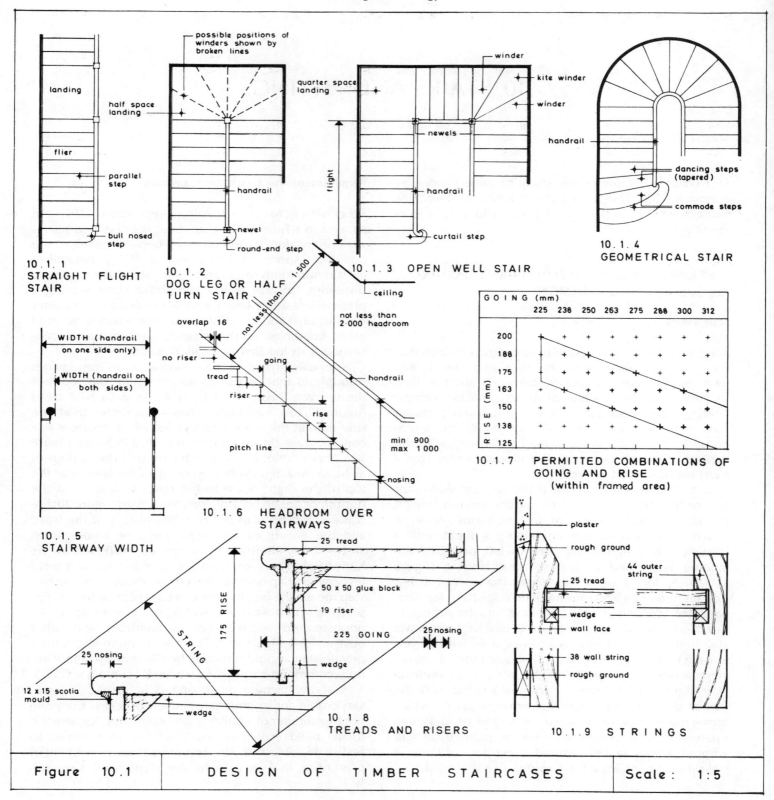

10.1.1 STRAIGHT FLIGHT STAIR

10.1.2 DOG LEG OR HALF TURN STAIR

10.1.3 OPEN WELL STAIR

10.1.4 GEOMETRICAL STAIR

10.1.5 STAIRWAY WIDTH

10.1.6 HEADROOM OVER STAIRWAYS

10.1.7 PERMITTED COMBINATIONS OF GOING AND RISE (within framed area)

10.1.8 TREADS AND RISERS

10.1.9 STRINGS

Figure 10.1   DESIGN OF TIMBER STAIRCASES   Scale: 1:5

same rise and the same going within the prescribed dimensions. There are also minimum headroom requirements over the whole width of the stairway: not less than 2 m clear headroom measured vertically from the pitch line (figure 10.1.6). For loft conversions where there is not enough space to achieve this height, an acceptable headroom will be 1.9 m at the centre of the stair width, reducing to 1.8 m at the side of the stair as illustrated in Approved Document K1 (1991). Alternating tread stairs can also be used to save space, where alternate handed steps have part of the tread cut away. The nosing of any tread which has no riser below it (open risers), shall overlap (on plan) the back edge of the tread below it by not less than 16 mm (figure 10.1.6). Steps with open risers should be constructed so that a 100 mm diameter sphere cannot pass through the open risers, to prevent children being held fast between the treads.

The dimensions of the going and rise have to be in a satisfactory relationship to one another and ensure that the stairway is neither too flat nor too steep. Approved Document K1 (1991) recommends that on a stair the aggregate of twice the rise plus the going shall be not less than 550 mm and not more than 700 mm ($2R + G = 550$ to 700). This permits a number of combinations as shown in the hatched area in figure 10.1.7. A common arrangement for a private stair is a rise of 175 mm and 225 mm going, which gives a sum of $225 + 350 = 575$ and is within the range outlined. In stairs serving single dwellings (private stairs) the rise of a step should not exceed 220 mm and the going shall be at least 220 mm. For private stairs Approved Document K1 (1991) prescribes any rise between 155 and 220 mm used with any going between 245 and 260 mm, or any rise between 165 and 200 mm used with any going between 245 and 260 mm.

For institutional and assembly stairs the rise should not exceed 180 mm and the going shall not be less than 280 mm, and with other stairs the corresponding dimensions are 190 and 250 mm. The pitch for private stairs shall not exceed 42° and that for gangways for seated spectators in assembly buildings shall not be more than 35°. A stairway should not normally have more than 16 rises in any flight. Consecutive tapered steps shall have uniform goings measured at the centre of the length, where the flight is less than 1 m in width, otherwise it should be measured 270 mm from each side, and the narrow end of a tapered tread should be at least 50 mm.

Stairs should have a continuous handrail fixed securely at a height of between 900 mm and 1.0 m, measured vertically above the pitch line (figure 10.1.6). A handrail shall be fixed on each side of a stair if it is 1 m wide or more, and on one side of the stair in other cases.

Landings should be provided at the top and bottom of every flight. The width and depth of a landing should be as least as great as the smallest width of the flight. Part of the floor of the building can count as landing. Landings should have a level surface and be clear of any permanent obstruction. A door may swing across a landing at the bottom of a flight where it will leave a clear space of at least 400 mm across the full width of the flight.

Approved Document K2/3 (1991) recommends that flights and landings should be guarded at the sides in dwellings where there is a drop of more than 600 mm. Suitable guarding would include a wall, screen, railing or balustrade. The guarding to a flight should prevent children from being held fast by the guarding, be constructed so that a 100 mm diameter sphere cannot pass through any openings in the guarding and be constructed so that children will not readily be able to climb the guarding. The minimum height of the guarding to single family dwellings should be 900 mm on stairs, landings, ramps and edges of internal floors. The guarding should be able to resist a horizontal force at these heights of 0.36 kN/m. Any glazing below these heights should be of glass blocks, toughened glass or laminated safety glass. Wired glass should not be used.

An alternative way to satisfy the stairway requirements of the Building Regulations is to comply with the provisions of BS 5395.[2] This Standard also provides additional information which may be useful in setting out stairways. Further design and constructional information concerning domestic timber stairways is provided in BS 585.[3]

Approved Document K1 (1991) requires fixed ladders to have fixed handrails on both sides and should only be used in loft conversions and then only when there is not sufficient space to accommodate a stair and give access to only one habitable room. Retractable ladders are not acceptable for means of escape.

Ramps

Approved Document K1 (1991) to the Building Regulations recommends that ramps should not be steeper than 1:12 and should be clear of permanent obstructions, to permit safe passage. All ramps and landings should have a clear headroom of at least 2 m. Furthermore, the requirements with regard to the provision of landings, handrails and guarding are the same as for stairs.

Pedestrian Guarding

Schedule 1 to the Building Regulations (1991) (paragraph K2) requires that stairs, ramps, floors and balconies, and any roof to which people normally have access, shall be guarded with barriers where they are necessary to protect users from the risk of falling, while paragraph K3 deals with vehicle barriers.

Approved Document K2/3 (1991) to the Building Regulations recommends that guarding should be provided where it is reasonably necessary for safety to guard the edges of any part of a floor, (including an opening window), gallery, balcony, roof (including rooflights and other openings), any other place to which people have access (unless only for the purpose of maintenance and repair), and any light well, basement or similar sunken area next to a building.

Any wall, parapet, balustrade or similar obstruction may serve as guarding. Guarding to edges of internal floors in single family dwellings should be at least 900 mm high, and that to external balconies and edges of roofs at least 1100 mm high. The guarding should (in order to stop children under 5 years being held fast by it) be constructed so that a 100 mm diameter sphere cannot pass through any openings on it and so that children will not readily be able to climb it. Guarding should be capable of resisting a horizontal force of 0.36 kN/m at a height of 900 mm and 0.74 kN/m at a height of 1100 mm.

The requirements with regard to protective barriers can also be met by following the relevant recommendations of BS 6180.[4]

TIMBER STAIRCASE CONSTRUCTION

Terminology

A considerable number of technical terms are used to describe component parts of staircases and the more important ones are listed and defined.

Tread. The upper horizontal surface of a step (figure 10.1.8).

Riser. The vertical front face of a step between two consecutive treads (figure 10.1.8).

Step. A combined tread and riser (figure 10.2.1).

Nosing. The front edge of a tread projecting beyond the face of the riser and includes the edge of a landing (figure 10.2.1).

Rise. The vertical distance between the upper surfaces of two consecutive treads (figure 10.2.1).

Going. The horizontal distance between the nosing of a tread and the nosing of the tread, ramp or landing next above it or the depth of the tread less any overlap with the next tread, or the horizontal dimensions from front to back of a tread less any overlap with the next tread above (figure 10.2.1).

Flier. A normal parallel step in a straight flight of stairs (figure 10.1.1).

Winder. A tapering tread where the stair changes direction, radiating from a newel (figure 10.1.3).

Kite winder. The middle step of three winders at a quarter turn, with a shape resembling a traditional flying kite (figure 10.1.3).

Round-end step. A step at the bottom of a flight of stairs with a semicircular end the width of the tread (figure 10.1.2).

Half-round step. A step with a semicircular end occupying the width of two treads.

Bullnose step. A step with a quadrant or quarter-round end (figure 10.1.1).

Curtail step. A step with a scroll end matching the finish to the handrail (figure 10.1.3).

String. An inclined member supporting the ends of treads and risers (*wall* string is fixed to a wall and *outer* string is away from it) (figure 10.1.9).

Cut or *open string*. An outer string with its upper edge cut to the shape of the treads and risers (figure 10.2.2).

Close string. An outer string with parallel edges into which the ends of treads and risers are housed (figure 10.2.2).

Newel. The post at the end of a flight to which the strings and handrail are framed; the cap is the head of the newel and may be applied or worked on solid, while the drop or pendant is the lower end projecting below a floor, which often forms a decorative feature (figure 10.2.2).

Baluster. The small bars or uprights between the handrail and string (figure 10.2.2).

Balustrade. This normally refers to solid panelling between handrail and string, but could also be applied to a balustrade wall or an 'open' balustrade (framework of

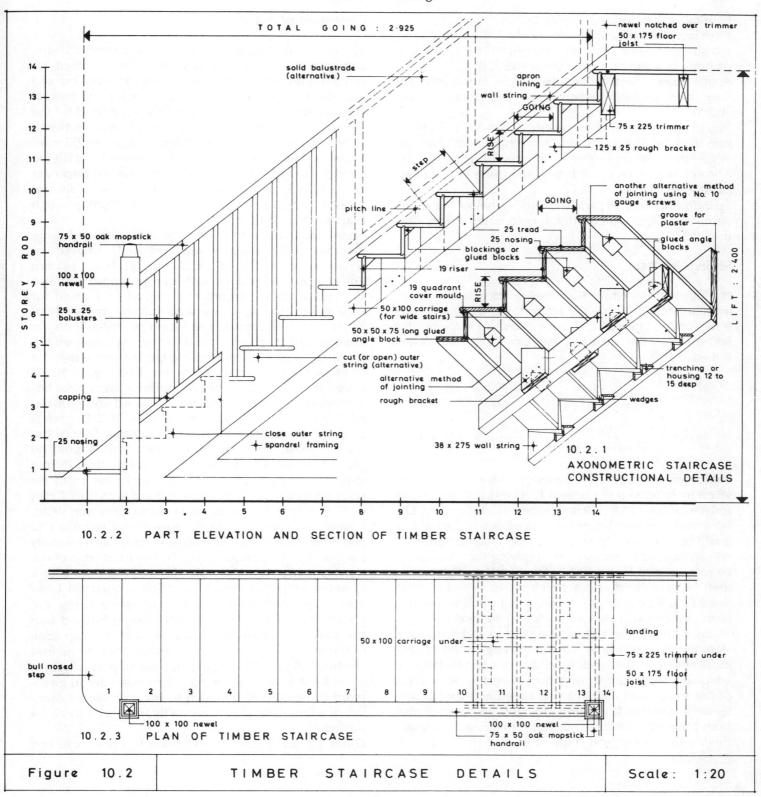

TOTAL GOING : 2·925

solid balustrade (alternative)

75 × 50 oak mopstick handrail

100 × 100 newel

25 × 25 balusters

capping

25 nosing

STOREY ROD

pitch line

cut (or open) outer string (alternative)

alternative method of jointing

rough bracket

close outer string
spandrel framing

step

newel notched over trimmer

50 × 175 floor joist

apron lining

wall string

GOING

RISE

75 × 225 trimmer

125 × 25 rough bracket

another alternative method of jointing using No. 10 gauge screws

groove for plaster

25 tread

25 nosing

blockings or glued blocks

19 riser

19 quadrant cover mould

50 × 100 carriage (for wide stairs)

50 × 50 × 75 long glued angle block

glued angle blocks

RISE

LIFT : 2·400

trenching or housing 12 to 15 deep

wedges

38 × 275 wall string

10.2.1 AXONOMETRIC STAIRCASE CONSTRUCTIONAL DETAILS

10.2.2 PART ELEVATION AND SECTION OF TIMBER STAIRCASE

bull nosed step

50 × 100 carriage under

landing

75 × 225 trimmer under

50 × 175 floor joist

100 × 100 newel

100 × 100 newel
75 × 50 oak mopstick handrail

10.2.3 PLAN OF TIMBER STAIRCASE

| Figure 10.2 | TIMBER STAIRCASE DETAILS | Scale: 1:20 |

handrail and balusters) (figure 10.2.2).

Carriage. An inclined timber placed halfway between the strings to give additional support to the underside of steps in a wide stairway (figure 10.2.1).

Angle blocks. Small triangular blocks glued into the underside angle between riser and tread (figure 10.2.1).

Rough brackets. Shaped pieces of wood, usually 25 mm thick, nailed to the side of carriages with the top edge fitting the underside of the tread (figure 10.2.1).

Apron lining. A lining or facing to a trimmer at a landing (figure 10.2.2).

Spandrel framing. A panelled infilling below the outer string of a flight which often incorporates a door (figure 10.2.2).

Setting Out Stairs

The normal procedure for drawing stairs is illustrated in figure 10.2.2. Firstly the lift or total rise (2.400) and travel or total going (2.925) are drawn to scale on the section. Next the tread and riser dimensions are determined.

$$\text{rise} = \frac{\text{lift}}{\text{number of risers}} = \frac{2.400}{14} = 172 \text{ mm}$$

$$\text{going} = \frac{\text{travel}}{\text{number of treads}} = \frac{2.925}{13} = 225 \text{ mm}$$

These are checked for suitability using the formula given in Approved Document K1 (1991) ($2R + G = 550$ to 700), $344 + 225 = 569$ mm, which is satisfactory. On occasions the only information available is the lift and it will be necessary to select a suitable combination of rise and going dimensions by applying the formula or by reference to figure 10.1.7. The number of steps can then be calculated and a check made to ensure that there is sufficient length available to accommodate the stairs, otherwise a steeper pitch may be needed.

Draw the bottom step on the section and produce a sloping line (pitch line) from its nosing to the uppermost nosing at the landing, and check to see there is adequate headroom throughout. Then the outer lines of the treads and risers can be speedily drawn by extending upwards from the graduated base and across from the vertical storey rod. The remaining details will then follow in a logical sequence.

Constructional Aspects

The essential parts of a wood stair are treads, risers, strings and newels, although risers are omitted on some modern stairs. The treads and risers are housed into the strings, the strings are tenoned and pinned into the newels and the newels are supported by the floors. The two strings may vary in thickness between 32 and 50 mm, and with a minimum depth of 225 mm. Where a stair is between 900 and 1220 mm in width the tread should be increased to 27 mm finished thickness and a rough carriage incorporated to give additional support to treads by means of interconnecting rough brackets (figure 10.2.1).

Steps. These are formed by framing treads and risers together with tongued and grooved, housed or screwed joints. Risers should be at least 14 mm finished thickness wood or 9 mm plywood, and also the tops of risers should be tongued or housed for their full thickness, 6 mm deep into the underside of the treads. It is desirable to locate the tongue on the inner face of the riser to avoid weakening the nosing of the tread unduly. The lower edges of risers may be fixed to treads (minimum 20 mm thick) with Nr 10 gauge screws not less than 32 mm long, at centres not exceeding 230 mm. In practice, tongued and grooved joints are often used (figures 10.1.8 and 10.2.1), preferably with the bottom tongue on the outer face of the riser to mask the joint. Treads and risers should be glued-blocked with angle blocks not less than 75 mm long and 38 mm wide, with two blocks per tread for stairs up to 900 mm wide, three for stairs 900 to 990 mm wide and four for wider stairs up to 1200 mm wide. Treads and risers should be housed not less than 12 mm deep into tapered housings in strings and securely wedged and glued (figure 10.2.1). The front edge of the tread projects over the outer face of the riser with a rounded edge (nosing). In addition a quadrant bead may be planted on the riser immediately below the nosing (figure 10.2.1) or a scotia mould housed into the underside of the tread, in which case the top edge of the riser may or may not be tongued or housed into the tread (figure 10.1.8). The bottom step of a flight is often finished differently from the remainder to give a more commodious approach to the stairs and a number of alternatives are shown in figures 10.1.1 to 10.1.4 inclusive.

Outer strings. These are two main forms – close and

cut or open. Close strings have both top and bottom edges parallel to the inclinations of the stairs with the ends of treads and risers housed into the strings (figure 10.1.9). Cut strings have their top edge cut to the shape of the steps, the bottom edge following the rake of the stairs (figure 10.2.2). The wall string can be grooved to receive the plaster at the top edge (figure 10.2.1) or be fixed to grounds with plaster extending behind the string (figure 10.1.9). Strings should have a minimum thickness of 32 mm.

Changes of direction in a stairway can be obtained by using landings or winders (figures 10.1.2 and 10.1.3). A half-space landing is formed by fixing a trimmer across the width of the stairway, to support the ends of bridging joists, and the newel is notched over it. The winders are supported by bearers built into the wall at one end and framed into the newel at the other, while the narrow ends of winders are housed to newels.

A *newel* is placed at every change of direction of flights. Outer newels normally have a minimum cross-section of 75 × 75 mm and wall newels (half-section plugged to walls) at least 75 × 38 mm. Treads, risers, winders and shaped ends of steps should be housed into newels to depth of at least 12 mm. Newels should be mortised and draw-bored to receive strings and mortised for handrails. Handrails, balusters and balustrades should be designed and constructed to secure a high degree of safety and rigidity. The handrail should preferably be of hardwood and be designed to afford a good grip. The balustrade may consist of vertical balusters, rails parallel to the inclination of the stairs, panelled infilling or a solid balustrade of studding, breeze blocks or bricks plastered on both faces. The solid balustrade will result in two wall strings and is often finished with a flat hardwood capping about 38 mm thick.

Defects in Staircases

Every year there are about 200 000 accidents on domestic stairs. BRE site surveys have found many defects, including newels and top nosings dangerously insecure, flights not rigidly fixed, handrails not sanded smooth, handrail brackets presenting sharp obstructions, wood spacers between handrail and apron linings obstructing passage of hand, and mouldings on apron linings fixed so that they wedged the hand between moulding and handrail. On over a quarter of the sites inspected, unprotected stairs were damaged after installation and loosened items, such as nosings, may not be re-fixed.[5]

STAIRWAYS IN OTHER MATERIALS

Apart from the traditional wood newel staircase, stairs can be constructed in various other materials or combinations of them. Some of the more common forms are examined.

Open-riser Stairways

These consist of treads supported by strings or carriages without risers and can be constructed in a number of ways. One method, as illustrated in figure 10.3.1, incorporates hardwood treads supported on hardwood brackets bolted to a pair of hardwood strings, and with steel balustrading and hardwood handrails. Another variation is to house the ends of the treads to the strings. Alternatively, the steel brackets may be housed and screwed to the treads and welded to steel channel strings. Metal or precast concrete treads may be used in external situations. This form of stairway gives an impression of spaciousness and facilitates cleaning and the distribution of light and heat.

Stone Stairways

These generally consist of stone steps supported at their ends by brick walls. *Skeleton* stairs incorporate stone slabs usually 50 mm thick for 750 mm wide stairs and 75 mm thick for 900 mm wide stairs, with open risers. *Built-up* stairs give a stronger job with stone slab risers between the treads. *Solid* rectangular steps about 300 × 150 mm in size with a 25 mm overlap (figure 10.3.3) give a very strong job, and the inclusion of a rebated joint provides a fire check. *Spandrel* steps with splay rebated joints, similar to the precast concrete steps illustrated in figure 10.3.6, are less strong than solid steps but give a better appearance, regular soffit and increased headroom below.

Reinforced Concrete Stairways

These can be formed of beams acting as strings with the steps spanning between them, but the more common arrangement consists of a reinforced concrete slab with projections forming the steps (figure 10.3.2).

Curbs may be formed *in situ* with the stairs to receive balustrades. The steps may be finished with granolithic (cement and granite chippings) or terrazzo (cement and marble chippings) placed in position after the concrete has set (figure 10.3.4). Precast inserts should be set in terrazzo faced treads, parallel with the nosings, for reasons of safety. The inserts can be formed of grooves filled with carborundum and cement or cubes of carbite, alundum or similar materials in various colours, arranged in chequerboard or other patterns.

Precast Concrete Stairways

These are generally constructed of spandrel steps faced with terrazzo, granolithic or cast stone with splay rebated joints (figure 10.3.6). They have non-slip finishes incorporating carborundum or alundum, or suitable precast inserts. Another arrangement is to bed precast concrete treads and risers either separately or as combined units onto an *in situ* reinforced concrete stairway (figure 10.3.5).

Pressed Steel Stairways

These often consist of steel strings, 250 to 300 mm deep, with steel treads and risers riveted to them. Landings may be formed of flat or dovetailed sheets supported on steel angles and channels. Treads may be covered with various finishes.

WOOD TRIM

Wood trim can be defined as 'products of uniform profile manufactured by linear machining only', and encompasses architraves, skirtings, picture rails, dado rails, cover fillets, and quadrant, half-round and scotia moulds. It might be helpful at this stage to describe the more common items of wood trim even though they will follow plastering on the site as part of joinery second fixings.

Architrave. A moulding or fillet around an opening applied to the face to cover the joint between joinery and the adjoining work (figures 10.3.7 to 10.3.9).
Skirting. A finishing member fixed to a wall where it adjoins the floor to cover the joint and protect the wall finish (figure 6.3.2).
Picture rail. A plain or moulded rail fixed to walls and

from which pictures and other decorative features can be hung (figure 10.3.13).
Cover fillet. A fillet to cover a joint in joinery or between joinery and adjoining work (figure 10.3.14).

The standard finishes for architraves, skirtings and picture rails are chamfered and rounded (figures 10.3.7 and 10.3.13), rounded (figure 10.3.8) and bevel-rounded (figure 10.3.9). All three mouldings are simple in form and reduce dust collection to a minimum. Quadrant and half-round moulds (figure 10.3.11) are frequently used to mask the joint between plaster and joinery. Typical sizes given in BS 1186, Part 3[6] are as follows and show significant reductions in recent decades.

> architraves – 13 × 45 and 20 × 70 mm
> picture rails – 13 × 45 and 13 × 70 mm
> skirtings – 13 × 70 to 20 × 120 mm
> scotia (figure 10.3.10) – 13 × 13 to 27 × 27 mm
> quadrant (figure 10.3.11) – 11 × 11 to 20 × 20 mm
> half-round (figure 10.3.11) – 13 × 33 and 20 × 45 mm
> cover fillet (figure 10.3.14) – 11 × 33 and 11 × 45 mm

More elaborate mouldings can be applied to specially designed members. Figure 10.3.12 shows an architrave incorporating a cyma reversa, ovolo and fillet band, while another (figure 10.3.15) embraces cavetto and cyma reversa moulds. Cornices at the junction of walls and ceilings generally contain a number of mouldings (figures 10.3.16 and 10.3.17), while figures 10.3.18 to 10.3.20 show some of the more ornate treatments for skirtings.

SIMPLE JOINERY FITTINGS

The design and construction of some of the more common domestic joinery fittings are now considered, with particular reference to kitchen fitments and built-in wardrobes.

Kitchen Fitments

Some measure of standardisation was essential if kitchen fitments were to be produced cheaply and efficiently. The pioneering work stems from the introduction of BS 1195[7] in 1948, which recommended a standardised range of different cupboard units, which could be assembled to produce the varying requirements for storage spaces and yet, at the same time, suit a particular kitchen plan. It was necessary to co-

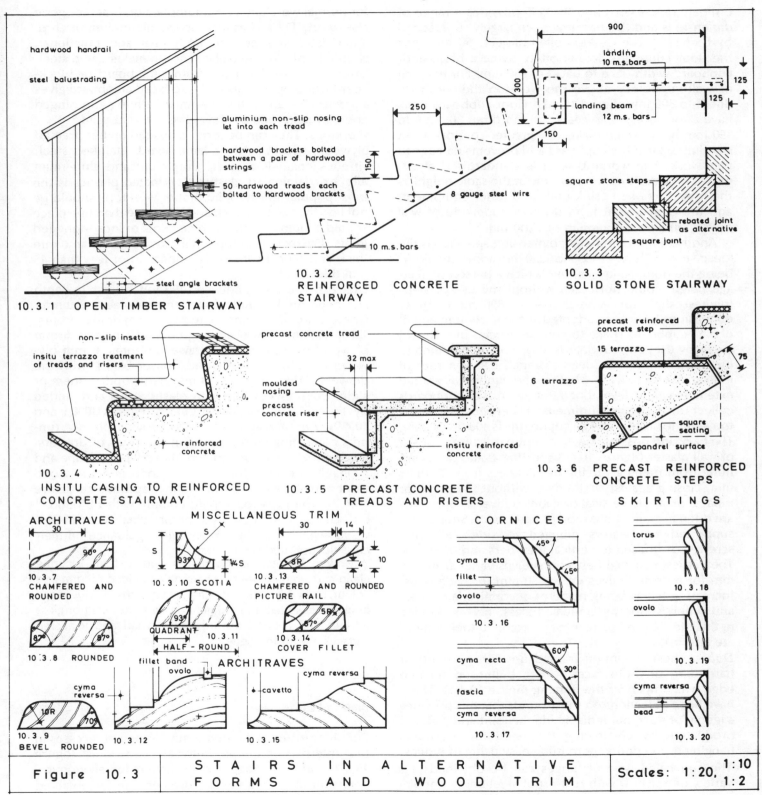

10.3.1 OPEN TIMBER STAIRWAY

hardwood handrail
steel balustrading
aluminium non-slip nosing let into each tread
hardwood brackets bolted between a pair of hardwood strings
50 hardwood treads each bolted to hardwood brackets
steel angle brackets

10.3.2 REINFORCED CONCRETE STAIRWAY

900
landing
10 m.s. bars
300
landing beam
12 m.s. bars
250
125
125
150
150
8 gauge steel wire
10 m.s. bars

10.3.3 SOLID STONE STAIRWAY

square stone steps
rebated joint as alternative
square joint

10.3.4 INSITU CASING TO REINFORCED CONCRETE STAIRWAY

non-slip insets
insitu terrazzo treatment of treads and risers
reinforced concrete

10.3.5 PRECAST CONCRETE TREADS AND RISERS

precast concrete tread
32 max
moulded nosing
precast concrete riser
insitu reinforced concrete

10.3.6 PRECAST REINFORCED CONCRETE STEPS

precast reinforced concrete step
15 terrazzo
6 terrazzo
75
square seating
spandrel surface

SKIRTINGS

ARCHITRAVES
MISCELLANEOUS TRIM
CORNICES

30
90°

10.3.7 CHAMFERED AND ROUNDED

S
S
93°
¼S

10.3.10 SCOTIA

30   14
3R
4
10

10.3.13 CHAMFERED AND ROUNDED PICTURE RAIL

cyma recta
fillet
ovolo
45°
45°

10.3.16

torus

10.3.18

87°   87°

10.3.8 ROUNDED

93°
QUADRANT
HALF-ROUND

10.3.11

5R
87°

10.3.14 COVER FILLET

cyma recta
fascia
cyma reversa
60°
30°

10.3.17

ovolo

10.3.19

cyma reversa
10R
70°

10.3.9 BEVEL ROUNDED

fillet band
ovolo

ARCHITRAVES

cavetto

10.3.12

cyma reversa

10.3.15

cyma reversa
bead

10.3.20

Figure 10.3   STAIRS IN ALTERNATIVE FORMS AND ALTERNATIVE WOOD TRIM   Scales: 1:20, 1:10, 1:2

ordinate dimensions and BS 1195 provides a range of dimensions and this has now been largely superseded by BS 6222.[8] The worktop often projects 50 mm over the floor cupboard below, often giving a basic floor cupboard width (face to back) of 550 mm. For hanging or wall cupboards above the worktop the width reduces to 300 mm (figure 10.4.2). Floor-cupboard units have a toe recess or recessed plinth often 50 × 80 to 150 mm high, which makes for greater comfort when standing against the cupboard and prevents damage to paintwork. The worktop level is standardised at 900 mm to produce a continuous line at the same height as the tops of cookers, sink units and other equipment, and the clear height from floor to underside of wall units should be a minimum of 1300 mm.[9]

Another important height dimension specified is the 'dead storage' level at 1950 to 2250 mm above the floor, being the nominal level above which a person of average height could not reach without the use of steps (highest shelf for general use – 1800 mm). These dimensions have been adopted as the maximum height of wall-cupboard space to contain articles in frequent everyday use, any cupboards above being confined to long-term storage or 'dead storage'. It is a matter of individual choice as to whether the cupboards should extend to ceiling level, but if not the tops of cupboards collect dust. The usual dimension between the worktop and the bottom of wall cupboards is 400 or 450 mm. The back and possibly sides of cupboards often consist of wall plaster (figures 10.4.3 and 10.4.4) and doors are hung from framing varying in thickness from 25 to 38 mm, closing against shelf edges without the need for rebates or stops. Vertical divisions between cupboards are often formed of plywood or hardboard. Shelves are supported on bearers plugged to side walls and screwed or housed to vertical members and divisions. The vertical distance between shelves varies from 150 mm for narrow shelves up to 450 mm for wide ones. Increasing use is being made of decorative laminates, and simply supported annealed glass shelves, varying in thickness from 6 to 12 mm according to their length, are also used.[6]

Doors. These are often framed up with plywood or hardboard on both faces with a bullnosed rebated edge which fits over the framing (figure 10.4.2). There have been objections on the grounds that this provides a ledge for dust, but it does hide the joint and it assists production by eliminating the need for individual hanging of the doors, permitting easy fixing of hinges. The concealed type of hinges can be obtained in a variety of finishes, with rust-proof steel being popular for normal work and chromiumplate on brass for good class work. D handles in anodised aluminium or chromiumplate on brass are both neat and functional, coupled with double ball catches in nickelplated steel. Doors may also be made from laminboard or blockboard with lipped edges. Plastic-faced chipboard gives a serviceable and attractive finish. Sliding and hinged doors of patterned or clear glass are also used.[6]

Worktops. They can be formed of solid timber, framed plywood, blockboard or laminboard, stainless steel, vitreous enamelled steel, aluminium and aluminium alloys or laminated plastics. Laminated plastics is the most popular finish. Solid timber worktops should be not less than 20 mm thick and if in more than one piece should be joined with machine joints or loose-tongued joints. Plywood worktops shall not be less than 6 mm thick and blockboard or laminboard not less than 15 mm thick.[6]

Drawers. These vary between 100 and 250 mm in depth and need double pulls if the length exceeds 500 mm. Drawers are often constructed of 25 mm drawer fronts, 12 or 19 mm drawer sides grooved to receive 6 mm plywood or 9 mm wood drawer bottoms, 6 mm plywood or 12 or 19 mm wood drawer backs, with the drawers sliding on 32 × 19 mm hardwood runners or slides possibly covered with plastic strips and guided by 12 × 15 mm hardwood guides (figures 10.4.1 and 10.4.2). Another approach is to provide plastics runners coinciding with grooves in the sides of drawers. Drawer fronts may be rebated to overlap the frame and probably the best form of joint between the wood front and sides of a drawer is the lap dovetail (figure 10.4.2). The plywood should comply with BS 6566.[10] BRE Digest 394[11] describes the principal types of plywood available in the UK and gives guidance on their classification and selection.

Modern kitchen design ranges widely from securing clean and simple lines to a harmonious blend of form and colour, in a wide variety of attractive designs. A wide range of laminate surfaces in different designs with matt or gloss finishes and a trend towards solid wood and veneered surfaces are all in vogue.

Built-in Wardrobes

Built-in wardrobes are frequently formed in recesses in the internal partitions between bedrooms (figure 10.5.1), often in staggered formation for economical provision. A strong vertical division can be obtained

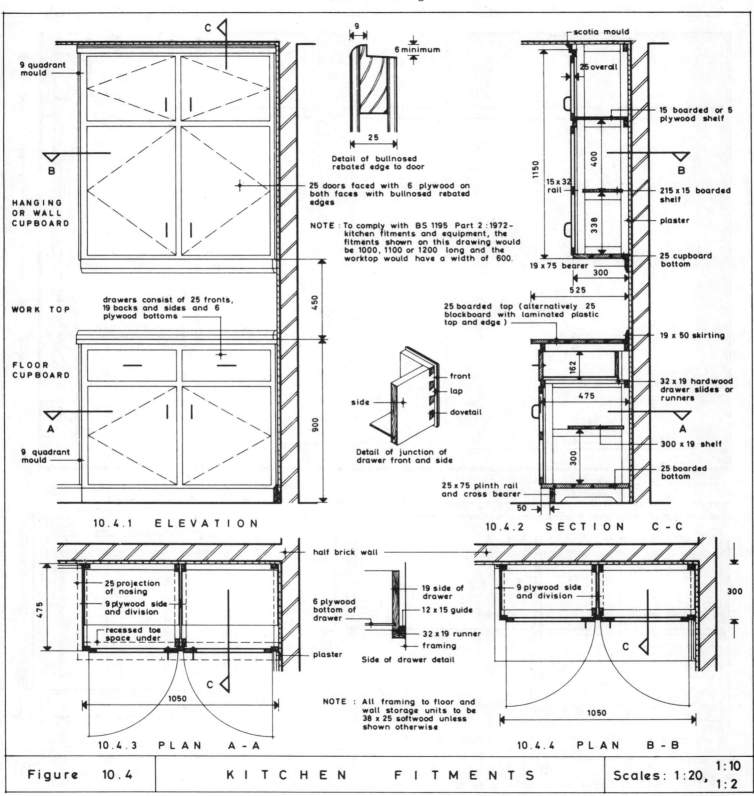

9 quadrant mould

HANGING OR WALL CUPBOARD

B

C

9

6 minimum

25

Detail of bullnosed rebated edge to door

25 doors faced with 6 plywood on both faces with bullnosed rebated edges

NOTE : To comply with BS 1195 Part 2 :1972 – kitchen fitments and equipment, the fitments shown on this drawing would be 1000, 1100 or 1200 long and the worktop would have a width of 600.

WORK TOP

drawers consist of 25 fronts, 19 backs and sides and 6 plywood bottoms

FLOOR CUPBOARD

9 quadrant mould

A

450

900

front
lap
side
dovetail

Detail of junction of drawer front and side

scotia mould

25 overall

15 boarded or 5 plywood shelf

400

B

1150

15 x 32 rail

338

215 x 15 boarded shelf

plaster

25 cupboard bottom

19 x 75 bearer

300

525

25 boarded top (alternatively 25 blockboard with laminated plastic top and edge)

19 x 50 skirting

162

32 x 19 hardwood drawer slides or runners

475

A

300 x 19 shelf

300

25 boarded bottom

25 x 75 plinth rail and cross bearer

50

10.4.1 ELEVATION

10.4.2 SECTION C - C

half brick wall

25 projection of nosing

9 plywood side and division

recessed toe space under

475

19 side of drawer

6 plywood bottom of drawer

12 x 15 guide

32 x 19 runner

framing

Side of drawer detail

plaster

C

1050

9 plywood side and division

300

C

1050

NOTE : All framing to floor and wall storage units to be 38 x 25 softwood unless shown otherwise

10.4.3 PLAN A - A

10.4.4 PLAN B - B

| Figure 10.4 | KITCHEN FITMENTS | Scales: 1:20, 1:10 1:2 |

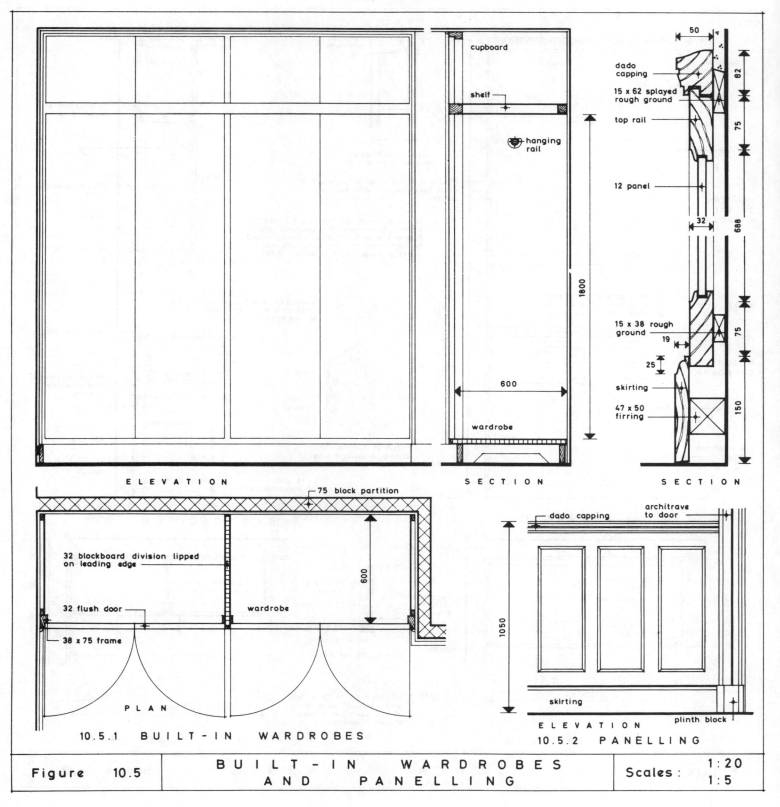

E L E V A T I O N

cupboard

shelf

hanging rail

1800

600

wardrobe

S E C T I O N

50

dado capping

15 x 62 splayed rough ground

top rail

12 panel

32

688

15 x 38 rough ground

19

25

skirting

47 x 50 firring

82

75

75

75

150

S E C T I O N

75 block partition

32 blockboard division lipped on leading edge

32 flush door

38 x 75 frame

600

wardrobe

P L A N

10.5.1   B U I L T - I N   W A R D R O B E S

dado capping

architrave to door

1050

skirting

plinth block

E L E V A T I O N

10.5.2   P A N E L L I N G

| Figure   10.5 | B U I L T - I N   W A R D R O B E S
A N D   P A N E L L I N G | Scales : | 1:20
1:5 |

with blockboard 32 mm thick, lipped on the leading edge. Flush doors may be hung to a 75 × 38 mm frame with quadrant bead masking the joint between the frame and the plaster. In figure 10.5.1, the base has been raised 150 mm to provide storage for shoes. Alternatively, a recessed plinth could be introduced. The upper cupboards are for storage of cases and similar little-used objects. A chromiumplated hanging rail is provided at a height that provides adequate clearance for clothes (men's suits: 1000 mm, men's overcoats: 1400 mm, and ladies' coats or dresses: 1750 mm). Shoe racks and drawers for underclothes should be 100 to 150 mm deep and a hat shelf could be 300 mm wide with 200 mm clearance above it.

The use of sliding doors on runners could be valuable if floor space is restricted. Wardrobe sliding-door gear is supplied by several manufacturers and generally incorporates overhead alloy track which supports hangers with nylon wheels. The bottoms of the doors are held in position by nylon guides screwed to the cupboard floor or have retractable bolts which run in suitable track. Wardrobe doors, framing and sides often have a plastic laminated finish for improved appearance and ease of cleaning.

PANELLING

Polished hardwood panelling, possibly up to dado or chair rail level, can form a very attractive albeit expensive internal wall finish, and is more suited for board rooms and prestige offices than domestic work. A typical detail is given in figure 10.5.2. It consists of 12 mm panels let into grooves in 32 mm rails and stiles. A 105 mm wide top rail is needed to obtain a 75 mm width of exposed flat surface, with 15 mm concealed within the dado capping and a further 15 mm moulded to give emphasis to the panelling. The dado capping or chair rail is a fully moulded member which provides an effective and attractive finish to the top of the panelling. A skirting, 19 × 150 mm, has a 25 mm moulding to provide a continuous horizontal line at the foot of the panelling. Rough grounds and firrings provide fixings for the panelling and associated members, with a splayed edge to the top ground to give a key for the plaster. A large and extremely ornate architrave is used around the door opening to blend in with the panell-

ing, terminating in a plinth block to frame the end of the skirting. The panels do not fit tightly into the grooves to allow for movement arising from shrinkage or expansion. Rails and stiles are secured by trenails or wood pins at intersecting joints.

It is important to guard against attack by dry rot fungus or other forms of decay. This is best accomplished by treating all grounds and the back of panelling with a suitable preservative, using adequately seasoned timber and providing ventilation through grilles at top and bottom.

REFERENCES

1. *The Building Regulations 1991 and Approved Document K (1991)*. HMSO
2. *BS 5395*: Stairs, ladders and walkways; *Part 1: 1977 (1984)* Code of practice for the design of straight stairs; *Part 2: 1984 (1989)* Code of practice for the design of helical and spiral stairs
3. *BS 585*: Wood stairs, *Part 1: 1989* Specification for stairs with closed risers for domestic use, including stairs with straight and winder flights and quarter and half landings; *Part 2: 1985* Specification for performance requirements for domestic stairs constructed of wood-based materials
4. *BS 6180: 1982* Code of practice for protective barriers in and about buildings
5. *BRE Defect Action Sheet 54*: Stairways: safety of users – installation (1984)
6. *BS 1186* Timber for and workmanship in joinery, *Part 3: 1990* Specification for wood trim and its fixing
7. *BS 1195*: Kitchen fitments and equipment; *Part 2: 1972* Metric units
8. *BS 6222*: Domestic kitchen equipment, *Part 1: 1982* Specification for co-ordinating dimensions; *Part 3: 1988* Specification for performance requirements for durability of finish; *Part 4: 1988* Code of practice for protection, storage and installation of fitted kitchen units
9. *DOE Design Bulletin 24, Part 2*: Spaces in the home – kitchens and laundering spaces. HMSO (1972)
10. *BS 6566*: Plywood; *Part 1: 1985* Specification for construction of panels and characteristics of plies including marking
11. *BRE Digest 394*: Plywood (1994)

11 FINISHINGS

This chapter is concerned primarily with finishings to walls and ceilings. Floor finishings were considered in chapter 6. In selecting wall and ceiling finishings, probably the two most important considerations are appearance and maintenance costs. In particular situations other factors may also be important, such as resistance to condensation, acoustic properties and provision of a smooth, even surface.

INTERNAL WALL FINISHINGS

Plastering

The great range of plasters and of background to receive them that are currently available has tended to complicate the choice of plaster.[1] Before making a choice it is advisable to list the functions of plaster. It is required to conceal irregularities in the background and to provide a finish that is smooth, crackfree, hygienic, resistant to damage and easily decorated. It may also be required to improve fire resistance, to provide additional thermal and sound insulation, to modify sound absorption or mitigate the effects of condensation. These latter aspects are generally of secondary importance, and the selection is influenced mainly by the surface finish desired. The stages in the selection of a plastering specification are now listed.

Finish. Whether it should be hard or soft, smooth or textured, the choice of decorative finish and the time available for drying out of plaster before decorations are applied. Table 11.1 lists various plaster finishes and considerations affecting their choice.

Number of coats. Until quite recently, three-coat plasterwork was applied to most backgrounds, with a first levelling undercoat followed by a second undercoat to provide suction for the finishing coat. Two-coat work is generally satisfactory for use with brickwork and block-

work, provided extreme variations of suction are avoided. Some backgrounds, such as plasterboard and smooth concrete, have level surfaces and uniform suction permitting single-coat plastering.

Plaster undercoats. These need to be selected to match the suction of the background, its surface irregularities and other important properties. Undercoats based on gypsum plasters unlike those based on cement do not shrink appreciably on drying. A background of very high suction, such as aerated concrete or insulating bricks, may absorb water so rapidly from the plaster mix that its adhesion to the background is weakened and it may severely limit the time available for levelling the coat. Hence it is advisable to use mixes which retain their water in contact with backgrounds of high suction. The undercoat must have a good key or bond with the background. With backgrounds offering little or no key, such as smooth concrete, bonding agents may be used prior to plastering, as described in BS 5270.[2] Backgrounds liable to drying shrinkage should be allowed to dry out completely before plastering. Thermal movements of backgrounds are rarely sufficiently serious to cause problems, except with concrete roofs and heated concrete floors.

Type of finish. There is a wide range of plasters available and the main types with their more important characteristics are scheduled in table 11.1.

Lime plaster is weak, easily indented, slow hardening and shrinks on drying, and for these reasons is now little used. It does however help in absorbing condensation. A gauged lime plaster (cement-lime-sand) has improved qualities.

Gypsum plasters should comply with BS 1191[3] and have a number of advantages over lime plaster, requiring fewer and thinner coats, which set hard in a few hours without shrinking, providing a hard, strong surface which does not cause alkali attack on paintwork. There are four classes of gypsum plaster.

Table 11.1 Characteristics of plaster finishes

Plaster finish	Surface hardness and resistance to impact damage	Other surface characteristics	Restrictions on early decoration	Shrinkage or expansion	Remarks
Lime plasters					
lime	Weak and very easily indented	Open-textured (depending on sand), absorbs condensation	Initially only suitable for permeable finishes that are unaffected by alkali	Shrink on drying, but shrinkage is reduced by addition of fine sand	Slow hardening; only apply on dry undercoats
gauged lime	Resistance to damage increases with proportion of gypsum plaster	Similar to above, but smoother finishes obtainable		Shrinkage is restrained by the gypsum content provided that over-trowelling is avoided	Only apply on dry undercoats
Gypsum plasters					
class D (Keenes)	Very hard and resistant to damage	Very level and smooth; particularly suitable for low-angle lighting conditions	None, except on undercoats containing cement or lime, or unless lime is added to the finishing coat	Expand during setting. Subsequent movements usually small, but too rapid drying can lead to delayed expansion	Set slowly and so allow ample time for finishing to a smooth surface; should not be allowed to dry too quickly
class C (anhydrous)	Hard and resistant to damage	Slightly less smooth than class D			
class B (hemihydrate)	Sufficiently hard and resistant for most normal purposes, but weakened by additions of lime	Sufficiently smooth and level for most purposes		Expand during setting, though extremely slightly with board finish plasters; subsequent movements are small	Set quickly; should be allowed to dry as soon as possible
lightweight	Surface hardness similar to class B plasters. Ease of indentation varies with the type of lightweight undercoat, but resilience tends to prevent serious damage	Sufficiently smooth and level for most purposes	None, but the higher water content of lightweight undercoats makes these somewhat slower to dry than sanded gypsum plaster undercoats	Expand during setting; subsequent movements usually small and easily restrained by background	Maximum fire resistance; lightweight plaster surfaces warm up more quickly than others and so help to prevent temporary condensation
Cement-lime-sand					
1:0–¼:3	Very strong and hard	Wood float finish	Initially only suitable for permeable finishes that are unaffected by alkali	Shrink on drying, but surface cracking can be minimised by avoiding over-working	Suitable for damp conditions
1:1:6	Strong and hard	Wood float finish			
1:2:9	Moderate	Wood float finish			
Single-coat finishes					
board finish gypsum plasters (class B)	Surface hardness similar to class B above, but resistance depends on background	On suitable backgrounds, similar to class B above	None; finish dries very quickly	Extremely small expansion on setting; subsequent movements small	
Thin-wall finishes					
based on gypsum	Softer than board finishes	Smooth and level on sufficiently level backgrounds	Dries very quickly: no restrictions when dry	Extremely small expansion on setting; subsequent movements small	
based on organic binders	Moderately hard, resistance depends on background	Matt surface, closely following the level of the background	Dries very quickly; no restrictions when dry	The very thin coats are restrained by the background	
gypsum projection plasters	Properties intermediate between class B and class C gypsum plasters				

Source: BRE Digest 213[1]

(1) *class A* (plaster of Paris) sets so rapidly that it is unsuitable for most work apart from small repairs;

(2) *class B* (retarded hemihydrate) plasters include board finishes and give a hard surface which is sufficiently resistant to impact for normal purposes, set quickly and expand during setting;

(3) *class C* (anhydrous) plasters give a surface harder than class B plasters and are slower setting;

(4) *class D* (Keene's) provides a very hard, smooth surface, and is slow-setting allowing ample time to bring to a fine finish that is particularly suitable for decoration with high gloss paints (the priming coat must be alkali resistant and applied when the work is dry).

Lightweight gypsum plasters are premixed, consisting of a lightweight aggregate, such as expanded perlite or exfoliated vermiculite and a class B plaster. Surface hardnesses are similar to class B plasters, but they are slower drying and so help to prevent temporary condensation.

Thin wall and *special finishing plasters* are based on gypsum or organic binders, and are generally applied by spraying and finishing with a broad spatula.

Although gypsum plasters give excellent results, care is needed in handling, mixing and applying. In particular, plasters belonging to different classes must not be mixed; plasters must be kept dry before use; only clean water must be used for mixing; mixes must be of the correct proportions, as too much sand in an undercoat will seriously reduce its strength; sand should comply with BS 1199,[4] cement should never be mixed in a mix with gypsum plasters; and the manufacturer's instructions should be followed regarding the suitability of plasters for different backgrounds.

Gypsum plaster mixes vary with the type of background, for instance thistle browning plaster (retarded hemihydrate gypsum plaster) is applied in proportions of one part of plaster to three parts of sand by volume to normal clay brick surfaces, but in the proportions of one to two to lightweight concrete blocks with lower suction. On the other hand Carlite browning plaster which is a premixed retarded hemihydrate gypsum plaster incorporating lightweight mineral aggregate is used neat, thus ensuring a uniform mix. The work is carried out in two coats in both cases; the floating coat normally being 11 mm thick and this is lightly scratched to form a key for the finishing coat which is normally about 2 mm thick, giving a total thickness of 13 mm.

The plaster should be finished truly vertical and be free from cracks, blisters and other imperfections. It is customary to round both internal and external angles of plaster to a radius of about 13 mm. Further guidance on plastering is given in BS 5492,[5] and includes mechanical projection plastering techniques.

Plastering on Building Boards

Plaster can be applied satisfactorily to a number of types of board, namely: insulating board, gypsum wallboard, gypsum lath, gypsum baseboard, gypsum plank and expanded plastics. Joints between all boards except gypsum lath should be reinforced with a strip of jute scrim with a minimum width of 90 mm to prevent cracking of the plaster. All internal and external angles should be reinforced with scrim around the angle between the boards and the plaster. Alternatively, galvanised mild steel corner beads may be used to give maximum protection at external angles complying with BS 6452,[6] and which details a range of profiles. Joints between walls and ceiling can be similarly scrimmed or, alternatively, a straight cut can be made through the plaster along the line of the junction. Small movements between walls and ceiling can be concealed by cornices, such as plasterboard cove.

A gypsum plaster should be used for single-coat work, preferably neat class B type b2 (retarded hemihydrate), often called board finish plaster, and for the undercoat in two-coat plastering, preferably sanded class B type a1 on building boards. The final coat could consist of neat gypsum plaster or a gauged lime plaster. Single-coat work is satisfactory when the boards have been fixed to give a true and level surface. Two-coat work normally finishes about 8 mm thick and single-coat work about 5 mm thick.

In rehabilitation work, the replacement of internal plaster is a common and often expensive item. BRE Good Building Guide 7[7] shows how to prepare commonly encountered background surfaces and how to select appropriate plaster systems to achieve a sound, durable finish.

Dry Linings

The use of plasterboard and similar sheet materials to form a dry lining to dwellings is quite popular, as it simplifies the work of finishing trades and can lead to earlier completions. Excellent guidance on the provision of dry linings and partitioning using gypsum plasterboard is given in BS 8212.[8] A permanent decoration can be selected in the knowledge that shrinkage cracks will not occur and painting or wallpapering can commence shortly after the lining is fixed. Several methods of fixing plasterboard are now described.

Timber battening This is a straightforward but rather expensive method, and is particularly well suited for counteracting damp penetration through the walls of a building. The battens are usually 40 mm wide and 15 mm thick, spaced at not greater than 450 mm centres, and fixed with chrome-steel nails. The plasterboard is fixed to the battens with taper-headed nails, taking care that nail heads do not puncture the plasterboard. When battens are fixed to walls which are damp or likely to become so, they should be pressure-impregnated with a timber preservative, to avoid the risk of staining the plasterboard and decorations. When the dampness is likely to be permanent, it is advisable to treat the board with a fungicidal wash before fixing, such as a one per cent solution of sodium pentachlorphenate. Plasterboard linings also provide a convenient method of improving the thermal insulation of external walls and to reduce the risk of condensation, preferably using aluminium foil-backed plasterboard.

Plasterboard battens. These consist of strips of plasterboard, stuck to the wall with a special adhesive, and can be substituted for the timber battens. Plasterboard grounds are first fixed horizontally at ceiling and floor level, and vertical grounds set between them at about 450 mm centres. The plasterboard is both stuck and nailed to the grounds.

Plaster dab and *fibre pack*. This is another fixing method. Pieces of bitumen impregnated fibreboard, about 75 × 50 mm, are stuck to the walls with plaster at the top, bottom and centre of each joint, with intermediate lines of packs to support the edges of boards. After about two or three hours, plaster is applied to the wall by trowel as dabs of about 100 mm diameter spaced at 450 mm centres in both directions. As boards are tamped into position, the dabs spread to give an effective contact surface of about 200 mm diameter. The boards are then nailed to the fibre packs with corrosion-resistant hardened nails which hold the boards into position until the plaster sets.

Plaster ribbon. In this method ribbons of plaster secure the boards to the wall. The ribbons, usually about 50 mm wide and varying in thickness from 12 to 38 mm, are applied to the back of boards through a spreading box and are laid parallel to the longest edge.

Plasterboards. These are produced with square, rounded, bevelled or tapered edges, and the joint finishes need to be considered in relation to the decoration. With square-edged boards it is difficult to fill the fine butt joint so as to produce a continuous plain surface to receive paint or distemper. Wallpaper with a slight texture or pattern will however mask the discrepancies. Other forms of edge can accommodate reinforcement to provide invisible joints and so give a satisfactory base for paint or distemper.

Plasterboard is essentially a building board consisting of a core of set gypsum plaster enclosed between and bonded to paper liners to form flat rectangular boards and it should comply with BS 1230.[9] The core can contain additives to impart additional properties. There are five basic types of plasterboard as follows

(1) *gypsum wallboard*, which is used primarily for linings to framed construction, with one self-finished surface, or designed to receive decoration direct;

(2) *gypsum lath*, which is a narrow-width plasterboard with rounded longitudinal edges, less expensive than wallboard and providing an ideal base for gypsum plaster;

(3) *gypsum baseboard*, either square or round edged plasterboard supplied in wider sheets than a lath as a base for gypsum plaster; unlike gypsum lath, all joints are scrimmed;

(4) *gypsum plank*, which is a relatively narrow-width board of greater thickness than other types of plasterboard; it is used as linings to framed construction and as casings to beams and columns; two types of board are available – one with two grey surfaces designed as a plaster base, and the other with one ivory surface intended for direct decoration; the former type has square edges and the latter has tapered edges for flush jointing. There is also a moisture resistant vapour check wallboard.

(5) *insulating gypsum plasterboard*, which has a bright metal veneer of low emissivity on one side.

BS 1230[9] describes five types of wallboard and two types of baseboard. Apart from gypsum wallboard and gypsum baseboard previously described, there is gypsum base wallboard to receive veneer finishes; gypsum moisture resistant wallboard; gypsum moisture repellant wallboard; gypsum wallboard F and gypsum baseboard F, both giving improved fire protection.

Gypsum plasterboard is obtained in thicknesses of 9.5, 12.5, 15 and 19 mm, with widths varying from 406 to 1200 mm and lengths from 1200 to 3000 mm, the actual sizes depending on the purpose and type of board used and the distance between fixings.

Fixing. Plasterboard is fixed to timber studs with 2 mm galvanised nails with small, flat heads, 30 mm long for 9.5 mm boards and 40 mm for 12.5 mm boards. No nailing should take place within 13 mm of edges of boards. The recommended spacing of studs is 450 mm centres for 9.5 mm board, 600 mm for 12.5 mm and 800 mm for 19 mm board. Boards should be fixed with a 3 mm gap between cut ends and they should be nailed to every support at 150 mm centres. It is desirable to avoid joints in line with door or window jambs. Plasterboard can also be fixed to metal studs.

Jointing. The method of jointing depends on the edge design of the boards. For instance, tapered edge boards normally have smooth seamless jointing composed of joint filler often reinforced with joint tape and finished with a slurry coat. A cover strip joint is often used with square-edged boards, covering the joint with embossed or corrugated paper strips, or paper-faced cotton tape, fixed with suitable adhesive. When a more pronounced panel effect is required, plain or patterned cover strips of wood, metal or plastic may be fixed over the joints. With bevelled-edge boards, the base of the V-joint may be partially filled with joint filler to bridge the joint between adjacent boards. This treatment improves the appearance of the V-joint and facilitates even decoration.

Angles. Most internal angles will have at least one cut edge and should desirably be reinforced with joint tape bedded in joint finish. External angles may be reinforced with corner tape bedded in joint filler and covered with joint finish. For maximum reinforcement a metal angle bead may be used, bedded in joint filler and covered with joint finish. A gypsum cove may be used to cover the junction of walls and ceilings and improve appearance. It consists of a gypsum core encased in ivory-coloured millboard and is normally fixed by sticking with adhesive or nailing or screwing to timber grounds or plugs.

Coves and cornices. The *gypsum cove* is the present day version of the much ornate and expensive plaster cornice of former days. Small *plaster cornices* are usually run in gauged plaster or a zinc mould. Where the projection exceeds 150 to 200 mm, the plaster must be firred out on brackets to suit the profile of the moulding. The brackets are lathed and rendered and the cornice finished with the help of a zinc mould mounted on a wooden horse running between guides fixed to wall and ceiling. Another alternative is to use *fibrous plaster* made up of plaster, glue, oil, wood wire and canvas. Fibrous plaster can be used to provide plain and moulded work and can cover large surfaces showing savings in time and weight compared with ordinary plaster. It is normally formed in gelatine moulds and fixed to grounds by specialist firms.

Plastics-surfaced Sheet Materials

These provide attractive finishes to wall surfaces on grounds of coverage, appearance, resistance to wear and water, durability, hygienic and low maintenance costs. Probably one of the best known types are melamine laminated thermosetting decorative sheets generally known as *decorative laminates*. They consist of layers of a fibrous material, such as paper, which has been impregnated with thermosetting synthetic resins and consolidated under heat and pressure. Upon the base lies the *print paper*, which is a layer of cellulose paper bearing the desired colour or printed pattern and impregnated with melamine formaldehyde (MF) resin. This may be covered by an over-lay of MF-treated alpha-cellulose paper which becomes transparent under heat-pressure treatment. Decorative laminates are often 1.5 mm thick and are already veneered to, or prepared for veneering to, a rigid substrate such as chipboard or plywood, to produce a composite sheet material suitable for interior cladding or free-standing constructions. The quality of decorative laminated plastics sheet is prescribed in BS 3794[10] and the normal thickness for domestic use is about 1.5 mm.

Decorative laminate surface finishes range from fine gloss to marked texture or raised patterns. They can normally withstand surface temperatures of up to 180°C for short periods without blistering or appreciable surface damage. They have been used successfully as wall linings to kitchens, bathrooms, halls, staircases, clubrooms, hospital operating theatres, public baths, lift cars, toilet cubicles and fitted furniture. Some melamine plastic laminates contain woven linens and fine quality hessians to provide the decorative feature. Plastic-faced plasterboard is also available.

Hygienic plastic cladding sheets are available in several colours and can be slotted or fitted to walls using adhesive. They can be supplied in polypropylene (3 mm thick), PVC-U (1, 2.5 and 3 mm thick) and glass fibre reinforced polyester (GRP) (2 and 2.4 mm thick).

Glazed Ceramic Wall Tiles

These are used extensively in kitchens and bathrooms to produce smooth, impervious and durable surfaces. There are two kinds of glaze: earthenware glazes and coloured enamels. The earthenware glazes are white and cream, while coloured enamels can be obtained in a variety of colours, either plain or mottled, and can have either a glossy or a matt surface. BS 6431[11] specifies European Standard requirements for sizes, dimensional tolerances and physical and chemical properties. Predominant sizes for tiles are 152 × 152 × 5 mm and 100 × 100 × 4 mm. There are three main types of fittings for use at angles, cappings and the like – round edge tiles which are the cheapest, attached angle tiles and angle beads. Alternatively, plastic trim may be used. To withstand wet conditions and chemicals, it is advisable to use tiles with low porosity and a thick glaze. Wall tiles should be fixed in accordance with BS 5385,[12] using adhesives meeting the requirements of BS 5980.[13]

BRE Defect Action Sheet 137[14] describes the main defects causing ceramic wall tiles to lose adhesion to internal walls as unsuitable adhesive, adhesive incorrectly spread, tiling insufficiently tamped, fixing delay, surface incorrectly prepared or primed, poor substrate bond and lack of movement joints in large tiled areas. It also provides extensive sound, practical advice on how to avoid failure.

INTERNAL CEILING FINISHINGS

The most common domestic ceiling finish is *gypsum lath plasterboard*, sometimes faced with aluminium foil on its upper face to give improved thermal insulation, all in accordance with BS 1230.[9] Each board should be nailed to each support using not less than four 40 mm clout-headed nails equally spaced across the width and driven no closer than 13 mm from the edges. End joints should be staggered in alternate courses and cut ends should be located at supports. The joints between boards are normally filled with gypsum plaster and the boards finished with a single coat of suitable neat gypsum plaster 5 mm thick.

Many years ago, *plaster* in three coats was applied to wood laths to form ceilings. The laths were usually of sawn timber 25 mm wide and 5 to 10 mm thick, in lengths varying from 900 to 1800 mm. They were spaced about 10 mm apart and fixed with flat-headed galvanised nails, butt jointed at their ends against supporting timbers and breaking joint frequently. Three-coat plasterwork in this situation was described as 'lath, plaster, float and set'. A pricking-up coat, about 10 mm thick of coarse stuff containing hair, adhered to the laths and also passed between them to form keys. The second or floating coat often contained hair and, after scratching, provided a good base for the thin final or setting coat, which was finished with a wood float or steel trowel according to the finish required. The plaster coats often consisted of lime and sand (1:3). Today *steel lathing* may be used to provide a key for plaster. The steel may be zinc coated or manufactured from austenitic stainless steel to give adequate protection against corrosion, and the mass varying from 0.89 to 2.62 kg/m^2 depends on the type of plaster and the spacing of the supports. It can be expanded metal, corrugated expanded metal or ribbed lath as described in BS 1369.[15] The metal lathing is fixed to timber with galvanised nails or staples at 100 mm centres.

Ceilings below *solid floors*, which are mainly of reinforced concrete or precast concrete slabs and beams, are often finished with traditional plaster. One, two or three-coat work can be applied according to the shape, texture and suction of the background. Board or lath and plaster finishes can be used by fixing timber battens to the underside of the concrete, with timber fixing pads or clips embedded in the concrete when it is poured. With precast concrete slabs the fixings are usually provided in the joints.

Plastic compound finishes are sometimes applied to the decoration side of square-edge wallboard or on taper-edge board. This type of ceiling finish is applied with a brush and the texture or pattern is obtained using a comb, lacer or stippler.

Ceiling linings are available in a variety of forms ranging from plywood, blockboard, laminboard, chipboard and fibreboard to softwood and hardwood boarding. Joints often form a feature by V-jointing or rebating, or, alternatively, by covering with metal, plastic or timber strip. Some manufacturers produce panels and strips ready for use, including some faced with thin plastic or metal sheet.

In buildings with taller storey heights, ceilings may be *suspended* using timber or metal framing. In this way services can be accommodated above the ceiling and it may also contain and distribute lighting, heating and ventilation. A variety of ceiling *tiles* are available made from various materials and with a number of textures and finishes. They are often designed to absorb sound and increase thermal insulation. Common types of thermal insulating tile are made of soft mineral fibre, and thin box metal or open-textured board backed with glass or mineral wool.

Before leaving ceiling finishings mention can usefully be made of *pattern staining* on ceilings, whereby dark and light patterns appear on plaster surfaces. Where the whole surface is not at a uniform temperature because of varying rates of heat transmission, dust is deposited on the various parts of the surface at different rates, depending on the temperature. Consequently the cooler areas become coated with darker patches of dust, such as beneath ceiling joists, because of the slower passage of heat. One of the best remedies is to provide about a 25 mm thickness of fibreglass, mineral wool or other suitable insulating material above the ceiling between the joists.

EXTERNAL RENDERINGS

An external solid wall of bricks or blocks may permit moisture penetration through haircracks between the mortar joints and the bricks or blocks and some form of external treatment then becomes necessary. Various weatherproofing applications were described in chapter 4, and in this chapter the characteristics of external renderings are examined in more detail. In selecting a suitable rendering attention should be paid to weatherproofing, durability and appearance. Renderings normally consist of two coats, the undercoat being about 13 mm thick and the final coat varying from 5 mm upwards. Brick joints should be raked out to form a key for the undercoat and the surface of the undercoat scratched to form a key for the final coat. The main types of rendering, as listed in BRE Digest 196,[16] are as follows.

Pebbledash or *drydash*. A rough finish of exposed small pebbles or crushed stone, thrown on to a freshly applied coat of mortar.

Roughcast or *wetdash*. A rough finish produced by throwing on a wet mix containing a proportion of coarse aggregate.

Scraped or *textured finish*. Treatment of the surface of the final coat, at the appropriate stage of setting, with different tools can produce a variety of finishes, depending on the artistry and skill of the craft operative.

Plain coat. A smooth or level surface finished with a wood, felt, cork or other suitably faced float.

Machine applied finish. The final coat is spattered or thrown on by a machine. The texture is determined by the type of machine and the mix, which is often a proprietary material such as Tyrolean.

Special finishes. These are mainly prebagged or premixed materials which have binders of cement resins, high strength acrylic resins or vinyl acetate copolymer emulsions, with aggregates of graded natural stone and glass. They have various surface textures created by the use of a spray, roller, stippler, brush or hand trowel and possess good wearing properties.

Smooth finishes tend to suffer surface crazing, and may be patchy in appearance. Mixes rich in cement, containing fine sands or finished with a steel trowel increase the risk of crazing. Pebbledash and roughcast are least liable to change in appearance over long periods and their good watershedding properties make them well suited for use in exposed coastal areas. Dirt from atmospheric pollution shows up more on white and pale coloured finishes than on natural grey cement finishes and the periodic painting of external rendered surfaces is a costly maintenance item. Dirt also appears worse on heavily textured finishes, although rain streaks produced by projecting features will be more prominent on smooth surfaces. The three most suitable *mixes* for external renderings are

(A) 1 part Portland cement: ½ part lime: 4 to 4½ parts sand by volume;

(B) 1 part Portland cement: 1 part lime: 5 to 6 parts sand by volume;

(C) 1 part Portland cement: 2 parts lime: 8 to 9 parts sand by volume.

Masonry-cement-sand mixes may be used as alternatives to mixes B and C. The higher sand contents should only be used if sand is well graded and a higher

Table 11.2 Mixes suitable for rendering (Source: BRE Digest 196[16])

Mix type	Cement : lime : sand	Cement : ready-mixed lime : sand		Cement : sand (using plasticiser)	Masonry cement : sand
		Ready-mixed lime : sand	Cement : ready-mixed material		
I	1 : ¼ : 3	1 : 12	1 : 3	–	–
II	1 : ½ : 4 to 4½	1 : 8 to 9	1 : 4 to 4½	1 : 3 to 4	1 : 2½ to 3½
III	1 : 1 : 5 to 6	1 : 6	1 : 5 to 6	1 : 5 to 6	1 : 4 to 5
IV	1 : 2 : 8 to 9	1 : 4½	1 : 8 to 9	1 : 7 to 8	1 : 5½ to 6½

NOTE. In special circumstances, for example where soluble salts in the background are likely to cause problems, mixes based on sulphate-resisting Portland cement may be employed.

Table 11.3 Recommended mixes for external renderings in relation to background materials, exposure conditions and finish required (Source: BRE Digest 196[16])

NOTE. The type of mix shown underlined is to be preferred.

Background material	Type of finish	First and subsequent undercoats			Final coat		
		Exposure			Exposure		
		Severe	Moderate	Sheltered	Severe	Moderate	Sheltered
(1) Dense, strong, smooth	Wood float	II or III	II or III	II or III	III	III or IV	III or IV
	Scraped or textured	II or III	II or III	II or III	III	III or IV	III or IV
	Roughcast	I or II	I or II	I or II	II	II	II
	Dry dash	I or II	I or II	I or II	II	II	II
(2) Moderately strong, porous	Wood float	II or III	III or IV	III or IV	III	III or IV	III or IV
	Scraped or textured	III	III or IV	III or IV	III	III or IV	III or IV
	Roughcast	II	II	II	as undercoats		
	Dry dash	II	II	II			
(3) Moderately weak, porous*	Wood float	III	III or IV	III or IV			
	Scraped or textured	III	III or IV	III or IV			
	Dry dash	III	III	III	as undercoats		
(4) No fines concrete[†]	Wood float	II or III	II, III or IV	II, III or IV	II or III	III or IV	III or IV
	Scraped or textured	II or III	II, III or IV	II, III or IV	III	III or IV	III or IV
	Roughcast	I or II	I or II	I or II	II	II	II
	Dry dash	I or II	I or II	I or II	II	II	II
(5) Woodwool slabs*[‡]	Wood float	III or IV	III or IV	III or IV	IV	IV	IV
	Scraped or textured	III or IV	III or IV	III or IV	IV	IV	IV
(6) Metal lathing	Wood float	I, II or III	I, II or III	I, II or III	II or III	II or III	II or III
	Scraped or textured	I, II or III	I, II or III	I, II or III	III	III	III
	Roughcast	I or II	I or II	I or II	II	II	II
	Dry dash	I or II	I or II	I or II	II	II	II

* Finishes such as roughcast and dry dash require strong mixes and hence are not advisable on weak backgrounds.
[†] If proprietary lightweight aggregates are used, it may be desirable to use the mix weaker than the recommended type.
[‡] Three-coat work is recommended, the first undercoat being thrown on like a spatterdash coat.

cement content is preferable when applied under winter conditions. The mix for a following coat should not be richer in cement than the preceding coat.

Table 11.2 gives the different mixes and table 11.3 shows how the choice of mix is influenced by the nature of the background, the type of finish and the degree of exposure. The mixes that are underlined are generally to be preferred. Dense, strong and smooth materials include dense clay bricks or blocks or dense concrete. Moderately strong and porous materials embrace most bricks and blocks and some medium-density concretes. Moderately weak and porous materials include lightweight concretes, aerated concretes and some low-strength bricks. Further guidance on external rendered finishes is contained in BS 5262.[17]

BRE Defect Action Sheets 37 and 38[18] describe how external renderings can fail through cracking or detachment and allow rain penetration into the structure. The main causes are inadequate bond or key to the wall, a continuous rendering over zones where relative movement occurs in the background, the rendering is stronger than the background or preceding coats, is too weak to exclude rainwater adequately or too rich or too wet to avoid cracking.

PAINTING

Types of Paint and Ancillary Materials

Paint consists essentially of a pigment, a binder and a solvent or thinner to make the mixture suitable for application by brush, roller or spray. After application, the paint undergoes changes which convert it from a fluid to a tough film which binds the pigment. The nature of these changes varies with different types of paint. Some such as size-bound distemper and chlorinated rubber paint lose the thinner by evaporation. With most paints containing drying oils, part of the change on drying results from reaction of the oil with oxygen from the air. In emulsion paints and oil-bound distempers, the binding material is emulsified or dispersed as fine globules in an aqueous liquid. After application the water evaporates and the globules coalesce to form a tough, water resistant film. Materials used in the painting of buildings fall into three main categories

(1) pigmented coatings, such as paints and wood stains;

(2) clear coatings, such as varnishes and lacquers; and

(3) ancillary materials, such as fillers, stoppers and other materials used in the preparation of surfaces.

BS 6150[19] gives general guidance on the painting of different materials and the selection of paint types. Standard paint colours are detailed in BS 4800.[20] The distinguishing features between solvent-borne, water-borne and microporous paints are described in the section dealing with 'painting woodwork'.

Painting Walls

Walls may be painted to provide colour or surface texture, to waterproof them, to reflect or absorb light, and to facilitate cleaning and hygiene (BRE Digest 197[21]). Problems may arise through dampness; drying out of new construction often takes an unacceptably long time and there is often a reluctance to accept temporary decoration. A coat of emulsion paint is an economical temporary decoration which, unlike distemper, need not be removed later. On internal work, heat and ventilation can accelerate the drying process. Difficulties may also arise through the presence of salts or alkali, although neat gypsum plasters are only slightly alkaline and do not usually affect paints. Surfaces to be painted must be sound and certainly not powdery or crumbly. The drying of paints is retarded by low temperatures, high humidity and lack of ventilation (BRE Digest 198[21]).

Matt finishes are often preferred because they avoid reflection of light sources and minimise surface irregularities. They are however less resistant to wear, more difficult to clean, and tend to absorb more condensation than glossy surfaces. Gloss finishes provide maximum washability and exterior durability and hard-gloss paints based on alkyd resins are popular. Chlorinated rubber paints provide a moderately glossy finish which is alkali-resistant. Emulsion paint, typically based on vinyl or acrylic polymers, is very suitable for application to walls and ceilings but normally avoiding impervious surfaces, kitchens and bathrooms, and is better able to withstand scrubbing and weathering than oil-bound distemper, and is less costly than oil paint. Special paints are available for specific purposes, such as anti-condensation paints with insulating or absorptive properties, fire-retardant paints to reduce fire risk, heat-resistant paints on surfaces that become hot, pigmented and two-pack epoxy and polyurethane coatings giving high resistance to abrasion and anti-graffiti treatments.[19]

Asbestos cement sheeting and masonry can be painted with emulsion paints based on alkali-resistant polymers, emulsion based masonry paints, and chlorinated rubber paints, but asbestos cement sheets require back painting (BRE Digest 197[21]).

Wallpaper also forms an attractive finish to walls and ceilings. A variety of finishes including hand or machine printed, embossed and washable papers are available. Lining papers are generally applied to walls prior to hanging heavy-weight or handprinted wallpapers. In addition to traditional wallpapers, there are metal foils, woven fabric coverings and cork panels. Suitable adhesives are specified in BS 3046.[22]

Painting Woodwork

BS 6952, Part 1[23] provides guidance to the user on the selection of wood coating systems for external use which are intended to meet their end user requirements in terms of both appearance and performance, accompanied by a suitable nomenclature and framework. It identifies and explains the significance of key properties such as permeability which has an important bearing on performance (durability) but it does not prescribe performance criteria.

On new wood, a *four-coat system* has been accepted as good practice, with a potential life of at least five years; although early-failure at susceptible areas, such as sills and bottom rails is common.[24] However, the life of the painting system may be reduced in conditions of severe exposure, such as in coastal areas up to 3 km inland and industrial areas with significant atmospheric pollution.[19] BRE Digest 354[24] recommends a water-repellant preservative treatment prior to painting, while BS 6150[19] suggests that this may permit the use of three coats (primer, undercoat and finish). Furthermore, *three-coat systems* are generally acceptable for internal woodwork in moderate/mild environments. The preparatory treatment includes knotting, priming and stopping. The *sealing of knots* to prevent resin bleeding through the paint is done with shellac knotting to BS 1336[25] or leafing aluminium primer. A good priming coat is essential for a durable paint

system and should preferably be applied in a paint-shop or factory to ensure thorough treatment of all surfaces.

The traditional pink *primer* for softwood contains a high proportion of white lead and some red lead, and has now been withdrawn from general sale because of objections to its use where children can lick or chew it. Aluminium wood primers to BS 4756[26] are durable and are generally preferred for woods which are resinous or have been treated with metallic naphthenate preservatives, and solvent-borne primer to BS 5358[27] for other timbers. Water-borne primers to BS 5082[28] have been useful for factory priming.

Stopping (deep holes) and *filling* (surface defects) should follow priming and traditionally single pack hard oil stoppers based on oil or cellulose media have often been used for deep cavities, but they do tend to shrink and more sophisticated two pack stoppers based on polyester, polyurethane or epoxy technology offer improved durability.[24] Water-mixed powder fillers based on soluble cellulose, gypsum plaster or cement can produce bad results in external woodwork and it is better to use ready mixed paste fillers based on emulsions to fill shallow flaws and fine cracks. Linseed oil putty is not satisfactory as it usually shrinks and can require 7 to 21 days to harden sufficiently to take paint.

Before applying undercoat, the primer should be checked and if thin, a further coat of primer applied. Undercoats should be of the same brand as the finishing coat. Better protection is often provided to exterior woodwork by using two finishing coats, but on interior work where rubbing down is specified, a smoother finish and better appearance will be obtained with two undercoats. Not all gloss finishes are suitable for use as consecutive coats and the manufacturer's instructions must be followed.

The *traditional paints* for use on exterior woodwork were lead-based with good durability. Concerns about the toxicity of the lead pigmentation have led to a decline in the availability of lead-based paints, and their use is now virtually prohibited except for historic buildings.[24]

The *traditional approach* to painting exterior wood embraces a three-part system of primer, undercoat and gloss coat. If these systems fail it is usually because the balance of properties has been inadequate to prevent or control the operative substrate conditions. However, it remains popular and still has a role, especially in the routine maintenance of existing conventional paint which is largely sound.[24]

In principle, good protection can be obtained by gloss coats alone but it is difficult to secure the desired quality of finish when applying gloss on gloss. The undercoat provides not only film build, opacity and colour, but also a surface with a good key to receive the finish. Undercoats and gloss finishing coat should be obtained from the same manufacturer to ensure compatibility. Most of the traditional undercoats based on varnish or alkyd resins are too hard and brittle to produce maximum durability on exterior woodwork, and emulsion-based primer/undercoats, applied in two consecutive coats, are more flexible and potentially more durable.[24]

Exterior quality paints are classified in BRE Digest 354[24] as solvent-borne and water-borne types. Some of these have a rather higher level of moisture permeability than the general-purpose paint systems and they are described as *'microporous'*, *'breathing'* or *'ventilating'*. They are generally claimed to resist the passage of liquid water but to allow it to escape freely from the wood as vapour and this, it is argued, maintains the wood throughout service at low moisture content. Microporosity is sometimes claimed to be a source of improved durability, and many of these paints do perform well in service. However, work at BRE has shown that the higher level of permeability built into microporous coatings has only at best very minor benefits in allowing softwood joinery to dry to slightly lower moisture contents, during warmer summer months, than they might otherwise do. There is no evidence that the long term overall moisture content in the wood differs significantly from that under a conventional general-purpose paint system in similar conditions. Hence preservation pretreatment remains essential.[24]

Solvent-borne exterior paints can be either gloss or low sheen finish. Exterior gloss paints are generally based on flexible types of alkyd resins and give a similar finish to general-purpose full gloss paints. There are exterior gloss systems designed as separate primer, undercoat and gloss, each applied in one coat. Other systems consist of two coats of primer/undercoat followed by one of gloss or one coat of primer followed by two of gloss. Exterior paints that dry to a low sheen (normally eggshell) are generally applied as coat-on-coat systems and thus offer the convenience and simplicity of one-can products.[24]

Water-borne exterior paints are available in both low sheen and gloss finish but the former are little used. Gloss types tend not to give as high a gloss level as solvent-borne (alkyd gloss) types. All water-borne exterior paints have a higher level of moisture permeability than equivalent solvent-borne exterior paints or conventional paint. This is a property of the acrylic emulsion, or alkyd-acrylic emulsion used as the resins for these paints. These paints are generally more durable than solvent-borne alkyd paints.[24]

There are two main types of *natural finishes* for exterior timber as described in BRE Digest 286,[29] namely clear varnishes and exterior wood stains. Varnishes give an initial fine appearance but this is often short lived externally and the treatment is expensive in both initial and maintenance costs. It becomes brittle by weathering and ultra-violet rays can bleach or degrade the underlying wood surface. Polyurethanes of the moisture-curing and two-pack types give good service indoors.

Exterior wood stains aim to provide durable coloured translucent coatings which allow the grain pattern of the wood to show through. They consist of dilute solutions of polymers containing added pigments and film fungicides. They are classified as low-solids and high-solids and both types combine some characteristics of a surface coating with those of a water-repellant preservative. Low-solid stains are generally of low viscosity and leave the texture of the wood unchanged, but they erode rapidly in fully exposed conditions. High-solid stains produce a noticeable film on the surface and give better resistance to water vapour transfer. They usually weather by erosion and will embrittle and flake if not regularly maintained, usually requiring two or three brush coats every three years. Softwood joinery requires suitable pretreatment with a water-repellant preservative. Imperfections in the timber such as knots and splits will remain visible after treatment and higher quality timber should advisably be specified. Glazing with conventional linseed oil putty is not satisfactory and bead glazing or a sealant or neoprene glazing system should be used.

BRE Digest 354[24] emphasises that present painting costs rarely allow sufficient time for the proper sanding down of wood and undercoats or use of fillers. Hence joinery specifications should require freedom from joinery imperfections. Extensive thinning of paint still occurs on site, to ease brush application and speed up the work, resulting in loss of durability and standard of finish. The moisture content of joinery at time of painting should not exceed 18 per cent for exterior work and 12 per cent for interior work, and exterior painting must not be undertaken on damp surfaces, and should preferably be carried out between mid-April and mid-September.[19]

On repainting it is ill-advisable to strip paint which is adhering well, chalking only slightly and which is free from other defects. It should be washed with a detergent solution or proprietary wash and rubbed with wet abrasive paper. Small areas of loose or defective paint must be scraped down to primer, if sound, or bare wood, and brought forward with primer and undercoat as necessary. On sound existing paint, one undercoat and one finishing coat is normally sufficient. Decayed timber should be cut out and replaced, with both new and old timber being treated with preservative. The blow lamp is the quickest and most effective way of removing defective paint from wood, but where impracticable, for instance near glass, paint removers of the organic solvent type may be used, taking care to remove all traces of paint remover or hot air strippers.[24]

The main causes of failure in paintwork are

(1) adhesion failure: application to damp, dirty, powdery, friable, dense or unprimed substrates;

(2) cracking: undercoat has not hardened sufficiently before finishing coat applied, and/or stresses within coating film;

(3) chalking: often results from slow erosion on lengthy exposure;

(4) blistering: usually results from liquid or vapour beneath the coating;

(5) discolouration: may result from use of paint in unsuitable conditions, chemical attack, or possibly lengthy exposure to bright sunlight;

(6) loss of gloss: after lengthy exposure.

Other causes of paint failure and suggested remedies are contained in BRE Digest 198,[21] and BS 6150.[19]

Painting Metalwork

Metals in building work are painted mainly for protection against corrosion and only secondarily for decora-tion. Sites subject to sea spray or in industrial areas with heavily polluted atmospheres require special attention. All metal surfaces to be painted must be free of mill scale, most rust, oil, grease, moisture and dirt. Suitable cleaning processes are

(1) blast cleaning: most effective method but requires priming within 4 hours of cleaning;

(2) flame-cleaning: particularly suitable for maintenance painting, but will not remove all rust and scale;

(3) acid pickling: little used for structural steelwork;

(4) manual cleaning: least satisfactory method.[30]

The paint system normally consists of primer, undercoat(s) and finishing coat(s). Guidance as to the suitable choice of a protective system for steel is given in BS 6150.[19] On rough surfaces, primers are best applied by brush. The best primers contain pigments which chemically inhibit the corrosion of iron and steel. Examples of such pigments include zinc chromate, zinc phosphate, zinc-rich to BS 4652,[31] and calcium plumbate. Priming paints are each normally intended for particular metals. Where a single primer is to be used on both steel and aluminium, a zinc chromate primer is preferable. For zinc and steel together, calcium plumbate or zinc chromate is best. Zinc-rich primer is especially well suited for steelwork to be left exposed on site. Advice on the protection of iron and steel structures against corrosion is given in BS 5493.[30] The lead-based paints listed in BS 6150[19] have intentionally been omitted because of the arguments over toxicity. The use of metal windows with a polyester powder coating is now quite common and provides a durable finish.

The undercoats and finishes provide additional film thickness, water resistance and possibly decoration. Bituminous paints, often used for cast iron rainwater gutters, give good inexpensive protection against water, salt and some chemicals, although they are not particularly durable in sunlight.

Paint Maintenance Practice

The author carried out a paint maintenance survey on a national basis in 1983[32] and found that standards of painting maintenance in general were deteriorating significantly, creating serious problems, and a follow up survey in 1985 showed increasing deterioration, particularly in the public sector. The situation worsened still further in the subsequent severe recession. For example, it is generally recognised that the frequency of repainting is influenced by climatic conditions, atmospheric pollution, degree of exposure and condition of substrate. Most property managers aimed for the repainting of external gloss surfaces two to three years from the initial painting and at four to five year intervals thereafter. In practice the period between painting cycles had often been extended to 6 to 9 years. Local authorities suffered from substantial budget cuts and many private property owners afforded painting a low priority, as their primary concern was profits and productivity. These delays in painting resulted in expensive bills for joinery repairs and replacements preparatory to painting, which in badly neglected situations amounted to as much as six to fifteen times the cost of the painting.

In many cases the failure of components at an early stage in the life of the building, which had been attributed to paint failure, was in fact due to the poor initial quality of the substrates and the lack of necessary protective measures. Another major weakness was the failure to monitor and enforce the specified painting cycles in full repairing leases of properties. To prevent the decay of building components, it is essential that repainting takes place before the existing paint film begins to break down. Saving on painting and decoration is frequently false economy as it is merely storing up much greater and more costly problems for the future.

One of the greatest weaknesses in painting work was found to be the general lack of attention to surfaces to be painted and satisfactory application methods. All too common were failures to remove all loose and flaking paint, to burn off existing wood surfaces where paint had broken down, to clean adequately the surfaces to be painted, the omission of sealing to knots, the lack of a good coat of suitable primer on all bare surfaces, the failure to seal all cracks and holes with an appropriate filling or stopping material, painting in damp conditions and on wet surfaces, the failure to lightly rub down between coats where appropriate, the mixing of incompatible paints, the omission of a specified undercoat or finishing coat and the excessive use of thinners. These faults, which demonstrate an overall decline in standards, drastically diminish the effectiveness of painting systems and result in premature paint failures. The importance of maintaining effectively the fabric of the nation's buildings appears to often go unrecognised.

REFERENCES

1. *BRE Digest 213*: Choosing specifications for plastering (1978)
2. *BS 5270* Bonding agents for use with gypsum plasters and cement, *Part 1: 1989* Specification for polyvinyl acetate (PVAC) emulsion bonding agents for indoor use with gypsum building plasters
3. *BS 1191* Gypsum building plasters, *Part 1: 1973* Excluding premixed lightweight plasters; *Part 2: 1973* Premixed lightweight plasters
4. *BS 1199: 1976 (1986)* Sands for external renderings and internal plastering with lime and Portland cement
5. *BS 5492: 1977 (1990)* Code of practice for internal plastering
6. *BS 6452* Beads for internal plastering and dry linings, *Part 1: 1983* Specification for galvanised steel beads
7. *BRE Good Building Guide 7*: Replacing failed plasterwork (1991)
8. *BS 8212: 1989* Code of practice for dry lining and partitioning using gypsum plasterboard
9. *BS 1230* Gypsum plasterboard, *Part 1: 1985* Specification for plasterboard excluding materials submitted to secondary operations
10. *BS 3794* Decorative laminated high pressure laminates (HPL) based on thermosetting resins, *Part 1: 1986* Specification for performance; *Part 2: 1986* Methods of determination of properties
11. *BS 6431* Ceramic floor and wall tiles, *Part 1: 1983* Specification for classification and marking, including definitions and characteristics; *Part 2: 1984* Specification for extruded ceramic tiles with low water absorption; *Part 6: 1984* Specification for dust-pressed ceramic tiles with a low water absorption
12. *BS 5385* Wall and floor tiling, *Part 1: 1990* Code of practice for the design and installation of internal ceramic wall tiling and mosaics in normal conditions
13. *BS 5980: 1980* Adhesives for use with ceramic tiles and mosaics

14. *BRE Defect Action Sheet 137*: Internal walls: ceramic wall tiles – loss of adhesion (1989)
15. *BS 1369* Steel lathing for internal plastering and external rendering, *Part 1: 1987* Specification for expanded metal and ribbed lathing
16. *BRE Digest 196*: External rendered finishes (1976)
17. *BS 5262: 1976 (1990)* Code of practice for external rendered finishes
18. *BRE Defect Action Sheet 37*: External walls: rendering – resisting rain penetration (1983); *Sheet 38*: Ditto – application (1983)
19. *BS 6150: 1982* Code of practice for painting of buildings
20. *BS 4800: 1989* Specification for paint colours for building purposes
21. *BRE Digest 197*: Painting walls; Part 1: Choice of paint (1982); *BRE Digest 198*: Painting walls; Part 2: Failures and remedies (1984)
22. *BS 3046: 1981* Specification for adhesives for hanging flexible wall coverings
23. *BS 6952* Exterior wood coating systems, *Part 1: 1988* Guide to classification and selection
24. *BRE Digest 354*: Painting exterior woodwork (1990)
25. *BS 1336: 1971 (1988)* Knotting
26. *BS 4756: 1971* Ready mixed aluminium priming paints for woodwork
27. *BS 5358: 1986* Specification for solvent-borne priming paints for woodwork
28. *BS 5082: 1986* Specification for water-borne priming paints for woodwork
29. *BRE Digest 286*: Natural finishes for exterior timber (1984)
30. *BS 5493: 1977 (1984)* Code of practice for protective coating of iron and steel structures against corrosion
31. *BS 4652: 1971 (1981)* Metallic zinc-rich priming paint (organic media)
32. I.H. Seeley. *Blight on Britain's buildings: a survey of paint and maintenance practice*. Paintmakers Association (1984)

12 WATER SERVICES, SANITARY PLUMBING AND SOLID WASTE STORAGE

This chapter examines the methods of supplying water to buildings, cold and hot water supply arrangements, sanitary appliances, waste systems, hot water heating and solid waste storage. These aspects all relate to 'sanitation and services' but are included here as it was felt that a study of 'building technology' would not be complete without them.

COLD WATER SUPPLY

A local authority has a duty to ensure that every dwelling in its area is provided with, or has reasonably available, a sufficient supply of wholesome water for domestic purposes. Water authorities/companies make byelaws for preventing waste, excessive consumption, misuse or contamination of water supplied by them. Sources of water include rainwater, surface water, lakes and rivers, and underground water in the form of springs, wells and boreholes. Some purification of the water is often necessary to remove organic or inorganic impurities and possibly to reduce the hardness of the water. Public water supplies are sterilised before being passed to the consumer by treatment with chlorine or chlorine and ammonia.

The connection to the water main is made by the water authority who normally drill and tap the main and lay a 'communication' pipe, often 13 mm in size, up to the boundary of the site, finishing with a stopvalve or stopcock in a suitable box or chamber often fitted with a hinged cast iron cover. This part of the water supply installation remains the property of the water authority/company, even although the property owner may contribute towards it. The supply pipe is laid from the stopvalve into the building and should be at least 750 mm deep for adequate protection from frost. A second stopvalve should be fitted at the first accessible position after the service enters the building, and is usually accompanied by a draincock to permit the cold water system to be drained down. The service pipe should enter the building through 100 mm diameter clay drain pipes and a 90° bend.

Cold Water Supply Systems

There are two principal cold water supply systems although variations of either are possible. They are often described as *direct* and *indirect*, but have also been referred to as *storage* and *non-storage* and as *downfeed* and *upfeed*. Typical arrangements for both systems are illustrated in figures 12.1.1. and 12.1.2. In the direct system all cold water drawoff points are fed directly from the rising service pipe with the cistern, where provided, serving the sole purpose of a feed cistern (supplying cold water to the hot water system). The latter system is found mainly in older properties and is unlikely to be accepted by water authorities in new dwellings. In the indirect system the rising service pipe normally serves only one drawoff point, often over the kitchen sink, from which fresh water can be obtained. All other cold water drawoff points are supplied from a storage cistern. The cistern may serve a dual purpose – feeding a hot water system and supplying cold water for other purposes.

With the direct system, potable water is available at all outlets, and it reduces the storage capacity and often the size of cold water distribution pipes and the length of pipe runs. To be satisfactory, there must be a constant supply of water of sufficient pressure during periods of peak demand. Furthermore, where the mains pressure is high it may result in excessive noise and damage to seatings of taps and valves.

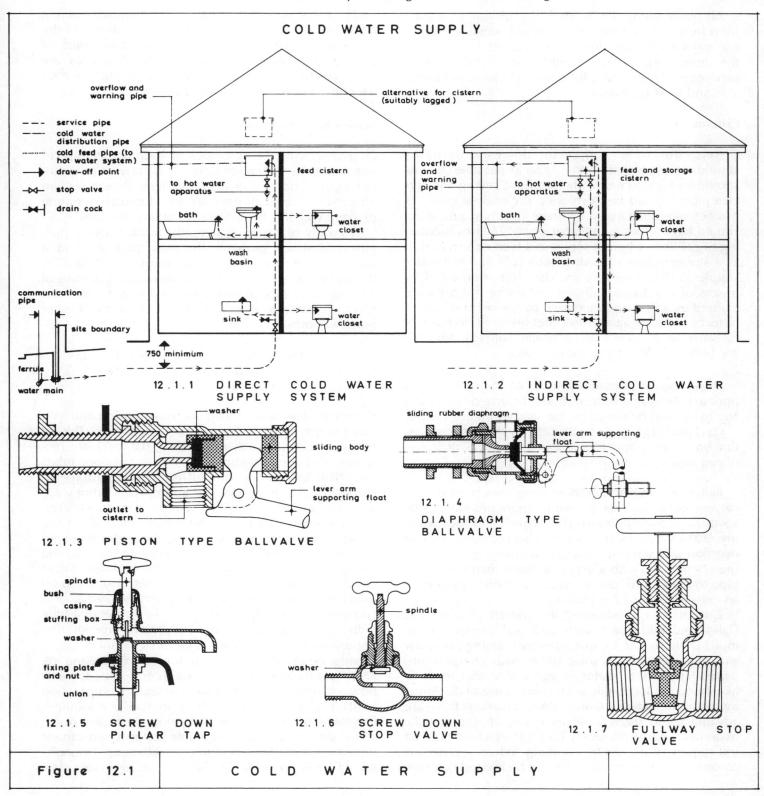

COLD WATER SUPPLY

overflow and warning pipe

alternative for cistern (suitably lagged)

--- service pipe
—·— cold water distribution pipe
········ cold feed pipe (to hot water system)
→ draw-off point
⋈ stop valve
⋈ drain cock

feed cistern

to hot water apparatus

bath

water closet

wash basin

communication pipe

site boundary

sink

water closet

ferrule

water main

750 minimum

12.1.1 DIRECT COLD WATER SUPPLY SYSTEM

overflow and warning pipe

feed and storage cistern

to hot water apparatus

bath

water closet

wash basin

sink

water closet

12.1.2 INDIRECT COLD WATER SUPPLY SYSTEM

washer

sliding body

lever arm supporting float

outlet to cistern

12.1.3 PISTON TYPE BALLVALVE

sliding rubber diaphragm

lever arm supporting float

12.1.4 DIAPHRAGM TYPE BALLVALVE

spindle
bush
casing
stuffing box
washer

fixing plate and nut

union

12.1.5 SCREW DOWN PILLAR TAP

spindle

washer

12.1.6 SCREW DOWN STOP VALVE

12.1.7 FULLWAY STOP VALVE

Figure 12.1 COLD WATER SUPPLY

All pipes should be located in positions protected from frost or be adequately insulated, suitable insulating materials are detailed in BS 5422.[1] BS 6700[2] covers the design, installation, testing and maintenance of services supplying water for domestic use within buildings and their curtilages.

Cisterns

Cisterns for water supply for domestic purposes should comply with BS 7181.[3] As far as possible cisterns should be located below the top floor ceiling and service pipes should be positioned on internal walls, as protection against frost. A storage cistern in a dwelling should have a minimum capacity of 114 litres increasing to 228 litres where it serves as a feed cistern as well as a storage cistern (serving both cold and hot water supplies). Water level in the cistern is maintained by means of a suitable ballvalve, whereby the pressure exerted by a float and increased by a lever arm is used to force a washer against the inlet orifice. Two types of ballvalve are in general use, although more efficient types are being developed to conserve water.

(1) Piston or Portsmouth type of BS 1212[4] with a horizontally sliding body of brass, bronze or gunmetal and nylon seat holding a rubber washer (figure 12.1.3).
(2) Diaphragm type to BS 1212[4] with brass or plastics body and copper or plastics spherical float (reduced noise and wear) (figure 12.1.4).

Ballvalves are made in three categories to withstand varying conditions of pressure: high pressure (1400 kN/m^2), medium pressure (700 kN/m^2) and low pressure (300 kN/m^2). Each cistern is also provided with an overflow and warning pipe with a diameter of not less than 19 mm and with a larger diameter than the supply pipe to the cistern, discharging in a visible position on an external wall of the building.

Cisterns can be obtained in a variety of materials. Galvanised mild steel cisterns to BS 417[5] are quite common. Their life can be extended by painting internally with one coat of non-toxic bituminous composition, particularly where water is aggressive and there is likelihood of electrolytic action due to use of dissimilar metals in the same system. In these situations there are advantages in using cisterns made of non-metallic materials, such as plastics to BS 4213[6] which do not rust but require more careful handling. Where a cistern is placed in the roof space, it should be located centrally

and be cased in timber or rigid sheet materials, leaving a 50 mm gap for a layer of insulating material to the sides of the cistern. Preformed insulating slabs of polystyrene, made to fit cisterns of different sizes, are also available. The cistern should be provided with a well-fitting cover, adequately insulated.

Service and Distribution Pipes

Stopvalves should be provided on each branch pipe to enable repairs to drawoff points to be carried out without closing down the whole system. Pipe sizes are affected by the water pressure and the total maximum demand which is likely to occur at any one time.

A variety of materials is available for water supply pipes, of which probably the most popular is *light gauge copper* tube, largely on account of its durability, flexibility, smooth bore, neat appearance and ease of jointing. The pipe dimensions in BS 2871[7] are now given in external diameters in place of the former nominal bores.

external diameter in mm	15	22	28
nearest equivalent internal bore in inches	½	¾	1

Common sizes are 15 mm for wash basins and WC flushing cisterns and 22 mm for sinks and baths. Thicker pipes in accordance with BS 2871[7] (table Y) are recommended for underground services. Copper tubes and fittings are jointed either with capillary or compression joints, and the appropriate fittings are detailed in BS 864.[8] *Capillary* joints are made with solder contained in a groove formed in the wall of the socket. After cleaning and fluxing, the solder is melted by a blowlamp and fills the space between the tube and fitting. In 1994 capillary joints were under threat because of dissatisfaction with solder. *Compression* joints are more expensive and generally consist of a compression ring compressed into the socket of a tube with a coupling nut. The bore of the compression ring corresponds to the outside diameter of the tube.
Plastics pipes. Polythene pipes to BS 6730[9] and BS 3284[10] are flexible and their smooth bore speeds water flow and prevents the formation of scale. Compression joints probably form the best method of jointing. Polythene pipes are rather soft, not completely resistant to ground gases, need ample support and cannot be used for earthing electrical installations. Unplasticised (PVC-U) pipes to BS 3505,[11] supplied in four clas-

ses resisting pressures ranging from 6 bar to 15 bar (600 to 1500 kN/m²), are stronger and more rigid than polythene, but neither is recommended for hot water systems.[12] Small diameter pipes can be hot bent but it is a skilled operation. Pipe joints are usually solvent welded.Useful guidance on the installation of thermoplastics pipes and associated fittings for use in domestic hot and cold water services and heating systems is given in BS 5955, Part 8[13], and detailed information on thermoplastic pipes and fittings is provided in BS 7291[14] and these include polybutylene (PB), cross-linked polyethylene (PE-X) and chlorinated polyvinyl chloride (PVC-C) pipes. The requirements for polyethylene pipes for general purposes are given in BS 6437[15] and for blue polyethylene pipes below ground use for potable water in BS 6572.[16]

Lead pipes. These are now little used on account of their high cost, weight and unsuitability with soft or acid water, and their use in new dwellings is not permitted. Soldered joints are normally used with lead pipes.

Mild steel pipes. They are relatively strong and inexpensive, and should comply with BS 1387,[17] they are made in three categories – light, medium and heavy. The most common form of joint is screwed and socketed. Small bore stainless steel tubes can be used as a feasible alternative costwise to copper, are free from corrosion and tarnishing and should comply with BS 4127.[18] They are jointed with capillary and compression fittings similar to those used for light gauge copper tubes and advice on installation is provided in BRE Digest 83.[19] Stainless steel tubes suitable for screwing with BS 21 pipe threads are covered by BS 6362.[20] Copper, polythene and steel pipes are usually supported with metal or plastics clips.

Noise in pipes (water hammer) may arise from any of a number of causes, including inadequately fixed pipework, sharp bends, defective valve seatings and large volumes of water entering the storage cistern. A Portsmouth ballvalve can be very noisy in operation unless provided with a dip pipe, although this is not always acceptable to water authorities/companies because of the risk of back siphonage.

Taps and Stopvalves

A pillar tap suitable for use with sinks, baths and wash basins is illustrated in figure 12.1.5, although the form of handle varies considerably and some are very sophisticated. Stopvalves on branches to WC flushing cis-

terns and the like are often of the screwdown variety (figure 12.1.6), complying with BS 1010,[21] but those on the main distribution pipes may, if permitted, be of the fullway type (figure 12.1.7), complying with BS 5154,[22] as screwdown valves offer some resistance to flow.

HOT WATER SUPPLY

Before choosing a hot water system, various criteria must be considered including required performance, characteristics of local water and need, if any, for space heating. The heating source for domestic hot water supply is often an independent boiler burning solid fuel, gas or oil. Alternatives are gas or electric hot water heaters (storage or instantaneous). In smaller dwellings some use is still made of back boilers behind open fires. The essential features of a hot water supply system should be to produce sufficient hot water to meet all demands, be economical in running costs, be suitable for the type of building and relatively easy to install and maintain. There are two main systems of hot water supply – direct and indirect.

Direct System

In a direct system (figure 12.2.1), hot water circulating between the boiler and storage cylinder or tank is drawn off as required for domestic use and replaced by fresh, cold water fed directly into the same circuit. A totally enclosed cylinder or tank, often placed in an airing cupboard, is located as close as possible to the boiler, and connected by primary flow and return pipes. A short primary circuit ensures rapid discharge of the hot water and reduces the heat loss from pipework. The cold water feed is taken from a cold water storage cistern to the hot water cylinder or tank located below it. The water to drawoff points is delivered from the top cylinder or tank where the hottest water is stored, often through an expansion pipe. This minimises interference with the primary flow and return during periods of drawoff. The main drawback is that distribution pipes are non-circulatory and so their water content has to be run to waste before hot water is discharged at the taps. Hence water authorities/companies normally restrict the length of non-circulatory delivery pipes to 12 m for pipes not exceeding 19 mm, 7.5 m for 25 mm pipes and 3 m for pipes exceeding 25 mm internal diameter.

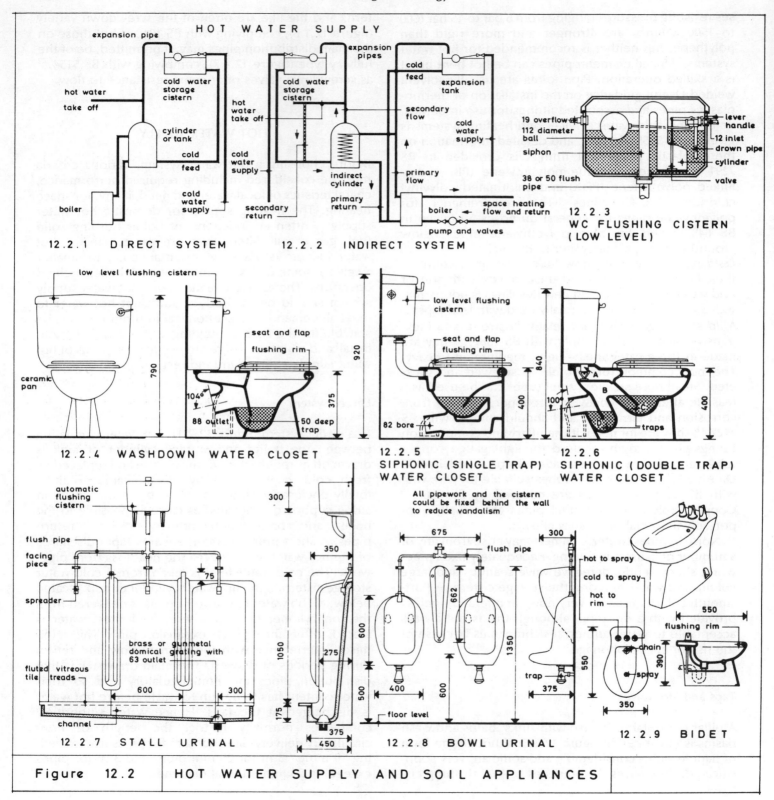

HOT WATER SUPPLY

expansion pipe

hot water take off

cold water storage cistern

cylinder or tank

cold feed

cold water supply

boiler

12.2.1   DIRECT   SYSTEM

expansion pipes

cold water storage cistern

hot water take off

cold water supply

secondary return

indirect cylinder

primary flow

primary return

boiler

pump and valves

space heating flow and return

12.2.2   INDIRECT   SYSTEM

cold feed

expansion tank

secondary flow

cold water supply

12.2.3 WC FLUSHING CISTERN (LOW LEVEL)

19 overflow
112 diameter ball
siphon
38 or 50 flush pipe

lever handle
12 inlet
drown pipe
cylinder
valve

low level flushing cistern

ceramic pan

790

104°
88 outlet
50 deep trap
375

seat and flap
flushing rim

12.2.4   WASHDOWN   WATER   CLOSET

low level flushing cistern

920

82 bore
100°
400

seat and flap
flushing rim

12.2.5   SIPHONIC (SINGLE TRAP) WATER CLOSET

840

A
B
C
b
c

traps
400

12.2.6   SIPHONIC (DOUBLE TRAP) WATER CLOSET

automatic flushing cistern

flush pipe
facing piece
spreader

brass or gunmetal domical grating with 63 outlet

fluted vitreous tile treads

channel
600   300

75

12.2.7   STALL   URINAL

300
350
275
1050
175
375
450

All pipework and the cistern could be fixed behind the wall to reduce vandalism

675
662
flush pipe
600
500
400   600

floor level

12.2.8   BOWL   URINAL

300
1350
550

hot to spray
cold to spray
hot to rim
trap
375

chain
spray
550

flushing rim
390
350

12.2.9   BIDET

| Figure   12.2 | HOT WATER SUPPLY AND SOIL APPLIANCES |

Indirect System

In an indirect system (figure 12.2.2), hot water also circulates between the boiler and storage cylinder or tank, but the storage vessel (an indirect cylinder or calorifier) is so designed that the hot water in the primary circuit from the boiler is used only to raise the temperature of the stored water; it does not mix with it nor is it drawn off for domestic use. Hot water for domestic use is drawn from the secondary side of the system, which may be a complete circuit, and is replaced by cold water fed into the secondary circuit. Apart from the replacement of water lost by expansion, the same water circulates continuously in the primary circuit.

In some parts of the country the cold water supply contains calcium and magnesium bicarbonates and when heated above 68°C, the bicarbonates are converted into insoluble carbonates, which form a hard lime deposit called 'fur'. With recirculation of hot water, as in the indirect system, there will be an initial production of scale but it will not be repeated. Water will expand by about 4½ per cent (1:23) when heated from 10°C to 93°C, hence a small expansion tank is necessary on an indirect system to absorb the expansion water from the primary circuit and to make up the voids when it contracts on cooling.

The water in the boiler is heated first and becomes less dense, permitting the colder, heavier water from the storage cylinder or tank to fall and force the warmer, lighter water up the primary flow into the top of the cylinder. Similarly, because of the difference in weight of the water in the feed and secondary flow, circulation of water in the secondary circuit will occur once the contents of the cylinder become heated. Loss of heat from the circulatory pipework to the surrounding air will create a temperature difference between the flow and return pipes, ensuring continuous movement.

Although both systems can be extended to serve hot water radiators for space heating, there are disadvantages with the direct system, as radiators tend to run cold when domestic water is drawn from the storage cylinder or tank and the rate of circulation of heated water is slow. Hot water for space heating in an indirect system is taken from the primary circuit and is less likely to cool when water is drawn from the secondary side of the hot water cylinder.

Cistern and Pipe Sizes (internal diameters)

Common sizes for domestic hot water supply installations are as follows

cold water storage cistern	250 litres
hot water storage cylinder or tank	125 litres
flow and return pipes (primary circuit)	25 mm or 32 mm on a direct system
cold feed pipe	25 mm
delivery pipes	19 mm minimum with 19 mm branch to bath and 13 mm branches to sink and wash basin
expansion and vent pipe	19 mm

(Note: pipe sizes in different materials are expressed in different ways)

Boiler Mountings

The most common mountings for hot water supply boilers are now listed.

Thermostat. The most common type is the vapour pressure thermostat, which consists of a small metal bulb containing a highly volatile oil, such as paraffin, which is immersed in the water of the boiler. In the case of oilfired, gasfired or gravity feed boilers, the thermostat will either shut off the gas or oil supply, or stop the air fan on the gravity feed boiler.

Drain valve. This enables the boiler and pipework to be emptied when required.

Thermometer. Fitted to show whether the boiler is operating effectively.

Check-draught valve. A simple pivoted valve inserted in the flue near the boiler, which opens when the draught in the flue becomes excessive due to high winds or gusty conditions, permitting the draught to be maintained at a reasonably constant level.

Condensing Boilers

BRE Digest 339[23] describes the nature and use of condensing boilers which constitute a new generation of heating appliance for use in both old and new commercial and domestic buildings, producing valuable savings in energy, by recovering extra heat from the flue gases with an increased heat exchange surface. The main departures from conventional practice are the provision of a condensate drain and careful consideration of flue arrangements and the installation of modern smaller, lighter boilers.

Immersion Heaters

Immersion heaters are often fitted in hot water cylinders for heating water when the boiler is not functioning. They consist of a resistance element connected to the electricity supply and contained in a protective material or sheath, the whole being immersed in the water to be heated. The immersion heater is usually provided with thermostatic control and lengths vary from 300 to 1050 mm with loadings of 1, 2 and 3 kW. They can be fitted vertically or horizontally in cylinders. Vertical fixing has the advantage of speedy recovery after some water has been drawn off, making more hot water quickly available. Water below an immersion heater will not be heated so that there would be merit in fixing one horizontal heater about one-third down the cylinder for quick heating of a small amount of water and another near the bottom for slow heating of all the water. These heaters should be controlled by a change-over switch so that they are not both working at the same time.

Unvented Hot Water Supply Systems

As an alternative to the two vented hot water supply systems previously described, Schedule 1 to the Building Regulations 1991[24] (paragraph G3) covers unvented hot water supply systems, which have been used extensively in Germany, the United States and other countries. It requires that if hot water is stored and the storage system does not incorporate a vent pipe to the atmosphere, it shall be installed by a person competent to do so, and there shall be adequate precautions to: (a) prevent the temperature of the stored water at any time exceeding 100°C; and (b) ensure that the hot water discharged from safety devices is safely conveyed to where it is visible but will not cause danger to persons in or about the building.

A typical unvented hot water system based on German practice embraces the following arrangements. All cold water outlets are supplied direct from the mains through a check valve, a meter and, if mains pressure is high, a pressure-reducing valve. The hot water supply is directly fed, and a second check valve is fitted in the feed pipe. Because this check valve prevents water expanding into the cold feed pipe when a charge of water is heated, it is necessary to provide a pressure relief valve and a drain to allow expansion water to leak away. Similarly a temperature-operated energy cut-out is normally used to disconnect the supply of energy. This ensures there will be no explosion if a thermostat failure results in uncontrolled heating of the water.[25]

Approved Document G3 (1991) of the Building Regulations contains the following recommendations with regard to an unvented hot water storage system with a storage vessel capacity of not more than 500 litres and a heat input not exceeding 45 kW, heated directly or indirectly. Any unvented hot water storage system should be in the form of a proprietary unit or package which is approved by a member body of the European Organisation for Technical Approvals (EOTA) which includes the British Board of Agrément (BBA). A unit is a storage water heater factory fitted incorporating a minimum of two temperature activated devices operating in sequence; a non self-setting thermal cut-out to BS 3955[26] and one or more temperature relief valves to BS 6283,[27] in addition to any thermostatic control which is fitted to maintain the temperature of the stored water. A package is supplied by the manufacturer with a kit containing the devices described for a unit and these are fitted to the system by the installer.

Approved Document G3 (1991) prescribes that electrical non self-resetting thermal cut-outs should be connected to a direct or indirect heat source in accordance with the IEE Regulations. In an indirectly heated system the thermal cut-out should be wired up to a motorised valve or some other device approved by a member of EOTA to shut off the flow to the primary heater. In both directly and indirectly heated systems, the temperature relief valve(s) should be located directly on the storage vessel, such that the stored water does not exceed 100°C.

There are also recommendations for the discharge pipe from the tun dish to be laid to a continuous fall, be generally no longer than 9 m, unless the bore is increased, and to discharge into a visible but safe place, such as a gully, where there is no risk to persons in the vicinity of the discharge and be of metal. Further provisions relate to the connection of electrical work.

Approved Document G3 (1991) also covers, in section 4, hot water storage systems of more than 500 litres capacity or having a heat input of more than 45 kW, which will generally be individual designs for spe-

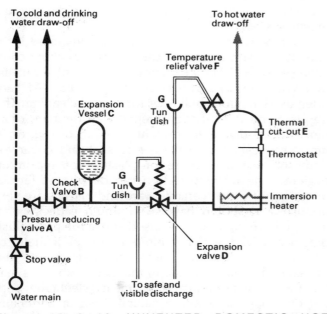

To cold and drinking water draw-off

To hot water draw-off

Temperature relief valve F

Expansion Vessel C

G Tun dish

Thermal cut-out E

Thermostat

Check Valve B

G Tun dish

Immersion heater

Pressure reducing valve A

Stop valve

Expansion valve D

Water main

To safe and visible discharge

Figure 12.2.10 UNVENTED DOMESTIC HOT WATER STORAGE SYSTEM

cific projects and not systems for which EOTA (European Organisation for Technical Approvals) or NACCB (National Accreditation Council for Certification Bodies) approval is appropriate. These systems should have safety devices in accordance with BS 6700,[2] with an appropriate number of temperature relief valves, non self-resetting thermal cut-outs and discharge pipes to convey any discharges from safety devices.

BRE Defect Action Sheet 139[28] describes how cold water pipes to the building may be taken off the supply either before or after a pressure reducing valve A as shown in figure 12.2.10. A check valve B is installed downstream of the cold water branch to prevent hot water from flowing back into the cold water pipework. Certain control devices are needed to guard against excessive pressure in the system. An expansion vessel C is required to accommodate the increased volume of heated water. Failure of this vessel will result in a discharge from expansion valve D, which also maintains the system working pressure.

To protect against over-heating resulting from failure of the control thermostat, a non-self-resetting thermal cut-out E is provided. If this should fail explosion protection is still provided by a temperature relief valve F. The expansion valve and temperature-relief valves discharge via tun dishes to a safe and visible

place, such as over a gully. If over-pressure develops in an unvented hot water cylinder, the system may rupture, and if the temperature exceeds 100°C the rupture will be an explosive one. Hence the Building Regulation requirement that there shall be adequate precautions to prevent the temperature of the stored water at any time exceeding 100°C.

BRE Defect Action Sheet 140[29] describes how failure of heating controls can lead to an explosion of the cylinder, scalding from steam and hot water, and wastage of water, in consequence of the system being incorrectly installed, commissioned and maintained. Further details of unvented hot water storage units and packages are contained in BS 7206.[30]

SANITARY APPLIANCES

While Approved Document G1 (1991) prescribes that any dwelling should have at least one closet and one washbasin, BS 6465[31] requires that every new dwelling shall be supplied with a minimum of one WC, one bath or shower, one washbasin and one sink. Sanitary appliances should be selected and sited with the object of reducing noise in living rooms and bedrooms. A partition between a bedroom and a bathroom or WC should be not less than 75 mm of insulating blockwork and the door should be as heavy as practicable and well-fitting. A WC cistern and pan should be isolated from the wall and floor respectively by resilient pads, sleeves and washers. The quietest WC suites are those with double siphonic traps and a close-coupled cistern, but these are also the most expensive.

The Building Regulations 1991[24] in paragraph G1 of Schedule 1 require that

(1) Adequate sanitary conveniences shall be provided in rooms provided for that purpose, or in bathrooms. Any such room or bathroom shall be separated from places where food is prepared.

(2) Adequate washbasins shall be provided in

(a) rooms containing water closets, or

(b) rooms or spaces adjacent to rooms containing water closets.

Any such room or space shall be separated from places where food is prepared.

(3) There shall be a suitable installation for the provision of hot and cold water to washbasins provided in accordance with sub-paragraph (2).

(4) Sanitary conveniences and washbasins to which this paragraph applies shall be designed and installed so as to allow effective cleaning.

In Approved Document G1 (1991) 'sanitary accommodation' is defined as a room containing closets or urinals whether or not it also contains other sanitary appliances. Sanitary accommodation containing one or more cubicles counts as a single space if there is free circulation of air throughout the space.

BS 6465[31] recommends that in houses and maisonettes accommodating 5 or more persons, there shall be two WCs of which one may be in a bathroom. An additional washbasin is to be provided in every separate WC compartment which does not adjoin a bathroom. Where there are two WCs in a house, it is desirable to locate them on separate floors.

The essential qualities of good sanitary appliances are

(1) cleanliness – strong, smooth, non-absorbent and non-corroding surfaces, largely self-cleansing and permitting easy cleaning;
(2) durability – withstanding hard wear;
(3) simplicity of design and construction;
(4) accessibility;
(5) economical in initial and maintenance costs;
(6) satisfactory appearance.

Traditionally, water closets, urinals and bidets were classified as 'soil appliances', but in more recent times the term 'sanitary appliances' has been used to describe all appliances including sinks, baths and wash basins.

Waterclosets

Approved Document G1 (1991) of the Building Regulations[24] requires that a watercloset receptacle or pan shall have a smooth, non-absorbent and easily cleaned surface, discharging through a trap and branch pipe into a discharge stack or drain. The flushing apparatus shall be capable of cleansing the receptacle effectively. No part of the receptacle should be connected to any pipe, other than a flush pipe or branch discharge pipe. A closet fitted with a macerator and pump may be connected to a small bore drainage system discharging to a discharge stack, provided they comply with a current EOTA approval, such as that issued by BBA and

there is also access to a closet discharging directly to a gravity system.

A typical washdown watercloset is illustrated in figure 12.2.4 with a ceramic pan of vitreous china complying with BS 5503[32] with a normal height of about 400 mm. Wall hung WC pans complying with BS 5504[33] are often used in public buildings for ease of cleaning. The trap is 'S' or 'P' in shape, integral with the pan, of 88 mm internal diameter and with a depth of seal of 50 mm. The seat is normally a ring seat and cover for domestic use and made of plastics for maximum hygiene. The flushing cistern can be of a variety of materials, of which ceramic ware and plastics are very popular and should have a minimum capacity of 9 litres. Where the cistern is raised above the pan a flush pipe is required, with a minimum internal diameter of 38 mm for a low-level suite, discharging into the flushing rim around the top of the pan. Flush pipes may be of galvanised or porcelain enamelled steel tube; copper or copper alloy with a natural finish or chromium plated; or PVC-U in accordance with BS 1125.[34]

Siphonic waterclosets are more efficient and quieter than the washdown variety but there is greater risk of blockage if misused. In the single trap variety (figure 12.2.5), the waterway is designed to produce full bore flow and the siphonic action assists in cleansing the pan. The double trap is illustrated in figure 12.2.6. When flushed the patent device A reduces the air pressure in chamber B; seal C is broken allowing the escape of flush water and setting up siphonage. Siphonic waterclosets are sometimes provided with flushing cisterns with capacities of 11 and 14 litres.

The older type high level flushing cisterns were usually of the siphon pipe or the dome pattern.[34] In the dome type, the dome is lifted by a handle forcing water over the open top of the flush pipe, relieving air pressure and starting siphonic action which draws all the water out of the flushing cistern. In low level WC suites, flushing cisterns of the type illustrated in figure 12.2.3 in ceramic ware or plastics are often used. The bent siphon tube, one leg of which is connected to the flush pipe, has an enlarged cylinder at the bottom of the other leg. The lever handle lifts a valve in the cylinder, which forces the water in the cistern over the siphon and into the flush pipe.

The installation of a Skevington/BRE controlled flush valve enables the WC user to control the amount of water flushed and can result in water savings of up to 40 per cent. As the flushing handle is depressed, a link pulls the valve shut. As soon as the hand pressure is removed, the spring opens the valve thereby allowing air through a tube into the siphon. In this way any volume of water from a minimum of about 0.5 litres to a maximum of the full cistern can be selected to flush the WC pan, depending on its contents.[35]

Urinals

Approved Document G1 (1991) of the Building Regulations[24] requires that a urinal shall have a smooth, non-absorbent and easily cleaned surface. A urinal fitted with a flushing apparatus should discharge through a grating, a trap and a branch pipe into a discharge stack or drain. Any flushing apparatus should be capable of cleansing the receptacle effectively. No part of the receptacle shall be connected to any pipe other than a flush pipe or branch discharge pipe.

Urinals can be formed of ceramic materials to BS 5520,[36] stainless steel to BS 4880[37] or moulded plastics. Stainless steel has advantages over vitreous china in that it is lighter and is not subject to breakage or chipping. A typical stall urinal is illustrated in figure 12.2.7 with a width of 600 mm per stall and heights of 900 and 1050 mm. A 50 or 62 mm diameter end outlet to the channel is adequate for four stalls, a central outlet for five to seven, and a minimum of two outlets where there are more than seven stalls. The flushing capacity is normally 4.5 litres per stall. Automatic flushing cisterns, often set to discharge at twenty-minute intervals, should comply with BS 1876.[38] A proprietary flushing mechanism such as the 'Cistermiser' ensures that flushing only takes place when required, and this can save substantial quantities of water. The cistern is fed with water at a steady rate and siphonic action takes place automatically and rapidly when water reaches the designed level. The water passes through 25 or 32 mm flush pipes and spreaders to thoroughly flush each stall. The bowl type urinal (figure 12.2.8) has a smaller surface to be fouled and flushed and the bowls are normally fixed at 600 to 675 mm centres, each with its own separate trap. The slab type urinal has a flat wall surface, ends and a channel, and a common example is the stainless steel urinal.

Bidets

Bidets are used for cleansing the lower excretory organs of the body in a thorough and convenient manner by sitting astride the appliance. Hot and cold water can be delivered either to the rim or to the ascending spray as shown in figure 12.2.9. Bidets have a secondary use as a footbath. They are normally formed of ceramic materials with 32 mm outlet and separate trap, and can be either the pedestal or wall-hung variety.[39]

Washbasins

Approved Document G1 (1991) prescribes that a washbasin shall be located in or adjacent to the room containing the water closet, or in a room or space giving direct access to the room containing the closet (provided it is not used for the preparation of food), should have a supply of hot water, which may be from a central source or from a unit water heater, and a piped supply of cold water. A washbasin shall discharge through a grating, a trap and branch discharge pipe to a discharge stack or may, where the washbasin is located on the ground floor, discharge into a gully or direct to a drain.

Most washbasins are in white or coloured ceramic ware; BS 1188[40] recognises both fireclay and vitreous china, each with and without a back skirting. Washbasins provide facilities for personal ablutions in bathrooms, dressing rooms, bedrooms and cloakrooms. They can be supported by a pedestal or by cantilever brackets built into or screwed to a wall. The normal size of washbasin is 635× 455 mm (figure 12.3.1) although 560 × 405 mm may be used in confined spaces. A common waste size is 32 mm and a separate trap is provided. BS 1188[40] prescribes requirements for tap holes, soap sinkings, chain stay and overflow. BS 5506[41] gives details of dimensions of pedestal and wall hung washbasins and other relevant particulars. A typical overflow detail is shown in figure 12.3.3 and a corner washbasin in figure 12.3.2. The waste pipe will be taken from a ground floor appliance to connect with a back-inlet gully. Metal washbasins are available in stainless steel to BS 1329,[42] which gives good service under most conditions, porcelain enamelled sheet steel, which is suitable for light duty, and porcelain enamelled cast iron, which is well suited for normal and heavy duty.[31] Dimensions of wall hung hand rinse basins with a width not exceeding 530 mm, whatever material is used in manufacture are provided in BS 6731.[43]

In recent years it has become quite common practice to incorporate steel or ceramic washbasins in bedrooms and ladies' powder rooms into a melamine-faced phenolic laminate surround, when the fittings are called *vanitory units*. This arrangement permits a counter-type surface for make-up and similar materials to be placed around the washbasin without the risk of being splashed. The laminate can be bent into two dimensional curves under heat, permitting the provision of vanitory units with rounded front edges and swept back upstands. Some vanitory units also incorporate shallow drawers.

Baths

Approved Document G2 (1991) prescribes that any dwelling (house, flat or maisonette) should have at least one bathroom with a fixed bath or shower. A house in multi-occupation should have at least the same provision as a dwelling and this should be accessible to all occupants. A bath or shower should have a supply of hot water, which may be from a central source or from a unit water heater, and a piped supply of cold water. A bath or shower should discharge through a grating, a trap and branch discharge pipe to a discharge stack, or may, if it is on the ground floor, discharge into a gully or directly to a foul drain. There is also provision for connection to a macerator and pump small bore drainage system subject to a current European approval issued by a member body of the European Organisation for Technical Approvals (EOTA), such as the British Board of Agrément (BBA).

The majority of baths are made of porcelain enamelled cast iron in accordance with BS 1189[44] or vitreous enamelled pressed steel conforming to BS 1390,[45] although baths are also manufactured in cast acrylic sheets.[46] Bath panels, for enclosing the bath are available in a variety of materials, including plastics faced hardboard, and the more durable and costly moulded plastics. A typical overall size for a rectangular bath is 1694 × 697 × 500 mm high, as illustrated in figure 12.3.5. Other bath types, specified in BS 1189,[44] include the rectangular shallow pattern and the tub (parallel) pattern. The bottom of the bath should be as flat as possible and hand grips are desirable for reasons of safety. The normal rectangular bath has a capacity of about 120 litres when filled to 225 mm above the waste fitting, a common waste size being 38 mm. A suitable bath trap is shown in figure 12.3.6; the depth of seal varies from 38 to 75 mm. A bath with a stepped bottom to form a seat may be useful where floor space is restricted or for use by handicapped persons.

Sinks

Sinks used for culinary, laundry and other domestic purposes are normally made either of white glazed fireclay to BS 1206[47] or of vitreous porcelain enamelled cast iron or pressed steel, hard wearing stainless steel to BS 1244[48], or GRP (glass fibre reinforced polyester) in a wide range of colours. Fireclay sinks can take various forms and sizes, with or without back shelves and with and without integral fluted drainers. A typical reversible sink (610 × 455 × 255 mm deep), without shelves (Belfast type), is illustrated in figure 12.3.10 and a suitable S-trap for use with it in figure 12.3.9, although most of this outdated variety have now been replaced. Outlets are similar to baths, while plain sinks have weir-type overflows (figure 12.3.10) and sinks with shelves have slot overflows (figure 12.3.8). A combined sink and drainer unit (combination unit) is illustrated in figure 12.3.8. Metal sinks are made in round bowls, normally 430 mm diameter, as well as the rectangular pattern. More attractive forms of exposed trap include the chromium-plated or stainless steel 'bottle' trap illustrated in figure 12.3.7. Butler's crockery or 'wash-up' sinks used for cleaning china and glasses may be made of teak, timber lined with lead or, more usually, stainless steel.

Waste Disposers

Waste disposers are fitted to sinks with a 88 mm diameter outlet and are designed to dispose of organic food waste quickly, hygienically and electro-mechanically and to flush the residue into the drain. Metal, rags and plastic objects should not be placed in a waste disposer. A typical disposer is illustrated in figure 12.3.11 and is operated by turning on cold water, switching on the disposer and feeding waste into the unit. The waste falls on to a high speed rotor and is flung against a stationary cutting ring with great force, shredding the waste into very small particles. The partially liquefied waste filters through the rotor into the waste pipe, joining the flow of water which keeps the apparatus clean and free from unpleasant smells. A thermal overload device cuts off the power in the event of jamming.

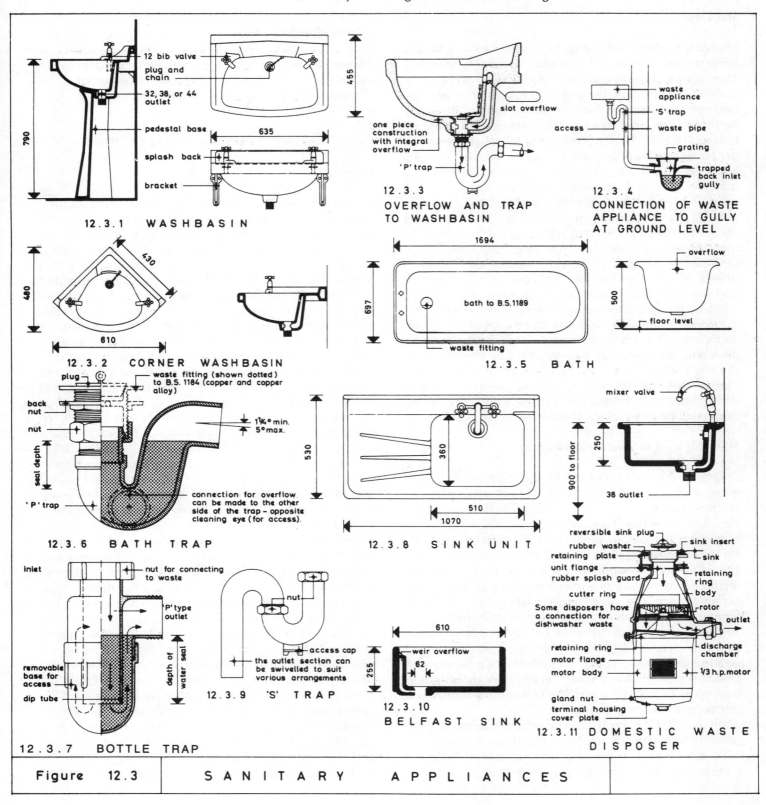

12 bib valve
plug and chain
32, 38, or 44 outlet
pedestal base
splash back
bracket

455
635
790

12.3.1   WASHBASIN

slot overflow
one piece construction with integral overflow
'P' trap

12.3.3   OVERFLOW AND TRAP TO WASHBASIN

waste appliance
'S' trap
access
waste pipe
grating
trapped back inlet gully

12.3.4   CONNECTION OF WASTE APPLIANCE TO GULLY AT GROUND LEVEL

430
480
610

12.3.2   CORNER WASHBASIN

1694
697
bath to B.S.1189
waste fitting

12.3.5   BATH

overflow
500
floor level

plug
back nut
nut
seal depth
'P' trap
waste fitting (shown dotted) to B.S. 1184 (copper and copper alloy)
1¼° min. 5° max.
connection for overflow can be made to the other side of the trap – opposite cleaning eye (for access).
530

12.3.6   BATH TRAP

360
510
1070

12.3.8   SINK UNIT

mixer valve
250
900 to floor
38 outlet

inlet
nut for connecting to waste
'P' type outlet
removable base for access
dip tube
depth of water seal

12.3.7   BOTTLE TRAP

nut
access cap
the outlet section can be swivelled to suit various arrangements

12.3.9   'S' TRAP

610
255
weir overflow
62

12.3.10   BELFAST SINK

reversible sink plug
sink insert
rubber washer
retaining plate
sink
unit flange
retaining ring
rubber splash guard
cutter ring
body
Some disposers have a connection for dishwasher waste
rotor
outlet
retaining ring
discharge chamber
motor flange
motor body
⅓ h.p. motor
gland nut
terminal housing cover plate

12.3.11   DOMESTIC WASTE DISPOSER

| Figure   12.3 | S A N I T A R Y   A P P L I A N C E S | |

Showers

Showers enable washing to be performed quickly, in limited space and with the minimum quantity of water. They should be selected and installed in accordance with BS 6340[49] and be sited in rooms that are adequately ventilated to reduce condensation. They may be fitted in separate enclosures or be used in conjunction with baths. They normally consist of an overhead or shoulder height rose or spray nozzle, usually attached to a flexible hose. The shower base or tray may be in porcelain enamelled cast iron, vitreous enamelled sheet steel, glazed ceramic, acrylic plastics or GRP; the most common sizes being 750 and 900 mm square by 150 and 175 mm deep overall, drained by a 38 mm trapped gully with bars not exceeding 6 mm apart. One-piece cubicles in acrylic plastics and built-up panels of ceramic tiles are available, and waterproof curtains or plastics doors may be used to give privacy.

Spray nozzles are of various patterns. Some contain a control tap to vary the shape or volume of the spray, and all may be mounted on a swivel joint to vary the direction. Showers may be controlled by stopcocks on the hot and cold supplies, by mechanical mixers or by thermostatic mixing valves, with a variety of mixing arrangements. Thermostatic mixing valves are to be preferred as they reduce the danger from scalding.

Materials used in Sanitary Appliances

Some of the more important characteristics of the more common materials used in sanitary appliances are now listed.

Ceramic materials. Fireclay is semi-porous yellow or buff refractory clay of great strength especially suitable for large appliances. Vitreous china is a white non-porous clay of very fine texture, which is very strong with high resistance to crazing and staining. The final coating of ceramic glaze resembles glass coating fused at high temperature and can be transparent or coloured.

Vitreous enamelled cast iron or *steel*. This is sometimes referred to as porcelain enamel and consists of opaque glass fused to the metal forming a permanent bond. It produces a smooth, even, very durable and high-gloss finish in a variety of colours which are permanent and non-fading.

Stainless steel. A high alloy steel containing a large proportion of chromium. Satin finish has a high resistance to marking, scratching and corrosion. Stainless steel is light in weight and very durable.

Plastics. Perspex, polypropylene, nylon and glass fibre reinforced plastics require no protective coating and are generally self-coloured, homogeneous, free from ripples or blemishes, very tough, lightweight, warm to the touch and resilient. They can however be damaged by abrasion, lighted cigarettes and hot utensils.

SANITARY PIPEWORK ABOVE GROUND

Traps

Approved Document H1 (1989) to the Building Regulations 1991[24] requires that all points of discharge into the system should be fitted with a water seal (trap) to prevent foul air from the system entering the building. Under working and test conditions traps should retain a minimum seal of 25 mm. These aspects are reiterated in BS 5572.[50] The entry of foul air from the drainage system into the building is prevented by the installation of suitable traps which should be self-cleansing. A trap which is not an integral part of the appliance should be attached to and immediately beneath its outlet and the bore of the trap shall be smooth and uniform throughout. All traps shall be accessible and be provided with adequate means of cleaning. There is advantage in providing traps which are capable of being readily removed or dismantled. Typical traps are illustrated in figures 12.3.3, 12.3.6, 12.3.7 and 12.3.9.

Waste outlets and traps should have a minimum internal diameter of 32 mm for washbasins and bidets, and 40 mm for sinks, baths and shower trays. Minimum depths of seals should be 50 mm for water closets and 75 mm for other appliances.

The traps to the various appliances must remain sealed in all conditions of use, otherwise there is a risk of unpleasant smells entering the building. A seal will be broken if the pressure changes in the branch pipe are of sufficient intensity and duration to overcome the head of water in the trap itself. One way of restricting the changes in air pressure during discharge is to provide an extensive system of vent piping with the object of equalising pressures throughout the system. Research conducted by the Building Research Establishment has shown that this may also be achieved by appropriate design of the pipework using the single-stack system. Another but less certain alternative is to use special resealing traps. The main causes of loss of seal are *self-siphonage* due to full bore flow in the branch waste pipe to the discharge stack, and *induced siphonage* due to air and water flow down the discharge stack and *back pressure* due to flow of air and water down the discharge stack in conjunction with restrictions to flow due to offsets and the bend at the base of the stack. Other causes are *wind effect* across the top of the stack, surcharging of the underground drainage pipework and the effect of intercepting traps.[51]

Discharge Pipes

BS 5572[50] introduced the term *discharge pipe* in place of 'soil and waste pipes'. A discharge pipe may also convey rainwater. The primary function of discharge pipes is to convey discharges from appliances to the underground drains through branch discharge pipes and discharge stacks, unless the appliances are on the ground floor, and they can be arranged in three different ways: two-pipe, one-pipe and single-stack systems. An efficient system should satisfy the following requirements.

(1) effective and speedy removal of wastes;
(2) prevention of foul air entering building;
(3) ready access to interior of pipes, including provision for cleaning where necessary;
(4) protection against extremes of temperature;
(5) protection against corrosion and erosion of pipes;
(6) restriction of siphonage and avoidance of liability to damage, deposition or obstruction;
(7) obtaining economical and efficient arrangements, which are assisted by compact grouping of sanitary appliances.

Approved Document H1 (1989) of the Building Regulations 1991[24] prescribes that the installation shall be capable of withstanding an air or smoke test of positive pressure of at least 38 mm water gauge for at least 3 minutes, with every trap maintaining a water seal of at least 25 mm. Smoke testing is not however recommended for PVC-U pipes.

Two-pipe System

This was the traditional system in the United Kingdom whereby soil (discharge from water closets, urinals, slop-hoppers, and similar appliances) is conveyed through soil pipes directly to the drain, and waste (discharge from sinks, washbasins, baths and similar appliances) is conveyed to the drain by a waste pipe discharging through a trapped gully or directly into a drain. This arrangement is straightforward and effective but costly, as it results in a large number of pipes. It is especially suitable, however, where the sanitary appliances are widely dispersed. Approved Document H1 (1989) to the Building Regulations[24] recommends that discharge stacks should not have any offsets in any wet portion; if they are unavoidable in a building of not more than three storeys there should be no branch connection within 750 mm of the offset. In a building over three storeys a ventilation stack may be needed with connections above and below the offset, and discharge stacks should be located inside the building. A typical arrangement of a two-pipe system for a two-storey house is shown diagrammatically in figure 12.4.1, with separate soil and waste pipes serving first floor sanitary appliances and with both pipes extended above roof level as ventilating pipes. Approved Document H1 (1989) prescribes that discharge stacks serving not more than one siphonic WC with a 75 mm outlet should not be less than 75 mm diameter. Ventilating pipes shall terminate so as not to become prejudicial to health or a nuisance, finishing at least 900 mm above any opening into the building within 3 m, and shall be fitted with a durable wire cage or other suitable cover, which does not restrict the flow of air. Branches may need anti-siphonage pipes if unsealing is otherwise possible. Ground floor appliances are best taken into the drains provided the drop is less than 1.5 m; the water closet is shown with a direct connection and the washbasin discharges between the grating and the top of the water seal in a back inlet gully.

An unventilated stub stack may be used if it connects above ground into a ventilated discharge stack or a drain free from surcharging subject to certain height limits, namely 2 m above invert of connection or drain and no branch serving a closet is more than 1.5 m from the crown of the closet trap to the invert of the connection or drain, as shown in diagram 5 of Approved Document H1 (1989). The length of branch drain from an unventilated stub stack should not be more than 6 m where a single appliance is connected and 12 m for a group of appliances.

One-pipe System

The one-pipe system is illustrated in figure 12.4.2, whereby all sanitary appliances discharge into one main stack or discharge pipe, relying upon appliance trap seals to act as the foul air barrier. The risk of loss of trap seal was overcome by using deeper seal traps and fully ventilating the system. This system requires the close grouping of appliances around the main stack to be economical and has now been largely superseded by the simpler and more economical single-stack system, which dispenses with the need for anti-siphonage pipes. It does however have installational and operational advantages over the two-pipe system, particularly for high-rise buildings, provided that a high standard of design and workmanship is secured.

Single-stack System

Research has shown that the unsealing of traps occurred far less frequently than was originally supposed and that in certain circumstances ventilating pipes could be omitted, except for the venting of the main stack. All appliances on upper floors can discharge into a single discharge pipe as shown in figure 12.4.3. The design of the pipework is important and relies upon close grouping of single appliances, each with a separate branch, around the stack. It produces the simplest possible system and is particularly well suited for high-rise housing where considerable savings in cost are possible.

The stack normally has a diameter of at least 100 mm, except for two-storey housing where 75 mm may be satisfactory. For buildings of more than five storeys, ground floor appliances should be connected separately to the drain. To obviate back pressure at the lowest branch to the stack and the build-up of detergent foam, the bend at the foot of the stack should be of large radius (at least 200 mm at the centre line), or alternatively two 135° bends may be used (figure 12.5.3). The vertical distance between the lowest branch connection and the invert of the drain should be at least 750 mm or 450 mm for three-storey houses with 100 mm stack and two-storey houses with 75 mm stack (figure 12.5.2). To guard against self-siphonage washbasins should have 75 mm P-traps. The maximum slope of a 32 mm diameter branch pipe can be determined from figure 12.4.4 according to the length of waste pipe, with no bends less than 75 mm radius, although they are avoided if possible. Branch pipes to washbasins longer than the recommended maximum length of 1.70 m (figure 12.5.1) should be provided with a 32 mm diameter trap with a short 32 mm tail pipe

discharging into a 40 or 50 mm branch pipe.[52] Another alternative is to ventilate the branch pipe. Bath and sink wastes should have 40 mm traps and branch pipes with 75 mm deep seals to guard against self-siphonage, when the length and slope of the branch pipe were not normally regarded as critical (figure 12.5.1). However, Approved Document H1 (1989) recommends maximum lengths of 3 m for 40 mm branch pipes and 4 m for 50 mm branch pipes with slopes between 18 to 90 mm/m for branches to baths and sinks, and 6 m maximum length for a branch to a single WC with a minimum slope of 9 mm/m. Watercloset connections should be swept in the direction of flow to a minimum radius of 50 mm (figure 12.5.1). Bath wastes must not be connected to the stack within 200 mm below the centre line of a WC branch (figure 12.5.1) to avoid the discharge from the WC branch backing up the bath branch.[52] Approved Document H1 (1989) recommends that branch ventilating pipes to branch pipes serving one appliance should be at least 25 mm diameter and be connected to the branch discharge pipe within 300 mm of the trap and not below the spillover level of the highest appliance served. Air admittance valves fitted to discharge stacks terminating inside a building must not interfere with the open stack ventilation to the below ground system and should comply with the current British Board of Agrément Certificate. All pipes should be reasonably accessible for repair.

Pipes and Joints

A variety of materials is available for discharge pipes and branches listed.

(1) *Galvanised steel* to BS 1387[17] with screw joints on small pipes or spigot and socket joints on large pipes, or pre-fabricated stack units to BS 3868.[53]

(2) *Copper* to BS 2871[7] (table X), with bronze welded joints on main stack pipes and capillary or compression joints on smaller pipes, complying with BS 864.[8]

(3) *Cast iron* to BS 416,[54] jointed with yarn and molten lead or a cold caulking compound, or a patent flexible jointing system.

(4) *Unplasticised* PVC to BS 4514[55] with solvent welded push-fit joints, with provision for expansion, or increasingly polypropylene to BS 5254[56] or plastics to BS 5255[57] which incorporate ABS (acrylonitrile butadiene

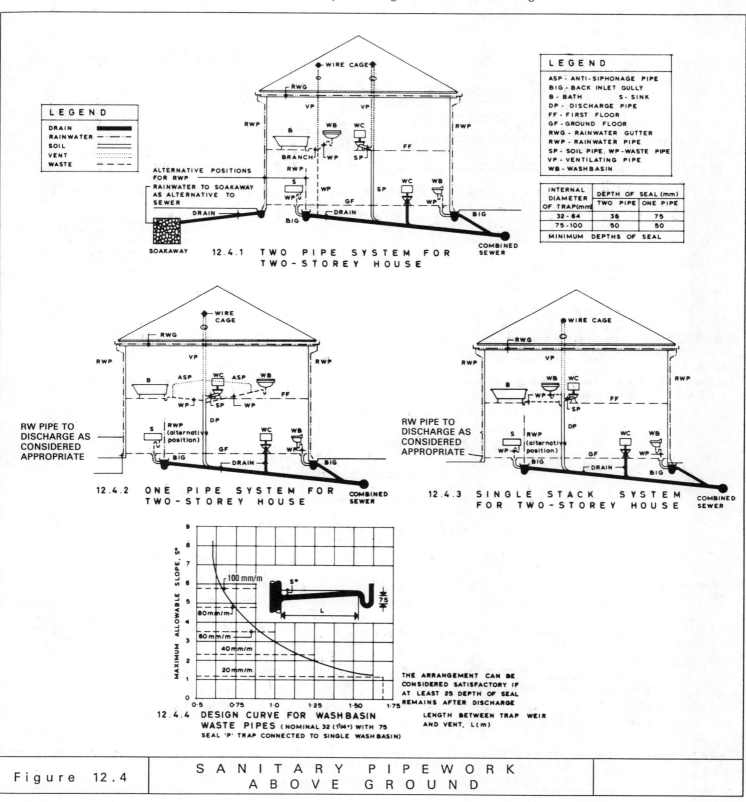

LEGEND

DRAIN	━━━
RAINWATER	─ ─ ─
SOIL	───────
VENT	··········
WASTE	─ · ─ · ─

12.4.1 TWO PIPE SYSTEM FOR TWO-STOREY HOUSE

LEGEND

ASP - ANTI-SIPHONAGE PIPE
BIG - BACK INLET GULLY
B - BATH   S - SINK
DP - DISCHARGE PIPE
FF - FIRST FLOOR
GF - GROUND FLOOR
RWG - RAINWATER GUTTER
RWP - RAINWATER PIPE
SP - SOIL PIPE. WP-WASTE PIPE
VP - VENTILATING PIPE
WB - WASHBASIN

INTERNAL DIAMETER OF TRAP (mm)	DEPTH OF SEAL (mm)	
	TWO PIPE	ONE PIPE
32-64	38	75
75-100	50	50
MINIMUM DEPTHS OF SEAL		

12.4.2 ONE PIPE SYSTEM FOR TWO-STOREY HOUSE

RW PIPE TO DISCHARGE AS CONSIDERED APPROPRIATE

12.4.3 SINGLE STACK SYSTEM FOR TWO-STOREY HOUSE

RW PIPE TO DISCHARGE AS CONSIDERED APPROPRIATE

12.4.4 DESIGN CURVE FOR WASHBASIN WASTE PIPES (NOMINAL 32 (1¼") WITH 75 SEAL 'P' TRAP CONNECTED TO SINGLE WASHBASIN)

THE ARRANGEMENT CAN BE CONSIDERED SATISFACTORY IF AT LEAST 25 DEPTH OF SEAL REMAINS AFTER DISCHARGE

LENGTH BETWEEN TRAP WEIR AND VENT, L(m)

| Figure 12.4 | SANITARY PIPEWORK ABOVE GROUND |

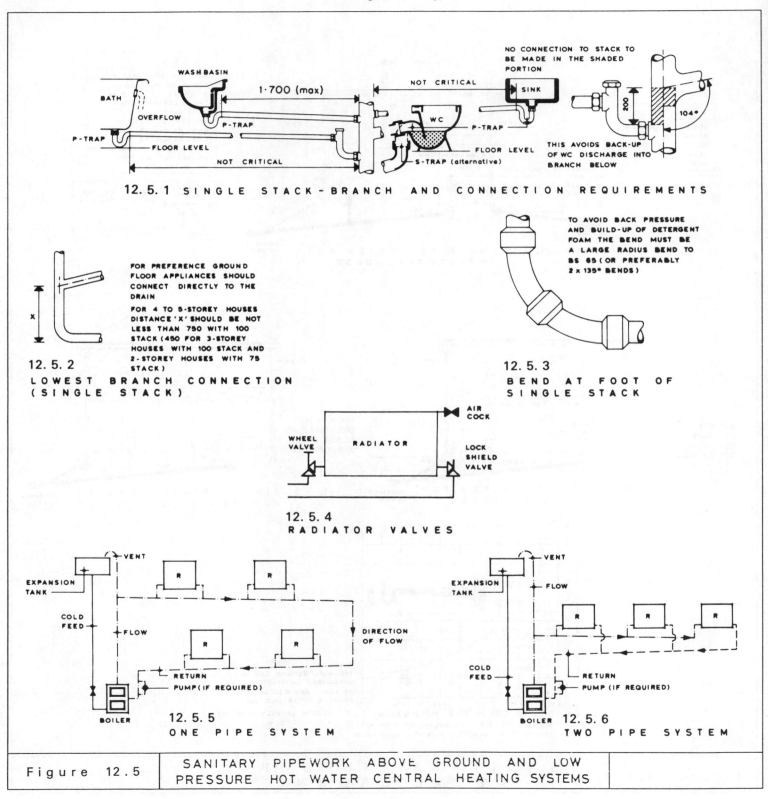

NO CONNECTION TO STACK TO
BE MADE IN THE SHADED
PORTION

WASH BASIN

NOT CRITICAL

BATH

1·700 (max)

SINK

OVERFLOW

P-TRAP

WC

P-TRAP

200

P-TRAP

FLOOR LEVEL

FLOOR LEVEL

104°

NOT CRITICAL

S-TRAP (alternative)

THIS AVOIDS BACK-UP
OF WC DISCHARGE INTO
BRANCH BELOW

12.5.1 SINGLE STACK - BRANCH AND CONNECTION REQUIREMENTS

FOR PREFERENCE GROUND
FLOOR APPLIANCES SHOULD
CONNECT DIRECTLY TO THE
DRAIN

FOR 4 TO 5-STOREY HOUSES
DISTANCE 'X' SHOULD BE NOT
LESS THAN 750 WITH 100
STACK (450 FOR 3-STOREY
HOUSES WITH 100 STACK AND
2-STOREY HOUSES WITH 75
STACK)

x

TO AVOID BACK PRESSURE
AND BUILD-UP OF DETERGENT
FOAM THE BEND MUST BE
A LARGE RADIUS BEND TO
BS 65 (OR PREFERABLY
2 x 135° BENDS)

12.5.2
LOWEST BRANCH CONNECTION
(SINGLE STACK)

12.5.3
BEND AT FOOT OF
SINGLE STACK

AIR
COCK

WHEEL
VALVE

RADIATOR

LOCK
SHIELD
VALVE

12.5.4
RADIATOR VALVES

VENT

EXPANSION
TANK

R

R

COLD
FEED

FLOW

R

R

DIRECTION
OF FLOW

RETURN

PUMP(IF REQUIRED)

BOILER

12.5.5
ONE PIPE SYSTEM

VENT

EXPANSION
TANK

FLOW

R

R

R

COLD
FEED

R

R

RETURN

PUMP (IF REQUIRED)

BOILER

12.5.6
TWO PIPE SYSTEM

| Figure 12.5 | SANITARY PIPEWORK ABOVE GROUND AND LOW PRESSURE HOT WATER CENTRAL HEATING SYSTEMS | |

styrene), MUPVC (modified unplasticised PVC), polyethylene and polypropylene.

LOW PRESSURE HOT WATER CENTRAL HEATING SYSTEMS

Many domestic central heating systems are of low pressure hot water served directly by a boiler fired by gas, oil or solid fuel. The pressure is provided by the static head of water in the cold water feed tank, hence the term 'low pressure'. Gravity systems operate by means of a natural convection current. Nevertheless, it is common to install a circulating pump connected into the return pipe near the boiler to provide a more vigorous or 'accelerated' circulation. There are two principal pipe arrangements: one pipe (figure 12.5.5) and two pipe (figure 12.5.6). The one-pipe system is easier and cheaper to install but the radiators become progressively cooler from the first to the last and individual control of radiators is largely ineffective. The two-pipe system is well suited to maximum individual control as the hot water is conveyed directly from the boiler to each radiator and the temperature of each is approximately the same; similarly the cold water from each radiator is returned directly to the boiler.

With taller buildings it is customary to use a drop-pipe system with radiators on each floor fed from drop pipes running vertically through the building. Another development has been the *small bore* system using 15 mm pipes and a circulating pump in a series of loops with about three medium-sized radiators on each loop. Radiators are normally controlled by a wheel valve on the flow pipe and a lockshield valve on the return pipe, with an aircock to release any air trapped in the top of the radiator (figure 12.5.4). Guidance on the design, planning, installation, selection and testing of low pressure hot water heating systems is given in BS 6880.[58] Advice on the planning, design and installation of smallbore and microbore domestic central heating systems is provided in BS 5449.[59] Useful guidance on the installation of gas fired hot water boilers is given in BS 6789,[60] and on the use, selection and installation of expansion vessels and ancillary equipment in BS 7074.[61] Details of relevant thermal insulation methods are provided in chapter 15.

BRE Digest 254[62] analyses the reliability and performance of solar collecting systems and gives guidance on the methods of checking system operation, in view of the failures that have occurred in practice.

SOLID WASTE STORAGE

Approved Document H4 (1989) requires solid waste storage to be

(1) designed and sited so as not to be prejudicial to health;

(2) of sufficient capacity having regard to the quantity of solid waste to be removed and the frequency of removal;

(3) sited so as to be accessible for use by people in the building and of ready access from a street for emptying and removal.

In low rise developments any dwelling should have, or have access to, a movable container with a capacity of not less than 0.12 m^3 or a communal waste container of between 0.75 m^3 and 1 m^3, each with a close fitting lid. This capacity assumes an output of refuse of 0.09 m^3 and collection at weekly intervals, otherwise larger capacity containers will be required. In multi-storey domestic developments dwellings up to the fourth floor may each have their own container or may share a container, and for taller buildings a chute is normally provided, terminating in a refuse storage chamber where the refuse generally falls into a large movable container.

Chutes should have a smooth non-absorbent surface and close fitting access doors at each storey which has a dwelling and be ventilated at top and bottom. Containers need not be enclosed but if they are, the enclosure should allow room for filling and emptying and provide a clear space of 150 mm between and around the containers and for communal containers be a minimum of 2 m high. The enclosure should be permanently ventilated at top and bottom.

Containers and chutes should be sited so that householders are not required to carry refuse further than 30 m. Containers should be within 25 m of the vehicle access. Containers in new buildings should be sited so that they can be collected without being taken through a building, unless it is a garage, carport or other open covered space. Further recommendations and information on solid waste storage can be obtained from BS 5906.[63]

REFERENCES

1. *BS 5422: 1990* Thermal insulating materials on pipes, ductwork and equipment (in the temperature range −40°C to +700°C)
2. *BS 6700: 1987* Specification for design, installation, testing and maintenance of services supplying water for domestic use within buildings and their curtilages
3. *BS 7181: 1989* Specification for storage cisterns up to 500 litres actual capacity for water supply for domestic purposes
4. *BS 1212* Float operated valves (excluding floats), *Part 1: 1990* Specification for piston type float operated valves (copper alloy body) (excluding floats); *Part 2: 1990* Specification for diaphragm type float operated valves (copper alloy bodies) (excluding floats); *Part 3: 1979* Diaphragm type (plastics body) for cold water services
5. *BS 417* Galvanised mild steel cisterns and covers, tanks and cylinders, *Part 2: 1987* Metric units
6. *BS 4213: 1986* Cold water storage and feed and expansion cisterns (polyolefin or olefin copolymer) and cistern lids
7. *BS 2871* Copper and copper alloys. Tubes, *Part 1: 1971 (1977)* Copper tubes for water, gas and sanitation
8. *BS 864* Capillary and compression tube fittings of copper and copper alloy, *Part 2: 1983* Specification for capillary and compression fittings for copper tubes
9. *BS 6730: 1986* Specification for black polyethylene pipes up to nominal size 63 for above ground use for potable water
10. *BS 3284: 1967 (1984)* Polythene pipe (type 50) for cold water services
11. *BS 3505: 1986* Unplasticized polyvinyl chloride (PVC-U) pressure pipes for cold potable water
12. *CP 312* Plastics pipework (thermoplastics material); *Part 1: 1973* General principles and choice of material; *Part 2: 1973* Unplasticized PVC pipework for the conveyance of liquids under pressure; *Part 3: 1973* Polyethylene pipes for the conveyance of liquids under pressure
13. *BS 5955* Plastics pipework (thermoplastic material), *Part 8: 1990* Specification for the installation of thermoplastics pipes and associated fittings for use in domestic hot and cold water services and heating systems
14. *BS 7291* Thermoplastics pipes and associated fittings for hot and cold water for domestic purposes and heating installations in buildings, *Part 1: 1990* General requirements; *Part 2: 1990* Specification for polybutylene (PB) pipes and associated fittings; *Part 3: 1990* Specification for cross-linked polyethylene (PE-X) pipes and associated fittings; *Part 4: 1990* Specification for chlorinated polyvinyl chloride (PVC-C) pipes and associated fittings and solvent cement
15. *BS 6437: 1984* Specification for polyethylene pipes (type 50) in metric diameters for general purposes
16. *BS 6572: 1985* Specification for blue polyethylene pipes up to nominal size 63 for below ground use for potable water
17. *BS 1387: 1985* Screwed and socketed steel tubes and tubulars and for plain end steel tubes suitable for welding or for screwing to BS 21 pipe threads
18. *BS 4127* Light gauge stainless steel tubes, *Part 2: 1972 (1986)* Metric units
19. *BRE Digest 83*: Plumbing with stainless steel (1980)
20. *BS 6362: 1990* Stainless steel tubes suitable for screwing in accordance with BS 21 pipe threads for tubes and fittings, where pressure type joints are made on threads
21. *BS 1010* Draw-off taps and stopvalves for water services (screwdown pattern), *Part 2: 1973 (1985)* Draw-off taps and above ground stopvalves
22. *BS 5154: 1983* Copper alloy globe, globe stop and check, check and gate valves for general purposes
23. *BRE Digest 339*: Condensing boilers (1990)
24. *The Building Regulations 1991 and Approved Documents G (1991) and H (1989)*. HMSO
25. *BRE News 58*: Research on water supply: unvented hot water systems (1982); *BRE Report 125*: Unvented domestic hot water systems (1988)
26. *BS 3955: 1986* Specification for electrical controls for household and general purposes
27. *BS 6283* Safety and control devices for use in hot water systems, *Part 2: 1991* Specification for temperature relief valves at pressures from 1 bar to 10 bar; *Part 3: 1991* Specification for combined temperature and pressure relief valves for pressures from 1 to 10 bar.
28. *BRE Defect Action Sheet 139*: Unvented hot water storage systems – safety (1989)
29. *BRE Defect Action Sheet 140*: Unvented hot water storage systems – installation and inspection (1989)
30. *BS 7206: 1990* Specification for unvented hot water storage units and packages
31. *BS 6465* Sanitary installations, *Part 1: 1984* Code of

practice for scale of provision, selection and installation of sanitary appliances

32. *BS 5503* Specification for vitreous china washdown WC pans with horizontal outlet, *Part 1: 1977* Connecting dimensions; *Part 2: 1977* Materials, quality, performance and dimensions, other than connecting dimensions

33. *BS 5504* Specification for wall hung WC pan, *Part 1: 1977* Wall hung WC pan with close coupled cistern. Connecting dimensions; *Part 2: 1977* Wall hung WC pan with independent water supply. Connecting dimensions; *Part 3: 1977* Wall hung WC pan. Materials, quality and functional dimensions other than connecting dimensions

34. *BS 1125: 1987* WC flushing cisterns (including dual flush cisterns and flush pipes)

35. *BRE Information Paper IP 12/83*: Water economy with the Skevington/BRE controlled flush valve for WCs (1983)

36. *BS 5520: 1977* Specification for vitreous china bowl urinals (Rimless type)

37. *BS 4880* Urinals, *Part 1: 1973* Stainless steel slab urinals

38. *BS 1876: 1990* Automatic flushing cisterns for urinals

39. *BS 5505* Specification for bidets, *Part 3: 1977* Vitreous china bidets over rim supply only. Quality, workmanship and functional dimensions other than connecting dimensions

40. *BS 1188: 1974 (1989)* Ceramic washbasins and pedestals

41. *BS 5506* Specification for washbasins, *Part 1: 1977* Pedestal washbasins. Connecting dimensions; *Part 2: 1977* Wall hung washbasins. Connecting dimensions; *Part 3: 1977* Washbasins (one or three tap holes), materials, quality, design and construction

42. *BS 1329: 1974 (1989)* Metal hand rinse basins

43. *BS 6731: 1988* Specification for wall hung hand rinse basins. Connecting dimensions

44. *BS 1189: 1986* Baths from porcelain enamelled cast iron

45. *BS 1390: 1990* Baths made from vitreous enamelled sheet steel

46. *BS 4305* Baths for domestic purposes made of acrylic material, *Part 1: 1989* Specification for finished baths; *Part 2: 1989* Specification for connecting dimensions

47. *BS 1206: 1974 (1989)* Fireclay sinks: dimensions and workmanship

48. *BS 1244* Metal sinks for domestic purposes, *Part 2: 1988* Specification for sit-on and inset sinks

49. *BS 6340* Shower units, *Part 1: 1983* Guide on choice of shower units and their components for use in private dwellings; *Part 2: 1983* Specification for the installation of shower units

50. *BS 5572: 1978 (1988)* Code of practice for sanitary pipework

51. *BRE Digest 248*: Sanitary pipework: Part 1: Design basis (1981)

52. *BRE Digest 249*: Sanitary pipework: Part 2: Design of pipework (1981)

53. *BS 3868: 1973* Prefabricated drainage stack units: galvanised steel

54. *BS 416: 1973 (1988)* Cast iron spigot and socket soil, waste and ventilating pipes (sand cast and spun) and fittings

55. *BS 4514: 1983 (1988)* Specification for unplasticized PVC soil and ventilating pipes, fittings and accessories

56. *BS 5254: 1976 (1984)* Polypropylene waste pipes and fittings (external diameter 34.6 mm, 41.0 mm and 54.1 mm)

57. *BS 5255: 1989* Thermoplastics waste pipes and fittings

58. *BS 6880* Code of practice for low temperature hot water heating systems of output greater than 45 kW, *Part 1: 1988* Fundamental and design considerations; *Part 2: 1988* Selection of equipment; *Part 3: 1988* Installation, commissioning and maintenance

59. *BS 5449* Code of practice for central heating for domestic premises, *Part 1: 1977 (1981)* Forced circulation hot water systems

60. *BS 6789: 1987* Specification for installation of gas-fired hot water boilers of rated input not exceeding 60 kW

61. *BS 7074* Application, selection and installation of expansion vessels and ancillary equipment for sealed water systems, *Part 1: 1989* Code of practice for domestic heating and hot water supply; *Part 2: 1989* Code of practice for low and medium temperature hot water heating systems

62. *BRE Digest 254*: Reliability and performance of solar collector systems (1981)

63. *BS 5906: 1980 (1987)* Code of practice for the storage and on-site treatment of solid waste from buildings

13 DRAINAGE AND SMALL TREATMENT WORKS

This chapter is concerned with building drainage from initial design and statutory requirements to the constructional techniques and materials used and methods of disposal of the effluent. As in previous chapters reference will be made to appropriate British Standards.

DESIGN OF DRAINS

Drainage Systems

Drainage systems must be designed to provide an efficient and economical method of carrying away waterborne waste, in such a way as to avoid the risk of pipe blockage and the escape of effluent into the ground. A drainage system normally consists of a network of pipes laid from a building to fall to a local authority sewer, although in some cases it may be necessary to install a plant to treat the effluent or a pump to raise it to a sewer at a higher level.

The layout of the drainage system is also influenced by the sewer arrangements, which can be one of three types.

(1) *Combined system* whereby foul water from sanitary appliances and surface water from roofs and paved areas discharge through a single drain to the same combined sewer. This simplifies and cheapens the house drainage system, ensures that the drains are well flushed in time of storm and that the house drain cannot be connected to the wrong sewer. On the other hand silting may occur in large pipes and it may entail storm overflows on sewers and high costs of pumping and sewage treatment. The Building Regulations 1991,[1] in paragraph H1 of Schedule 1, describes foul water as waste water which comprises or includes (a) waste from a sanitary convenience or other soil appliance;

(b) water which has been used for cooking or washing.

(2) *Separate system* in which foul wastes pass through one set of drains to a foul sewer, whereas surface water is conveyed to a separate surface water sewer or soakaways. This arrangement reduces pumping and sewage treatment costs to the main drainage authority but results in additional expense with the house drainage system, eliminates the flushing action of the surface water in foul drains and permits the possibility of an incorrect connection.

(3) *Partially separate system* in which the foul sewer takes some of the surface water (possibly that from the rear roof slopes and paved areas) and another sewer takes surface water only. This arrangement is a compromise and lessens both the advantages and disadvantages of the previous systems.

The layout of a drainage system should be as simple and direct as possible, and drains should be laid in straight lines between points where changes of direction or gradient occur, and these should be kept to a minimum. Connections to other drains or sewers should be made obliquely in the direction of flow. The drains should be of sufficient strength and be constructed of sufficiently durable materials with flexible watertight joints to minimise the effects of any differential settlement. Approved Document H1 (1989) to the Building Regulations 1991[1] requires that a drain under a building should be surrounded with at least 100 mm of granular or other flexible filling, and further precautions taken where excessive subsidence is possible. It further provides that a drain may run through a wall or foundation if either (1) an opening is formed to give at least 50 mm clearance all round and the opening masked with rigid sheet material to prevent ingress of fill or vermin, or (2) a short length of pipe is built in with its joints as close as possible to the wall faces (150 mm maximum) and connected on each side to rocker pipes with a length not exceeding 600 mm and with flexible joints.

Pipe Sizes and Gradients

Pipes must have sufficient capacity to carry the flow and be laid to falls which ensure a self-cleansing velocity and so prevent the settlement of solids, which might lead to a blockage. The pipe size and gradient selected should ensure as far as practicable that the pipe will not run full which might, by suction, unseal traps either within the building or in gullies. BS 8301[2] recommends that at peak flow, foul drains should be designed to permit adequate air movement in the pipe by ensuring that the depth of flow does not exceed 0.75 of the pipe diameter.

Approved Document H1 (1989) of the Building Regulations 1991[1] and BS 8301[2] both prescribe that a drain carrying waste water should have a diameter of at least 75 mm and a drain carrying soil water or waste water containing trade effluent at least 100 mm. The flow rates of appliances are detailed in BS 8301[2] as 2.3ℓ/s for 5 s for a WC, 0.6 ℓ/s for 10 s for a washbasin (32 mm outlet), 0.9ℓ/s for 25 s for a sink (40 mm outlet) and 1.1 ℓ/s for 75 s for a bath (40 mm outlet). Probable discharge factors and unit ratings can be calculated from these flow rates and are tabulated in BS 8301,[2] and a useful graph showing the capacities of foul drains running at 0.75 proportional depth at varying gradients is contained in diagram 7 of Approved Document H1 (1989) of the Building Regulations.[1]

For normal house drains serving two or three dwellings, the flow in dry weather is very small and usually intermittent, and to be self-cleansing the pipe diameter should be kept to the smallest practicable (100 mm). The volume of surface water is usually calculated on the basis of a rainfall intensity of 50 mm per hour.

BS 8301[2] prescribes the following guidelines on gradients

(1) For flows of less than 1ℓ/s, pipes not exceeding 100 mm nominal bore at gradients not flatter than 1:40 have proved satisfactory.

(2) Where the peak flow is more than 1ℓ/s, a 100 mm nominal bore pipe may be laid at a gradient not flatter than 1:80, provided that at least one WC is connected.

(3) A 150 mm nominal bore pipe may be laid at a gradient not flatter than 1:150, provided that at least five WCs are connected.

(4) Experience has shown that for gradients flatter than those given in items (1) to (3), a high standard of design and workmanship is necessary if blockages are to be minimised. Where this has been achieved, gradients of 1:130 for 100 mm pipes and 1:200 for 150 mm pipes have been used successfully.

Where the available fall is less than that necessary to achieve the recommended gradient, increasing the pipe diameter at low flows is not a satisfactory solution. It will lead to a reduction in velocity and depth of flow and an increase in the tendency for deposits to accumulate in the pipes.

Structural Design of Drains

The design of bedding for pipes is based on the principle that the ability of a pipe to carry a load may be increased by the provision of a suitable bedding. A rigid pipe has inherent strength but by providing a degree of encasement higher loads may be carried. A flexible pipe, on the other hand, will deform under the application of loads and requires support from surrounding material, and thus from the sides of the pipe trench, in order to avoid excessive deformation of the pipe. Rigid pipes include asbestos, vitrified clay, concrete and grey iron, while PVC-U is classified as a flexible pipe.

The load on a pipeline depends on the diameter of the pipe, the depth at which it is laid, the trench width, the traffic or other superimposed loading and the prevailing site conditions. Detailed recommendations for the different classes of pipes in a variety of situations are given in BS 8301[2] and Approved Document H1 (1989) of the Building Regulations.[1] Arrangements for bedding and backing of pipes are described in more detail later in the chapter under Pipe Laying.

Approved Document H1 (1989) of the Building Regulations[1] also prescribes the following special protective measures against ground loads.

(1) Where rigid pipes of less than 150 mm diameter have less than 0.3 m of cover, or pipes of 150 mm or more have less than 0.6 m of cover, the pipes should be surrounded with concrete with a thickness of at least 100 mm, although many prefer 150 mm (as shown in figure 13.1.7), and with movement joints formed with compressible board at each socket or sleeve joint face.

(2) Where flexible pipes are not under a road and have less than 0.6 m of cover, they should be surrounded with concrete as previously described or concrete paving slabs should be laid as bridging above the pipes and on at least 75 mm of granular or other flexible filling, as shown in diagram A4 of Approved Document H1 (1989) of the Building Regulations.[1]

(3) Where flexible pipes are under a road and have less than 0.9 m of cover, reinforced concrete bridging should be used instead of paving slabs, or a reinforced concrete surround.

Access to Drains

Access to drains is required for testing, inspection, maintenance and removal of debris. Rodding is normally done with flexible canes, or metal or plastic rods or tubes to which a suitable ram is attached. The traditional means of access is the inspection chamber (figure 13.3.5), which becomes a manhole when it provides working space at drain level. They are mainly provided at changes of direction and gradient, and are described in more detail later in this chapter.

Manholes and inspection chambers are, however, expensive to build and rodding eyes may therefore be substituted particularly at the head of shallow branch drains. A rodding eye consists of a vertical or inclined length of pipe with a suitable cover or cap at the top end at ground level, while at its base it makes a curved junction with the drain. A typical rodding eye as produced by Marley is illustrated in figure 13.1.2. The alternative to the rodding eye is an access fitting as illustrated in BRE Digest 292,[3] and is a pipe fitting or small chamber having a removable, sealed cover, similar to a rodding eye, but it enables the drain to be tested or rodded in more than one direction. It has minimum internal dimensions of 150 x 100 mm or 150 mm diameter, coupled with a maximum depth to invert of 0.6 m. For greater depths, internal sizes of 225 x 100 mm are recommended in Approved Document H1 (1989) of the Building Regulations 1991.[1]

Approved Document H1 (1989) of the Building Regulations[1] prescribes that sufficient and suitable access points should be provided for clearing blockages from any drain runs which cannot be reached by any other means. The siting, spacing and type of the access points will depend on the layout, depth and size of the runs. It further recommends that access should be provided at the following points: (1) on or near the

Table 13.1 Maximum spacing of access points to drains in metres (Source: Approved Document H1 (1989) (Building Regulations))

| | Access fitting to | | | | |
From	Small	Large	Junction	Inspection chamber	Manhole
Start of external drain*	12	12	–	22	45
Rodding eye	22	22	22	45	45
Access fitting					
small: 150 × 100 mm					
or 150 mm diameter	–	–	12	22	22
large: 225 × 100 mm	–	–	22	45	45
Inspection chamber	22	45	22	45	45
Manhole	22	45	45	45	90

* Connection from ground floor appliances or stack
See also paragraphs 1.9 and 1.26 of Approved Document H1 (1989)

head of each drain run; (2) at a bend and at a change of gradient; (3) at a change of pipe size; and (4) at a junction unless each run can be cleared from an access point.

Access should be provided to long runs. The distances between access points depend on the types of access used but should not be more than shown in table 13.1 for drains up to and including 300 mm.

CHOICE OF PIPES

All the pipes subsequently described are suitable for use below ground, but the beam strength of a pipeline may become a limiting factor under difficult loading or ground conditions. In these situations, rigid pipes with flexible joints should be used, and short lengths of pipe in ground subject to severe settlement. Where pipes are laid above ground, special attention should be paid to structural support and protection against mechanical damage, frost and corrosion. Approved Document H1 (1989) of the Building Regulations[1] recommends that all rigid pipes should have flexible joints. All pipes require careful handling and jointing. Different metals should be separated by non-metallic materials to prevent electrolytic corrosion. The various pipe materials are now described and compared.

Rigid Pipes

Vitrified clay pipes. Manufactured to BS 65[4] with nominal bores of 100 to 1000 mm and lengths of 0.3 m

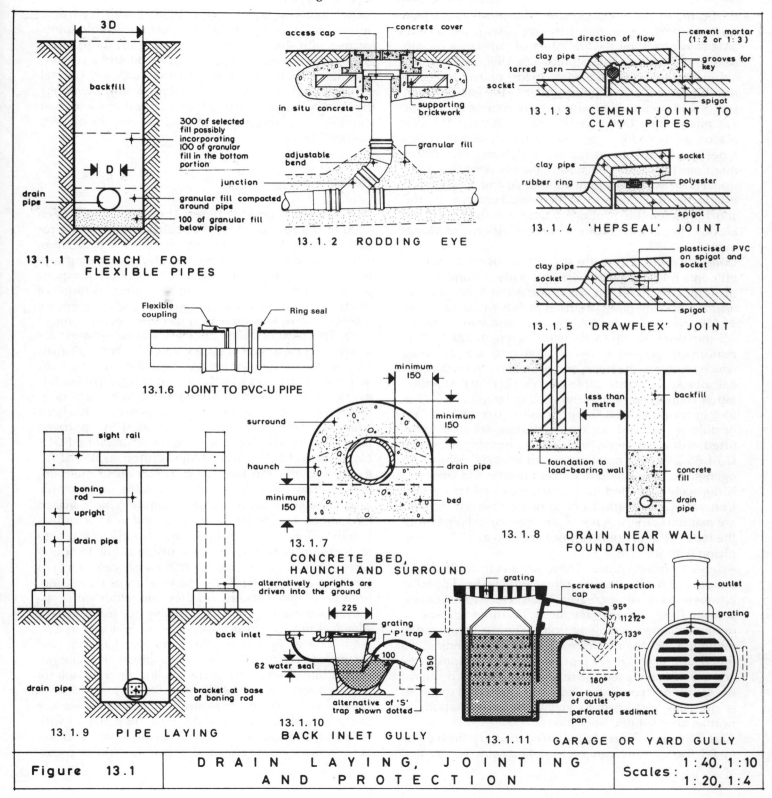

13.1.1 TRENCH FOR FLEXIBLE PIPES

- 3D
- backfill
- drain pipe
- 300 of selected fill possibly incorporating 100 of granular fill in the bottom portion
- D
- granular fill compacted around pipe
- 100 of granular fill below pipe

13.1.2 RODDING EYE

- access cap
- concrete cover
- in situ concrete
- supporting brickwork
- adjustable bend
- granular fill
- junction

13.1.3 CEMENT JOINT TO CLAY PIPES

- direction of flow
- cement mortar (1:2 or 1:3)
- clay pipe
- tarred yarn
- socket
- grooves for key
- spigot

13.1.4 'HEPSEAL' JOINT

- clay pipe
- rubber ring
- socket
- polyester
- spigot

13.1.5 'DRAWFLEX' JOINT

- clay pipe
- socket
- plasticised PVC on spigot and socket
- spigot

13.1.6 JOINT TO PVC-U PIPE

- Flexible coupling
- Ring seal

13.1.7 CONCRETE BED, HAUNCH AND SURROUND

- minimum 150
- surround
- haunch
- drain pipe
- minimum 150
- bed

13.1.8 DRAIN NEAR WALL FOUNDATION

- less than 1 metre
- backfill
- foundation to load-bearing wall
- concrete fill
- drain pipe

13.1.9 PIPE LAYING

- sight rail
- boning rod
- upright
- drain pipe
- drain pipe
- bracket at base of boning rod
- alternatively uprights are driven into the ground

13.1.10 BACK INLET GULLY

- 225
- grating
- back inlet
- 'P' trap
- 62 water seal
- 100
- 350
- alternative of 'S' trap shown dotted

13.1.11 GARAGE OR YARD GULLY

- grating
- screwed inspection cap
- 95°
- 112½°
- 133°
- 180°
- various types of outlet
- perforated sediment pan
- outlet
- grating

| Figure 13.1 | DRAIN LAYING, JOINTING AND PROTECTION | Scales: 1:40, 1:10 1:20, 1:4 |

to 4.0 m, in three classifications of normal, surface water and extrachemically resistant. Clay pipes are resistant to attack by a wide range of substances, both acid and alkaline. These pipes are still very popular although the traditional joint made of two rings of tarred yarn and with the remainder of the gap between spigot and socket filled with cement mortar to a 1:2 or 1:3 mix (figure 13.1.3) is becoming increasingly displaced by mechanical or flexible joints, of which two types are illustrated in figures 13.1.4 and 13.1.5. Cement mortar shrinks on setting thus making the traditional form of cement joint vulnerable and, being rigid, it is liable to damage by settlement. Furthermore the short pipe lengths produce a large number of joints. Flexible joints are recommended in Approved Document H1 (1989).[1]

Concrete pipes. These are suitable for use with normal effluents but may be attacked by acids or sulphates in the effluent, or in the surrounding soil or groundwater. Concrete pipes are used mainly for larger pipes of 150 mm diameter and upwards, and with pipes of 225 mm diameter upwards, external wrappings of glass reinforced polyester resin laminates are available which reinforce the pipes and protect them from external attack. Concrete pipes to BS 5911[5] are supplied either reinforced or unreinforced in lengths of 0.45 m to 3 m for pipes of 600 mm nominal size or less with flexible spigot and socket or flexible rebated joints fitted with elastomeric ring seals, complying with type D of BS 2494.[6] Concrete pipes to BS 5911[5] with plain ogee joints are jointed in cement mortar and their use is normally restricted to the carriage of surface water. Concrete pipes made from sulphate resisting cement are marked with the letter 'S' and reinforced pipes with the letter 'R'. Prestressed pipes are also available complying with BS 5178.[7]

Asbestos cement pipes. These are made to BS 3656,[8] are provided with flexible joints, and can be dipped in bitumen to give increased resistance to chemical attack. They are manufactured in three classes in nominal diameters ranging from 100 to 1050 mm and in minimum lengths of 3 m for pipes not exceeding 200 mm diameter to 4 m for pipes exceeding 200 mm. They are occasionally used for drainage purposes but have the same shortcomings as concrete pipes and their use is normally restricted to gravity flow installations at normal atmospheric pressure.

Cast iron pipes. These are referred to as grey iron pipes in BS 8301.[2] They can be supplied with spigot and socket joints to BS 437[9] for caulking with lead or a proprietary material. BS 6087[10] specifies flexible joints for use with these pipes. The coating on these pipes gives good protection against corrosion and a reasonable life with average ground conditions and normal effluents, although care is needed during handling; they can be laid at any depth on account of their great strength. Cast iron pipes are made in lengths of 1.83, 2.74 and 3.66 m.

Flexible Pipes

Pitch-impregnated fibre pipes. These pipes have been used in the past as they were considered suitable for normal domestic use in nominal bores of 50 to 225 mm. They were more economical than clay pipes being supplied in longer lengths and better suited for poor ground conditions. There were two principal forms of joint: (1) The pipes have tapered ends and an internally tapered sleeve was driven on to the pipes to connect them. This gave flexibility but there was no provision for longitudinal expansion or contraction. (2) A better alternative was to use a polypropylene moulded coupling with snap ring rubber seals which provided longitudinal flexibility. However, the British Standard (BS 2760) which prescribed the minimum requirements for these pipes has been withdrawn and they were not included in Approved Document H1 (1989) of the Building Regulations 1991.[1]

Unplasticised PVC pipes. Manufactured to BS 4660,[11] to 110 and 160 mm nominal diameters (external), are golden brown in colour and are suitable for domestic installations and surface water drainage. Unplasticised PVC pipes are available in 1, 3 and 6 m standard lengths. They should not be used for effluents at high temperatures and they become brittle at low temperatures when they require handling with care. Push-in joints are available for PVC-U pipes either as loose couplings or integral sockets and these, together with elastomeric sealing rings, provide flexibility (see figure 13.1.6). Although relatively expensive, they are available in long lengths, are lightweight and readily handled and assembled. Larger diameter PVC-U pipes, with integral sockets fitted with elastomeric sealing rings or solvent cement joints, should comply with BS 5481.[12]

Glass fibre reinforced polyester (GRP) pipes. These are made from glass reinforced thermosetting plastics with or without an aggregate filler, having nominal diameters ranging from 100 to 4000 mm for use at pressures up to 25 bar and can convey either foul or surface water. They should comply with BS 5480[13] and have effective lengths of 3 m and 6 m with either rigid or flexible joints.

PIPE LAYING

Pipes should always be laid with the sockets pointing uphill, commencing from the point of discharge (sewer connection or treatment works). Painted sight rails, as illustrated in figure 13.1.9, should be fixed across the trench, usually at manholes or inspection chambers, at a height equal to the length of the boning rod to be used above the invert level of the drain. A line sighted across the tops of two adjacent sight rails will represent the gradient of the drain at a fixed height above invert level. At any one time there should desirably be at least three sight rails erected on a length of drain under construction.

Wooden pegs or steel pins are driven into the trench bottom at intervals of at least 900 mm less than the length of straight-edge in use. The use of a boning rod (figure 13.1.9) will enable each peg or pin to be driven until its head represents the pipe invert at that point. The underside of the straight-edge resting on the tops of the pegs or pins will give the levels and gradient of the pipe. The pegs or pins are withdrawn as the pipes are laid.

To obtain a true line in a horizontal plane, a side line is strung tightly between steel pins at half pipe level, with the pipe sockets just free of the side line. Pins will normally be located at each manhole or inspection chamber, but intermediate pins will also be needed on very long lengths.

In firm ground, rigid pipes may be laid upon the carefully trimmed trench bottom or formation, with socket and joint holes formed where necessary to give a minimum clearance of 50 mm between the socket and the formation. Where accurate hand trimming is not possible the trench should be excavated below the pipe invert to give a minimum thickness of 100 mm of granular bed. When laying 100 mm pipes in deep trenches exceeding 5.5 m, the granular fill should be extended up to half the depth of the pipe. In all cases, the selected fill should be used to surround the pipes and extend for a depth of 150 mm above them. The selected fill should be free from stones larger than 40 mm, lumps of clay over 100 mm, timber, frozen ma-

terial and vegetable matter as recommended in Approved Document H1 (1989) of the Building Regulations 1991.[1] The granular fill shall conform to BS 8301,[2] Appendix D, and be free from stones larger than 40 mm.

Where rigid pipes are to be laid on a concrete bed, pipes should be supported clear of the trench bottom by placing blocks or cradles under each pipe, after which concrete is placed solidly under the barrel of each pipe to give continuous support. The concrete bed and surround are normally 150 mm thick, and the bed or haunch extending 150 mm on each side of the pipe (figure 13.1.7), but Approved Document H1 (1989) of the Building Regulations[1] illustrates concrete encasement not less than 100 mm thick and having movement joints formed of 13 mm compressible board at each socket or sleeve joint face.

Where a pipe trench is less than one metre from a loadbearing wall foundation, the trench shall be filled with concrete to the level of the underside of the foundation (figure 13.1.8), and where the trench is one metre or more from the building, it shall be filled with concrete to a level below the lowest level of the building equal to the distance from the building, less 150 mm, as illustrated in Approved Document H1 (1989) of the Building Regulations.[1]

In general, flexible pipes should be laid on a granular bed not less than 100 mm thick, and with the granular material extending upwards to the top surface of the pipe. It is also necessary to place selected fill free from stones larger than 40 mm up to a level of 300 mm above the top of the pipe, and often incorporating a 100 mm layer of granular fill in the bottom portion as illustrated in figure 13.1.1, for trench depths not exceeding 10 m.

Approved Document H1 (1989) of the Building Regulations 1991[1] also contains provisions in paragraph A13 for protection of the building where a drain is liable to surcharge and the protective measures are described in BS 8301.[2] Where anti-flood devices are used additional ventilation may be needed to maintain trap seals. Paragraph A14 deals with special precautions to secure rodent control by providing sealed drainage or interceptor traps, each of which can pose problems in operation. Special precautions should also be taken to accommodate the effects of settlement where pipes run under or near a building, on piles or beams, in common trenches or on unstable ground (paragraph 2.7).

DRAINAGE LAYOUT

When designing a drainage layout the primary objectives should be to secure an efficient and economical arrangement, with pipes laid in straight lines to even gradients, and access points provided only if blockages could not be cleared without them. Wherever practicable, sanitary appliances should be grouped together to reduce the lengths of drain and to obtain steeper gradients. A drainage layout for a bungalow and ancillary buildings is illustrated in figure 13.2.1, and consists of separate foul and surface water drains each connected to a separate sewer. The foul drain picks up the sinks, washbasins and bath through back inlet gullies of the type illustrated in figure 13.1.10, which act as a water seal and the pipes discharge below gratings to prevent fouling, whilst waterclosets are connected direct to the drains and are normally fitted with a 'S' trap. Gullies prevent undesirable matter entering drains but gratings and traps need cleansing periodically. A ventilating pipe must be provided at the head of the foul drainage system, and at the head of any branch longer than 6 m serving a single appliance or 12 m serving a group of appliances.

Rainwater pipes discharge through back inlet gullies or direct into the surface water drains. These drains also pick up the yard gullies in the drive and paved patio. Care has to be taken to obtain adequate difference in level between the two drains to allow surface water connections to cross the line of the foul drain, to provide adequate falls to ensure self-cleansing velocities and yet, at the same time, to avoid excessive excavation. Inspection chambers are provided at the head of each main pipe run, at changes of direction and gradient and at main intersections. It would be possible to reduce the number of inspection chambers by inserting some rodding eyes and/or access fittings as advocated in BRE Digest 292.[3] Other forms of disposal of surface water are soakaways (figure 1.1.6 in chapter 1) and watercourses, where permitted by Water Authorities/Companies to drain large areas and where there are impervious soils, with a non-return flapvalve at the outlet end of the pipe to prevent backflow or debris blocking it in times of flood.

Sewer Connection

Except where junctions have been provided on the sewer, the drain connection is made with a saddle of the type illustrated in figure 13.2.3. A hole is formed in the top half of the sewer and carefully trimmed to fit the saddle. The saddle, incorporating a socket piece, is jointed all round in cement mortar and finally surrounded in 150 mm of concrete after the mortar has set. The saddle connection should be carried out by the appropriate authority's employees or under their supervision. Where the sewer and drain are of similar size or the drain is laid to a very steep gradient, it is advisable to construct a manhole at the point of connection. Another alternative when connecting to a 150 mm sewer is to take out three sewer pipes and replace with two straight pipes and a junction pipe, or to use a double spigot pipe and loose collar.

INTERCEPTORS

An interceptor or intercepting trap may be used to intercept, by means of a water seal at least 62 mm deep, the foul air from a sewer or cesspool from entering a house drainage system. The trap is located in an inspection chamber or manhole and is provided with a rodding arm to give access to the section of drain between the trap and the sewer or cesspool. The stopper fitted in the rodding arm should be firmly seated but easily removable, and is best fitted with a lever locking device supported by a rust-resistant chain attached to the wall.

Unfortunately, interceptor traps are often responsible for blockages in house drains, and their additional cost, together with that of the extra ventilation work involved is not justified. Occasions arise when their use may be beneficial, such as when one or two new houses are being connected to an existing sewerage system, where all existing house connections are intercepted. In the absence of interceptor traps, the sewer would be ventilated through the stacks of the new houses, possibly resulting in an accumulation of sewer gas at one or two points.

The primary aim of drain ventilation is to ensure that the air in the drainage system is always at about atmospheric pressure. If there is no interceptor trap, the house drain and sewer will be adequately vented by means of soil and vent pipes, which normally have an internal diameter of not less than 75 mm. Where an interceptor trap is provided, an additional fresh air inlet must be installed near the interceptor trap and these fittings are easily damaged.

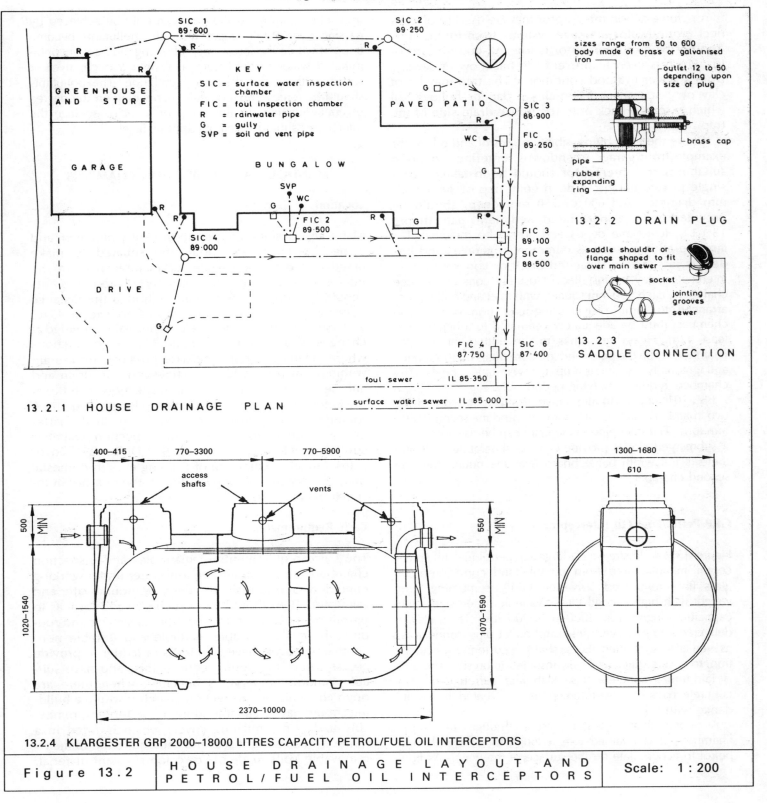

13.2.1 HOUSE DRAINAGE PLAN

K E Y

SIC = surface water inspection chamber
FIC = foul inspection chamber
R   = rainwater pipe
G   = gully
SVP = soil and vent pipe

13.2.2 DRAIN PLUG

sizes range from 50 to 600
body made of brass or galvanised iron
outlet 12 to 50 depending upon size of plug
brass cap
pipe
rubber expanding ring

13.2.3 SADDLE CONNECTION

saddle shoulder or flange shaped to fit over main sewer
socket
jointing grooves
sewer

13.2.4 KLARGESTER GRP 2000–18000 LITRES CAPACITY PETROL/FUEL OIL INTERCEPTORS

access shafts
vents

| Figure 13.2 | H O U S E   D R A I N A G E   L A Y O U T   A N D P E T R O L / F U E L   O I L   I N T E R C E P T O R S | Scale: 1 : 200 |

A reverse action interceptor may be used to disconnect groundwater drainage systems from foul drains. Another type of interceptor is the anti-flood interceptor which is used to prevent the backflow of sewage from a sewer in flood conditions. The trap may be of clay or cast iron and contains a flap or hollow ball which is forced back onto the house drain side of the trap.

Where there is any possibility of petrol and oil, as for example from garage washdowns, entering a drain, a suitable petrol interceptor should be installed. For a single private garage, a deep gully trap of 300 or 375 mm diameter and 600 or 750 mm deep, should be provided with a perforated sediment pan (figure 13.1.11), to enable debris and grit to be removed. For larger premises, such as commercial garages and bus depots, a specially designed brick or concrete petrol interceptor may be installed. It usually consists of three chambers, each 900 mm square, with inlet and outlet pipes arranged to form a seal using cast iron bends. The chambers must be adequately ventilated to a height of at least 4 m. Heavy grit sinks to the bottom of the first chamber, while the petroleum mixture floats on the surface and gradually evaporates up the ventilating pipes. The chambers require emptying periodically.

BS 8301[2] shows an alternative arrangement comprising two chambers, the first 900 mm long and the second 1800 mm long. The inlet pipe to the first chamber is submerged, a submerged slot is provided in the separating wall and there is full width baffle board near the outlet from the second chamber.

GRP Petrol/Fuel Oil Interceptor

Figure 13.2.4 illustrates a Klargester petrol/fuel oil interceptor manufactured from durable and corrosion-proof glass fibre reinforced polyester (GRP), combining light weight with high strength. It is available in ten sizes with capacities ranging from 2000 to 18 000 litres. The units are delivered complete with inlet and outlet pipe connectors, as well as factory-fitted access shafts to customer specification, to ensure quick and easy installation on site. The unit is laid on a concrete base slab and surrounded with concrete backfill. The interceptors are vented in accordance with BS 8301.[2]

As contaminated water passes through the three chambers, it is retained long enough to allow oils and petrol to accumulate on the surface of the water. As flow from one chamber to the next can only be achieved by passing over a full length weir, the pollutants become trapped within the interceptor, allowing the now uncontaminated water to discharge safely into a ditch, river or stream without the risk of oil pollution. The pollutants should be removed periodically depending on the degree of contamination. In the event of a major spillage the interceptor should be emptied immediately.

MANHOLES AND INSPECTION CHAMBERS

Location

Manholes, designed to permit the entry of a man, and inspection chambers should be situated to make lengths of drain accessible for maintenance. In the interests of economy the number of access points should be kept to a minimum, although the distance between inspection chambers should not exceed 45 m, and manholes 90 m. They are also normally provided at changes of direction and gradient, at drain junctions where cleaning is not otherwise possible, on a drain within 12 m from a junction between that drain and another drain unless there is an inspection chamber at the junction, and at the head of each length of drain. A rodding eye may however be sufficient in the latter position, provided there is an inspection chamber sufficiently close downstream. BRE Digest 292[3] advocates a much greater use of rodding eyes in domestic drainage systems, thereby reducing the number of inspection chambers.

Basic Requirements

The basic requirements of manholes and inspection chambers are to contain the foul water under working conditions and to resist the entry of groundwater and rainwater. They should be of such size and form as to permit ready access for inspection, cleansing and rodding; have a removable, suitable and durable non-ventilating cover; have step irons or ladder to provide access where the depth requires this; and have suitable, smooth impervious benching when there are open channels. An inspection chamber within a building should have a suitable, durable, watertight, removable and non-ventilating cover, which is fitted in a frame with an airtight seal and secured to the frame by removable bolts made of corrosion-resistant material.

Table 13.2   Minimum internal dimensions for manholes and inspection chambers

Type	Depth to invert	Minimum dimensions
Inspection chamber (figure 13.3.5)	Not exceeding 1.0 m Not exceeding 0.6 m	450 × 450 mm or 450 mm diameter, if circular, with a minimum cover opening of 450 × 450 mm or 450 mm diameter, if circular 190 mm diameter with 190 mm diameter cover
Manhole (shallow)	Not exceeding 1.5 m	Chamber must be large enough to permit entry of a man and have minimum internal dimensions of 1200 × 750 mm or 1050 mm diameter, if circular, and a minimum cover opening of 600 × 600 mm, or 600 mm diameter, if circular
Manhole (moderate depth)	Exceeding 1.5 m and not exceeding 2.7 m	Man can stand up in chamber 1200 × 750 mm in size, or 1200 mm diameter, if circular, covered by slab; cover size of 600 × 600 mm or 600 mm diameter, if circular
Manhole (deep) (figure 13.3.2)	Exceeding 2.7 m	Economical to provide working chamber with an access shaft, with height of chamber not less than 2 m above top of benching; chamber to be not less than 1200 mm × 840 mm or 1200 mm diameter for circular chambers; minimum size of access shaft to be 900 × 840 mm or 900 mm diameter, if circular; the clear cover opening shall be not less than 600 × 600 mm or 600 mm diameter, if circular

Notes: (1) In the case of clayware and plastics inspection chambers, the clear opening may be reduced to 430 mm in order to provide proper support for the cover and frame.
(2) The dimensions should be increased at junctions if they do not allow enough space for the branches.

A cast iron access chamber of the type illustrated in figure 13.3.6 would be suitable for this situation.

Minimum Internal Dimensions

The dimensions of manholes and inspection chambers will be largely determined by the size and angle of the main drain, the position and number of branch drains, and the depth to invert. Table 13.2 is based on data contained in Approved Document H1 (1989) of the Building Regulations 1991.[1]

Form of Manhole Construction

Manholes and inspection chambers can be constructed in brickwork (figures 13.3.2 and 13.3.5), *in situ* concrete, precast concrete sections (figure 13.3.3) or cast iron (figure 13.3.6), PVC-U, glass fibre reinforced polyester (GRP) and vitrified clay.

Brick manholes. These are normally built of 215 mm walls in English bond, preferably using class B engineering bricks to BS 3921[14] in cement mortar, finished fair face internally, as internal rendering may fail and result in blockages. In granular soils above the water table, inspection chambers not exceeding 900 mm deep may be built in half-brick walls.[2] Chambers are normally roofed with a precast or *in situ* reinforced concrete cover slab, 125 or 150 mm thick (figure 13.3.2). The access shaft may be corbelled inwards or be surmounted by a perforated reinforced concrete slab to receive the cover frame. The concrete base is normally 150 mm thick and need not normally extend beyond the walls as this results in needless extra cost. When pipes of 300 mm diameter are built into walls, half-brick rings should be turned over them for the full thickness of the brickwork or concrete lintels inserted to divert loads from the pipes. Brick manholes are particularly suitable for shallow depths and offer considerable flexibility in design.

Concrete in situ *manholes.* These have walls of no less thickness than comparable brick manholes, and the concrete shall conform to table 1 of BS 8301.[2] This form of construction is not used extensively although it could be particularly advantageous for irregularly shaped shallow manholes containing large diameter pipes.

Precast concrete manholes and inspection chambers. These have the advantage of speedy construction and are usually built in circular sections or rings which may be connected with ogee joints or rebated joints sealed

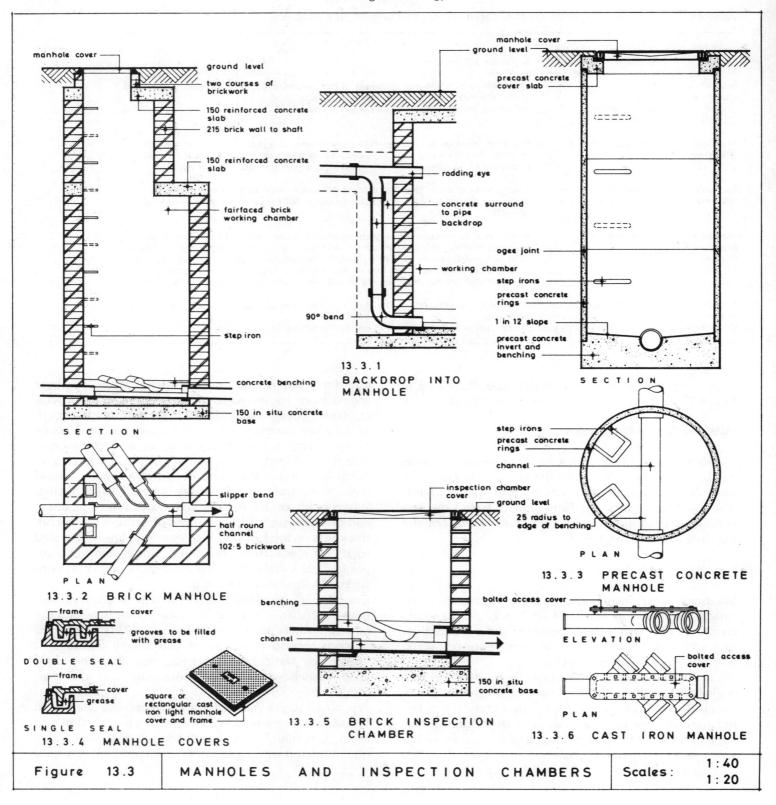

manhole cover

ground level

two courses of
brickwork

150 reinforced concrete
slab

215 brick wall to shaft

150 reinforced concrete
slab

fairfaced brick
working chamber

step iron

concrete benching

150 in situ concrete
base

SECTION

slipper bend

half round
channel

102·5 brickwork

PLAN

13.3.2   BRICK MANHOLE

frame   cover

grooves to be filled
with grease

DOUBLE SEAL

frame

cover

grease

SINGLE SEAL

square or
rectangular cast
iron light manhole
cover and frame

13.3.4   MANHOLE COVERS

manhole cover

ground level

rodding eye

concrete surround
to pipe
backdrop

working chamber

90° bend

**13.3.1
BACKDROP INTO
MANHOLE**

inspection chamber
cover

ground level

benching

channel

150 in situ
concrete base

**13.3.5   BRICK INSPECTION
CHAMBER**

manhole cover

ground level

precast concrete
cover slab

ogee joint

step irons

precast concrete
rings

1 in 12 slope

precast concrete
invert and
benching

SECTION

step irons

precast concrete
rings

channel

25 radius to
edge of benching

PLAN

**13.3.3   PRECAST CONCRETE
MANHOLE**

bolted access cover

ELEVATION

bolted access
cover

PLAN

13.3.6   CAST IRON MANHOLE

| Figure   13.3 | MANHOLES   AND   INSPECTION   CHAMBERS | Scales: | 1:40 1:20 |

with cement mortar in accordance with BS 5911, Part 200.[5] In waterlogged ground, the joints may be sealed with a mastic sealant or a rubber ring joint. In unstable ground, a 150 mm surround of at least mix design C20P concrete should be provided.[2] The concrete base can be *in situ* or precast concrete and the cover is supported on a precast concrete cover slab (figure 13.3.3), possibly with one or two intervening courses of bricks to make up to the required finished level. Step irons are provided already built into the precast sections or rings. With deep manholes a taper section and access shaft, often 675 mm diameter, will be incorporated.

Cast iron manholes. These are formed of bolted cast iron sections for use in bad ground or cast iron access chambers for use in buildings. Their cost is very high.

Plastics manholes and inspection chambers. These are available moulded in thermoplastic and thermosetting materials such as polypropylene, PVC-U or GRP (glass fibre reinforced polyester), either as integral bases or as complete chamber units. They offer the same advantages as described for GRP petrol/fuel oil interceptors.

Vitrified clay inspection chambers. These are available in vitrified clay conforming to BS 65.[4] The integral base provides a variety of inlet and outlet connection positions. Flexible joints incorporating elastometric sealing rings, connect raising pieces to the base and allow for depth variations.[2]

Channels and Benchings

The open channel in the bottom of a manhole or inspection chamber may be formed of half-round channel pipes, although with precast concrete chambers the base containing the channels may be precast. Side branches should be in the form of three-quarter section standard branch bends, where the angle of the branch is more than 45°, bedded in cement mortar and discharging in the direction of flow in the main channel, and at or above the level of its horizontal diameter. Large diameter channels may be formed of *in situ* concrete finished with a granolithic rendering to a mix of 1:1:2.

The benching should rise vertically from the top edge of the channel pipe to a height not less than that of the soffit of the outlet and then be sloped upwards to meet the wall of the manhole at a gradient of 1 in 12, to control the flow and to allow a man to stand on the benching. It is usually floated to a smooth, hard surface with a coat of cement mortar (1:2), laid monolithic with the benching. The benching should be rounded at the channel with a radius of at least 25 mm.

Access to Manholes

The most usual form of access to manholes deeper than 1 m is galvanised ferrous or stainless steel steps to BS 1247,[15] and may be general-purpose pattern single or double steps or round bar double steps. They are normally built into manhole walls at 300 mm vertical intervals and set staggered in two vertical runs at 300 mm centres horizontally. The top step should be not more than 750 mm below the top of the manhole cover and the lowest not more than 300 mm above the benching, to be of practical use. On very deep manholes it is customary to provide mild steel ladders, often made up of 65 x 12 mm stringers supporting 22 mm shouldered diameter rungs at 300 mm centres.

Approved Document H1 (1989) of the Building Regulations 1991[1] prescribes that inspection chambers and manholes should have removable non-ventilating covers of durable material, such as cast iron, cast or pressed steel, precast concrete or PVC-U, and be of suitable strength. Those located in buildings should have mechanically fixed airtight covers unless the drain itself has watertight access covers.

The majority of manhole covers and frames are made of cast iron complying with BS 497[16] and coated with hot applied coal tar based material or cold applied black bitumen-based composition. They are supplied in four grades: grade A for use in carriageways, grade B1 for use in minor residential roads, grade B2 for use in pedestrian precincts (occasional vehicular access) and grade C in positions inaccessible to wheeled vehicles. Covers and frames are also made in various shapes – mainly circular (500 to 600 mm diameter) for grades A and B and rectangular for grade C. Light-duty cast iron covers and frames with both single and double seals are illustrated in figure 13.3.4. Sizes of cover frame openings vary between 450 x 450 and 600 x 600 mm. The covers are usually bedded in grease and the frames in cement mortar. Covers are also manufactured in cast or pressed steel, a steel frame and steel mesh bottom filled with concrete and surfaced with a material to match the surrounding finish, or precast reinforced concrete with a minimum thickness of 50 mm to BS 5911.[5]

Backdrop Manholes

These are sometimes referred to as drop-pipe manholes and provided to accommodate a significant difference in invert levels, which may arise where a domestic drain approaches a sewer. The usual method is to construct a vertical drop pipe outside the manhole, surrounded in concrete or granular fill, as in figure 13.3.1. The upper drain is carried through the manhole wall to form a rodding eye and the vertical drop terminates at its lower end with a bend discharging into an open channel at the bottom of the manhole. Alternatively, the vertical drop pipe may be supported on brackets within a suitably enlarged manhole. The backdrop arrangement economises in drain trench excavation and avoids the need for excessively steep drain gradients. Where the difference in invert levels is less than 1 m, a ramp may be formed by increasing the gradient of the last length of the upper drain to about 45°.[2]

TESTING OF DRAINS

All drains after laying, surrounding with concrete, where appropriate, and backfilling of trenches, should withstand a suitable test for watertightness. Soil, waste and ventilating pipes should be able to withstand a smoke or air test for a minimum period of three minutes at a positive pressure equivalent to a head of not less than 38 mm of water. The principal objective in testing a length of drain is to ensure that it is watertight, as defective drains can create serious problems.

Water Test

The most effective test is the water test whereby suitably strutted expanding drain (or inflatable canvas or rubber bags) plugs of the type illustrated in figure 13.2.2 are inserted in the lower ends of lengths of drain, which are then filled with water before the trenches are backfilled. For house drains, a knuckle bend and length of vertical pipe may be temporarily jointed at the upper end to provide the required test head. A drop in the level of water in the vertical pipe may be due to one or more of the following causes:

(1) absorption by pipes or joints;
(2) sweating of pipes or joints;
(3) leakage from defective pipes, joints or plugs;
(4) trapped air.

Hence it is advisable to fill the pipes with water for two hours before testing and topped up, and then to measure the loss of water over a 30-minute period,

normally applying a test pressure of 1.5 m head of water above the invert at the upper end and not more than 4 m at the lower end, to prevent damage to the drain. It may thus be necessary to test steeply graded drains in stages, to avoid exceeding the maximum head. Approved Document H1 (1989) of the Building Regulations[1] recommends that the leakage over the 30-minute period for drains up to 300 mm nominal bore should not exceed 0.05 litres for each metre run of drain for a 100 mm drain – a drop in water level of 6.4 mm/m, and 0.08 litres for a 150 mm drain – a drop in water level of 4.5 mm/m. Where there is a trap at the upper end of a branch drain, a rubber or plastics tube should be inserted through the trap seal to draw off the confined air as the pipes are filled with water. A second and similar test is generally undertaken after the completion of bedding or concrete surround, backfilling, compaction and reinstatement.

Air Test

An air test may be used to test the watertightness of drains, in which an appreciable drop in pressure will indicate a defective drain. The length of drain is plugged in the manner previously described and air pumped into the pipes, usually by means of a hand pump, until a pressure of 100 mm head of water is shown in an attached manometer U-tube gauge. The air pressure should not fall to less than 75 mm head of water (25 mm loss of head) during a period of five minutes after a period for stabilisation. Where a 50 mm gauge is used, the equivalent loss of head would be 12 mm. Where there is an appreciable drop in pressure, it may still be difficult to detect the location or rate of leakage. Hence failure to satisfy this test should be followed by a water test to provide more conclusive information. An alternative to the water test is the smoke test with the smoke generated by a smoke-testing machine consisting of a double-action leather bellows and a copper firebox enclosed within a copper tank and floating dome, but this is now little used.

Infiltration Test

This test should be applied after backfilling where any part of the soffit of the drain (underside of crown of pipe) is 1.2 m or more below the water table. All inlets are sealed and inspection of manholes or inspection chambers will show any flow.

Tests for Straightness and Obstruction

Tests for line, uniform gradient and freedom from obstruction can be carried out using a mirror at one end of the drain and a lamp at the other.

CESSPOOLS

Schedule 1 to the Building Regulations 1991[1] (paragraph H2) requires any cesspool, septic tank or settlement tank to be

(a) of adequate capacity and so constructed that it is impermeable to liquids;

(b) adequately ventilated; and (c) so sited and constructed that–

(i) it is not prejudicial to the health of any person,

(ii) it will not contaminate any underground water or water supply, and

(iii) there are adequate means of access for emptying.

A cesspool is an underground chamber constructed for the reception and storage of foul water from the building until it is emptied. Approved Document H2 (1989) of the Building Regulations 1991[1] requires that cesspools should be constructed so as to prevent leakage of the contents and ingress of subsoil water, with adequate ventilation, be sited so as not to be prejudicial to health nor to contaminate water supplies, to permit satisfactory access for emptying, and have a minimum capacity below the level of the invert of 1800 litres (18 m³). Cesspools, if they are to be emptied using a tanker, should be sited within 30 m of a vehicle access and at such levels that they can be emptied or desludged and cleaned without hazard to the building occupants or the contents being taken through a dwelling or place of work. Access may be through an open covered space. When siting a cesspool attention should be paid to the slope of the ground, direction of the prevailing wind, access for emptying and possibility of future connection to sewer. It should be sited a minimum of 15 m from any inhabited building. It should be noted that cesspools are not permitted in Scotland.

In general cesspools should only be used when no other alternative is available because of the problems associated with ensuring watertightness and emptying; an average household of three persons will produce 7 m³ (the capacity of a typical tanker) in about three weeks.[17] A small sewage treatment works, if practicable, will be cheaper in the long-term and much more satisfactory. Capacity limitations will restrict the use of cesspools to single properties with not more than eight inhabitants. BS 6297[17] recommends a capacity of not less than 45 days storage which may be based on 150 litres per head per day. Constructional considerations will probably limit the economic capacity of a single tank cesspool to a maximum of about 50 m³.

The depth below cover should not normally exceed 4 m on a flat site to facilitate pumping, and the most satisfactory shape for a traditional type cesspool is circular with a diameter equal to the depth below the incoming drain. The walls may be of 215 mm engineering brickwork in cement mortar (1:3), minimum of 150 mm concrete of C25P mix in accordance with BS 5328,[18] glass reinforced concrete or large diameter precast concrete pipes surrounded with a minimum of 100 mm of concrete, built from a concrete base. The drain inlet enters through a bend and a 100 mm fresh air inlet (750 mm above ground) and 100 mm vent pipe (about 800 mm above ground) must be provided to ensure adequate ventilation. The access should not have any dimension less than 600 mm and the cover should be durable and lockable. Cesspools can also be factory made of glass fibre reinforced polyester (GRP) polyethylene or steel, each covered by a BBA certificate.

GRP Cesspools

The ENTEC cesspool comprises a glass fibre reinforced polyester (GRP) rib-stiffened cylinder with hemispherical ends and is available in two grades (standard and heavy duty) and three capacities (18 000, 27 000 and 36 000 litres); it is illustrated in figure 13.6.1. The tanks are manufactured in two sections, upper and lower, which are jointed with a flanged horizontal joint sealed with bonding paste and bolted together. Each cesspool is provided with a 600 mm diameter sliding access shaft provided with a mastic sealant. The shaft incorporates a 110 mm diameter inlet socket. The access cover must be of durable quality and be lockable with a strength equivalent to BS 497, Part 1. In the opinion of the British Board of Agrément (BBA), this product when suitably installed will have a life in excess of 30 years. It is quickly and easily installed in poor ground conditions or below the water table, and there is the further advantage of the guaranteed integrity of a factory manufactured unit.

In a dry, well drained site where the water table is always below the bottom of the tank, the tank is sited on a minimum of 200 mm thick grade 20 concrete and a minimum of 150 mm carpet of pea shingle (nominal 10 mm rounded aggregate) is laid on the concrete base before the tank is lowered into position to provide a good bearing. A backfill of pea shingle surrounds the tank as shown in figure 13.6.1. In a wet or poorly drained site, the concrete base is laid on a 200 mm thick layer of hardcore and separated from it by a polythene membrane, and topped with pea shingle as before. The tank is subsequently surrounded by concrete.

SMALL DOMESTIC SEWAGE TREATMENT WORKS

General Background

In country areas where no public sewers are available, sewage from single houses or groups of houses, with a total population up to 1000, can be treated in small sewage treatment plant, consisting of a septic/settlement tank and a biological filter. In the septic tank heavier solids settle at the bottom as sludge and lighter solids rise and form a scum which acts as a surface seal and permits anaerobic decomposition by bacteria. Oxidation of the organic matter contained in the tank effluent takes place by aerobic bacteria in the filter medium. BS 6297[17] prescribes a minimum distance of 25 m from any dwelling and this should be progressively increased for larger treatment works. Other siting requirements are similar to those listed for cesspools.

Septic/Settlement Tanks

Sufficient capacity is required to enable the breakdown and settlement of solid matter in the foul water from buildings. This can be calculated in litres as ($180P$ + 2000) where P is the design population, with a minimum value of 4, where desludging is carried out at not more than 12-monthly intervals. Approved Document H2 (1989) prescribes a minimum capacity of 2700 litres (2.7 m^3) and BS 6297[17] recommends 2720 litres. A simple tank is illustrated in figure 13.4.2 consisting of a rectangular chamber of from 1.2 to 1.8 m water depth

with its length about three times its width. Precast concrete slabs, set slightly apart for ventilation purposes and fitted with lifting rings, make suitable covers. Tee-junction dip pipes form suitable inlets and outlets for tanks up to 1.2 m in width to prevent disturbance to the surface scum or settled sludge. For tanks wider than 1.2 m, the arrangements shown in figure 13.4.3 should be adopted with two pipes feeding a baffle inlet and with the outlet formed of a weir extending across the full width of the tank and with the scum retained by a scumboard fixed 150 mm from the weir.[17] Approved Document H2 (1989) of the Building Regulations[1] advocates minimising turbulence by limiting the flow rate of the incoming foul water, and for steeply laid drains up to 150 mm minimum diameter the velocity may be reduced by laying the last 12 m of the incoming drain at a gradient of 1 in 50 or flatter.

BS 6297[17] advocates the use of two tanks in series, either by using two separate chambers (figure 13.4.3) or by dividing a tank into two parts with a partition. The first chamber should have a capacity of about two-thirds of the total. A suitable installation for more than 30 persons is shown in figure 13.4.3, but for populations over 60 duplicate tanks should be provided and operated in parallel. Walls may be built of 215 mm brickwork in class A engineering bricks to BS 3921[14] in cement mortar (1:3) or be in concrete or glass reinforced concrete not less than 150 mm thick to mix C25P (BS 5328).[18] Factory made septic tanks are also available in glass fibre reinforced polyester (GRP) polyethylene and steel, complying with BBA certificates. Approved Document H2 (1989) of the Building Regulations 1991[1] prescribes that septic and settlement tanks shall be constructed so as to prevent leakage of the contents and ingress of subsoil water, and be adequately ventilated and covered or fenced in. Access to tanks should not have any dimension less than 600 mm, and access covers should be durable having regard to the corrosive nature of the contents and be lockable. Emptying of sludge normally takes place at six-month intervals.

GRP Septic Tanks

Modern septic tanks are designed to promote the formation of micro biological cultures in an anaerobic environment. Furthermore, a by-product of the biological activity is the generation of gases which require efficient ventilation for their satisfactory dispersal.

Klargester have developed a highly efficient glass fibre reinforced polyester (GRP) three-stage septic tank as illus-

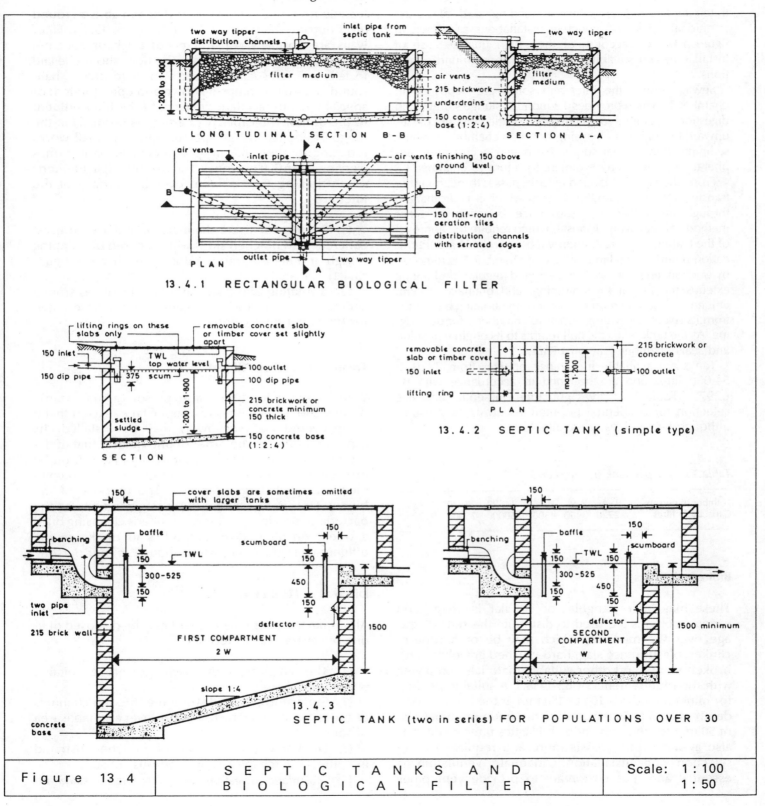

LONGITUDINAL SECTION B-B

SECTION A-A

PLAN

13.4.1 RECTANGULAR BIOLOGICAL FILTER

SECTION

13.4.2 SEPTIC TANK (simple type)

FIRST COMPARTMENT
2 W

SECOND COMPARTMENT
W

13.4.3 SEPTIC TANK (two in series) FOR POPULATIONS OVER 30

| Figure 13.4 | SEPTIC TANKS AND BIOLOGICAL FILTER | Scale: 1:100 1:50 |

trated in figure 13.5, which is tending to displace the traditional brick and concrete installations as shown in figure 13.4, on account of their high quality, ease of installation and suitability for use in poor ground conditions.

Sewage enters the inlet and solids are deposited in chamber 1, which provides 12 months' sludge storage and digestion capacity. From chamber 1, settled effluent passes upward through self-cleansing slots into chamber 2, where sedimentation of finer suspended matter occurs and peripheral slots allow any sediment to return to chamber 1.

From chamber 2, clarified effluent passes through further transfer slots into chamber 3, from which it is discharged through the outlet into subsurface irrigation or other methods of soakaway disposal, which require the approval of the National Rivers Authority (NRA) or the Water Purification Board in Scotland. Sludge in chamber 1 is removed by suction emptier once a year and regular desludging extends the life of the adjoining land drainage system. Effluent is first drawn off chamber 3, the floating ball drops from its seating providing access to chamber 1 for desludging. As the tank refills the ball returns to its original position and seals chamber 1 automatically.

Tanks are available in a range of sizes from 2720 to 54 000 litres and are designed in accordance with BS 6297.[17] Table 13.3 gives guidance on septic tank size selection for residential populations up to 38 persons, although larger sizes are available.

Table 13.3 Septic tank size selection

Number of persons	1–4	5–9	10–13	14–22	23–30	31–38
Capacity in litres	2720	3750	4500	6000	7500	9000

Biological Filters

These may be rectangular or circular in shape and various methods are used to distribute the settled sewage over the medium, which may be of hardburnt clinker, blastfurnace slag, hard-crushed gravel or hard-broken stones and other suitable materials complying with the requirements of BS 1438.[19] A suitable grading for mineral media is 100 to 150 mm at the bottom for a depth of about 150 mm, with a maximum nominal size of 50 mm for the remainder.[17] Plastics filter medium is also available but it costs more and requires careful selection and installation. Adequate ventilation is essential and it is customary to provide vent pipes

leading from underdrains up to 150 mm above ground level outside the filter as shown in figure 13.4.1. Filter walls are normally constructed of brick or concrete supported on a concrete base. The floor should be laid to falls and is usually covered with field drains, half-round channels laid upside down and open-jointed, or special tiles, to facilitate drainage of the filter effluent to the outlet. The capacity of a filter is normally in the range of about 1 m³ per person for very small works reducing to about 0.6 m³ per person for larger ones (serving up to 300 persons). The two principal methods of distributing tank effluent over the surface of the filter are

(1) a series of fixed channels of metal or precast concrete with notched sides which are fed by a tipping trough or other intermittent dosing mechanism (figure 13.4.1);

(2) a rotating arm distributor which rotates from a central column by the head of liquid and is more suited for the larger installations.

Treatment of Filter Effluent

With works serving 100 or more persons a humus tank, designed on similar lines to a septic tank, except that it is uncovered and shallower, may be installed. Its capacity is often in the order of one-quarter that of the septic tank with a baffle or weir inlet and a weir outlet protected by a scumboard, although BS 6297[17] recommends a much larger capacity. The purpose of the humus tank is to remove the waste products of the bacterial action in the filter and it needs cleansing once a week. An alternative method is to discharge the effluent over grassland, using about 3 m² per person.

Disposal of Final Effluent

After final treatment the effluent may be disposed of in one of several way as follows

(1) The simplest is by discharge into a suitable watercourse.

(2) Surface irrigation distributing through channels or carriers controlled by handstops over a suitable area of soil.

(3) Discharge to a soakaway of the type illustrated in figure 1.1.6 (chapter 1), in a porous subsoil.

(4) Subsurface irrigation using perforated clayware,

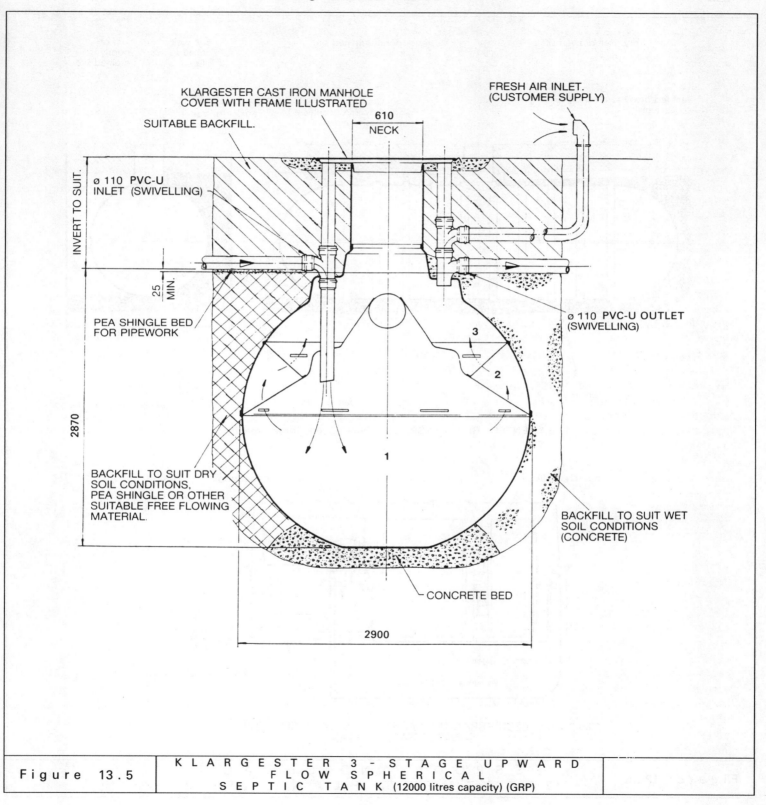

KLARGESTER CAST IRON MANHOLE
COVER WITH FRAME ILLUSTRATED

FRESH AIR INLET.
(CUSTOMER SUPPLY)

SUITABLE BACKFILL.

610
NECK

INVERT TO SUIT.

ø 110 PVC-U
INLET (SWIVELLING)

25 MIN.

PEA SHINGLE BED
FOR PIPEWORK

ø 110 PVC-U OUTLET
(SWIVELLING)

2870

3

2

1

BACKFILL TO SUIT DRY
SOIL CONDITIONS,
PEA SHINGLE OR OTHER
SUITABLE FREE FLOWING
MATERIAL.

BACKFILL TO SUIT WET
SOIL CONDITIONS
(CONCRETE)

CONCRETE BED

2900

| Figure 13.5 | KLARGESTER 3-STAGE UPWARD FLOW SPHERICAL SEPTIC TANK (12000 litres capacity) (GRP) | |

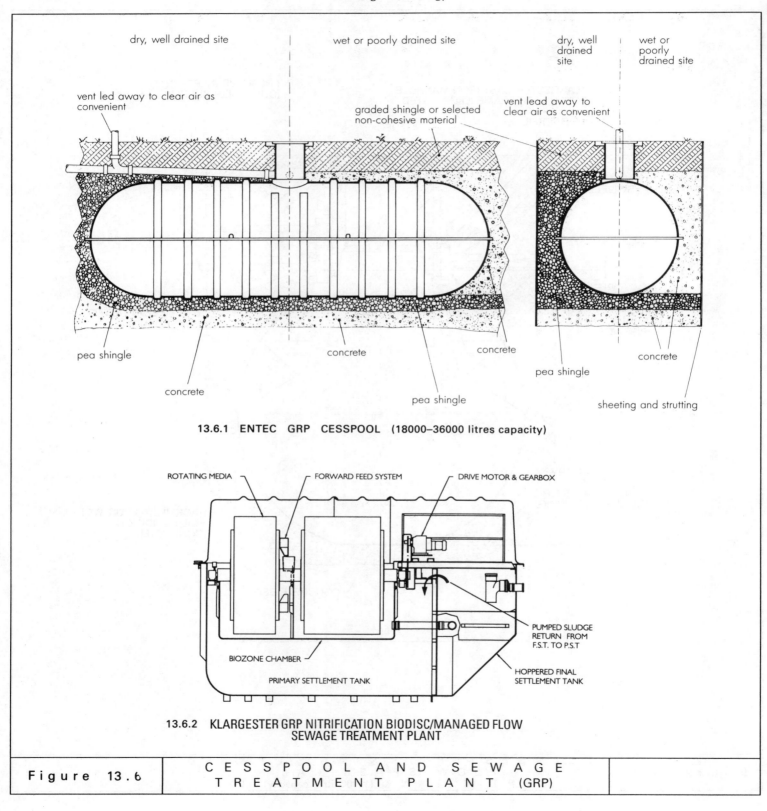

dry, well drained site

wet or poorly drained site

dry, well drained site

wet or poorly drained site

vent led away to clear air as convenient

graded shingle or selected non-cohesive material

vent lead away to clear air as convenient

pea shingle

concrete

concrete

concrete

pea shingle

concrete

pea shingle

sheeting and strutting

13.6.1 ENTEC GRP CESSPOOL (18000–36000 litres capacity)

ROTATING MEDIA

FORWARD FEED SYSTEM

DRIVE MOTOR & GEARBOX

PUMPED SLUDGE RETURN FROM F.S.T. TO P.S.T

BIOZONE CHAMBER

PRIMARY SETTLEMENT TANK

HOPPERED FINAL SETTLEMENT TANK

13.6.2 KLARGESTER GRP NITRIFICATION BIODISC/MANAGED FLOW SEWAGE TREATMENT PLANT

| Figure   13.6 | C E S S P O O L   A N D   S E W A G E   T R E A T M E N T   P L A N T   (GRP) | |

plastics or pitch fibre pipes, normally 100 mm diameter and about 500 mm deep to gradients not exceeding 1 in 200, and laid on a 150 mm layer of 30–50 mm shingle rejects or similar material, filled to a level 50 mm above the pipe and covered with strips of permeable sheeting to prevent the entry of silt.[17]

Klargester GRP Nitrification BioDisc Units

Increasing environmental pollution control standards have been introduced in the UK to protect, maintain and improve the quality of water in inland watercourses. Controlling bodies such as the National Rivers Authority (NRA) are insisting on more stringent standards for final effluent discharges. Often these standards require a significant reduction in the organic and suspended solids content of discharges, frequently with ammoniacal nitrogen limits in the final effluent of not more than 5 mg/l. In consequence, Klargester introduced a range of GRP single piece Nitrification BioDisc units, in addition to the series of Carbonaceous BioDisc units, conforming to BS 6297. These can serve a wide range of developments including single dwellings, housing estates, hotels, country clubs, offices and industrial estates, leisure developments, schools, nursing homes and motorway service stations.

The N-Range Nitrification units, as illustrated in figure 13.6.2 comprise an integral primary settlement tank, two-stage biozone, staged rotating biological contactor and hopper-bottomed final settlement tank with sludge return system. Each unit is powered by a small electric drive motor with associated control panel.

Primary settlement is achieved within the BioDisc, with the incoming crude sewage being stilled initially by a deflector box. The heavier solid matter is settled out and retained in the primary settlement tank. This accumulated primary sludge is drawn off at regular intervals, with partially clarified liquor containing fine suspended solids passed to the biozone above.

A rotating biological contactor comprises a circular steel zinc sprayed tube containing polypropylene media on to which a biomass film will adhere. The contactor assembly is driven by an integral motor, with the shaft rotated through a reduction gear box and chain drive assembly.

The biozone chamber is suspended above the primary settlement zone and is divided into two sections by a fixed baffle. Partially clarified liquor enters the first stage of the biozone though a submerged slot in the biozone base or by a submerged transfer pipe in the smaller units. The second-stage biozone is hydraulically isolated from the first and maintains a constant water level. The shaft mounted media banks are rotated within the biozone, and humus from the first stage media bank will return to the primary settlement tank. Humus from the second stage will remain in suspension and be settled in the final settlement tank. The second-stage biozone is fed liquor from the first through the forward feed transfer system.

Treated effluent from the second stage biozone, containing excess humus (active biological culture) from the rotating biological contactor, enters the final settlement tank through a transfer pipe and diffuser baffle. The final clarifier is designed with a hopper bottom to facilitate the consolidation of sludge, which is removed at regular intervals. A desludge pump within the unit assists hydrostatic removal of consolidated active sludge, which is returned to the primary settlement zone where it will assist with biological oxygen demand (BOD) reduction. The fine sludge is returned to the primary settlement tank at predetermined intervals which are electrically controlled.

Useful further guidance on septic tanks and small treatment works is given in CIRIA Technical Note 146.[20]

REFERENCES

1. *The Building Regulations 1991 and Approved Document H (1989)*. HMSO
2. *BS 8301: 1985 (1989)* Code of practice for building drainage
3. *BRE Digest 292:* Access to domestic underground drainage systems (1984)
4. *BS 65: 1988* Vitrified clay pipes, fittings, joints and ducts
5. *BS 5911* Precast concrete pipes and fittings for drainage and sewerage, *Part 1: 1981* Specification for pipes and fittings with flexible joints and manholes; *Part 2: 1982* Specification for inspection chambers and street gullies; *Part 3: 1982* Specification for pipes and fittings with ogee joints; *Part 100: 1988* Specification for unreinforced and reinforced pipes and fittings with flexible joints; *Part 101: 1988* Specification for glass composite concrete (GCC) pipes and fittings with flexible joints; *Part 120: 1989* Specification for reinforced jacking pipes with flexible joints; *Part 200: 1989* Specification for unreinforced and reinforced manholes and soakaways of circular cross section

6. *BS 2494: 1990* Specification for elastomeric seals for joints in pipework and pipelines
7. *BS 5178: 1975* Prestressed concrete pipes for drainage and sewerage
8. *BS 3656: 1981 (1989)* Specification for asbestos-cement pipes, joints and fittings for sewerage and drainage
9. *BS 437: 1978 (1988)* Specification for cast iron spigot and socket drain pipes and fittings
10. *BS 6087: 1990* Specification for flexible joints for grey or ductile cast iron drain pipes and fittings (BS 437) and for discharge and ventilating pipes and fittings (BS 416)
11. *BS 4660: 1989* Unplasticized polyvinyl chloride (PVC-U) pipes and plastic fittings of nominal sizes 110 and 160 for below ground gravity drainage and sewerage
12. *BS 5481: 1977* Specification for unplasticized PVC pipes and fittings for gravity sewers
13. *BS 5480: 1990* Glass fibre reinforced plastics (GRP) pipes, joints and fittings for water supply or sewerage
14. *BS 3921: 1985* Specification for clay bricks
15. *BS 1247* Manhole steps, *Part 1: 1990* Specification for galvanised ferrous or stainless steel manhole steps
16. *BS 497* Manhole covers, road gully gratings and frames for drainage purposes, *Part 1: 1976 (1986)* Cast iron and cast steel
17. *BS 6297: 1983* Code of practice for design and installation of small sewage treatment works and cesspools
18. *BS 5328: 1981 (1986)* Methods for specifying concrete, including ready-mixed concrete
19. *BS 1438: 1971 (1981)* Media for biological percolating filters
20. CIRIA. *Technical Note 146:* Septic tanks and small sewage treatment works – a guide to current practice and common problems (1994)

14 EXTERNAL WORKS

This chapter is concerned with the various external works provided in connection with buildings, apart from drainage and other underground services. They consist principally of roads, paved areas, footpaths, landscape work, fencing and gates, and lighting.

ROAD DESIGN

The normal single 2-lane residential/access road should desirably have a minimum width of 5.50 m, although this is sometimes reduced to 5.00 m for 2-lane back or service roads used occasionally by heavy vehicles. For the principal distribution roads to large residential estates, widths of 6.00 or 6.75 m would be more appropriate. At the other end of the scale, the shorter service roads with development on one side only and culs-de-sac often have a width of 4.0 m, although this is inadequate to accommodate two wide vehicles alongside one another. Private drives and accesses to garages may be 2.50 to 3.00 m wide.

Longitudinal gradients should be kept within reasonable limits, such as 1 in 20 to 1 in 200. If a road is too flat it will be difficult to remove surface water, while if it is too steep it will become difficult to negotiate in snowy or frosty weather. Vertical curves should be designed to provide a suitable parabolic curve linking the two gradients, with the levels normally determined at 6 m intervals. Roads can be constructed with a camber (figure 14.1.3) or with a single crossfall, normally of 2.0 to 2.5 per cent (1 in 50 to 1 in 40).

The selection of the form of road construction will be influenced by a number of factors, including type of subgrade, liability to subsidence, initial costs, maintenance costs, appearance, resistance to wear and non-skid qualities.

Footways should ideally be physically separated from roads by verges up to 3 m wide, although these may have to be omitted in some urban areas through lack of space.

ROAD CONSTRUCTION

The construction methods can be broadly subdivided into two main groups

(1) flexible (figure 14.1.1) consisting of a stone base with a surfacing of tar or bitumen-coated stones;
(2) rigid (figure 14.1.2) constructed of a concrete road slab.

Flexible Roads

Flexible roads or pavements usually contain a sub-base of clinker, crushed slag, crushed rock, hardcore or hoggin, ranging from 100 to 250 mm thick. The sub-base reduces stress in the subgrade (natural formation), protects the subgrade against frost and constructional traffic, and prevents mud entering the road structure. The road base may be formed of various materials, such as (1) dry bound macadam (crushed slag or rock) to BS 63,[1] spread in layers 75 to 100 mm thick and rolled, normally with a 2.5 tonne roller, covered with a 25 mm layer of 4.7 mm down material rolled with a 8 tonne roller; (2) cement bound granular base with an aggregate of gravel, rock or slag mixed with cement to produce a strength of not less than 3.35 MN/m^2; (3) dense road base macadam to BS 4987;[2] and (4) hot rolled asphalt to BS 594.[3] Road bases normally vary from 75 to 100 mm in thickness.

Bituminous surfacings which often incorporate a base course and a wearing course, provide a smooth, non-skid, abrasion-resistant and jointless surface which possesses colour and texture, and has good cleansing

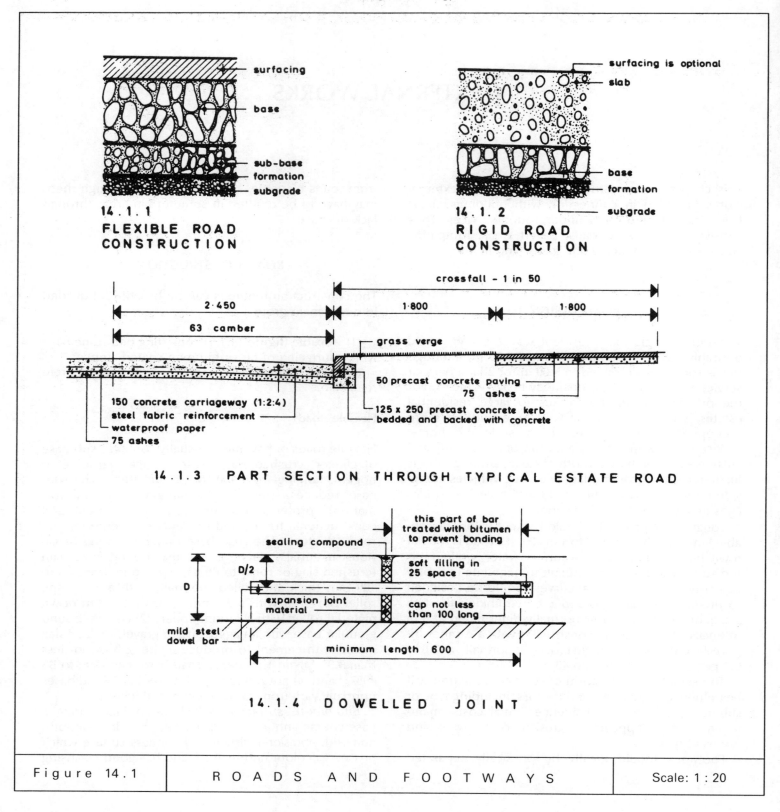

surfacing

base

sub-base
formation
subgrade

**14.1.1
FLEXIBLE ROAD
CONSTRUCTION**

surfacing is optional
slab

base
formation
subgrade

**14.1.2
RIGID ROAD
CONSTRUCTION**

crossfall – 1 in 50

2·450

1·800   1·800

63 camber

grass verge

150 concrete carriageway (1:2:4)
steel fabric reinforcement
waterproof paper
75 ashes

50 precast concrete paving
75 ashes

125 x 250 precast concrete kerb
bedded and backed with concrete

14.1.3   PART SECTION THROUGH TYPICAL ESTATE ROAD

this part of bar
treated with bitumen
to prevent bonding

sealing compound

soft filling in
25 space

D/2

D

expansion joint
material

cap not less
than 100 long

mild steel
dowel bar

minimum length 600

14.1.4   DOWELLED   JOINT

| Figure 14.1 | R O A D S   A N D   F O O T W A Y S | Scale: 1 : 20 |

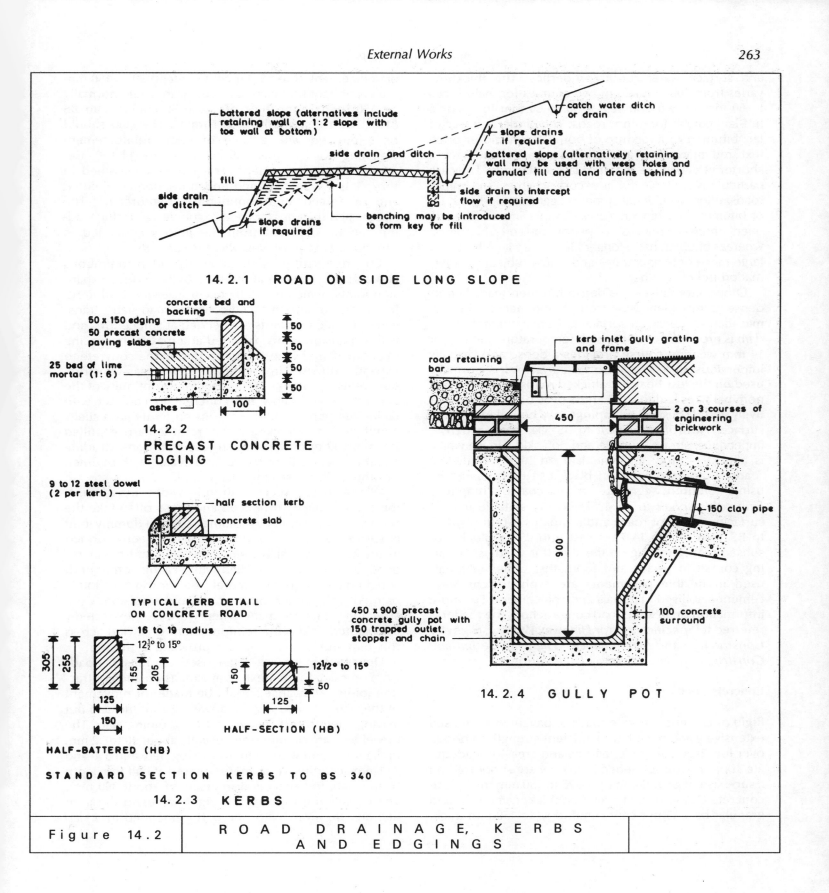

14.2.1 ROAD ON SIDE LONG SLOPE

battered slope (alternatives include retaining wall or 1:2 slope with toe wall at bottom)

catch water ditch or drain

slope drains if required

side drain and ditch

battered slope (alternatively retaining wall may be used with weep holes and granular fill and land drains behind)

fill

side drain or ditch

slope drains if required

side drain to intercept flow if required

benching may be introduced to form key for fill

concrete bed and backing

50 x 150 edging

50 precast concrete paving slabs

25 bed of lime mortar (1:6)

ashes

50
50
50
50
50

100

14.2.2 PRECAST CONCRETE EDGING

9 to 12 steel dowel (2 per kerb)

half section kerb

concrete slab

TYPICAL KERB DETAIL ON CONCRETE ROAD

16 to 19 radius

$12\frac{1}{2}^\circ$ to 15°

305
255
155
205

125
150

HALF-BATTERED (HB)

$12\frac{1}{2}^\circ$ to 15°

150
50

125

HALF-SECTION (HB)

STANDARD SECTION KERBS TO BS 340

14.2.3 KERBS

kerb inlet gully grating and frame

road retaining bar

2 or 3 courses of engineering brickwork

450

150 clay pipe

900

450 x 900 precast concrete gully pot with 150 trapped outlet, stopper and chain

100 concrete surround

14.2.4 GULLY POT

Figure 14.2   ROAD DRAINAGE, KERBS AND EDGINGS

and surface water runoff properties. The thickness varies from 100 mm laid in two courses for main roads to 60 mm in two courses or 50 to 60 mm in a single hot-laid course for minor roads. The binders consist of tar, bitumen or a mixture of both. Warm-laid or open textured macadams are usually cheaper but have a shorter effective life than hot-laid surfacings. They are particularly suitable for low-cost schemes and minor roads, often using local stone as aggregate and with tar or bitumen as a binder. Tar is cheaper in first cost and offers more resistance to petrol and oil droppings, whereas bitumen has a longer life, can be laid through a large range of temperatures and is less subject to deformation in hot weather.

Other alternatives are dense bitumen macadam or dense tarmacadam base course and dense bitumen macadam or dense tar surfacing. *Dense bitumen macadam* is produced at a very high temperature and a 10 or 14 mm wearing course provides a dense and virtually impervious surface. *Dense tar surfacing*[4] is normally used on the less heavily trafficked roads in 10 or 14 mm nominal sizes laid to a thickness of 30 or 40 mm respectively and, if desired, chippings may be rolled into the surface. *Hot-rolled asphalt* to BS 594[3] gives a dense, impervious, strong, durable and non-skid surface which is almost entirely machine laid on highly trafficked roads. Colours other than black can be obtained by using coloured aggregate or pre-coated chippings. *Fine cold asphalt* to BS 4987[2] is very suitable for the surfacing of minor roads and footpaths. *Mastic asphalt* to BS 1446[5] or BS 1447[6] is cast into 25 kg blocks for subsequent reheating on the site. It is used as a wearing course for roads and footpaths. Other materials used in flexible pavements are asphaltic concrete, cement-stabilised materials and soil cement. For more information on flexible road construction, the reader is referred to *Specification for Highway Works*,[7] *Highway Construction and Maintenance*,[8] *Highway Design and Construction*,[9] and *Highway Engineering*.[10]

Concrete Roads

Rigid or cement concrete roads or pavements are used extensively and must have sufficient strength to bridge over localised subgrade failures and areas of inadequate support without deflection. They are supported on a sub-base of granular material 75 to 150 mm thick, lean concrete (1:15/24) 100 to 200 mm thick, cement-bound granular base (5 per cent cement) 90 to 150 mm thick,

or soil cement (5 to 15 per cent cement) 90 to 150 mm thick. For light traffic the concrete slab would normally be 150 to 200 mm thick of grade 30 concrete to BS 8110[11] with a strength of 30 N/mm^2. The slab should be reinforced with a layer of fabric reinforcement weighing not less than 2.5 kg/m^2 (figure 14.1.3). Cement concrete pavements are generally classified in three basic types: jointed and unreinforced, jointed and reinforced, and continuously reinforced.[10] The slab design thickness related to the design traffic loading can be obtained from the graphs contained in *Structural Design of New Road Pavements*.[12]

Concrete road slabs are generally laid in continuous construction, broken at intervals by transverse expansion joints, which normally occur at every third joint. In the case of unreinforced concrete road slabs, transverse contraction joints are needed at 5 m centres and with expansion joints provided at a maximum spacing of about 40 m. Expansion joints permit the concrete to expand with rises in temperature, whereas construction joints are breaks in the structural continuity of the concrete which allow the concrete to contract with falling temperatures. The expansion joints are usually about 20 mm in thickness with rounded arrises, filled to within 12 m of the road surface with non-extruding material such as fibre board, and sealed with bitumen or other suitable sealing compound.

Where roads exceed 4.50 m in width, longitudinal joints are usually introduced and these often take the form of dummy joints. A common form of dummy joint is a groove about 5 mm wide and 50 mm deep – formed in the top of the slab before the concrete has hardened, and the space is filled with sealing compound. Longitudinal joints may merely be vertical butt joints painted with hot tar, bitumen or cold emulsion. Construction joints are made when work is unexpectedly interrupted, when the reinforcement is carried across the joint and a sealing groove provided at the top.

Dowelled joints are often used to transmit loads from one slab to the next, as at expansion and contraction joints. Allowance should be made for movement of the slabs by securing the dowel bars in one slab but leaving them free to move in the adjoining slab. The dowel bars are of mild steel, usually about 20 to 32 mm in diameter and with a length varying between 550 and 750 mm according to diameter, placed at half the depth of the slab and spaced at intervals of about 300 mm. The dowels must be fixed at right angles to the joint line and be level; with one part embedded in a slab

Table 14.1   Comparison of constructional methods for roads

Characteristics	Concrete roads	Coated macadam	Asphalt roads
Liability to subsidence damage	Considerable	Reasonably flexible	Reasonably flexible
Initial cost	Reasonably priced in most areas; requires little skilled labour	Depends on proximity of production plant, but not excessively high	More expensive than coated macadam
Maintenance costs	If well constructed should be low for many years after construction	Fairly high as periodic surface dressing required about every five years and carpet coat at longer intervals	Carpet coat required at about eight to ten-year intervals at much greater cost than coated macadam
Appearance	Glare from sun	Reasonable	Good
Non-skid properties	Slightly corrugated surface gives reasonable resistance	Good	Fairly good
Resistance to wear	Good if concrete quality is carefully controlled	Fairly good to good, depending on type of filler and surface dressing stone	Good
Ease of reinstatement	Costly, difficult, requires curing time	Comparatively simple	Comparatively simple

and the other part painted with bitumen, wrapped in paper or fitted in a sleeve, to prevent it adhering to the concrete. Figure 14.1.4 illustrates a typical dowelled joint. All these joints are well described and illustrated in *Highway Construction Details*[13] and *Highway Design and Construction*.[9]

The concrete can be tamped off the kerbs or side forms of steel or timber, often using vibrating tampers. On large contracts spreading machines or pavers will probably be used. Concrete must not be permitted to dry out quickly or it will lose part of its strength and surface cracks will appear, particularly in hot weather or with cold drying winds. One protective method is to spray a suitable aluminised curing compound on to the finished surface of the concrete at the rate of about 4 m² per litre. As an additional precaution the concrete may be protected against the effects of sun and rain during setting by covers of opaque and waterproof material of white colour externally, supported on a suitable framework. No vehicular traffic should be permitted on the finished surface of the concrete within twenty days of completion when ordinary Portland cement is used, or ten days when rapid hardening cement is incorporated.

Comparison of Constructional Methods

Table 14.1 provides a comparative study of the three main forms of road construction.

Kerbs

The main functions of kerbs are to resist the lateral thrust of the carriageway, to define carriageway limits, to direct the flow of surface water to gullies, and to support and protect footpaths and verges. The most common form of kerb is in precast concrete to BS 7263,[14] and may incorporate Portland or blastfurnace cement and natural stone or slag aggregate, and it may be coloured. Manufacturing processes include cast, vibrated and hydraulically pressed – the latter giving the strongest kerbs which are recognisable by their patterned faces. Less popular kerbs include natural granite and sandstone.

Precast concrete kerbs are made in 915 mm lengths and in three standard sections – bullnosed (with 16 to 19 mm radius edge), splayed (75 mm wide × 75 mm deep) and half-battered (angle of 12½° to 15° to vertical). Two half-sections are illustrated in figure 14.2.3 and the most commonly used sizes are 125 × 255 mm and 125 × 150 mm respectively. Radius kerbs are made to a variety of radii ranging from 1 to 12 m. Bullnosed and half-battered kerbs are used extensively for urban roads, whereas splayed kerbs are more often used on dual carriageways and rural roads.

With flexible roads, it is usual to lay full-section kerbs (125 × 255 to 150 × 305 mm) to take the lateral thrust. They are usually bedded on a 12 to 25 mm layer of cement mortar on a concrete foundation (1:3:6), 100 to 150 mm thick, and backed with similar quality concrete, 100 to 125 mm thick, to within 75 mm of the top of the kerb. Alternatively, the kerbs may be laid on a fresh semi-dry concrete foundation without an intervening mortar bed. The kerbs may be butt jointed without any mortar, or have mortared and pointed joints.

With rigid roads, it is better to use half-section kerbs on a mortar bed, 12 to 25 mm thick, laid directly on the concrete road slab. To give adequate support to the kerbs it is advisable to provide two 9 to 12 mm diameter steel dowel bars, 125 to 150 mm long, to each length of kerb and driven 75 mm into the concrete while it is still green and then backed with concrete (figure 14.2.3). This form of construction prevents water penetrating between the kerb face and the edge of the road slab, but prevents the tamping of the concrete carriageway from the kerbs.

Channels

Channels help surface water runoff, define the limits of the carriageway, facilitate road sweeping and strengthen the edges of flexible roads. The most commonly used channels are precast concrete to BS 7263,[14] of 255 × 125 mm rectangular section, laid on a concrete foundation. In concrete roads channels are normally formed about 200 or 250 mm wide with a wood or steel float. Precast concrete edging (figure 14.2.2) varying in size from 50 × 150 to 50 × 250 mm, with tops finished to round, flat or bullnosed sections, may be used to support and define the edges of drives in place of kerbs or channels.

ROAD DRAINAGE

Groundwater Drainage

Water entering the subgrade can cause a substantial loss in strength in the soil foundation to a road. Hence the road formation should be kept a minimum of 1.20 m above the highest groundwater level. With permeable soils it is possible to reduce the water table by installing longitudinal land drains on each side of the road, with the drain trenches filled with coarse sand or other suitable material and sealed with clay at the top of the trench. Where there is a risk of surface water from adjoining higher land penetrating into the subgrade this must be intercepted by side

drains or ditches as shown in figure 14.2.1. With embankments it is customary to provide slope drains, often constructed as French drains, running down the slope to connect with a side drain or ditch at the side of the road.

Surface Water Drainage

Road surfaces are laid to specified cambers or straight crossfalls to shed surface water to channels. Minimum crossfalls to road surfaces vary with the form of construction ranging from 1:33 to 1:50 (3 to 2 per cent) for relatively smooth surfaces such as mechanically laid asphalt and dense bitumen macadam or concrete, to 1:30 to 1:40 (3.33 to 2.5 per cent) for rougher surfaces such as hand laid bituminous materials. Drainage channels consisting of preformed units bedded on concrete should have a gradient of at least 1:150, whereas those of reinforced *in situ* concrete, bituminous materials or brick pavers should have a minimum gradient of 1:50 on the channel line. Roads in rural areas are often drained through trenches or 'grips' about 375 mm wide and 150 mm deep, leading across grassed roadside areas to ditches. Access roads normally have a crossfall of 1 in 40 and paved areas a minimum of 1 in 60.[16]

Gullies

In urban areas surface water runoff from roads is almost invariably discharged into gullies of the type illustrated in figure 14.2.4. These are often spaced at about 45 m intervals, with each gully draining about 220 m^2 of road surface. Gully pots are made in a variety of materials including precast concrete, clayware, cast iron, brickwork and *in situ* concrete. Precast concrete pots to BS 5911[17] are the most commonly used. They can be trapped or untrapped and supplied with or without rodding eyes, in a variety of sizes ranging, for instance, from 300 mm diameter and 600 mm deep to 600 mm diameter and 1200 mm deep for precast concrete pots. The outlets are usually 150 mm in diameter, although the smallest gullies have 100 mm diameter outlets. A typical gully pot is illustrated in figure 14.2.4, which shows brickwork above the pot to permit some flexibility in the fixing of the cover and frame.

Gully gratings, through which surface water enters the pot, are of two main forms – channel and kerb inlet in cast iron, cast steel or a mixture of both to BS 497.[18] The kerb inlet type are only really suitable for light duty but have a number of advantages – they eliminate ob-

Table 14.2   Comparison of constructional methods for footpaths

Form of construction	Advantages	Disadvantages
Precast concrete paving flags (slabs)	Good appearance; hard wearing; non-slip; fairly easily reinstated	Expensive in initial and maintenance costs; can soon become dangerous with slight settlement; easily damaged by vehicles mounting kerb
In situ concrete	Reasonably cheap; can be coloured; reasonably hard wearing if concrete of good quality; reasonably non-slip	Needs frequent expansion joints and time for curing; unattractive appearance and difficult to reinstate satisfactorily
Bitumen macadam or tarmacadam	Reasonably priced and hard wearing; non-slip; flexible; fairly easily maintained; reasonable appearance	Periodic surface dressing required; generally needs path edging (figure 14.2.2) at back of path
Asphalt	Good appearance; hard wearing; reasonably non-slip; flexible; fairly easily maintained	Fairly expensive; generally needs path edging (figure 14.2.2) at back of path

structions in the carriageway, facilitate road construction and resurfacing and there is less risk of blockage. On the other hand they obstruct a section of the path or verge, need anti-deflector plates in the channel where the gradient exceeds 1 in 50, and have a smaller hydraulic efficiency. The channel type gratings are made in two grades: grade A capable of bearing wheel loads up to 11.50 t and grade B for wheel loads up to 5.00 t for use in culs de sac and minor residential roads.

FOOTPATHS AND PAVED AREAS

Footpaths and paved areas can be constructed of a variety of materials, and the choice will be determined largely by such factors as initial cost, maintenance cost, appearance, wearing qualities and non-skid properties. An examination of the comparative schedule in table 14.2 may prove helpful in this connection. The width of a footpath varies from about 1.80 to 2.00 m on housing estates to permit the satisfactory passing of prams or wheelchairs, and may increase to as much as 6 m in shopping centres. They are normally laid to a crossfall of between 1 in 30 to 1 in 40 towards the kerb and there may be a grass verge between the path and kerb. A tree-planted grass verge improves the appearance of residential development but it increases maintenance costs.

Precast Concrete Paving Flags/Slabs

Precast concrete paving flags or slabs are used extensively in the built-up areas of towns and to a lesser extent on residential estates. The best flags/slabs are hydraulically pressed, as these are the densest and strongest, and all should comply with BS 7263.[14] They can be made with a variety of aggregates, of which the most common are natural aggregates to BS 882,[19] and to various colours and surface finishes. For example, tactile flags have domes on the wearing surfaces as illustrated in BS 7263.[14] Standard sizes are 600 × 450, 600 × 600, 600 × 750, 600 × 900, 450 × 450, 400 × 400 and 300 × 300 mm, in both 50 and 63 mm thicknesses for the four larger sizes of flag.

The slabs/flags are usually laid on a 75 or 100 mm bed of ashes or hardcore consolidated with a 2.5 tonne roller or 300 to 500 kg vibratory roller, and a 25 mm bed of sand, lime mortar or gauged mortar. Where the slabs/flags may be subject to vehicular traffic they should be laid on concrete. Slabs/flags are often laid to a 150 mm bond and a crossfall of 1 in 30. After laying, slabs/flags are usually grouted with lime or cement mortar to give a solid job. Although expensive, precast concrete paving slabs/flags are attractive, durable and relatively easily reinstated.

Alternatives to precast concrete slabs are natural stone or reconstructed stone.

Natural stone. Natural stone for paving can have a riven sawn, tooled or ribbed surface. For example, York stone, which is a durable sandstone with an attractive creamy white/brown colour provides flags of fairly even thickness, but they are very expensive, and require a thoroughly consolidated base overlaid with a 25 mm bed of 1:6 to 1:5 cement sand mortar or 1:3 to 1:4 lime sand mortar. Lisney and Fieldhouse[20] have described the alternative materials of granite, slate and limestone.

Reconstructed stone slabs. These contain natural stone both as fine and coarse aggregates, giving colour and non-slip qualities. They are supplied in several shapes for patterned work, including hexagonal, triangular and circular, when they are often surrounded by asphalt, tarmacadam or cobbles, to provide an attractive combination of colours and textures.

Precast Concrete Paving Blocks

Precast concrete paving blocks complying with BS 6717[21] normally have a work size of 200 × 100 × 60 mm thick and are suitable for low speed roads and other paved surfaces subjected to vehicle loading and pedestrian traffic.

In situ Concrete

In situ concrete is very suitable for large paved areas and paths to houses. A common mix is 1:2:4 (grade 21 N/mm^2), using 19 mm graded aggregate and the concrete is often 75 mm thick, but may be thickened at the edges, on a clinker or hardcore base. It is advisable to provide expansion joints, often of impregnated sheet for the full depth of the slab, at about 6 m intervals. A variety of finishes is available including tamped, rolled with a crimping roller leaving pyramidal depressions, brushed aggregate, pigmented, and surface dressed with coloured chippings lightly tamped into the concrete. Interesting patterns can be obtained by using concrete with different aggregates in alternate bays and inserting bricks, setts or other small units as dividers for the bays, which also serve as contraction joints. It is relatively cheap but its main deficiency is the difficulty of reinstatement to gain access to underground services.

Bituminous Surfacings

A common footpath surfacing is coated macadam to BS 4987[2] using a crushed rock, gravel or slag aggregate and a binder of tar or tar-bitumen mixture, with a 6 mm nominal size medium-textured wearing course. It is laid warm on a bed of clinker, hardcore or other suitable material, 50 to 75 mm in thickness. Dense tar surfacing to BS 5273[4] may be laid in a single course, 30 mm thick, with a nominal size of coarse aggregate of 10 mm. It should be consolidated with a roller while hot to a crossfall of not less than 1 in 48. It provides a relatively economical, close-textured, impervious, non-slip and fairly easily reinstated surface, but it does require periodic surface dressing.

Another alternative, usually incorporating slag or limestone aggregate, is a single course of fine cold asphalt to BS 4987,[2] laid on hardcore, concrete, soil-cement or other suitable base. A tack coat of bitumen emulsion (1.8 to 2.7 m^2 per litre) should precede the asphalt, unless the asphalt is being laid on a recently laid base course. The average thickness is about 20 mm and the asphalt is consolidated with a 350 to 2500 kg roller. After initial consolidation, 10, 13 or 20 mm precoated chippings are sometimes added and rolled into the surface of the asphalt. Fine cold asphalt gives a dense and hard-wearing surface, and is used primarily for thin wearing courses or for patching.

Cobbles, Setts and Bricks

These smaller paving units are used to give interest, for demarcation purposes, to form drainage channels, to pave small irregular areas and to discourage the passage of vehicles and pedestrians.

Cobbles. These are obtained from river beds, beaches and gravel quarries, are hard wearing and have a rich diversity of colours and good texture. They are often about 50 to 75 mm in size but can be 125 mm or more in areas subject to vehicular traffic. For lightly trafficked areas cobbles are usually laid on a fine gravel or ash bed with the intervening spaces between cobbles filled with a dry cement and sand mix brushed in and sprayed with water. For heavily trafficked areas cobbles are usually laid on a semi-dry concrete bed about 100 mm thick, with the cobbles extending above the concrete for about one-third of their depth. They are attractive in appearance but difficult to keep clean, requiring a high standard of maintenance.

Setts. These are usually of granite to BS 435[22] and are generally either hammer dressed (left roughly rectangular) or nidged (chisel dressed to a true rectangle), although most supplies now available are salvaged materials. Typical sizes on plan vary from 100 × 100 to 150 × 150 and 100 × 200 mm with the thicknesses varying from 100 to 150 mm. They are extremely hard wearing with good textures and colours, and are usually laid to a crossfall of about 1 in 40 (1:40) on a concrete base with a 25 mm cement mortar bed, and with the joints between setts sometimes filled with 6 mm chippings and grouted with pitch or pitch–tar compound. Cement mortar can be used for jointing in pedestrian areas. Alternatively, they may be laid on a 25 mm bed of coarse sand on a 100 mm hardcore base, when confined to pedestrian use. The alternative of whinstone setts have a smoother finish.

Engineering bricks and brick pavers. These are suitable for paving in a variety of colours, bonds and patterns to give an attractive and durable surface. They are usually laid on a concrete or hardcore base with a 25 mm bed of semi-dry cement or lime mortar (1:4) with 6 to 9 mm joints between bricks. Typical bonds are stack, stretcher, basketweave and herringbone.[20]

BS 6677[23] lays down requirements for both clay and calcium silicate pavers for flexible pavements. Typical work sizes are 210 × 105 mm with work size thicknesses of 50 and 65 mm. This standard classifies PA pavers for use by pedestrians and light vehicles and PB pavers for public transport and commercial road vehicles. In flexible construction the pavers are laid on coarse sand with close butt joints (2 to 5 mm wide) and flexible paving must be contained on all sides by an edge restraint such as brick trims and concrete kerbs and channels bedded in concrete.

Gravel

Gravel is used for surfacing lightly used pedestrian areas in rural situations, landscaped areas, around trees in car parks and in shaded areas where grass will not thrive. The formation is usually treated with a weedkiller. The gravel can be 'all-in' material such as ballast or hoggin, or graded material such as river gravels and crushed rock (often 19 to 3 mm in size) in various colours. It is normally laid in two layers, well watered and rolled to a consolidated depth of about 150 mm to a crossfall of about 1 in 30.

Other Pavings

Another alternative is to use hoggin, a naturally occurring mixture of gravel and clay, with a maximum particle size of 50 mm, watered and rolled to a compacted thickness of 100 mm, possibly laid on a clinker bed. It has a relatively low initial cost but is expensive in maintenance. A number of proprietary products incorporate interlocking precast concrete slabs with rows of protuberances about 100 mm deep, usually bedded in sand on hardcore, and with the interstices soiled and seeded. They are useful in providing parking areas for occasional traffic, whilst retaining an attractive grassed appearance at all times.

LANDSCAPE WORK

This section is mainly concerned with grassed areas and the planting of trees and shrubs.

Grassed Areas

Grassed areas and roadside verges possess great amenity value and also perform other useful functions such as absorption of sound and reduction of glare, and accommodating public utility services. They are relatively cheap to establish but expensive in maintenance. It is undesirable to lay verges to gradients in excess of 1 in 4 and less than 1.35 m in width. The finished surface should project about 20 mm above adjoining surfaces such as kerbs, pavings and manhole covers, for ease of grass cutting. It is good practice to provide a strip of gravel or ballast, treated with weedkiller at least 220 mm wide, between grassed areas and walls.

Prior to soiling and seeding, the formation should be broken up and a layer of clinker or gravel provided for drainage purposes with heavy clay. The layer of soil should be at least 100 mm deep raked or harrowed (cultivated), removing all stones and debris exceeding 25 mm, to give 50 mm of fine tilth and have a pH value just over 5.0, followed by the application of a pre-emergent weedkiller. A suitable fertiliser is often distributed at the rate of 70 kg/ha, followed by a pre-seeding pesticide at about 70 kg/ha. A suitable mixture of grass is sown in two directions at an overall rate ranging from 50 to 150 kg/ha, depending on its function and the type of subsoil, and should be lightly raked in and lightly rolled if the surface is dry. The choice of mix is influenced by the type of soil, location, wear and expected maintenance. For moderate wear in residential areas the constituent grasses should include Chewings fescue, red creeping fescue, browntop bent and smooth-stalked meadow grass. Fescues and bents thrive on poor soils.[24]

Practical difficulties of the seasons when seeding can take place (April and mid-August to mid-September) and the time taken for grass to become established (complete growing season) may favour expensive turfing in some situations. Turves vary in size from 300 × 900 mm to 900 × 2100 mm and are laid from boards to break bond over well-prepared soil. They should be watered after laying in dry periods. Meadow turf is the most readily available and is the cheapest but

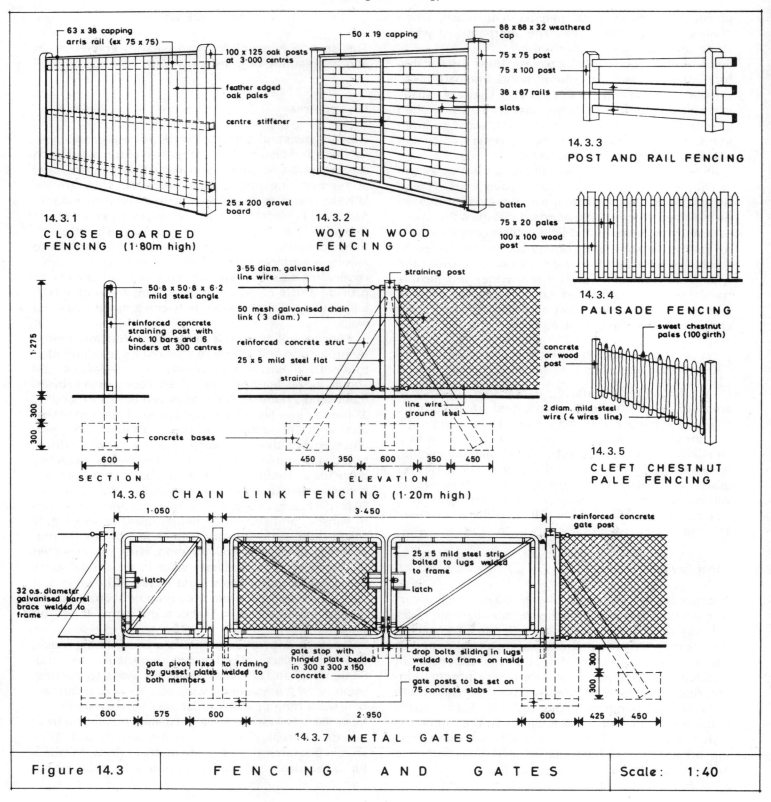

63 x 38 capping
arris rail (ex 75 x 75)
100 x 125 oak posts
at 3·000 centres
feather edged
oak pales
centre stiffener
25 x 200 gravel
board

50 x 19 capping
slats
batten

88 x 88 x 32 weathered
cap
75 x 75 post
75 x 100 post
38 x 87 rails

14.3.1
CLOSE BOARDED
FENCING (1·80m high)

14.3.2
WOVEN WOOD
FENCING

14.3.3
POST AND RAIL FENCING

75 x 20 pales
100 x 100 wood
post

14.3.4
PALISADE FENCING

3·55 diam. galvanised
line wire
straining post
50·8 x 50·8 x 6·2
mild steel angle
reinforced concrete
straining post with
4no. 10 bars and 6
binders at 300 centres
50 mesh galvanised chain
link (3 diam.)
reinforced concrete strut
25 x 5 mild steel flat
strainer
line wire
ground level
concrete bases

1·275
300 300
300
600
SECTION
450   350   600   350   450
ELEVATION
14.3.6 CHAIN LINK FENCING (1·20m high)

sweet chestnut
pales (100 girth)
concrete
or wood
post
2 diam. mild steel
wire (4 wires line)

14.3.5
CLEFT CHESTNUT
PALE FENCING

1·050   3·450
reinforced concrete
gate post
25 x 5 mild steel strip
bolted to lugs welded
to frame
latch
32 o.s. diameter
galvanised barrel
brace welded to
frame
latch
gate pivot fixed to framing
by gusset plates welded to
both members
gate stop with
hinged plate bedded
in 300 x 300 x 150
concrete
drop bolts sliding in lugs
welded to frame on inside
face
gate posts to be set on
75 concrete slabs
300
300
600   575   600   2·950   600   425   450
14.3.7 METAL GATES

| Figure 14.3 | F E N C I N G   A N D   G A T E S | Scale: 1:40 |

is likely to contain a high proportion of weeds and coarse grasses. Sea-washed turf and downland turf from southern England contain fine leaved grasses and their normal size is 300 × 900 × 25 to 35 mm thick. However, they need laying promptly once lifted to prevent drying out or rotting but can stand wear within two months or less of laying. Recommendations on grading and cultivation, grass seeding, turfing and planting of trees and shrubs are contained in BS 4428.[25]

Trees and Shrubs

The provision of trees and shrubs can do much to improve the appearance of a building project. Trees take many years to grow and so existing trees which are in good condition should be preserved wherever possible. However, new building developments frequently alter the water table, endangering the lives of existing trees and underground services must be positioned to avoid their roots. Furthermore, existing trees and shrubs give a good indication of what new species are likely to thrive as far as climate and soil are concerned. There is rarely sufficient space on residential developments for forest-type trees, but smaller trees such as whitebeam and mountain ash, as well as many types of flowering tree, with heights of 6 to 9 m, can be used to advantage.

Trees generally have great value in form as well as colour, and they should be chosen to blend the architecture and landscape into a single composition. They change with the season and give living quality to the hard surfaces and angular shapes of buildings and roads. In summer trees in leaf have great beauty in both form and colour, and in winter the silhouette and shadow of branches and twigs provide fascinating decoration to nearby walls. The irregular shapes and siting of trees can often be used deliberately as a contrast to architectural regularity. Trees are also useful as a barrier to noise and glare, and to give shade and shelter from winds. Consideration must be given to soil and climatic conditions, ultimate size of trees, and location of underground services and buildings, when selecting trees.[26]

Shrubs provide a variety of colours in flowers, berries and leaves, with differing flowering seasons and heights varying from 0.60 to 2.40 m. Plants commonly used in landscape work comprise the *Rosaceae* family which encompasses cotoneasters, hawthorns, potentillas, pyracanthas and spiraeas, and are best planted in bold, informal groups. The *Ericaceae* family includes many acid-tolerant plants which provide good ground cover, such as heaths and heathers. Hedges form attractive boundaries and a wide range of colourful species is available.

BS 3936[27] specifies trees and shrubs, including climbing plants and conifers, that are suitable to be transplanted and grown for amenity, as well as forest tree stock; while BS 3998[28] contains general recommendations for tree work, including treatment of wounds, cleaning out, pruning, repair of roots and bark wounds, bracing, irrigation, nutrition, felling and stump treatment. BS 4043[29] gives recommendations for transplanting trees which are to be moved with a ball of earth around their roots and this is becoming increasingly important as more semi-mature trees are transplanted on to housing sites.

Small trees require holes about 1.5 m in diameter which should be filled with topsoil, preferably with some leaf mould, peat or other suitable organic matter placed around the tree roots. In wet soils, brick rubble or clinker should be laid over the bottom of the hole. The normal planting season stretches from October to mid-March. The roots should be well spread out and the soil made firm around them. Young trees should be protected with stakes and guards and they need to be watered and mulched with grass cuttings or manure during dry periods for the first two or three years after planting.

FENCING

Fences of a variety of materials and types are used to form boundaries between land of different occupiers. The choice may be influenced by a number of factors including appearance, durability, initial and maintenance costs, and effectiveness.

Close boarded fences. These are both attractive and effective but have a high initial cost. They should comply with BS 1722[30] and a typical fence of this type is illustrated in figure 14.3.1. Posts may be of concrete or timber, preferably oak. Timber posts look better – they should ideally have weathered tops and mortices to receive rails, be 100 × 100, 100 × 125 or 100 × 150 mm in section according to height with 600 to 750 mm in the ground. The normal maximum spacing is 3 m (centre to centre). Rails are normally triangular in section with two cut from 75 × 75 mm timber – with two rails for fences up to 1.20 m high and three rails for higher fences. The boarding usually consists of vertical pales about 100 mm wide and feather edged approximately 13 to 6 mm (minimum) thick (two ex 100 × 22 mm), lapped approximately 18 mm when fixed, with a 65 × 38 mm twice-weathered capping to protect the top of the pales. The pales are nailed to the rails with 50 mm galvanised steel nails. A horizontal gravel board, often 25 or 32 × 150 mm, is often fixed below the pales to prevent their bottoms from being in contact with the ground with liability to decay.

Woven wood and lap boarded panel fences. These are reasonably attractive, fairly expensive and not exceptionally durable (figure 14.3.2). They should comply with BS 1722[30] and posts may be of concrete or timber. The maximum length of panels is 1.80 m and heights vary from 0.60 to 1.80 m. The panels consist of horizontal and vertical slats woven together within a frame of double vertical battens and either double horizontal battens or a single counter rail and gravel board, together with single or double vertical centre stiffeners. Battens are not usually less than 19 × 38 mm and slats not less than 75 × 5 mm. A twice-weathered capping, about 19 × 50 mm is usually fixed to the top of each panel. Wood posts should not be less than 75 × 75 mm with a weathered cap, and reinforced concrete posts are generally 100 × 85 mm with a weathered top.

Wooden post and rail fences. These are also specified in BS 1722[30] with posts often 75 × 150 mm possibly with pointed bottoms for driving (figure 14.3.3), and intermediate or prick posts may also be provided usually 38 × 87 mm in size. Rails are often rectangular, 38 × 87 mm in size. This type of fence is reasonably durable and attractive, of moderate cost, but does not provide a very effective division, as people, animals and objects can pass through it.

Wood palisade fences. These (figure 14.3.4) may consist of either concrete or wooden posts supporting two or three arris or triangular rails (ex 75 × 75 mm), according to the height of the fence (three rails when exceeding 1.20 m high), and vertical pales or palisades of 75 × 20 or 65 × 20 mm section, all in accordance with BS 1722.[30] This is a reasonably popular type of fence but has no particular merit.

Cleft chestnut pale fences. These (see figure 14.3.5) are also covered by BS 1722[30] with the pales normally hand-riven from sweet chestnut and a girth of not less than 100 mm, and roughly triangular or half round in section. The pales are supported by two or three lines

of 2 mm diameter zinc coated low carbon mild steel wire, with each line consisting of four wires twisted together between the pales. Straining and intermediate posts can be of concrete or wood with the spacing of intermediate posts varying between 2.00 and 3.00 m according to the material used, function and height of fence. This type of fence is useful for temporary work but its poor appearance and liability to damage prevent its more widespread use.

Chain-link fences. Made to BS 1722,[30] are widely used as they are reasonably economical and form a very effective boundary, although their appearance even using plastic-coated chain link is not very attractive (figure 14.3.6). The chain link consists of a diamond-shaped mesh with an average mesh size of 50 mm and 2.5 to 3 mm diameter finished with zinc coating and/or plastics coating. The chain link is tied with wire to low carbon mild steel line wires of 3.0 to 4.75 mm diameter (two lines for fences up to 900 m high and three lines for higher fences). Posts may be of reinforced concrete, steel or wood. The line wires are pulled tight with straining fittings at straining posts (figure 14.3.6) and intermediate posts are provided at not more than 3 m centres. Post sizes vary with the height of the fence, but with 1.20 m high fences typical sizes would be:

	straining posts	intermediate posts
	(mm)	(mm)
reinforced concrete	125 × 125	125 × 125
steel (angle)	50 × 50 × 6	40 × 40 × 5
wood	100 × 100	75 × 75

Straining posts are strutted in the manner shown in figure 14.3.6.

Other types of fence. Covered by BS 1722,[30] these include rectangular wire mesh and hexagonal wire netting; strained wire; mild steel (low carbon steel) continuous bar; mild steel (low carbon steel) fences with round or square verticals and flat posts and horizontals; anti-intruder fences in chain link and welded mesh; and steel palisade fences.

GATES

Domestic front entrance gates are covered by BS 4092[31] and are of two main types – wooden and metal. Standard wooden single gates have widths varying from 810 to 1020 mm, whereas pairs of gates may have overall widths of 2.13 to 2.64 m, with heights of not less than 840 mm. Minimum component sizes are specified as follows

hanging stiles for pairs of gates	95 × 45 mm
hanging stiles for single gates	70 × 45 mm
shutting stiles	70 × 45 mm
rails behind infilling	70 × 25 mm
other rails	70 × 45 mm
brace	70 × 25 mm
thickness of infilling	14 mm

Rails should be through-tenoned into stiles and each tenon should be pinned with a hardwood pin (not less than 10 mm diameter) or a non-ferrous metal star dowel (not less than 6 mm diameter).

For metal gates, the standard widths are 0.90, 1.0 and 1.1 m for single gates and 2.3, 2.4 and 2.7 m for pairs of gates, with the top of the gate framework a minimum of 0.90 m above ground. Metal gates may be constructed of either mild steel tubular frame or wrought iron flat frame, should have a continuous framework, and be truly square and welded all round at junctions. Minimum dimensions of framework and infilling are prescribed in BS 4092,[31] together with suitable hanging and latching fittings and protective treatments. Figure 14.3.7 shows suitable single and pairs of metal gates for use with chain link fencing. Alternative forms of construction are detailed in BS 1722.[30]

Gates are normally hung from steel band hinges which may be hot dip galvanised or sherardised. The type of gate latch depends on the gate height; low gates can be fitted with automatic gate latches fitted to the inside of the gate, whereas high gates can be fitted with a Suffolk type of thumb latch or a ring handle gate latch operated from either side of the gate. Double gates shall be provided with slam plates and drop bolts.

LIGHTING

Road Lighting Design

BS 5489,[32] Part 1 provides guidance on the general principles of road lighting and contains recommenda-

tions relating to the aesthetics and technical requirements of road lighting. Part 3 of this British Standard gives recommendations for the lighting of access roads, residential roads and associated pedestrian areas and it encompasses lighting design, general installation design and operation and maintenance.

By illuminating the road surface uniformly, obstructions are seen as a dark outline against a light background (silhouette) for the whole width of the road. The driver is thus able to interpret the whole scene from a distance and this enables him to react to any obstruction in sufficient time. To achieve this, the pools of light from adjacent lamps should overlap so that dark patches on the road surface are kept to a minimum. This objective can be nullified in practice by large overhanging roadside trees located between the lights.

The principal criteria of quality for road lighting to ensure good visual conditions are as follows:

(1) achievement of the appropriate level of road surface luminance;

(2) uniformity of luminance distribution across and along the road;

(3) adequacy of lighting to the immediate surroundings of the road;

(4) correct limitation of glare from the lighting equipment.

Road lighting in the United Kingdom is classified into two categories: group A on main roads and traffic routes and group B mainly on residential and side roads. In group B lighting the mounting height of lanterns may be 8 m, 6 m or 5 m, depending on the road width and use. The lamps range from 35 W to 150 W discharge lamps depending on the type and mounting height. The lamps are usually of the semi cut-off type but provide more light above the horizontal plane to illuminate adjacent buildings for security purposes. The lighting columns are generally sited in a staggered formation with particular emphasis on road junctions, with the actual siting influenced by the need to avoid entrances to properties, house windows and the like. The spacing of columns is generally 5 to 6 times the mounting height for 5 m and 6 m columns and 3 to 4 times the mounting height for 8 m columns, depending on the road width.

Light Fittings

Light fittings embrace the following three basic components

(1) The lamp or light source which determines the output and colour of light emitted. For example, filament lamps, such as tungsten in halogen, are used for low output applications and give good colour rendering by not changing the natural colours of the illuminated objects. By contrast, discharge lamps such as fluorescent, sodium and mercury give high quality of output at the expense of quality.

(2) The luminaire houses the lamp and its electrical supply and provides weatherproofing and protection. It forms an important design element visually and must be of durable and corrosion resistant materials, using acrylic and polycarbonates in preference to glass where vandalism may occur.

(3) The supporting structure may be a column of concrete, aluminium, cast iron, galvanised steel or reinforced plastics or the luminaire may be fixed to a wall of a building with a bracket.

On residential developments, lighting for pedestrians is important for reasons of safety and security. Particular attention needs to be paid to the lighting of steps, ramps, low walls, road crossings and other potentially hazardous features. Lighting with low level fittings is not very efficient and they are more vulnerable to vandalism.[20]

REFERENCES

1. *BS 63* Road aggregates, *Part 1: 1987* Specification for single-sized aggregate for general purposes; *Part 2: 1987* Specification for single-sized aggregate for surface dressing
2. *BS 4987* Coated macadam for roads and other paved surfaces, *Part 1: 1988* Specification for constituent materials and for mixtures; *Part 2: 1988* Specification for transport, laying and compaction
3. *BS 594* Hot rolled asphalt for roads and other paved areas, *Part 1: 1985* Specification for constituent materials and asphalt mixtures; *Part 2: 1985* Specification for the transport, laying and compaction of rolled asphalt
4. *BS 5273: 1975 (1985)* Dense tar surfacing for roads and other paved areas
5. *BS 1446: 1973* Mastic asphalt (natural rock asphalt fine aggregate) for roads and footways
6. *BS 1447: 1988* Specification for mastic asphalt (limestone fine aggregate) for roads, footways and pavings in building
7. Department of Transport. *Specification for highway works, Part 3: road pavements*. HMSO (1986)
8. J. Watson. *Highway Construction and Maintenance*. Longman (1989)
9. R.J. Salter. *Highway Design and Construction*. Macmillan (1988)
10. C.A. O'Flaherty. *Highways Vol. 2: Highway Engineering*. Arnold (1988)
11. *BS 8110* Structural use of concrete, *Part 1: 1985* Code of practice for design and construction
12. Department of Transport. *Structural Design of New Road Pavements*. HMSO (1987)
13. Department of Transport. *Highway Construction Details*. HMSO (1987)
14. *BS 7263* Precast concrete flags, kerbs, channels, edgings and quadrants, *Part 1: 1990* Specification; *Part 2: 1990* Code of practice for laying
15. Surrey County Council. *Roads and footpaths: A design guide for Surrey. 2* (1981)
16. *BS 6367: 1983* Code of practice for drainage of roofs and paved areas
17. *BS 5911* Precast concrete pipes and fittings for drainage and sewerage, *Part 2: 1982* Specification for inspection chambers and street gullies; *Part 200: 1989* Specification for unreinforced and reinforced manholes and soakaways of circular cross section
18. *BS 497* Manhole covers, road gully gratings and frames for drainage purposes, *Part 1: 1976* Cast iron and cast steel
19. *BS 882: 1983* Specification for aggregates from natural sources for concrete
20. A. Lisney and K. Fieldhouse. *Landscape Design Guide, Vol. 2: Hard Landscape*. Gower (1990)
21. *BS 6717* Precast concrete paving blocks, *Part 1: 1986* Specification for paving blocks; *Part 3: 1989* Code of practice for laying
22. *BS 435: 1975* Dressed natural stone kerbs, channels, quadrants and setts
23. *BS 6677* Clay and calcium silicate pavers for flexible pavements, *Part 1: 1986* Specification for pavers;

Part 2: 1986 Code of practice for design of lightly trafficked pavements; *Part 3: 1986* Method of construction of pavements

24. A. Lisney and K. Fieldhouse. *Landscape Design Guide, Vol. 1: Soft Landscape.* Gower (1990)
25. *BS 4428: 1989* Code of practice for general landscape operations (excluding hard surfaces)
26. *BS 5837: 1980* Trees in relation to construction
27. *BS 3936* Nursery stock, *Part 1: 1980* Specification for trees and shrubs; *Part 4: 1984* Specification for forest trees
28. *BS 3998: 1989* Recommendations for tree work
29. *BS 4043: 1989* Recommendations for transplanting root-balled trees
30. *BS 1722* Fences, *Part 1: 1986* Specification for chain link fences; *Part 2: 1989* Specification for rectangular wire mesh and hexagonal wire netting for fences; *Part 3: 1986* Specification for strained wire fences; *Part 4: 1986* Specification for cleft chestnut pale fences; *Part 5: 1986* Specification for close boarded fences; *Part 6: 1986* Specification for wooden palisade fences; *Part 7: 1986* Specification for wooden post and rail fences; *Part 8: 1978* Mild steel (low carbon content) continuous bar fences; *Part 9: 1979* Mild steel (low carbon content) fences with round or square verticals and flat posts and horizontals; *Part 10: 1990* Specification for anti-intruder fences in chain link and welded mesh; *Part 11: 1986* Specification for woven and lap boarded panel fences; *Part 12: 1990* Specification for steel palisade fences
31. *BS 4092* Domestic front entrance gates, *Part 1: 1966 (1971)* Metal gates; *Part 2: 1966 (1971)* Wooden gates
32. *BS 5489* Road lighting, *Part 1: 1987* Guide to general principles; *Part 3: 1989* Code of practice for lighting subsidiary roads and associated pedestrian areas

15 SOUND AND THERMAL INSULATION, DAMPNESS, VENTILATION AND CONDENSATION

This chapter deals with aspects of building science with important technological and constructional implications, and each of which encompasses more than one building element. Hence it was considered necessary to examine them on a wider scale to secure a more comprehensive approach than could be obtained by including them in earlier elemental chapters, and to enable the reader to appreciate the diverse applications and consequences.

SOUND INSULATION

General Characteristics of Sound

Sound is a form of energy and the energy content of a sound wave may be classified according to its power output and measured in watts (W). The simplest method of stopping unwanted sounds is to convert the vibrational energy of the sound waves into other forms of energy such as frictional heat. The rate at which sound waves vibrate is described as their *frequency*, measured in vibrations/second using the unit of the Hertz (Hz). The hearing system is most sensitive to sounds in a range of 1000 to 5000 Hz.[1]

Simple energy or pressure measurements of sound are converted to sound level values expressed in *decibels* (*dB*). Sound reduction values of different constructional components are also expressed in decibels. Zero dB represents the threshold of hearing, busy traffic would register about 80 dB and 120 dB has been described as the threshold of discomfort. Table 15.1 illustrates the effect on hearing of a range of sound level changes.

Table 15.1   Effect on hearing of sound level changes (Source: R. McMullan[1])

Sound level change	Effect on hearing
± 1 dB	Negligible
± 3 dB	Just noticeable
+ 10 dB	Twice as loud
− 10 dB	Half as loud
+ 20 dB	Four times as loud
− 20 dB	One-quarter as loud

When sound makes contact with a building surface such as a wall, floor or ceiling, part will be reflected back into the room, part will be absorbed by the materials of which the element is constructed and part will be transmitted through to the next room, with the respective proportions depending on the type and thickness of material and the frequency of the sound.

Sound is classified into the two following main types:

(1) *Airborne sound* is that emanating from such sources as human voices, musical instruments and radio and television sets, resulting in the generation of sound waves in the air which produce vibrations in walls and floors.

(2) *Impact sound* is caused by the direct impact of a solid object on a building element, such as footsteps, moving furniture or banging a door, and thereby causing vibration of the element.

The operation of equipment or plant within a building may generate both types of sound.

Noise has been aptly described as unwanted sound and has been categorised according to its source, as either outdoor or indoor. Outdoor sources of noise are mainly traffic oriented but can also include slamming of car doors, replacement of refuse bin lids and human voices, while indoor sources encompass conversation, singing, playing of musical instruments, radio and television sets, water cisterns and closets, movement of furniture and footsteps. Almost every activity is likely to produe noise of one kind or another and while it is possible to measure the energies and frequencies of sounds, it is not possible to assess with any accuracy the reactions of an individual person to them.

General Principles of Sound Insulation

As people generally have become more noise conscious, so the need for improved insulation in and between dwellings has assumed greater importance. A radio or television in a ground floor room of one dwelling can often be heard in the bedroom of an adjacent dwelling. Vibrations induced in the separating wall at ground floor level are transmitted up the separating wall and then radiate into the upper floor rooms. This is known as *flanking transmission*. This can still occur when the rooms adjoin, either horizontally or vertically, and often provides a serious addition to sound passing by the obvious path through the common wall or floor.[2]

In standard tests, described in BS 2750,[3] the insulation of airborne sound is measured in each of sixteen one-third-octave bands, the centre frequencies of which range from 100 to 3150 Hz. In order to determine whether a satisfactory standard has been achieved results are expressed in terms of *weighted standardised level difference*, calculated in accordance with BS 5821,[4] when checking the performance of existing walls. This entails the difference in decibels (dB) between the energy levels in the adjoining rooms suitably corrected to allow for a standard amount of absorption representative of normal furnished conditions. For example, the sound reduction of the brick or block components of an external wall is in the range of 45 to 50 dB, whereas the reduction with closed single windows is about 20 to 25 dB.

BRE Digest 337[2] describes how at each frequency the noise levels are measured in the source room and the adjoining receiving room and that the difference between these two levels is a measure of the sound insulation. The actual level in the receiving room is influenced by the following factors

(1) the sound insulation of the separating wall or floor;

(2) the area of the separating wall or floor;

(3) the volume of the receiving room;

(4) the amount of flanking transmission;

(5) the amount of absorbing material, such as furniture, in the receiving room.

The problem of sound insulation is almost entirely one of reflecting energy back into the source room, while the role of absorption is limited to supplementing reflection at high frequencies in some types of wall or floor. The transmitted energy is inversely proportional to the square of the mass of the wall. Thus by doubling the mass of the wall, transmission is reduced to a quarter.

Materials containing a large proportion of voids are very good at absorbing sound waves and have good thermal insulation properties. When used for lining wall and ceiling surfaces to a room they reduce the level of sound within the room significantly, but they have little effect upon the sound insulation properties of a dividing wall. For example, a 25 mm layer of mineral wool could absorb about 80 per cent of sound waves, but if used as a separating medium between two rooms, the sound passage through the wall would only be reduced by about 3 dB.[5]

With double leaf lightweight walls, there is often a large difference between the vibration levels in the two leaves and the insulation can be adversely affected by solid bridges. The insulation is almost certain to be inadequate unless the two leaves are attached to different studs. While heavy concrete and brick or concrete walls provide good insulation against airborne sound provided their critical frequency is near or below 100 Hz. However to obtain satisfactory insulation with less weight, it is necessary to use double leaf construction.[2]

As insulation against airborne sound is increased the presence of gaps assumes greater importance. For example, if a brick wall contains a hole or crack with an area equivalent to only 0.1 per cent of the total wall area, the average sound reduction index (SRI) of the wall will be reduced from about 50 to 30 dB. Serious air gaps can also arise through poorly constructed seals around partitions, particularly at the junctions with floors, ceilings, windows, doors, service pipes and ducts.[6] Other sources of flanking transmission are unsealed doors and windows and even the keyholes of doors.

Building Regulation Requirements

Part E of Schedule 1 to the Building Regulations 1991[7] prescribes the following requirements with regard to airborne and impact sound:

E1. *Airborne sound (walls)*
A wall which separates a dwelling from another building or from another dwelling, or separates a habitable room or kitchen within a dwelling from another part of the same building which is not used exclusively with the dwelling, such as a machinery room or tank room, shall resist the transmission of airborne sound. A habitable room is defined as a room used for dwelling purposes but not a kitchen or scullery.

E2. *Airborne sound (floors and stairs)*
A floor or a stair which separates a dwelling from another dwelling, or from another part of the same building which is not used exclusively with the dwelling, shall resist the transmission of airborne sound.

E3. *Impact sound (floors and stairs)*
A floor or a stair above a dwelling which separates it from another dwelling, or from another part of the same building which is not used exclusively with the dwelling, shall resist the transmission of sound.

Sound insulation of walls

Building Regulation Approved Document E (1991)[7] classifies walls into four types for the purpose of prescribing satisfactory standards of sound insulation and these are now examined in some detail.

Wall type 1: solid masonry

With this type of wall the resistance to airborne sound depends mainly on the mass of the wall. It is also important to fill the joints solid between the bricks or blocks with mortar, to seal the joints between the wall and other parts of the construction to achieve the mass and to avoid air paths, and to control sound paths around the wall (to reduce flanking transmission). Five forms of construction are specified as follows:

(1) Brick plastered on both room faces, with the mass of the wall, including plaster, being 375 kg/m^2, 13 mm plaster and the bricks laid in a bond which includes headers (215 mm wall).

(2) Concrete block plastered on both room faces, with the mass of the wall, including plaster, being 415 kg/m^2, 13 mm plaster and the blocks extending to the full thickness of the wall (215 mm).

(3) Brick with plasterboard on both room faces with the same mass and brick laying requirements as in (1) and with 12.5 mm thick plasterboard.

(4) Concrete block with plasterboard on both room faces with the mass of the wall, including 12.5 mm plasterboard, being 415 kg/m^2, using blocks which extend the full thickness of the wall (215 mm).

(5) Concrete (*in situ* or large panel) of similar mass to that prescribed for (4), but plaster is optional. Joints between panels filled with mortar and unplastered wall is 190 mm thick.

Precautions to be observed at key junctions in the construction of these walls are fully detailed in Approved Document E (1991). For example, the joint between wall and roof must be filled, normally using firestopping. Above ceiling level the mass of the wall may be reduced to 150 kg/m^2, and if lightweight aggregate blocks are used then one side should be sealed with cement paint or plaster skim. Ceilings should be 12.5 mm plasterboard or similar mass material. With timber floor construction joists at right angles to the wall should be fixed with a joist hanger, while concrete floors will be built into the wall.

Wall type 2: cavity masonry

The resistance to airborne sound depends on the mass of the leaves and the degree of isolation. The leaves should be 50 to 75 mm apart and be connected with butterfly pattern wall ties spaced as required by BS 5628. The following types of wall are specified

(1) Two 102 mm brick leaves with 50 mm cavity plastered on both room faces. The mass of the wall, including plaster, being 415 kg/m^2.

(2) Two 100 mm leaves of concrete block with 50 mm cavity plastered on both room faces. The mass of the wall, including plaster, being 415 kg/m^2.

(3) Two leaves of lightweight aggregate 100 mm concrete block with 75 mm cavity plastered or dry lined

on both room faces. The mass of the wall, including finish of 13 mm plaster or 12.5 mm plasterboard, being 300 kg/m^2.

The precautions to be taken at key junctions are similar to those outlined for type 1 walls, except that with concrete floors, the floor shall be carried through to the cavity face of each leaf.

Wall type 3: masonry between isolated panels

The resistance to airborne sound depends partly on the mass and type of core and partly on the isolation and mass of panels. The lightweight panels are supported only from floor and ceiling and shall not be tied or fixed to the masonry core to maintain isolation.

The masonry core may be formed of brick, concrete block or lightweight concrete block with at least 25 mm air space between the masonry core and the lightweight panels on either side. The lightweight panels may consist of two sheets of plasterboard either joined by a cellular core or with or without a supporting framework. The plasterboard has staggered joints between the sheets to avoid air paths. The gap between the ceiling and the masonry core should be sealed with a timber batten and the gap between the ceiling and lightweight panels with mastic, tape or a cove. The lightweight panels are fixed to a timber batten secured to the floor, while timber floor joists which are at right angles to the masonry core are fixed with joist hangers and the space between the joists sealed with a timber batten.

Wall type 4: timber frame with absorbent material

The resistance to airborne sound depends mainly on the use of two isolated frames and also on the absorption of sound in the air space. If the two frames need to be connected together, 14 to 15 gauge metal straps should be fixed below ceiling level and spaced 1.2 m apart.

The wall claddings as shown in figure 4.12 should be 200 mm apart, and each cladding shall consist of at least two sheets of plasterboard with or without plywood sheeting with a combined thickness of 30 mm, with the joints between the sheets staggered to avoid air paths. The construction also embraces an absorbent curtain of unfaced mineral fibre batts or quilt, which may be wire reinforced, with a density of at least 10 kg/m^3 and a thickness of 25 mm if the curtain is suspended in the cavity between the two frames, or 50 mm if fixed to one of the frames. The precautions to be taken at key junctions in the construction are well described and illustrated in Approved Document E1/2/3 (1991).[7]

Sound insulation of partitions

Sound proofing stud partitions

When installing stud partitions between rooms, it is often desirable to provide an adequate level of sound insulation in the partitions to permit the reasonable use and enjoyment of the adjoining rooms. A normal stud partition with plasterboard on each face provides only an average sound reduction of 29 dB.

One approach is to install 50 mm Rockwool slabs or other similar mineral fibre insulation as illustrated in figure 15.2.1, to give a reasonable level of sound reduction in dwellings or offices of about 42 dB.

Another alternative is to install steel faced partitions as illustrated in figure 15.2.2 comprising 1.25 mm steel sheet on both faces enclosing 40 × 60 mm timber studs at 900 mm centres and 60 mm thick Rockwool slab or other equivalent mineral fibre insulation. This type of construction should ensure a reasonable standard of privacy in offices, provide adequate sound control in studios and isolate noisy machines or processes in industrial situations, with an average sound reduction in the order of 44 dB.

A third approach is to provide double staggered timber stud partitions with two layers of 50 mm insulation as illustrated in figure 15.2.3, giving an average sound reduction of 55 dB, comparable to a masonry wall, and suitable for a wide variety of purposes.

Rock-Lam APS acoustic panel wall lining system 1

The Rockwool APS system 1 provides a relatively simple technique for the wall application of a sound absorbent lining protected by a perforated metal facing, offering optimum acoustic absorption with the advantage of high impact resistance. The system consists of lengths of galvanised steel retaining channel fixed horizontally at 1200 mm centres into which the absorber panels are held by cover strips, secured by self-tapping screws. The metal facing of the panel forms a tray into which a specially constructed Rockwool core is laminated, as illustrated in figure 15.2.4. No vertical supports are required other than the use of the same type of retaining channels at external corners or vertical terminations of the system. When fixed to timber battens, the average noise absorption coefficient is 0.90.

Another alternative is to use APS system 2, consisting of Rockwool slabs bonded to perforated hardboard to form panels which can be mounted on a simple timber battening system at 600 mm horizontal or vertical centres.

Sound insulation of floors

Building Regulation Approved Document E1/2/3 (1991)[7] classifies floors into three types for prescribing satisfactory sound insulation standards and these are now examined in some detail.

Floor type 1: concrete base with soft covering
The resistance to airborne sound depends on the mass of the concrete base and on eliminating air paths and the soft covering reduces the impact sound at source. The floor base can consist of a solid *in situ* concrete slab, solid concrete slab with permanent shuttering, concrete beams with infilling blocks, or hollow or solid concrete planks. The mass of the base, including any floor screed and any ceiling finish bonded to the concrete, shuttering, beams and blocks or planks shall be 365 kg/m². The soft covering consists of a resilient material or a material with a resilient base with an overall uncompressed thickness of at least 4.5 mm. A material is deemed to be resilient if it returns to its original thickness after it has been compressed.

At the junction of a floor with a sound resisting or internal solid wall which weighs less than 355 kg/m², the floor base but not the screed pass through the wall. If however the wall weighs more than 355 kg/m², either the wall or the floor base, excluding the screed, may pass through. If the wall passes through, the floor base is tied to the wall and the joint grouted. Where a pipe or duct penetrates the floor, the pipe or duct shall be wrapped with 25 mm of unfaced mineral wool.

Floor type 2: concrete base with floating layer
The resistance to airborne sound depends partly on the mass of the concrete base and partly on the mass of the floating layer, while a resilient layer resists impact sound by isolating the floating layer from the base and surrounding construction.

The four base specifications described for type 1 floors are applicable, except that the weight of the base, including the floor screed and ceiling finish if bonded to the concrete, beams and blocks or planks shall be 300 kg/m². The floating floor may consist of tongued and grooved timber boarding or wood based

board 18 mm thick fixed to 45 × 45 mm timber battens or alternatively a 55 mm cement sand screed reinforced with 20 to 50 mm wire mesh. The resilient layer incorporated in this type of construction can be mineral fibre 13 mm thick and with a density of 36 kg/m³, with rolls tightly butted to avoid air paths; or for use under screeds, boards of pre-compressed expanded polystyrene (impact sound duty grade), can be provided with the boards tightly butted.

Detailed information on the precautions to be taken at key junctions in the construction is contained in Approved Document E1/2/3 (1991).[7]

Floor type 3: timber base with floating layer
The resistance to airborne sound depends partly on the structural floor plus absorbent blanket or pugging, and partly on the floating layer. Resistance to impact sound depends mainly on the resilient layer isolating the floating layer from the base and the surrounding construction. Three floor specifications are included for this type of floor.

The platform floor with absorbent material is illustrated in figure 15.1.1 and consists of (i) a floating layer of 18 mm thick tongued and grooved boarding or wood based board on a substrate of 19 mm thick plasterboard; (ii) a resilient layer of 25 mm thick mineral fibre having a density between 80 and 100 kg/m³; (iii) a floor base of 12 mm thick timber or wood based board deck nailed to timber joists with a ceiling of 30 mm plasterboard in two layers with staggered joints; and (iv) an absorbent material of 100 mm thick unfaced rock fibre with a density of not less than 10 kg/m³ laid on the ceiling.

Ribbed floor with absorbent material is illustrated in figure 15.1.2 and consists of (i) a floating layer of 18 mm tongued and grooved timber or glue jointed wood based board on a substrate of 19 mm thick plasterboard nailed to 45 × 45 mm timber battens; (ii) resilient strips of 25 mm thick mineral fibre having a density between 80 and 140 kg/m³; (iii) a floor base of timber joists of 45 mm nominal width with a ceiling of 30 mm plasterboard in two layers with staggered joints; and (iv) an absorbent blanket of 100 mm thick unfaced rock fibre with a density of not less than 10 kg/m³ laid on the ceiling.

Ribbed floor with heavy pugging is illustrated in figure 15.1.3 and consists of (i) a floating layer of 18 mm tongued and grooved timber or glue jointed wood based board nailed to 45 × 45 mm timber battens; (ii) resilient strip of 25 mm thick mineral fibre having a density between 80 and 140 kg/m^3; (iii) a floor base of timber joists of 45 mm nominal width with a ceiling of either 19 mm of dense plaster on expanded metal or 6 mm plywood fixed under the joists plus two layers of plasterboard with total thickness of 25 mm; and (iv) pugging of mass 80 kg/m^2 laid on a polyethylene layer.

The main precautions to be taken at junctions of floors and walls are as follows
 (i) seal the gap between the skirting and floating layer with a resilient strip glued to the wall where appropriate;
 (ii) leave a 3 mm gap between the skirting and the floating layer, sealed with acrylic caulk or neoprene where appropriate;
 (iii) where appropriate use a suitable method of connecting the floor to the wall which will block air paths between the floor and the wall.

Practical Constructional Applications

There are a number of practical steps which can be taken to improve sound insulation and eliminate or reduce the transmission of unwanted sounds, and some of the most important approaches will now be described.

Layout. Habitable rooms should be kept away from noisy roads where practicable; adjacent dwellings can with advantage be staggered or stepped; it is desirable for adjoining rooms to have similar uses and the shorter dimensioned walls of rooms shared; entrance halls and passageways are best located alongside separating walls; and rooms generating noise should not be located adjoining bedrooms in an adjacent dwelling.
Services. Kitchens and bathrooms in flats should be grouped vertically above one another; services should be sited away from living rooms and bedrooms; boilers, pumps, telephones and other vibrating devices should not be mounted on separating walls; extractor fans should be mounted on rigid panels; and flush mounted fittings should not be mounted back to back on separating walls but kept at least 200 mm apart.
Doors and windows. Door closers should be fitted so as to avoid slamming; door surrounds and keyholes sealed to draughtproof standard; all glazing which is required to have sound proofing qualities should be as thick as possible; double glazing should incorporate

large air gaps between the panes; and opening windows should be adequately sealed to draught proof standard.
Walls. Heavy brick and block walls give good sound insulation, with frogs laid upwards and joints well filled with mortar; pores of concrete blocks should be sealed with plaster and cement based paint; wall ties to cavity walls of the butterfly type; plasterboard partitions should have staggered studs, absorbent quilts and gaps at junctions with floors and ceilings effectively sealed.
Floors. Heavy materials are good airborne sound insulants; a resilient layer is needed to resist impact sound; floating floors should not be connected to the structure beneath or surrounding walls; and skirting boards should not have a rigid connection with the floor.[1]

Sound insulation of flats. Many of the principles examined in this chapter can be applied to the implementation of effective sound insulation techniques to both new and existing flats, and these are well described in BRE Digests 333[8] and 334.[9] Many criticisms have been levelled at the poor sound insulation qualities of many large houses which have been converted into flats and BRE Information Paper IP 6/88[10] shows very clearly how this can be achieved with minimum annoyance to the occupants.

A typical timber joist floor will not provide adequate sound reduction and one satisfactory solution shown in BRE Information Paper IP 6/88[10] and illustrated in figure 15.1.4 is to provide an independent ceiling giving a high performance of sound insulation. If the original ceiling has to be removed the reduction in room height can be minimised by positioning the new joists between the original joists. Other alternatives are to reconstruct the floors to form either timber platform floors or timber raft floors as detailed in the same paper and illustrated in figures 15.1.5 and 15.1.6. BRE Digest 333[8] examines the factors affecting the performance of separating walls and recommends, in particular, that direct air paths through or round a separating wall, even in the roof space, should be kept to a minimum by the use of joist hangers where necessary and by careful sealing around any unavoidable penetrations of the wall.
Insulation against external noise. The main criterion for protection against external noise is by means of enclosure, entailing adequate surface mass, the con-

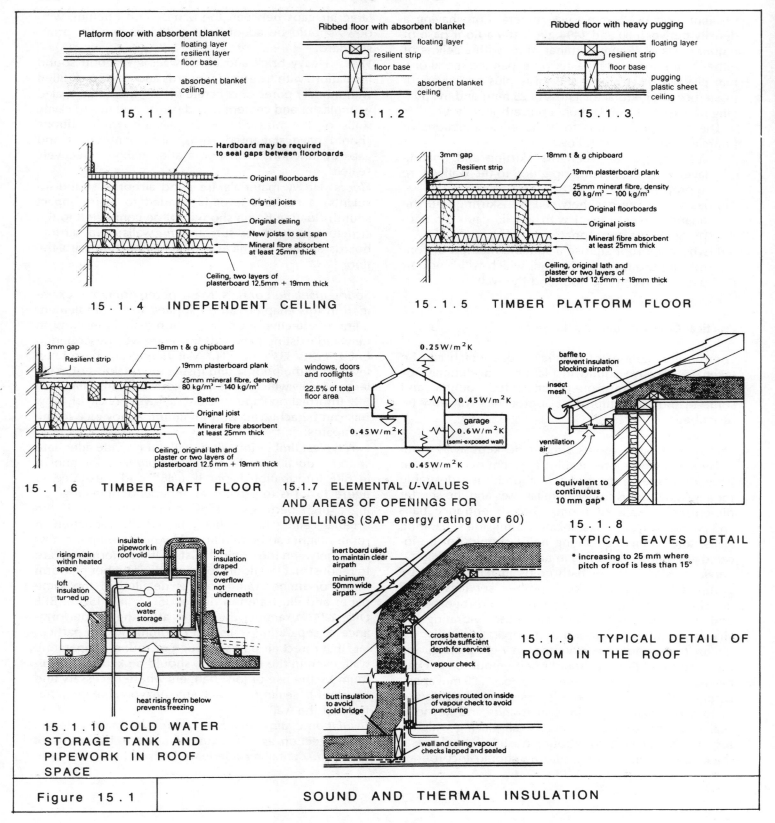

Platform floor with absorbent blanket

floating layer
resilient layer
floor base

absorbent blanket
ceiling

15.1.1

Ribbed floor with absorbent blanket

floating layer

resilient strip
floor base

absorbent blanket
ceiling

15.1.2

Ribbed floor with heavy pugging

floating layer

resilient strip
floor base

pugging
plastic sheet
ceiling

15.1.3

Hardboard may be required
to seal gaps between floorboards

Original floorboards

Original joists

Original ceiling

New joists to suit span

Mineral fibre absorbent
at least 25mm thick

Ceiling, two layers of
plasterboard 12.5mm + 19mm thick

15.1.4   INDEPENDENT CEILING

3mm gap
Resilient strip

18mm t & g chipboard

19mm plasterboard plank

25mm mineral fibre, density
60 kg/m³ – 100 kg/m³

Original floorboards

Original joists

Mineral fibre absorbent
at least 25mm thick

Ceiling, original lath and
plaster or two layers of
plasterboard 12.5mm + 19mm thick

15.1.5   TIMBER PLATFORM FLOOR

3mm gap
Resilient strip

18mm t & g chipboard

19mm plasterboard plank

25mm mineral fibre, density
80 kg/m³ – 140 kg/m³

Batten

Original joist

Mineral fibre absorbent
at least 25mm thick

Ceiling, original lath and
plaster or two layers of
plasterboard 12.5 mm + 19mm thick

15.1.6   TIMBER RAFT FLOOR

windows, doors
and rooflights
22.5% of total
floor area

$0.25 W/m^2 K$

$0.45 W/m^2 K$

garage
$0.6 W/m^2 K$
(semi-exposed wall)

$0.45 W/m^2 K$

$0.45 W/m^2 K$

**15.1.7 ELEMENTAL *U*-VALUES
AND AREAS OF OPENINGS FOR
DWELLINGS (SAP energy rating over 60)**

baffle to
prevent insulation
blocking airpath

insect
mesh

ventilation
air

equivalent to
continuous
10 mm gap*

**15.1.8
TYPICAL EAVES DETAIL**

*** increasing to 25 mm where
pitch of roof is less than 15°**

insulate
pipework in
roof void

rising main
within heated
space

loft
insulation
turned up

cold
water
storage

loft
insulation
draped
over
overflow
not
underneath

heat rising from below
prevents freezing

**15.1.10 COLD WATER
STORAGE TANK AND
PIPEWORK IN ROOF
SPACE**

inert board used
to maintain clear
airpath

minimum
50mm wide
airpath

cross battens to
provide sufficient
depth for services

vapour check

butt insulation
to avoid
cold bridge

services routed on inside
of vapour check to avoid
puncturing

wall and ceiling vapour
checks lapped and sealed

**15.1.9   TYPICAL DETAIL OF
ROOM IN THE ROOF**

| Figure 15.1 | SOUND AND THERMAL INSULATION |

tinuity, uniformity and completeness of the enclosure, and the extra insulation afforded by double leaf enclosure. The serious effect of single windows on the sound reduction of external walls has been described earlier. BRE Digest 338[11] describes how a flat concrete roof will give a sound reduction of about 45 dB, while a pitched slated or tiled roof with an underceiling and glass fibre or mineral wool quilt gives about 35 to 40 dB reduction.

Sound insulation of lightweight dwellings. BRE Digest 347[12] lists the essential requirements for good acoustic performance of dry lightweight construction as

(1) an independent structure for each house;
(2) a double leaf separating wall construction;
(3) an acceptable minimum weight in each leaf;
(4) wide cavity separation between the two leaves;
(5) good sealing of joints.

The principal constructional techniques have been described earlier and are well illustrated in figure 4.12. The sound insulation performance compares favourably with most types of masonry construction.

A comprehensive and authoritative study of sound control for homes has been issued jointly by CIRIA and BRE.[55]

THERMAL INSULATION

General Background

To maintain a constant temperature within a building it is necessary to reduce heat losses and thereby conserve energy and reduce heating costs. The importance of conserving energy is now fully recognised and statutory requirements have progressively increased the thermal insulation standards of new buildings, culminating in the Building Regulations Approved Document L1 (1989).[7]

The heat savings of a typical house where insulation of walls and roof, double glazing and draughtproofing of doors have been carried out effectively could approach 60 per cent, with the cost of the insulating measures recovered over a relatively short period by savings in energy costs and a reduction in the size of the heating plant. In addition, good thermal insulation reduces the risk of surface condensation and, if the ventilation is correctly controlled, the building will remain cooler in summer.

A thermal insulant is a material which restricts the transfer of heat between surfaces at different tempera-

tures and can take many forms, such as rigid preformed (aerated blocks), flexible (fibreglass quilts), loose fill (expanded polystyrene granules), site formed (foamed polyurethane), and reflective (aluminium foil).[6]

Thermal conductivity (k) is a measure of the rate at which heat is conducted through a material under specific conditions and the appropriate unit is W/mK, whereas *resistivity* (r) is the reciprocal of thermal conductivity and thus $r = 1/k$. The thermal conductivity varies with the moisture content and density, but a set of standard values has been formulated, as described later, for calculation purposes.

The thermal transmittance through a building element is measured by its *U-value* or thermal transmittance coefficient which is the rate of heat transfer in watts through 1 m² of a structure when the temperature on each side of it differs by 1°C, expressed in W/m²K.

Insulated buildings can however give rise to defects which did not occur in uninsulated buildings of traditional construction, such as interstitial condensation and damp penetration. Hence it is important to understand fully the physical principles involved and to take the necessary precautionary measures as fully described and illustrated in *Thermal insulation: avoiding risks*,[13] which will be considered later in the chapter.

Statutory Requirements and Procedures

Basic requirements and approaches

Schedule 1 to the Building Regulations 1991[7] in L1 requires that reasonable provision shall be made for the conservation of fuel and power in buildings, and this requirement applies to dwellings and other buildings whose floor area exceeds 30 m². This requirement will be met by the provision of the following energy efficient measures

(a) to limit the heat loss and where appropriate maximise the heat gains through the fabric of the building;
(b) to control as appropriate the output of the space heating and hot water systems;
(c) to limit the heat loss from hot water storage vessels, pipes and ducts.

Approved Document L1 (1995) of the Building Regu-

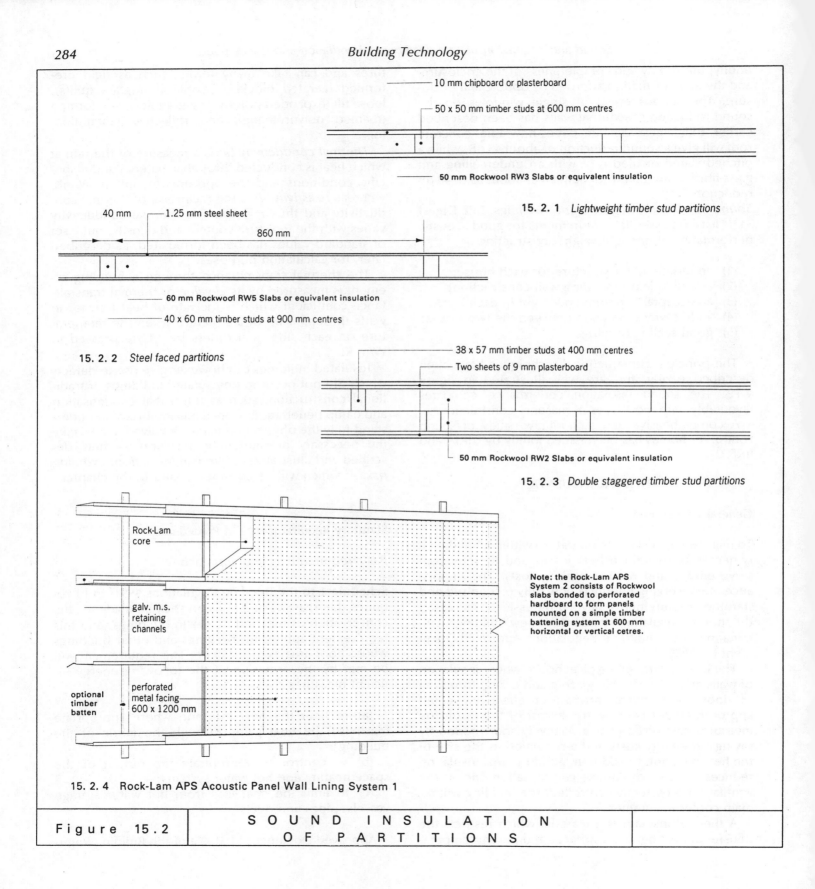

10 mm chipboard or plasterboard

50 x 50 mm timber studs at 600 mm centres

50 mm Rockwool RW3 Slabs or equivalent insulation

15. 2. 1   *Lightweight timber stud partitions*

40 mm   1.25 mm steel sheet

860 mm

60 mm Rockwool RW5 Slabs or equivalent insulation

40 x 60 mm timber studs at 900 mm centres

15. 2. 2   *Steel faced partitions*

38 x 57 mm timber studs at 400 mm centres

Two sheets of 9 mm plasterboard

50 mm Rockwool RW2 Slabs or equivalent insulation

15. 2. 3   *Double staggered timber stud partitions*

Rock-Lam
core

galv. m.s.
retaining
channels

**Note: the Rock-Lam APS
System 2 consists of Rockwool
slabs bonded to perforated
hardboard to form panels
mounted on a simple timber
battening system at 600 mm
horizontal or vertical cetres.**

optional
timber
batten

perforated
metal facing
600 x 1200 mm

15. 2. 4   Rock-Lam APS Acoustic Panel Wall Lining System 1

| Figure   15.2 | SOUND   INSULATION
OF   PARTITIONS | |

lations 1991[7] contains the following three methods of complying with the revised requirements:

1. The elemental method
2. The target *U*-value method
3. The energy rating method.

Elemental method

With this method the dwelling is constructed such that:

1. The *U*-values of walls, floors and roofs do not exceed the following:

Standard *U*-values (W/m²K) for dwellings

Element	For SAP energy ratings of:	
	60 or less	over 60
Roofs	0.2	0.25
Exposed walls	0.45	0.45
Exposed floors and ground floors	0.35	0.45
Semi-exposed walls and floors	0.6	0.6
Windows, doors and rooflights	3.0	3.3

Source: Approved Document L1 (1995)

2. The total area of windows, doors and rooflights does not exceed 22.5 per cent of the total floor area of the dwelling.

An 'exposed' wall or floor is a wall or floor directly exposed to the external air, while a 'semi-exposed' wall or floor is any wall or floor between a heated space and another space having one or more exposed walls not insulated to the required level (usually because this second space is unheated). This approach is illustrated in figure 15.1.7.

Any part of a roof having a pitch of 70° or more may have the same *U*-value as a wall, while for loft conversions in existing dwellings it would be reasonable to have a roof *U*-value of 0.35 W/m²K.

Approved Document L1 (1995) describes how some flexibility is built into the elemental approach to thermal insulation calculations. For example, double glazing may be provided with consequent changes to some of the insulation levels shown earlier. Thus the designer may wish to provide *U*-values of 0.6 W/m²K for walls and 0.35 W/m²K for roofs, when he can either double glaze half the total window area with the exposed walls at 0.6 W/m²K, or double glaze all

windows and have exposed walls at 0.6 W/m²K, roof at 0.35 W/m²K and floor uninsulated. Another approach would be to insulate the ground floor to a *U*-value of 0.35 W/m²K in order to reduce the roof insulation to 0.35 W/m²K. Furthermore a ground floor may achieve the specified *U*-value 0.45 W/m²K without the addition of insulating material, as shown in figure 6.1.4, if the floor is of sufficiently large area. With improved average *U*-values the area of windows and doors in a dwelling can be increased as illustrated in table 3 of Approved Document L (1995).

A common method of achieving the prescribed *U*-values is to provide insulation of a thickness calculated from the tables in Appendix A of Approved Document L (1995) covering walls, roofs and floors. When dealing with walls, roofs and exposed or semi-exposed floors, the procedure is to select the thermal conductivity value of the relevant insulating material and by reference to the appropriate table, read off the base level thickness for the prescribed *U*-value. Next refer to table A15, select the relevant constructional features, add up the thicknesses given and deduct their total from the base level thickness, which will give the minimum thickness that should be used.

Calculation of *U*-values

The *U*-values of a constructional element can be calculated by adding together the thermal resistances of the component parts of the construction and then taking the reciprocal.

The *thermal resistance* of a material is obtained by dividing the thickness in metres of the material by its thermal conductivity (W/mK), and using standard values for resistance of air spaces and surfaces, as shown in the following list.

Exposed walls:	outside surface	0.06 m²K/W
	inside surface	0.12 m²K/W
	air space (cavity)	0.18 m²K/W
Roofs	outside surface	0.04 m²K/W
	inside surface	0.10 m²K/W
	roof space (pitched)	0.18 m²K/W
	roof space (flat)	0.16 m²K/W
Exposed floors	outside surface	0.04 m²K/W
	inside surface	0.14 m²K/W

Table 15.2 gives the conductivity values of some of the more common building materials.

The following example shows how the U-value for an exposed wall can be calculated. The thermal conductivity values in W/mK are shown in brackets following each component. The wall is to be constructed of an outer brick leaf 102.5 mm thick (0.84), 50 mm cavity (0.18 thermal resistance), 50 mm thermal insulation (0.04), 100 mm concrete blocks (0.15), and 13 mm plasterboard (0.16). The overall width of cavity is 100 mm.

Table 15.2 Thermal conductivity of some common building materials (Source: Approved Document L1 (1995)

Material	Density (kg/m²)	Thermal conductivity (W/mK)
Walls (external and internal)		
Brickwork (outer leaf)	1700	0.84
Brickwork (inner leaf)	1700	0.62
Cast concrete (dense)	2100	1.40
Cast concrete (lightweight)	1200	0.38
Concrete block (heavyweight)	2300	1.63
Concrete block (mediumweight)	1400	0.51
Concrete block (lightweight)	600	0.19
Normal mortar	1750	0.80
Fibreboard	300	0.06
Plasterboard	950	0.16
Tile hanging	1900	0.84
Timber	650	0.14
Surface finishes		
External rendering	1300	0.50
Plaster (dense)	1300	0.50
Plaster (lightweight)	600	0.16
Calcium silicate board	875	0.17
Roofs		
Aerated concrete slab	500	0.16
Asphalt	1700	0.50
Felt/bitumen layers	1700	0.50
Screed	1200	0.41
Stone chippings	1800	0.96
Tile	1900	0.84
Wood wool slab	500	0.10
Floors		
Cast concrete	2000	1.13
Metal tray	7800	50.00
Screed	1200	0.41
Timber flooring	650	0.14
Wood blocks	650	0.14
Insulation		
Expanded polystyrene (EPS) slab	25	0.035
Mineral wool quilt	12	0.040
Mineral wool slab	25	0.035
Phenolic foam board	30	0.020
Polyurethane board	30	0.025

Note
If available, certified test values should be used in preference to those in the table.

The combined resistance of the wall is calculated as follows:

	m²K/W
resistance of outside surface	0.06
resistance of brick leaf = 0.1/0.84	0.12
resistance of cavity	0.18
resistance of insulation material = 0.05/0.04	1.25
resistance of block leaf = 0.1/0.15	0.67
resistance of plasterboard = 0.013/0.16	0.08
resistance of inside surface	0.12
total resistance (m²K/W)	2.48

U-value = 1/2.48 = 0.40 W/m²K

Target U-value Method

The Building Regulations requirement will be met if the calculated U-values do not exceed the following targets:

(a) For dwellings with SAP energy ratings of 60 or less:

$$\text{Target } U\text{-value} = \frac{\text{total floor area} \times 0.57}{\text{total area of exposed elements}} + 0.36$$

(b) For dwellings with SAP energy ratings of more than 60:

$$\text{Target } U\text{-value} = \frac{\text{total floor area} \times 0.64}{\text{total area of exposed elements}} + 0.4$$

The Approved Document L (1995) provides guidance on the calculation of areas. For instance, the dimensions for the areas of walls, roofs and floors are measured between finished internal faces of the external elements of the building including any projecting bays. Roofs are measured in the plane of the insulation, while floor areas include non-usable space such as builders' ducts and stair wells.

An exposed element is an element exposed to the outside air (including a suspended floor over a ventilated or unventilated void) or an element in contact with the ground.

The total area of exposed elements comprises:

(a) the area of fabric including windows, doors and rooflights exposed externally to outside air, plus

(b) the area of the ground floor.

Example calculations for target U-values and average U-values are given in Appendix F of Approved Document L (1995), and an outline example follows to show the approach used.

The example selected covers a two-storey detached dwelling measuring 7.7 m × 7.3 m to inside faces of external enclosing walls × 5 m high to eaves. It is proposed to use the target U-value method with U-values for the walls and roof slightly higher (worse) than would be required by the elemental method. The SAP energy rating is to be more than 60. The cavity walls are to have dry linings and the windows and doors are to have metal frames with thermal breaks and sealed double glazing with 12 mm gaps.

The relevant details for this dwelling are now listed.

Exposed element	Exposed surface area (m²)	U-value (W/m²K)	Rate of heat loss per degree (area × U-value) (W/K)
Floor	56.2	0.45	25.29
Windows	24.8	3.3	81.84
Doors	3.8	3.3	12.54
Walls	121.4	0.5	60.7
Roof	56.2	0.3	16.86
Totals	262.4		197.23

Using the formula given previously to obtain the target U-value for dwellings with SAP energy ratings of more than 60 and inserting the relevant data for the detached dwelling gives the target U-value.

$$\frac{112.4 \times 0.64}{262.4} + 0.4 = 0.67 \text{ W/m}^2\text{K}$$

The average U-value for the dwelling is given by the ratio of the two values:

$$\frac{\text{Total rate of heat loss per degree}}{\text{Total external surface area}}$$

The relevant figures from the previous details of the dwelling can now be interposed to give the average U-value.

$$\frac{197.23}{262.4} = 0.75 \text{ W/m}^2\text{K}$$

Hence the proposed design fails to meet the requirements and modifications are required. Possible courses of action include improving the thermal resistance of the exposed walls and windows and doors and incorporating the benefits of solar gain and more efficient space heating.

In this example, the optional method of accounting for solar gains can be implemented to advantage. The target U-value equations are based on a calculation assuming equal distribution of glazing on north and south elevations. Where the area of glazing on the south elevation exceeds that on the north, the total window area included in the calculation can be reduced to take account of the benefits

of solar gains. This is taken as the actual window area less 40 per cent of the difference in area of south-facing glazing ±30° and north-facing glazing ±30°. In this example there is a predominance of south-facing glazing and this results in a reduction of exposed surface area of windows in excess of 2 m².

In a similar manner it was possible to improve the thermal resistance of the walls and windows by improving the specification in each case to give lower U-values and much reduced heat losses. Another possible approach is to use a heating system with a higher efficiency. The target U-value is based on a calculation assuming a gas- or oil-fired central heating system with a seasonal efficiency of at least 72 per cent. If the dwelling is to be heated in a more efficient manner (having regard to both heating system efficiency and primary energy consumption) a proportion of the benefits so derived can be taken into account by increasing the target U-value by up to 10 per cent. For example, if the orthodox boiler were to be replaced by a condensing boiler, thus increasing the seasonal efficiency to 85 per cent, the target U-value could be increased by the full 10 per cent allowance.

After adjusting the figures to take cognisance of solar gain and improved thermal efficiency of walls and windows, the exposed surface area was reduced to 260.12 m² and the rate of heat loss per degree to 173.49 W/K. The average U-value for the revised proposals then became:

$$\frac{173.49}{260.12} = 0.67 \text{ W/m}^2\text{K}$$

In practice, it will be necessary to make all the detailed computations with possible alternatives, probably needing referral to the client where specification changes are involved, and then to prepare a schedule of revised data with the alterations suitably highlighted.

Energy Rating Method

Approved Document L (1995) requires that all dwellings provided as new construction or by way of material changes of use, which include building work, must be given energy ratings using the Government's Standard Assessment Procedure (SAP). There is however no obligation to achieve a particular SAP energy rating. Nevertheless, higher levels of insulation are justified for new dwellings having SAP ratings of 60 or less, whereas a SAP rating of 80 to 85 (depending on dwelling size) will demonstrate compliance with the energy conservation requirements.

The energy rating method entails a calculation for dwellings using the SAP as illustrated in Appendix G of Approved Document L (1995), which permits the use of any valid energy conservation measures. The procedure is very exhaustive and takes account of ventilation rate, fabric losses, water heating requirements, internal heat gains and solar gains. The operative energy conservation requirement will be met if the SAP energy rating for the dwelling (or each dwelling in a block of flats or converted building) is not less than the appropriate figure as shown in the following table.

SAP energy ratings to demonstrate compliance

Dwelling floor area (m²)	SAP energy rating
80 or less	80
more than 80 up to 90	81
more than 90 up to 100	82
more than 100 up to 110	83
more than 110 up to 120	84
more than 120	85

Source: Approved Document L (1995)

When using the calculation procedures in the target *U*-value and energy rating methods, it may be possible to achieve satisfactory solutions where the *U*-values of some elements are worse than those given in the table on page 285 (standard *U*-values for dwellings). However, as a general rule the *U*-values of exposed walls and exposed floors should not be worse than 0.7 W/m²K and the *U*-values of roofs should not be worse than 0.35 W/m²K.

Provision should be made to limit the thermal bridging which occurs around windows, doors and other wall openings. This is necessary in order to avoid excessive additional heat losses and the possibility of local condensation problems. When using the energy rating method the designs for lintels, jambs and sills should perform no worse than those described and illustrated in Approved Document L (1995).

The Approved Document L (1995) also provides much sound advice on methods of securing efficient controls of space heating and hot water storage systems, and insulation of vessels, pipes and ducts.

Appendix G of Approved Document L (1995) describes how SAP ratings are influenced by the following range of factors that contribute to energy efficiency.

- thermal insulation of the building fabric
- efficiency and control of the heating system
- ventilation characteristics of the dwelling
- solar gain characteristics of the dwelling
- the price of fuels used for space and water heating.

The same appendix contains a SAP worksheet for calculating the rating of a dwelling accompanied by a set of valuable tables, covering such aspects as hot water energy requirements, heating system efficiency, and lighting and heating factors. The SAP worksheet occupies four pages and is extremely comprehensive. Appendix G also provides examples of SAP ratings for selected dwelling types, and three typical examples are now given.

Example 15.1 Two-bedroom mid-terrace house with electric storage heaters

Element	Description	Area	*U*-value
Wall	Brick/cavity/dense block with 70 mm blown fibre cavity insulation	30.3	0.44
Roof	Pitched roof, 100 mm insulation between joists 50 mm on top	27.3	0.25
Ground floor	Suspended timber, 25 mm insulation	27.3	0.37
Windows and doors	Double glazed (6 mm gap), wooden frame	11.7	3.3
Heating	Electric storage heaters (efficiency 100%)		

SAP rating = 68

Example 15.2 Four-bedroom detached house with condensing boiler

Element	Description	Area	*U*-value
Wall	Brick/partial cavity fill/ medium density block	116.5	0.45
Roof	Pitched roof, 100 mm insulation between joists 50 mm on top	50	0.25
Ground floor	Suspended timber, 35 mm insulation	50	0.45
Windows and doors	Double glazed (6 mm gap), wooden frame	24.9	3.3
Heating	Central heating with gas condensing boiler (efficiency 85%)		

SAP rating = 85

Example 15.3 Two-bedroom bungalow with gas boiler

Element	Description	Area	*U*-value
Wall	Brick/cavity/aerated concrete block with insulated plasterboard	64.2	0.45
Roof	Pitched roof, 100 mm insulation between joists 50 mm on top	56.7	0.25
Ground floor	Concrete suspended beam and medium density concrete block, 25 mm insulation	56.7	0.45
Windows and doors	Double glazed (6 mm gap), PVC-U frame	13.4	3.3
Heating	Central heating with gas boiler (efficiency 72%)		

SAP rating = 68

Cavity Wall Insulation

Statutory requirements

The thermal insulation value of a cavity can be increased considerably by inserting a suitable insulant to assist in meeting the fuel conservation requirements of Part 1 of Schedule 1 to the Building Regulations 1991[7] and Approved Document L1 (1995),[7] whereby exposed walls need to have a *U*-value of 0.45. Part D of Schedule 1 to the same Regulations states that if insulating material is inserted into a cavity in a cavity wall, reasonable precautions shall be taken to prevent the subsequent permeation of any toxic fumes from that material into any part of the building occupied by people. This applies particularly to formaldehyde fumes given off by urea–formaldehyde (UF) foams.

Provisions Satisfying the Cavity Insulation Requirements

Approved Document D1 (1985) provides that a cavity wall may be insulated with UF foam where

(a) the inner leaf of the wall is built of masonry (bricks or blocks), and
(b) the suitability of the wall for foam filling is assessed before the work is carried out in accordance with BS 8208,[15] and
(c) the person carrying out the work holds or operates under a current BSI Certificate of Registration of Assessed Capability for the work, and
(d) the material is in accordance with the relevant recommendations of BS 5617,[16] and
(e) the installation is in accordance with BS 5618.[17]

Alternative Insulation Arrangements

Various alternative approaches are available for the provision of cavity wall insulation. One approach is to use 50 mm thick cavity insulation batts filling standard width cavities with materials such as glass fibre or rock fibre, in accordance with BS 6676.[18] Another method comprises partial cavity fills using materials such as 25 mm boards of expanded polystyrene, polyurethane foam, polyisocyanurate foam or mineral fibre. These are normally fixed against the cavity face of the inner leaf either with clips attached to the wall ties or with independent clips or nails.[19]

However, BRE Digest 358[20] recommends the use of rigid urethane foams, which include polyurethane and polyisocyanurate foams, that have been manufactured using HCFCs (hydrochlorofluorocarbons) instead of CFCs (chlorofluorocarbons) as blowing agents and, in the longer term, HFC (hydrofluorocarbon) alternatives containing no chlorine, in order to reduce and subsequently to eliminate the ozone depletion potential.

Insulating Materials and their Installation

BRE Digest 236[21] describes the principal materials used in cavity insulation and the problems that can arise in their installation, and the main findings are now summarised.

Glass and Rock Fibre Slabs

Although the slabs themselves are very effective barriers to water, the joints between them may be vulnerable if the slabs are not installed with care. Wide gaps between adjacent slabs must be avoided and any mortar droppings prevented or removed, necessitating good supervision.

Expanded Polystyrene Board

Insulating boards, usually 25 mm thick, are fixed to the inner leaf, leaving a clear cavity between the outer leaf and the boards; special fixings are available for this purpose. Again, good site supervision is important to make sure that the boards are restrained by the fixings against the inner leaf and do not lean across the cavity against the outer leaf. The polystyrene should ideally be produced using a HCFC as a blowing agent.[20]

Urea–formaldehyde (UF) Foam

This is the most commonly used form of cavity fill in the United Kingdom, and accounted for over half-a-million installations in 1980. The material is a low density cellular plastics foam which is produced by foaming together in a gun a mixture of water-based resin solution, a hardener-surfactant solution, and compressed air. The foam has a consistency of shaving soap and is injected into the cavity where it subsequently hardens and dries. As it dries it will normally shrink and this will lead to fissuring. Occasionally the fissures are able to provide a bridge which will allow water to cross from the outer leaf to the inner leaf.

When the foam is injected, and for some time after, it gives off fumes of formaldehyde gas. For a normal cavity wall of two leaves of masonry, plastered on the inner leaf, these fumes will rarely enter the house. However, there is a risk of fumes entering the dwelling when the inner leaf is vapour permeable, as for example with plasterboard or fair faced brickwork, particularly if the outer leaf is impermeable. Where the cavities to be filled are very wide (100 mm or more) the problem is more likely to occur. Formaldehyde vapour can cause irritation to the eyes and nose.

Man-made Mineral Fibre

This consists of fibres coated with a water repellant which are blown as tufts into the cavity where they form a water-repellant mat. This method is generally more expensive than urea-formaldehyde, and should comply with BS 6232.[22]

Polyurethane Granules/Beads

These are irregularly shaped granules usually between 5 mm and 20 mm across, made by chopping waste rigid polyurethane foam. The material should be kept away from hot surfaces such as flues built into or crossing the cavity.

Expanded Polystyrene (EPS) Loose Fills

Expanded polystyrene beads are white spheres with a diameter between 2 mm and 7 mm. They are extremely free running and so very few filling holes are necessary. The free running nature of this insulant can lead to an unnoticed escape of material where there are holes in the inner leaf. The fill should be kept away from hot flues and PVC-coated electrical cables.

Polyurethane Foamed *in situ*

This consists of two liquid components which are mixed and injected into the cavity, where they spontaneously foam and rise to fill the space. The foam adheres strongly to masonry and does not shrink. However, the material cost is very high.

Constructional Problems

Mortar in an insulated cavity not only affects the thermal properties of the wall but can also lead to rainwater reaching the inner leaf. With full cavity fill the major problems are mortar extrusions from the outer leaf squeezing into the joints between the batts, and mortar dropped onto the batts during bricklaying or scraped off the wall when the batts are placed in position. These can direct rainwater, which normally trickles down the cavity face of the external leaf, into the insulation and hence to the inner leaf. Essential principles for total cavity fill are to keep the insulation dry during installation, to keep it free from mortar debris and to design so that any water entering the cavity is directed away from the insulation and the inner leaf.[23]

With partial cavity fill, the bridging paths are the same as for an unfilled cavity, with the addition of a further route via any loose or displaced insulation boards. Mortar snots bridging the cavity can accentuate the problem.

Difficulties can also arise when cutting slabs to fit wall ties at non-standard centres and above and around window openings, and adjoining damp-proof courses. The precautions to be taken when installing mineral fibre batts are further described in BRE Defect Action Sheet 17.[24]

Controls for Space Heating and Hot Water Supply Systems

Building Regulations Approved Document L (1995), section 1.33, prescribes that for gas and oil fired hot water central heating systems for dwellings there shall be provided a room thermostat or thermostatic radiator valves or other equivalent form of sensing device, to control the output from the heating system.

Section 1.34 gives the following requirements for the controls for hot water storage vessels:

(a) heat exchanger in storage vessel to comply with BS 1566, BS 3198 or equivalent;

(b) a thermostat is to be provided which shuts off the supply of water when the storage temperature is reached; and

(c) a time switch is to be provided which will shut off the supply of heat when there is no demand for hot water.

Section 1.37 covers the insulation of hot water storage vessels, pipes and ducts, in the following manner:

Hot water storage vessels: these shall be insulated to limit the heat loss to 1W/litre by testing using the method in BS 1566, part 1. One way of achieving this is to provide a storage vessel complying with either BS 699[25], or BS 1566[26] or BS 3198[27] or a hot water storage cylinder insulating jacket complying with BS 5615.[28] Where an insulating jacket is used the segments of the jacket shall be taped together to provide an unbroken insulation cover for the storage vessel.

Pipes and ducts: unless the heat loss from a pipe or duct contributes to the useful heat requirement of a room or space, the pipe or duct should be insulated so that

(a) for pipes, the insulation material should have a thermal conductivity not greater than 0.045 W/mK and a thickness equal to the outside diameter of the pipe up to a maximum of 40 mm, or

(b) alternatively for pipes, and in the case of ducts, the insulation should meet the recommendations of BS 5422.[29] The relevant code of practice for the thermal insulation of pipework and equipment is contained in BS 5970.[30]

Avoidance of Risks in Thermal Insulation

The BRE Report *Thermal insulation: avoiding risks*[13] provides comprehensive advice on the technical problems arising from improved thermal insulation and how they can best be avoided. For example, certain parts of the construction remain colder and thus make interstitial condensation more likely, and changes to traditional forms of construction to improve thermal insulation can lead to damp penetration. Some of the more important aspects relating to the principal building elements will now be considered.

Roofs. With pitched tiled or slated roofs, whether insulation is placed at ceiling level, rafter level or a combination of the two, as with rooms in roof spaces, the technical risks are essentially the same, namely condensation in roof spaces, cold bridging at roof perimeters, fire hazards and freezing pipes and tanks. The roof space requires ventilating to the outside air in accordance with BS 5250,[31] care should be taken to ensure that ventilation is not blocked by insulation at the eaves or by sagging sarking felt, and 3 to 4 mm mesh should be fixed across ventilation openings to prevent the entry of insects, as shown in figure 15.1.8. When insulation is provided at rafter level, a vapour control layer often of 500 gauge polythene should be inserted on the warm side of permeable insulation for rooms in the roof space, as illustrated in figure 15.1.9. Further examples of good practice in insulating flat and pitched roofs are illustrated in figures 7.6, 7.7, 7.8 and 7.9.

All pipes should be located in heated spaces wherever possible, such as below loft insulation or below the ceiling, and all pipes, including overflows, in unheated spaces shall be adequately insulated. The top and sides of cold water storage tanks in roof spaces require insulating, but omitting insulation from directly under the tank to allow warmth from below to reach the bottom of the tank. The space below the tank base can be insulated by turning up the loft insulation against the tank insulation and the rising main should desirably be located within the insulated enclosure for the tank as illustrated in figure 15.1.10.

The precautions to be taken with the three main types of flat roof (cold deck, warm deck: sandwich, and warm deck: inverted) are well described and illustrated in the BRE Report.[13]

Walls. When filling the cavities of exposed walls with insulation, the precautions described and illustrated in the BRE Report[13] should be taken to avoid cold bridges at openings, spread of fire gases where cavities abut insulation, and interstitial condensation with timber frame construction. Furthermore, the selection of wall construction will be influenced by the local exposure to wind driven rain. Some examples of good cavity wall insulation practice are illustrated in figures 4.5 and 4.6.

Windows. The three main areas of technical risk with double glazing are

1. deterioration of edge seals in double glazed units;
2. deformation of undersized frames;
3. condensation within double windows.

The precautionary measures to be taken in each case are well described and illustrated in the BRE Report.[13]

Floors. A screeded finish may be used above any type of concrete slab, and where supported by insulation board the latter must be totally rigid and the screed shall be of adequate thickness and well compacted to resist bending. When screeds are being laid, there is a danger that the insulation board and the membrane above the insulation may be damaged.

With cavity insulated walls the cavity insulation should commence below the underside of the slab, and where the slab is built into the inner leaf of the wall this leaf should be constructed with blocks of high insulation value, and similarly below damp-proof course level where the slab is supported on the ground.

Floors incorporating insulation beneath chipboard and plywood panels are often used as 'floating floors' to control impact sound in flats, as described earlier in the chapter. When used for thermal insulation purposes, the insulant need not be resilient and there are positive advantages in it being rigid. An important technical aspect is to prevent moisture weakening the board finish, whether from within the structure or from spillages.

An insulated floor finish can be constructed with

1. composite panels using chipboard and a rigid insulant;
2. loose laid systems with the chipboard or plywood boards and the rigid insulant installed separately;
3. timber battens supporting boards with resilient insulant between the battens.[13]

The BRE Report[13] recommends that the surface of the sub-floor should be flat, and tongued and grooved chipboard panels be fitted tightly together and the joints glued using a PVA or mastic adhesive and temporary wedges at the perimeter in the gap provided for moisture expansion. Where the insulant supports the boarding, it must have sufficient compressive strength to avoid depressions or indentations under load.

In timber suspended floors with insulation inserted between the joists, the insulation is normally quilt supported in plastics mesh, or rigid boards supported on battens, corrosion resisting nails or clips.[13] Further examples of good practice in the insulation of floors are illustrated in figures 6.1, 6.2 and 6.3.

Domestic Draughtproofing

BRE Digest 319[32] describes how draughtproofing the doors, windows and other sources of excessive air leakage of a dwelling can be an effective and relatively inexpensive method of improving comfort and reducing heat loss by natural ventilation. The doors and windows being considered are existing components not originally designed for draughtproofing, which applies to many millions of dwellings in the United Kingdom. Hence the products to be used for draughtproofing have to fill a wide range of gap sizes, be durable and permit easy opening and closing. The digest describes their basic requirements, general characteristics and the situations where they can be cost effective.

The digest recommends that draughtproofing in a dwelling should be restricted to worthwhile and safe

measures. For instance, provided the basic ventilation requirements are satisfied, it will generally be safe to draughtproof doors, windows and loft hatches. In general, it is not worthwhile to draughtproof doors and windows where the average gap is below 0.5 mm, unless there are pronounced draughts. Single-glazed components, and metal frames, can create draughts by the local cooling of indoor air, and these can be reduced by providing close-fitting heavy curtains or double glazing, or fitting secondary glazing windows at the outset.

The main benefits gained from draughtproofing a dwelling are fewer draughts and a reduction in excess ventilation rates, the latter resulting in reduced heating energy whilst retaining the same temperatures. In 1991, the cost of draughtstrip materials for a dwelling was in the order of 65p to £1.25 per linear metre, and the average cost of draughtproofing the doors and windows in a house was about £65 on a DIY basis or about £260 if undertaken by a contractor, producing savings in heating costs of about £16 per annum.

Energy Efficiency in Dwellings

Calculating energy requirements

BRE Digest 355[33] describes how calculations of energy requirements in dwellings are used because measurements of energy performance are rarely practicable, and the actual consumption varies between households. Domestic energy calculations need to take account of many factors including climate, physical characteristics of the dwelling and its heating system, the way in which the house is used and the internal temperatures and use of domestic appliances. The BRE Domestic Energy Model (BREDEM) has been developed to make realistic estimates of annual needs simply and conveniently, based on experience gained from measurements made in a large number of occupied dwellings.

Factors affecting energy efficiency

Heat losses through the fabric of the building constitute a major factor, termed fabric heat loss rate. In addition heat is also lost through the replacement of heated air by fresh air drawn from outdoors (ventilation heat loss rate). Together, fabric and ventilation heat loss rates make up the specific heat loss rate, which is a good measure of the thermal performance of the building and for comparison with that of other buildings of similar size. Other factors affecting energy efficiency include activities such as cooking, water heating and the use of electrical appliances, in addition to fuel costs.

The efficiency of a heating system is the ratio of energy it produces to that it consumes, usually expressed as a percentage. The benefits of good insulation can be offset by an inefficient heating system while, conversely, a building which is difficult to heat can benefit from a very efficient heating system.

Securing energy efficiency in housing

For new buildings the Building Regulations set minimum standards based upon cost effectiveness calculated using typical fuel and building costs. These may need adjusting for higher fuel costs, higher temperatures, or unusual methods of construction.

With existing buildings there are many opportunities for improving energy efficiency. Some can be applied without disruption to the occupants and are cost effective, such as pitched roof insulation, hot water cylinder insulation, cavity wall insulation, and draughtproofing. Further opportunities arise when buildings are being refurbished or components replaced and these include double glazing when windows are replaced, solid wall insulation when re-rendering or rain screening, ground floor insulation when floors are renewed, and high efficiency boilers and controls when boilers are replaced or central heating installed.[33] The cost effectiveness of some common measures are shown in table 15.3, and further examples are given in *Building Economics*.[56]

Extensive guidance on securing maximum efficiency in timber frame housing can be obtained from *Energy efficient housing – a timber frame approach*.[34]

The Government's Standard Assessment Procedure (SAP) for energy efficiency ratings now forms an integral part of revised Building Regulations, Approved Document L (1995) with the following requirements:

(1) All Building Regulations applications for new residential development must include an energy rating.

Table 15.3 Energy efficiency in existing dwellings (Source: BRE Digest 355)

Measure	Typical payback period	Applications and comments
Hot water cylinder jacket	6 to 12 months	All dwellings with hot water storage
Loft insulation	1 to 3 years	All dwellings with accessible lofts
Condensing boilers	2 to 4 years	At replacement of worn-out boilers
Draughtproofing windows and doors	2 to 10 years	All dwellings — subject to need to maintain adequate ventilation
Cavity wall insulation	4 to 7 years	Dwellings with cavity walls not subjected to severe driving rain
Double glazing	10 to 12 years	Cost effective when windows need replacing

(2) The energy rating must follow the Government's Standard Assessment Procedure (SAP) on a scale of 1 to 100.

(3) If the proposed property fails to achieve a SAP rating not exceeding 60, a more stringent set of *U*-value targets must be used.

(4) If the proposed property achieves a SAP rating of 80 or more, it may not be necessary to meet the *U*-value targets.

(5) The 'energy rating' approach to show Approved Document L compliance is based on the SAP calculation.

These new arrangements were incorporated in the 1995 edition of Approved Document L and Appendix G to the Approved Document includes worked examples of this Method.

DAMPNESS

General Background

As outlined in chapter 4, damp penetration is one of the most serious defects in buildings, as apart from causing deterioration of the structure, it can also result in damage to finishings and contents and even adversely affect the health of the occupants. It can arise in a variety of different ways, including direct penetration through the structure, faulty rainwater disposal, faulty plumbing, rising damp and dampness in solid floors. It is vital that all new buildings shall be soundly designed and constructed with a view to making them entirely waterproof.

Paragraph C4 of Schedule 1 to the Building Regulations 1991[7] requires the walls, floors and roof of new buildings to resist the passage of moisture to the inside of the building, and Approved Document C4 (1991) describes and illustrates satisfactory technical solutions covering floors, walls and roofs, many of which have been examined in the relevant chapters of the book.

As highlighted by Addleson,[35] diagnosis is often difficult as there may often be more than one cause, and the point of emission of the dampness may be some distance from its source. For example, movement of a building can result in cracking and subsequent water penetration can cause internal dampness, but the dampness could have entered the building in ways other than through the cracks. Water from a leaking service water pipe can travel along the pipe and enter the interior of the building at a point some distance from the leak.

Richardson[36] has rightly emphasised that building designs should take account of normal sources of moisture such as penetrating, rising and condensing dampness, but that excessive moisture can arise in other ways which may not have been anticipated at the design stage, such as leaks and floods, and these will probably require special attention when they occur.

It should also be borne in mind that in a traditionally constructed house several tonnes of water are introduced into the walls during bricklaying and plastering, as described in chapter 4. The walls will often remain damp until a summer season has passed and as the moisture dries out from inner and outer surfaces it may leave deposits of soluble salts or translucent crystals.

The measurement of moisture in walling materials can be achieved using a carbide moisture meter, while an electrical resistance meter will give accurate measurements of the moisture content of timber within limited ranges and indications of moisture in other materials, such as stone, brick and concrete. Another approach is to secure the accurate measurement of moisture content and hygroscopicity levels in samples of brick, concrete, tiles, plaster and other materials by laboratory testing as described by Hollis.[37]

Walls

Building regulation requirements

A solid external wall in conditions of very severe exposure may be constructed of brickwork at least 328 mm thick, dense aggregate concrete blockwork at least 250 mm thick, or lightweight aggregate or aerated autoclaved concrete blockwork at least 215 mm thick, with in each case the exposed face rendered in two coats finishing not less than 20 mm thick with a scraped or textured face to satisfy the Building Regulation requirements for solid walls. The rendering should be 1:1:6 (cement:lime:sand), unless the blocks are of dense concrete aggregate when the mix way be 1:½:4, in accordance with Approved Document C4 (1991).

With external cavity walls the outer leaf may be built of masonry (bricks, blocks, stone or cast stone) and the cavity shall be at least 50 mm wide, and bridged only by wall ties or damp-proof trays, designed and constructed to prevent moisture being carried to the inner leaf. The inner leaf shall be of masonry or a frame with a suitable lining (Approved Document C4:1991).

Cavity walls

Cavity walls often fail to prevent water penetration in practice because of poor construction on site. The cavities are frequently bridged by mortar droppings or extrusions (snotting), damp-proof membranes are punctured and there are inadequate damp-proofing arrangements around door and window openings. It is also necessary to ensure that the horizontal damp-proof course in the wall links effectively with the water-proof membrane at ground floor level, and that the damp-proof course is laid at least 150 mm above finished ground level. Details of suitable damp-proof courses and the constructional details of cavity walls are described and illustrated in chapter 4.

Ground levels around existing houses tend to be raised by the addition of garden soil and the laying of new paving surrounding the building on top of the original. Where the clearance between external levels and the damp-proof course is insufficient, it is necessary to lower the ground levels or provide a suitable sized impervious channel laid to fall to surface water gullies.[38]

Problems of damp penetration can also arise if the masonry is highly permeable, the bed joints and perpends are not properly filled or there are small cracks between the mortar joints and the masonry units. Window openings provide a vulnerable point for the penetration of moisture and filling the cavities with insulation tends to highlight the weaknesses. The water may overflow from the ends of the cavity tray over a lintel and penetrate to the inner leaf or the cavity tray may be damaged during construction. Water may also penetrate between the window jamb and frame where the mastic pointing is insufficient or has lost its adhesion or the vertical damp-proof course is wrongly positioned or defective.

Solid walls

With solid walls it is essential to have adequate thickness to resist driving rain and the masonry should have a low moisture absorption rate. A generous overhang at the eaves will reduce water penetration into the walls and gutters and downpipes should be thoroughly cleaned periodically, particularly after the autumn falls of leaves, to avoid heavy discharges of rainwater during severe storms. Useful information on damp-proofing old solid walls and external waterproofing treatments are given in chapter 4.

Rising damp

In older buildings damp may rise up walls to heights in excess of 1 m because of the lack of damp-proof courses. The height of damp penetration depends on several factors, such as the pore structure of the wall, degree of saturation of the soil, rate of evaporation from wall surfaces and presence of salts in the wall. In newer buildings rising damp may occur through a defective damp-proof course, the bridging of the damp-proof course by a floor screed internally or by an external rendering or pointing, path or earth outside the building, or mortar droppings in the cavity. Damp may also penetrate a solid floor in the absence of a damp-proof membrane.[39] These sources are well illustrated in BRE Digest 245.[40]

BRE Digest 245[40] describes how measurements of surface moisture are in themselves no positive indication that a genuine rising damp problem exists. The electric meters commonly used by surveyors are responsive to both the amount of moisture present and to the salt concentration and cannot distinguish between the two, although they can identify areas where further investigation is necessary. This normally involves the taking of samples of bricks, blocks or mortar at some depth in the wall, from which an accurate measurement of the moisture content of the wall can be obtained, together with an indication of the influence of any hygroscopic salts that may be present, using the direct weighing and drying method or a carbide meter.

Damp-proofing Basements

Basement floors and walls are usually in direct contact with the ground and are therefore vulnerable to dampness, as described in chapter 3. Where houses are being rehabilitated and basements which were used originally as utility rooms are being converted to living areas, a careful assessment is required. A major criterion is that living areas must be free from dampness as this can damage finishes and decorations, allow

Table 15.4   Selection of damp-proofing system for basement (Source: BRE Good Building Guide 3)

System	Comparative cost of material and installation	Durability[1]	Use in high water pressure or high water table areas	Specialist contractor needed	Space requirement
Drained cavity	Medium–high	Life of building	Not recommended	No	High[2]
Mastic asphalt	High	Life of building	Yes	Yes	Medium[2]
Cementitious render or compound	Medium	Life of building	Yes	Recommended	Low[3]
Self-adhesive membrane	Medium	Life of building	Yes	Not usually	Medium
Liquid-applied membranes	Medium	Life of building	Yes	Not usually	Medium
Dry lining	Low	20 years	No	No	No

[1] If correctly applied and structure is stable (and provided the dpm is not accidentally punctured).
[2] Height penalty in addition to loss of area.
[3] Without addition of optional lining wall.

mould growth and may result in the deterioration of structural timbers.

There are several techniques and proprietary products available for damp-proofing basements. A traditional approach was to build a drained cavity in the wall, but the more usual current approach is to provide a damp-proof membrane envelope around the basement as illustrated in figure 3.5.1. A selection of common envelope treatments are described and illustrated in BRE Good Building Guide 3,[41] and they all need to be built into a stable structure and usually require to be restrained against possible water pressure in the ground and protected against accidental puncturing. Dry lining offers an alternative option where the dampness is minimal.

However, not all damp-proofing systems are equally suitable for a particular situation. For example, building type, internal layout, position of openings, drainage layout and height of water table may all influence the choice of method. Table 15.4 provides guidance on the selection of method to be used. BRE surveys have shown that a large number of damp-proofing failures were associated with poor detailing.

Pitched Roofs

The first sign of a leaking roof is often a damp patch on a ceiling or in the top corner of a wall. Common causes are defective flashings to upstands, cement fillets which have shrunk or broken away from the adjoining surfaces, choked or defective gutters or slipped or broken tiles or slates. Defective flashings usually need redressing but perished zinc flashings will require replacement, preferably with a more durable material. Defective cement fillets should be replaced with metal flashings. Choked gutters need cleaning and checking to ensure that they are satisfactory, while coating the internal surfaces of parapet and valley gutters with bituminous composition may extend their useful lives.[42]

Rain and snow may penetrate a pitched roof because the slates or tiles are laid at too flat a pitch. The problem is aggravated on exposed sites and the use of a flatter bellcast at eaves to improve appearance can create a vulnerable condition at the point of greatest rainwater run-off. For example, plain tiles should not be laid to a pitch flatter than 40° and in extreme cases may require stripping and replacing with single lap interlocking tiles. Poor quality tiles may be subject to

severe lamination and galvanised nails are unlikely to last the life of the slates or tiles. Problems of moisture penetration may also occur with sarking felt which sags excessively between the rafters or does not extend into the eaves gutter or over a barge board.[42]

Flat Roofs

Extensive publicity has been given to the numerous failures of flat roofs resulting in substantial water penetration. As emphasised in chapter 7, most of these failures could have been avoided if the design principles outlined in BRE Digest 312[43] had been implemented.

Whether constructed of timber or concrete, flat roofs need to be laid to adequate falls and to incorporate a vapour barrier (vapour control layer). The vapour barrier is designed to prevent moist air reaching parts of the external wall or roof construction which are cold enough to cause moisture vapour to condense, and it should be placed on the warm side of the insulation. The minimum fall should be 1 in 40 to comply with the recommendations of BS 6229,[44] to allow for any inaccuracies on the site and possible deflection of the roof structure.[42]

As to flat roof coverings, one-half of the bitumen roof failures investigated by BRE arose from splitting of the felt caused by its inability to withstand more than a slight amount of stretching without splitting or tearing apart, and this can be remedied by patching with a strip of felt reinforced with hessian bedded in bitumen. The next most common cause of failure resulted from differential movement at skirtings to parapets and at other peripheral weatherings, and felt skirtings should be turned up over an angle fillet at the base of the upstand, and ideally be masked by a metal or semi-rigid asbestos/bitumen sheet flashing. Blisters do not usually lead to leakage.[42]

A survey of 130 mastic asphalt covered flat roofs ·to Crown buildings showed a 28 per cent failure rate, resulting either from splitting and cracking of the asphalt because of movement of the substrate and absence of an isolating membrane, or peripheral cracking caused by differential movement between a roof deck and a non-integral parapet wall to which an asphalt skirting was fixed without any provision for movement. Cracked and blistered areas should be heated, cut out and made good with new asphalt without delay. Parapet walls must be constructed of durable masonry surmounted by a suitable weathered and throated coping laid on a damp-proof course, as illustrated in figures 4.7.1, 4.7.2 and 7.3.4.

Floors

Solid ground floors can suffer from rising damp and should always be provided with an adequate damp-proof membrane, except when finished with asphalt, even if the floor finish to be laid on it is unaffected by moisture. Furthermore, the damp-proof membrane preferably situated under the floor screed over the concrete slab shall be effectively jointed to the horizontal damp-proof courses in the adjoining walls, as illustrated in figures 6.1.1, 6.1.2 and 6.2.1. BRE Digest 364[45] shows how wood flooring should be fixed in a solid ground floor to avoid piercing the waterproof membrane and so rendering it ineffective.

With suspended timber ground floors the greatest danger is an outbreak of dry rot (Serpula lacrymans), as described in chapter 6, with its serious and expensive consequences. The softwood species used in modern building construction all have a low natural resistance to decay and, to be immune from attack, the moisture content of the timber must be maintained below 22 per cent. BRE Digest 364[45] recommends adherence to the following basic principles of good building practice to prevent the risk of decay in timber floors

(1) selection of timbers of suitable quality;

(2) storage of timbers in dry conditions on site prior to fitting;

(3) design to protect against ingress of rain or ground moisture;

(4) design to provide adequate ventilation of sub-floor voids.

Designs satisfying criteria (3) and (4) are illustrated in figure 6.3.2.

With timber upper floors no special precautions are necessary beyond ensuring that jobs ends are not built into masonry which could become damp in service.

External Joinery

Nature and types of decay

In recent years there has been a substantial increase in the number of instances of decay in wood windows in comparatively new houses. Decay occurs most frequently in ground floor windows and in the lower parts of the members concerned, such as the bottom rail of an opening light, the bottoms of jambs and mullions, and the sill itself, often at or near a joint.

In old buildings, decay in window joinery may be part of a widespread attack of dry rot fungus. However, most decay in window woodwork exposed to the weather is of the wet rot type (both brown rot and white rot), which will not spread to other timber in the building. The rot is almost certain to be of this variety where the decay is confined to relatively small localised pockets detectable only from outside the building or if no actual fungus growth can be found. The basic causes of decay are the low natural resistance of sap-wood and the presence in the wood of sufficient moisture to permit the growth of wood-destroying fungi, usually in excess of 20 per cent.[46]

Entry of moisture

Decay of existing joinery involves in the main preventing access of moisture to the wood and the following factors accentuate the problem

(1) flat surfaces on horizontal rails which do not effectively shed water;

(2) failure to seal joints and exposed end grain, resulting in capillary entry of water;

(3) use of animal and casein glues which fail under damp conditions;

(4) failure to cover joinery in transit and on site and placing too much reliance on pink shop primers which are of variable quality;

(5) failure to prime rebates or careless puttying leading to putty failure;

(6) failure to provide an effective seal under wood glazing beads;

(7) overstressing of joints in opening lights, causing joints to open and putty to come away from glass;

(8) poor paintwork maintenance resulting in excessive swelling and shrinkage with consequent opening of joints and putty lines, which new paint fails to seal;

(9) excessive condensation, particularly in bathrooms and kitchens, and aggravated by defective back putties.[46]

Remedial measures

Where there are indications of water penetration but no decay is detected, remedial measures should be undertaken during a dry period of the year. Paintwork should be stripped in suspect areas and extending about 100 mm around them, and loose or cracked putty removed. Horizontal wood surfaces should be cut to form a slope for drainage. After an adequate drying out period, follow with generous applications of wood preservative worked well into the joints of the woodwork. Strained joints should be strengthened by metal brackets. Putty can then be renewed and open joints carefully sealed.[39]

The easiest way to preserve sound but wet wood against decay is to drill and insert pellets at joints and other vulnerable positions. The pellets contain a fungicide based on boron compounds which dissolve in the moisture of the timber, diffuse into the wood and stop any decay. An alternative method is to inject wood preservative at the joints using specially designed plastic injectors.

Regular maintenance inspections ensure that decay is diagnosed at an early stage when minor repairs will suffice. Relatively inexpensive repairs will prolong the life of a window by several years adopting the following procedure

(1) strip paint and cut out decayed wood in late spring or early summer and allow to dry;

(2) inject preservative, often of organic solvent type, and allow to dry;

(3) apply priming coat of paint;

(4) fill holes using a proprietary hard filler – a large hole may be filled with a piece of wood, preservative-treated and primed after shaping, and sealed with filler;

(5) seal whole of repair area with primer;

(6) apply undercoats and finishing coat of paint.

Where the woodwork is in an advanced stage of decay, the whole part of an affected window may have to be replaced.

Leaking Water Services

Inadequate forethought to future maintenance requirements at the design stage frequently results in problems of access to service pipes, and this can cause delays and increased damage when leaks occur.

Damp patches may appear on walls and ceilings, often some distance from the external walls and may be spasmodic in occurrence. Copies of the original drawings, preferably 'as built', can help when attempting to trace the source of dampness. Where the services are not housed in ducts, the service runs require tracing. If the damp patches appear on ceilings, the floor above should be checked, particularly if it is a toilet or bathroom.[47]

In many cases the water will appear some distance from the point of origin, and diagnosis of the fault can be time-consuming, particularly when the water occurs only intermittently. Leaks from joints or defective pipes that run in ducts or under floorboards will allow water to pass down the pipes until it meets a joint in the pipework or the pipes come into contact with the enclosing structure. When access is severely restricted it may be good policy to reroute the pipes to make future maintenance easier.[47]

Continually leaking overflow pipes discharging on to external walls will result in saturated masonry, mould growth and staining and, in extreme cases, moisture penetration through the wall following the weakening of mortar joints and possible frost action.

VENTILATION

General aspects

Part F of schedule 1 to the Building Regulations 1991[7] requires that there shall be adequate means of ventilation for people in the dwelling. This requirement will be met if the ventilation provided under normal conditions is capable in use of restricting the accumulation of moisture which could lead to mould growth and pollutants from within the building which could form a hazard to the health of occupants.

Building Regulations Approved Document F1 (1995)[7] defines a *ventilation opening* as any means of ventilation, whether permanent or closable, which opens directly to external air, such as the openable parts of a window, a louvre, airbrick, progressively openable ventilator, or window trickle ventilator, and includes a door which opens directly to external air. It prescribes that ventilation openings shall have a smallest dimension of at least 5 mm for slots or 8 mm for square or circular holes, in order to minimise resistance to the flow of air.

Approved Document F2 (1995) states that the general objective of the requirement in F1 is to provide a means of

(1) extracting moisture from areas where it is produced in significant quantities, such as kitchens, utility rooms and bathrooms;
(2) achieving rapid ventilation for the dilution of pollutants and of moisture likely to produce condensation in occupied rooms and sanitary accommodation;
(3) making available over long periods a minimum supply of fresh air for occupants and to disperse where necessary residual water vapour without significantly affecting comfort.

Habitable rooms

Habitable rooms are rooms used for dwelling purposes but excluding kitchens, for which the ventilation requirement will be satisfied if there is

(1) for rapid ventilation one or more ventilation openings with a total area of at least 1/20th of the floor area of the room, and with a part of the ventilation opening, such as an opening window, at least 1.75 m above floor level; and
(2) for background ventilation a ventilation opening or openings having a total area of not less than 8000 m², as for example a trickle ventilator. The opening(s) shall be controllable and secure so as to avoid unacceptable draughts.

Kitchens

The ventilation requirement will be satisfied it there is both
(1) extract ventilation for rapid ventilation, with an extract rating of not less than 60 litres per second, or incorporated within a cooker hood with an extraction rate of 30 litres per second, which may be operated intermittently as for instance during cooking or passive stack ventilation (PSV); and
(2) background ventilation by a controllable and secure ventilation opening(s) with a total area of not less than 4000 m², located to avoid draughts, and probably using a trickle ventilator.

Passive stack ventilation (PSV) is a ventilation system using ducts from the ceilings of rooms to terminals on the roof, which operate by a combination of the natural stack effect, i.e. the movement of air due to the difference in temperature between inside and outside and the effect of wind passing over the roof of the building.

Bathrooms

Bathrooms include shower-rooms and the ventilation requirement can be met by providing mechanical extract ventilation capable of extracting at a rate not less than 15 litres per second, which may be operated intermittently, or passive stack ventilation (PSV).

Sanitary accommodation

Sanitary accommodation is defined as a space containing one or more closets or urinals and the ventilation requirement can be met by either

(1) providing rapid ventilation by one or more ventilation openings with a total area of at least 1/20th of the floor area of the room, and with part of the ventilation opening at least 1.75 m above floor level; or
(2) mechanical extract ventilation, capable of extracting air at 6 litres per second and background ventilation of not less than 4000 m².

Habitable rooms ventilated through other rooms and spaces

Two habitable rooms may be treated as a single room for ventilation purposes if there is an area of permanent opening between them equal to at least 1/20th of the combined floor areas. A habitable room may be ventilated through an adjoining space if

(a) the adjoining space is a conservatory or similar space; and
(b) there are one or more ventilation openings with a total area of at least 1/20th of the combined floor

area of the habitable room with some part of the ventilation opening at least 1.75 m above floor level; and

(c) for background ventilation one or more ventilation openings of at least 8000 m²; and

(d) there are openings, which may be closable, between the habitable room and the space for rapid and background ventilation as previously described.

Ventilation to a restricted external air space
(now omitted from the 1995 edition of Approved Document F1)

If a ventilation opening serving a habitable room faces a wall nearer than 15 m, the following minimum distances apply

(a) if there is a wall on each side of the opening, forming a closed court as shown in figure 8.1.1, then the vertical distance from the top of the opening to the top of the wall containing the opening, Dt, should be less than twice the horizontal distance from the opening to the facing wall, Df, or

(b) if there is a wall on only one side of the opening, forming an open court as shown in figure 8.1.1, and if the length of the facing wall, D1, is more than twice the horizontal distance from the opening to the facing wall, Df, then either

(i) the vertical distance from the top of the opening to the top of the wall containing the opening, Dt, or
(ii) the horizontal distance from the side of the opening to the open side of the court, Ds,

should be less than twice the horizontal distance from the opening to the facing wall, Df.

Alternative approaches

Alternative approaches to those provided in Approved Document F1 (1995) to satisfy the ventilation requirements are contained in the relevant recommendations of the following:

(a) BS 5925: 1991. Code of practice for ventilating principles and designing for natural ventilation; or

(b) BS 5720: 1979[48]. Code of practice for mechanical ventilation and air conditioning of buildings; or

(c) BS 5250: 1989[31]. Code of practice for the control of condensation in buildings; or

(d) BRE Digest 398. Continuous mechanical ventilation in dwellings: design, installation and operation.

CONDENSATION

Nature of Condensation

Condensation has tended to become a greater cause of dampness in post-war dwellings than rain penetration and rising ground moisture. Warm air can hold more water vapour than cold air and when moist air meets a cold surface it is cooled and releases some of its moisture as condensation. Air containing a large amount of water vapour has a higher vapour pressure than drier air and hence moisture from the wetter air disperses towards drier air. This has a special significance since

(1) a concentration of moist air as in a kitchen or bathroom readily disperses throughout a dwelling; and

(2) moist air at high pressures inside buildings tries to escape by all available routes to the outside, not only by normal ventilation exits but also through the structure, when it may condense within it.[49]

Condensation takes two main forms

(1) surface condensation arising when the inner surface of the structure is cooler than room air; and

(2) interstitial condensation where vapour pressure forces water vapour through slightly porous materials, which then condenses when it reaches colder conditions.

The term *relative humidity* (RH) expresses as a percentage the ratio between the actual vapour pressure of an air sample and the total vapour pressure that it could sustain at the same temperature (per cent RH at °C). Air is described as saturated when it contains as much water vapour as it can hold – it is then at 100 per cent RH. If moist air is cooled, a temperature will be reached at which it will become saturated and below which it can no longer hold all of its moisture. This temperature is the *dew point*.[49]

The occurrence, persistence, extent and level of condensation are influenced by many factors, of which the most important are probably

(1) number of occupants and use of property;
(2) type of dwelling, construction and layout;
(3) heat levels maintained in the property;
(4) type of heating;
(5) length of time the property remains unheated;
(6) degree of insulation;
(7) amount of ventilation;
(8) prevailing weather conditions.

A BRE report[50] estimated that two million dwellings in the UK suffer from serious dampness and a further two million suffer to a lesser extent and the main cause was diagnosed as condensation. This gives a clear indication of the enormity of the problem.

Causes of Condensation

There are two main reasons for the increase in frequency and severity of condensation – (1) changes in living habits and (2) changes in building techniques. More wives are now employed, often resulting in dwellings being left unoccupied, unventilated and unheated for much of the day. Moisture-producing activities such as cooking and washing of clothes tend to be concentrated into shorter periods of time. Furthermore, washing and drying of clothes are often carried out within the main area of the dwelling instead of in a separate washhouse or fairly isolated scullery as in older dwellings. Flueless paraffin and gas heaters produce large quantities of moisture up to 1 to 2 litres and unvented tumble driers 3 to 7.5 litres daily. Additionally, occupants have become more sensitive to slight dampness in their dwellings and often endeavour to maintain a high standard of decoration, so that local deterioration assumes greater importance.

Structurally, probably the most significant change is the disappearance of many open fires and air vents which provided valuable ventilation routes. Modern windows reduce ventilation flows and this may be further accentuated by draughtproofing by occupants. Solid floors without an insulating floor finish or screed are slow to warm, and modern wall plasters and paints are less absorptive. Flat roofs and newer forms of wall construction also need to be carefully designed if they are not to lead to increased condensation.[39]

Thermal insulation laid on top of the ceiling in the roof space is one of the most cost-effective methods of conserving energy, but it also increases the risk of condensation in the roof. BRE Digest 270[51] highlights the importance of adequate roof ventilation by the provision of openings in the eaves on opposite sides of the roof, and this aspect will be examined further later in the chapter, when considering the requirements of Building Regulations Approved Document F2 (1989). When modernising older buildings, it is common practice to install suspended ceilings supported on an insulating quilt on the top floor. This creates a cold roof where the temperature can be reduced below dew point, resulting in the depositing of moisture. Here again adequate cross ventilation is needed and this will also be considered in more detail later.

The use of factory components in industrialised building systems often contributes to condensation. The reinforced concrete panel system of building produces cold wall surfaces and low levels of air circulation.

Surface condensation can lead to unsightly and unpleasant blue, green and black mould growth on walls, ceilings, fabrics and furnishings, which produce many complaints from occupants. On paint it may show as pink or purple staining.[52] Condensation within the fabric is slower to show up but may be much more serious in the long term.

Diagnosis of Condensation

Rising damp can be distinguished from condensation by the pattern and positioning of staining, while moisture penetration through cavity walls across wall ties also shows pattern staining. Gutters and downpipes must be checked for cracks, defective joints, blockages and the resultant leakage and water penetration. Roofs should be checked for defects and here again the type and positioning of staining is often a useful guide. Less obvious causes of dampness are slight weeping at pipe joints and wastes, and pinhole leaks in pipes.[39]

Drying out of construction moisture can lead to defects similar to those resulting from condensation and it is desirable to allow drying out to finish before carrying out remedial measures. As this can take up to three years, occupants are only likely to accept this advice with reluctance.

Condensation frequently occurs as occasional damp patches in cold weather, although a sudden change from cold to warm humid weather may also cause condensation. Apart from investigating damp conditions, attention should be directed to the heating arrangements, possible use of portable gas or oil-fired appliances, ventilation, arrangements for drying clothes, means of dispersal of moisture from the kitchen, form of construction of floors, walls and roof, and whether there is any uninsulated pipework. Measurement of temperatures and humidities will show whether conditions favourable to condensation exist at the time of measurement. Suitable charts and useful calculations are contained in *Condensation in Dwellings, Part 1*.[49] A sling or whirling hygrometer is useful for this purpose and consists of wet and dry bulb thermometers. The interrelationship of the various factors can be shown on a psychrometric chart where the moisture content of the air is plotted against temperature as illustrated in BRE Digest 297.[53] Electric meters can help in indicating the amount of moisture held beneath the surface of any material, subject to the limitations described earlier. A surveyor also needs the capacity to assess the reliability of information supplied by occupants.[39]

Condensation problems will lead to damp patches that are more diffuse and without the definite edges that occur with other causes. Impermeable surfaces, such as gloss paint or vinyl wallpaper, can be covered with a film or droplets of water. Trouble starts in areas that are usually cold, such as inside exposed corners, wall to floor junctions or solid lintels, or poorly ventilated, such as kitchen cupboards, wardrobes or behind furniture. Spores from moulds and other fungi can germinate over a whole range of temperatures (0 to 20°C) given suitable conditions (supply of food, oxygen and liquid water), resulting in deterioration of decorations, a musty smell and possible health hazards.[39]

Remedial Measures

The principal remedial measures consist of improved ventilation, insulation or heating, or a combination of them. If the relative humidity is excessive, the amount of moisture must be reduced or temperatures raised. Alternatively the moisture vapour should be removed at source, preferably by mechanical means.

Ventilation is normally the cheapest solution and can be very effective provided it does not result in unpleasant draughts, otherwise it is likely to be rendered ineffective by occupants. This is particularly important in kitchens. A limited amount of ventilation is essential to keep relative humidities below 70 per cent. Ideal ventilation rates are between 0.5 and 1.5 air changes per hour.[50] This can be achieved by the installation of trickle ventilators in bedrooms and extractor fans in kitchens and bathrooms. Automatic controls, such as humidistats, can improve the effectiveness of fans at

little extra cost. Care should be taken to make the fans as unobtrusive as possible, both in visual and acoustic terms.[53] Ventilated hoods above cookers are valuable in quickly trapping and removing the steam generated by cooking.

Insulation is generally more expensive than ventilation but more acceptable to the occupier. The main aims are to keep surface temperatures above dew point, improve *U*-values and secure better value for money from heating and improved comfort conditions. This is normally done by filling cavities with suitable insulant and/or fixing a plasterboard–insulation composite board with an integral vapour check internally.[53]

Heating is the most effective measure of all but also the most expensive, and may be opposed by occupants faced with increased running costs. The aim is to raise air and surface temperatures and so reduce the relative humidity. Living rooms, even if only heated during evenings, seldom suffer from surface condensation, while bedrooms, which are frequently very poorly heated, often cause problems.[49] Marsh[54] recommends the attainment of a general air temperature of about 15°C, ideally with the system running throughout the night, and not allowing surface temperatures of the structure to fall below 10°C. Casual heat gains from other rooms can be important, such as, for example, a bedroom over a well-heated living room. Conversely, a bedroom at one end of a flat may receive very little benefit from other heating and may also have a high heat loss if it is an external corner room with two exposed walls. Similarly, top floor rooms may suffer high heat loss through poorly insulated roofs.[52]

Clothes drying cupboards should be heated and well ventilated. While *cupboards* on external walls, especially clothes cupboards in poorly heated bedrooms, often suffer from condensation. They will benefit from high and low internal vents and in extreme cases low wattage tubular heaters should be installed.

With *bathrooms*, rapid ventilation provided by opening windows after bathing is usually adequate, and particularly so if there is a heated towel rail. Bathroom doors should be well-fitting and kept closed. Internal bathrooms with fan ventilation rarely give trouble if the fan is functioning satisfactorily. Separate WCs are rarely heated and condensation often occurs on cold fittings. In extreme cases the provision of a low wattage tubular heater could be considered. In a similar manner, condensation may drip from cold storage tanks located in cupboards and may be intercepted by a drip tray, or have suitable insulation installed under them. Condensation on cold water pipes may also cause drips and the pipes should ideally be insulated.

There is normally sufficient air movement in halls, passages and stairways to prevent condensation, but additional heat may be needed in extreme cases. *Living rooms* which are heated to above 18°C for several hours a day rarely suffer from serious condensation. If it does occur, it may be caused by an adjoining kitchen without a fan, very poor ventilation of the living room or poor structural insulation. In the latter case some background heating for long periods should be provided or, alternatively, a low thermal capacity lining should be fixed to the wall face of the structure.[52]

The measures necessary to prevent *interstitial condensation* can be determined by calculation. Unless there is a vapour barrier (vapour control layer) on the room side of an external structure, water vapour will enter and condense when it reaches colder conditions towards the outside. With flat roofs, a vapour check at ceiling level may be formed of gloss paint or vinyl-faced paper. With some composite forms of walling, a vapour barrier is needed on the inside face, often in the form of polythene sheeting on impregnated battens with a dry lining, or an insulated lining with an integral vapour barrier. Anti-condensation paints can be used in certain situations but their use on a large scale is rarely justified. Finally, occupants of dwellings should be informed of what is and what is not reasonable so far as living patterns are concerned, as quite trivial changes in living habits may produce a major improvement.[53]

Electric dehumidifiers that operate on a closed refrigeration cycle both dry and heat the air. They are most effective in warmer dwellings when condensation problems are caused by high vapour pressures but they tend to be relatively obtrusive and noisy in operation.[53]

Condensation in Roofs

Basic statutory requirements

Part F2 of schedule 1 to the Building Regulations 1991[7] prescribes that adequate provision shall be made to prevent excessive condensation in a roof or in a roof void above an insulated ceiling, which could give rise to wet rot in the roof timbers. The requirements will be met if condensation in a roof and in spaces above

insulated ceilings is limited so that, under normal conditions, the thermal performance of the insulating materials and the structural performance of the roof construction will not be substantially and permanently reduced.

Approved Document F2 (1995) contains the following introductory points

(1) the requirement will be met by the ventilation of cold deck roofs;

(2) it is not necessary to ventilate warm deck roofs or inverted roofs;

(3) for the purposes of health and safety, it may not always be necessary to provide ventilation to small roofs, such as those over porches and bay windows;

(4) ventilation openings may be continuous or distributed along the full length and may be fitted with a screen, fascia or baffle;

(5) further guidance is given in *Thermal Insulation – Avoiding Risks*.[13]

Approved Document F2 (1995)[7] prescribes the following methods of satisfying the requirements.

Roofs with a pitch of 15° or more (pitched roofs)

(1) Pitched roof spaces should have ventilation openings at eaves level to promote cross ventilation, with an area on opposite sides at least equal to continuous ventilation running the full length of the eaves and 10 mm wide, as shown in figures 15.1.8 and 15.3.2.

(2) Purpose made components are available to ensure that quilt and loose fill insulation will not obstruct the flow of air where the insulation and roof meet.

(3) A pitched roof which has a single slope and abuts a wall should have ventilation openings at both eaves and high levels. The ventilation at high level may be located at the junction of the roof and the wall or through the roof covering, and if through the roof covering it should be placed as high as possible. The area at high level shall be at least equal to continuous ventilation running the full length of the junction and 5 mm wide.

(4) An alternative approach is to follow the relevant recommendations of BS 5250.[31]

Roofs with a pitch of less than 15° and those where the ceiling follows the pitch of the roof

(1) Roof spaces should have ventilation openings in two opposite sides to promote cross ventilation. These openings shall have an area at least equal to con-

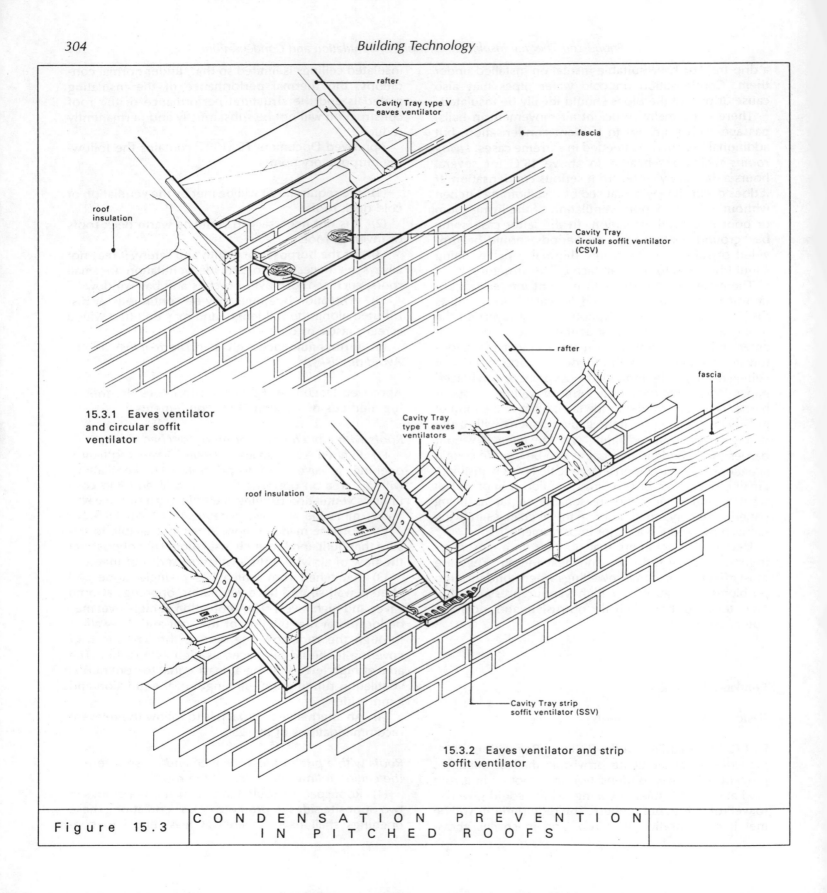

rafter

Cavity Tray type V
eaves ventilator

fascia

roof
insulation

Cavity Tray
circular soffit ventilator
(CSV)

**15.3.1 Eaves ventilator
and circular soffit
ventilator**

rafter

fascia

Cavity Tray
type T eaves
ventilators

roof insulation

Cavity Tray strip
soffit ventilator (SSV)

**15.3.2 Eaves ventilator and strip
soffit ventilator**

| Figure   15.3 | C O N D E N S A T I O N   P R E V E N T I O N
I N   P I T C H E D   R O O F S |

tinuous ventilation running the full length of the eaves and 25 mm wide.

(2)   Roofs with a span exceeding 10 m or being other than a simple rectangle in plan may require more ventilation, totalling 0.6 per cent of the roof area.

(3)   The void shall have a free air space of at least 50 mm between the roof deck and the insulation, as shown in figure 15.1.9. Where roof joists run at right angles to the flow of air a suitable air space may be formed by using counter battens.

(4)   Pitched roofs where the insulation follows the pitch of the roof also need ventilation at the ridge at least equal to continuous ventilation running the length of the ridge and 5 mm wide.

(5)   Where the edges of the roof abut a wall or other obstruction in such a way that free air paths cannot be formed to provide cross ventilation or the movement of air outside any ventilation openings would be restricted, an alternative form of roof construction should be used.

(6)   Vapour checks can reduce the amount of moisture reaching a void but they cannot be relied on as an alternative to ventilation, as a complete barrier to moisture is required.

(7)   An alternative approach is to follow the relevant recommendations of BS 5250.[31]

Avoidance of condensation in pitched roofs

Eaves ventilators

The Cavity Trays type V eaves ventilator, as illustrated in figure 15.3.1, is designed to allow the roof insulation to be laid over the wall plate, thereby avoiding and minimising cold bridging. A free-flow air channel is maintained under the felt into the roof space to satisfy the requirements of the Building Regulations. The ventilator is manufactured from galvanised steel to suit most popular truss spacings, and is positioned above the wall plate, with the top flanges secured to the upper faces of the trusses. The type V ventilators are of tension-formed shape to give rigidity and the V600 type, for use with 600 mm truss centres, is 180 mm × 580 mm; the V600 exceeds 25 000 mm² free air space per unit, equivalent to 10 mm continuous gap, when used along the entire length of rafters, which is suitable for roof pitches above 15°.

The Cavity Trays type T eaves ventilator is another preformed eaves ventilator which permits adequate airflow into roof spaces, satisfying the requirements of the Building Regulations. The type T eaves ventilator, illustrated in figure 15.3.2, is manufactured from polypro-pylene. The ventilator consists of a centre section which clamps on to the timber rafter or truss. From the centre section two wings project and create two air-flow voids, whereby incoming and outgoing air travels over the unobstructed wings. The free air space per unit exceeds 15 000 m², giving an equivalent of over 25 000 mm² per metre of length, when used at 600 mm truss centres.

Soffit ventilators

The introduction of increased insulation levels in existing properties makes it advisable to consider the ventilation requirements of the roof. The circular soffit ventilator (CSV), as illustrated in figure 15.3.1, may be inserted in existing soffit boards, thus permitting existing properties to be upgraded to meet Building Regulation requirements. The injection moulded polypropylene ventilator as manufactured by Cavity Trays can be white, brown or black and is 79/70 mm × 50 mm in size, providing the equivalent of a 10 mm gap when fitted at 200 mm centres along the soffit. It is designed to fit soffit board thicknesses of 4 mm upwards and it incorporates a 4 mm grid for fly screening. This product can also be used in new build situations.

The preformed strip soffit ventilator (SSV), as manufactured by Cavity Trays, is designed for new work situations and is illustrated in figure 15.3.2. It satisfies the roof ventilation requirements, thus alleviating the problems of condensation by providing defined air circulation, and embraces insect-resistant screening. The strip soffit ventilator of UV stabilised virgin polymer PVC, in white, brown or black, can be used with a standard soffit board. It permits air-flow through the soffit while also·providing support for soffit boards from 4 mm to 10 mm in thickness, and is pre-drilled for immediate fixing. It is provided in lengths of 2400 mm × 50 mm wide × 29 mm high, giving a free air-flow of 10 800 mm² per metre of length, equivalent to a 10 mm continuous gap, when installed along the whole length of the soffit. The SSV15 type is 80 mm wide to provide for pitches below 15°, producing 25 000 mm² free air-flow per metre of length, equivalent to a 25 mm continuous gap, when installed along the whole length of the soffit.

REFERENCES

1. R. McMullan. *Noise Control in Buildings*. BSP (1991)
2. *BRE Digest 337*: Sound insulation: basic principles (1988)

3. *BS 2750* Methods of measurement of sound insulation in buildings and of building elements, *Part 4: 1980* Field measurements of airborne sound insulation between rooms; *Part 7: 1980* Field measurements of impact sound insulation of floors

4. *BS 5821* Methods for rating the sound insulation in buildings and of building elements, *Part 1: 1984* Method for rating the airborne insulation in buildings and of interior building elements; *Part 2: 1984* Method for rating the impact sound insulation

5. J. Stephenson. *Building Regulations 1985 Explained*. Thomson Publishing (1989)

6. R. McMullan. *Environmental Science in Building*. Macmillan (1992)

7. *The Building Regulations 1991 and Approved Documents C(1991), D(1985), E(1991), F(1995) and L(1995)*. HMSO

8. *BRE Digest 333*: Sound insulation of separating walls and floors – Part 1: walls (1988)

9. *BRE Digest 334*: Sound insulation of separating walls and floors – Part 2: floors (1988)

10. *BRE Information Paper IP 6/88*: Methods for improving sound insulation between converted flats (1988)

11. *BRE Digest 338*: Insulation against external noise (1988)

12. *BRE Digest 347*: Sound insulation of lightweight dwellings (1989)

13. BRE. *Thermal insulation: avoiding risks*. HMSO (1994)

14. B.R. Anderson. *Building Regulations: conservation of fuel and power – the 'energy target' method of compliance for dwellings*. BRE (1989)

15. *BS 8208* Guide to assessment of suitability for external cavity walls for filling with thermal insulants, *Part 1: 1985* Existing traditional cavity construction

16. *BS 5617: 1985* Specification for urea–formaldehyde (UF) foam systems suitable for thermal insulation of cavity walls with masonry or concrete inner and outer leaves

17. *BS 5618: 1985* Code of practice for thermal insulation of cavity walls (with masonry or concrete inner and outer leaves) by filling with urea–formaldehyde (UF) foam systems

18. *BS 6676* Thermal insulation of cavity walls using man-made mineral batts (slabs), *Part 1: 1986* Specification for man-made mineral fibre batts (slabs); *Part 2: 1986* Code of practice for installation of batts (slabs) filling the cavity

19. *BRE News 60*: Built-in cavity wall insulation (1984)

20. *BRE Digest 358*: CFCs and buildings (1991)

21. *BRE Digest 236*: Cavity insulation (1984)

22. *BS 6232* Thermal insulation of cavity walls by filling with blown man-made mineral fibre, *Part 1: 1982* Specification for performance of installation systems; *Part 2: 1982* Code of practice for installation of blown man-made mineral fibre in cavity walls with masonry and/or concrete leaves

23. *BRE Digest 277*: Built-in cavity wall insulation for housing (1983)

24. *BRE Defect Action Sheet 17*: External masonry walls insulated with mineral fibre cavity-width batts: resisting rain penetration (1983)

25. *BS 699: 1984* Specification for copper direct cylinders for domestic purposes

26. *BS 1566* Copper indirect cylinders for domestic purposes, *Part 1: 1984* Specification for double feed indirect cylinders; *Part 2: 1984* Specification for single feed indirect cylinders

27. *BS 3198: 1981* Specification for copper hot water storage combination units for domestic purposes

28. *BS 5615: 1985* Specification for insulating jackets for domestic hot water storage cylinders

29. *BS 5422: 1990* Thermal insulating materials on pipes, ductwork and equipment (in the temperature range $-40°C$ to $+700°C$)

30. *BS 5970: 1981* Code of practice for thermal insulation of pipework and equipment (in the temperature range $-100°C$ to $+870°C$)

31. *BS 5250: 1989* Code of practice for control of condensation in dwellings

32. *BRE Digest 319*: Domestic draughtproofing: materials, costs and benefits (1987)

33. *BRE Digest 355*: Energy efficiency in dwellings (1990)

34. TRADA. *Energy Efficient Housing – A Timber Frame Approach* (1989)

35. L. Addleson. *Building Failures*. Butterworth Architecture (1989)

36. B.A. Richardson. *Defects and Deterioration in Buildings*. Spon (1991)

37. M. Hollis. *Surveying Buildings*. RICS (1991)

38. H.S. Staveley and P. Glover. *Building Surveys*. Butterworths (1990)

39. I.H. Seeley. *Building Maintenance*. Macmillan (1987)

40. *BRE Digest 245*: Rising damp in walls: diagnosis and treatment (1986)

41. *BRE Good Building Guide 3*: Damp proofing basements (1990)

42. I.H. Seeley. *Building Surveys, Reports and Dilapidations*. Macmillan (1985)

43. *BRE Digest 312*: Flat roof design: the technical options (1987)

44. *BS 6229: 1982* Code of practice for flat roofs with continuously supported coverings

45. *BRE Digest 364*: Design of timber floors to prevent decay (1991)

46. *BRE Digest 304*: Preventing decay in external joinery (1985)

47. PSA (DOE). *Defects in Buildings*. HMSO (1989)

48. *BS 5720: 1979* Code of practice for mechanical ventilation and air conditioning in buildings

49. DOE. *Condensation in Dwellings, Part 1: A Design Guide*. HMSO (1970)

50. J. Garratt and F. Nowak. *Tackling Condensation: A guide to the causes of and remedies for surface condensation and mould in traditional housing*. BRE (1991)

51. *BRE Digest 270*: Condensation in insulated domestic roofs (1983)

52. DOE. *Condensation in Dwellings, Part 2: Remedial Measures*. HMSO (1971)

53. *BRE Digest 297*: Surface condensation and mould growth in traditionally built dwellings (1990)

54. P. Marsh. *Thermal Insulation and Condensation*. Construction Press (1979)

55. BRE and CIRIA. *Sound Control for Homes* (1993)

56. I.H. Seeley. *Building Economics*. Macmillan (1995)

16 BUILDING IN WARM CLIMATES

Mainly because there are many overseas readers of this book but also because UK readers are increasingly requiring a knowledge of overseas building practice, it seemed desirable to add a further chapter describing the main problems experienced and the techniques used to overcome them.

FACTORS INFLUENCING OVERSEAS BUILDING

The functions of buildings – to provide shelter from the weather and a comfortable living and working environment – are basically the same worldwide, but there are very significant differences in the requirements for buildings in hot and temperate climates. Hence BRE Digest 302[1] emphasises that building designs and processes must be modified to suit local conditions.

(1) Buildings must be protected against heat as well as cold, and different sky and ground conditions affect the design of natural lighting. In a humid climate, extensive ventilation is needed, which may cause problems of noise control. In the wealthier hot countries, air conditioning is used extensively. Different social customs will also affect building needs.

(2) Building materials, components and equipment have to cope with different environmental conditions. Intense tropical heat and humidity, as experienced in Singapore, encourage mould and fungal growth, rot and insect attack. Strong sunlight places greater strain on paints, plastics, bitumens and jointing materials. High temperatures and contaminated aggregates cause difficulties with concrete. Close to the coast, hot, humid, salt-laden atmospheres hasten the corrosion of metals.

(3) Many tropical countries are under-developed both economically and technologically. However, the finding of oil has led to rapid change and the adoption of more sophisticated and expensive approaches to building in oil-producing countries, while in poorer countries the buildings must be simpler and cheaper, except where financed by external agencies. Many local building materials and products may be of poor quality, necessitating the importing of materials and particularly components on a considerable scale, as in Kenya and Tanzania, and involving the expenditure of scarce foreign currency.

(4) There are significant differences in the organisation of the building industry and in the training, skills and responsibilities of its members in different countries. Regulations, government procedures and commercial practice also vary significantly. Housing space requirements vary widely according to the social and cultural background and what is acceptable in one region is not necessarily operative in an adjoining territory.

BRE Overseas Building Note 160[2] describes how many developing countries have warm climates and people are accustomed to spending much of their time out of doors. In these regions, depending on local circumstances, both sanitary and cooking facilities could be conveniently located outside the main building. Cooking facilities might advantageously be enclosed in pigeon-holed blockwork, thereby providing adequate ventilation and avoidance of heat transfer into the dwelling.

Manufactured sanitary ware and similar appliances account for a large part of the cost of a small building and good design ensures that these costs are kept to a minimum to reduce the need for imported goods with the consequent drain on foreign currency. Good design in this context would include the close grouping of appliances to reduce the length of plumbing and drainage connections, the use of showers which can often be made locally in lieu of baths, and the skilful arrangement of electrical cables to reduce the use of copper or aluminium.

The housing deficit in developing countries was estimated by UNESCO in 1980 as 300 million units and is still rising. Hence the need to improve living conditions and reduce health hazards on a global scale is enormous, and yet every conceivable effort must be made to achieve this goal. Countless millions of people occupy land as squatters as they see little prospect of obtaining housing in the accepted sense in the foreseeable future. One intermediate step, which has been implemented in a number of areas, is to provide basic sanitation and water supplies on plots where the occupier can be assured of security of tenure. On these plots, people erect their own shelters and governments often offer some assistance by providing subsidised materials from nearby compounds and a range of standard designs capable of construction by the local handy man.

DESIGN OF BUILDINGS FOR WARM CLIMATES

Climate

Information on the climates of many overseas countries is obtainable from the UK Meteorological Office world tables. BRE Digest 302[1] shows the main characteristics of a range of warm climates that affect building design. The following examples give an indication of the wide range of variations that can occur.

(1) A warm, humid climate prevails in equatorial lowland in the tropical belt at or near sea level. It is never very hot nor very cold, with little seasonal variation in temperature. Rain falls on more than one-third of days throughout the year, sometimes being very heavy. Strong winds can occur during occasional rain squalls.

(2) In tropical inland areas away from the equator a composite monsoon climate is experienced. Two-thirds of the year is hot and dry, and one-third warm and humid. The total rainfall may vary from year to year, and mostly falls in the monsoon period when rain is heavy and prolonged. In the dry season it is hotter by day and cooler by night.

(3) Hot, dry climates occur in inland low-latitude deserts or semi-deserts located near and beyond the tropics. They are the hottest and most arid places in the world, with a meagre but variable rainfall. Settlements occur only where there are underground sources of water or oil. Large seasonal and diurnal changes in temperature are experienced. It is cold at night, especially in winter. Local winds may develop into dust storms.

Various factors influence local climate, including latitude, height above sea level, proximity to lakes or sea, hills or mountains and vegetation. Man changes the climate by cutting down forests, draining swamps, damming rivers and polluting the atmosphere. Agricultural irrigation and the planting of trees also have an effect.

In hot climates, buildings must be designed to counter the influence of heat, while in warm, humid climates it is also important to minimise the effects of humidity. East-west orientation and shading are advantageous in all areas. For hot, dry climates, a compact layout, heavy walls and roof structures, reflective external surfaces and small external openings are desirable. In warm, humid regions, air movement is important and buildings should be well spaced apart and generally be of lightweight construction, with reflective external surfaces and large external openings. For composite climates, where the construction programme is substantial, the most appropriate balance must be selected having regard to the duration and intensity of the different seasons and to the style of living. Hence climatic conditions have an important influence on the design of buildings.

Orientation

In the tropics, east and west-facing walls receive more direct radiation from the sun and should therefore be kept short. Openings in these walls, whether glazed or not, should be as few and as small as possible. Buildings should desirably have an east-west axis, with openings for access, ventilation and light located on the longer north and south walls. Moreover, buildings should be sited so that their openings do not face directly onto brightly sunlit surfaces of other buildings or boundary walls.

Shade

Buildings need to be shaded from the sun's heat in all warm climates, apart from highlands or outside the tropics in winter. Direct sunlight through doors and windows should be shaded after early morning, preferably outside the glass. North and south-facing walls are best shaded by horizontal overhangs and canopies, permitting windows to be kept open when it is raining. West and east-facing surfaces require sunbreakers and screens which reduce daylight but also obstruct the view. Shades should always be light in colour. In hot, dry climates, buildings should ideally be grouped closely together to provide some shade to each other and to create small shady spaces between them. In humid regions, wide spacing of buildings assists ventilation.[1]

Solar charts enable sun angles to be calculated for any position on the earth's surface, and for any time of the day and any day of the year. A shadow angle protractor, superimposed over the particular solar chart, shows shadow angles for buildings of any orientation.

Colour

BRE Overseas Building Note 158[3] describes how colour should be used with care as it affects an environment both physically and psychologically. The colour of an external surface influences the absorptivity of walls or roofs to the sun's radiation to a significant extent. For example, black absorbs about 85 to 100 per cent, dark greens and greys 70 per cent, light greens and greys 40 per cent, white oil paint 20 per cent, white emulsion paint 12 to 20 per cent, and new whitewash 12 per cent.

It is therefore advantageous to colour all roofs and unshaded walls of buildings as near to white as possible. Boundary and screen walls should be in dark colours, such as browns, greens and blues, to prevent the reflection of heat and glare, whereas shaded walls of buildings can be in bright cheerful colours to enliven and enrich the environment. Paving should preferably be in dark colours or, if light, should have broken surfaces to avoid reflecting heat and glare.

Internally, ceilings should ideally be white in order to reflect maximum light from the windows and spread it as evenly as possible throughout the room. White or light grey window walls will reduce the contrast between the light window aperture and the surrounding wall as seen from the inside, thus minimising glare.

Thermal Capacity and Insulation

Several factors have to be considered when deciding whether a building should have a low or high thermal capacity and the extent of insulation required

(1) the relative importance of day or night-time use; and when buildings are used day and night throughout the year, a compromise may be necessary when designing for comfort;

(2) the diurnal temperature range;

(3) the yearly seasonal range of weather, and changes over the seasons in the method of use; and

(4) whether or not the building is air conditioned.

In hot, dry areas buildings are best constructed traditionally with heavy walls and roofs, whereby they will gain and lose heat slowly and thus even out temperature fluctuations within the building. Insulation near the outer surface reduces solar gain in the daytime, although it will tend to retain heat at night. In warm, humid climates, where the variation between day and night temperatures is small, insulation can be a disadvantage because it will prevent the escape of heat. Walls and roofs of thin, light construction will heat and cool more quickly.

Air conditioning is now used extensively, particularly in the more prosperous countries, and this improves the quality of comfort significantly. Even where air conditioning is installed, it is advisable to adopt designs which are suitable for the climate, both to reduce running costs and to minimise discomfort in the event of a power failure, which occurs periodically in many regions. Furthermore, a building may be used differently in different seasons, and air conditioning may be run only in hot, humid periods. The internal temperature should be set at a level which does not cause thermal shock relative to the outside temperature.

Table 16.1, produced by the BRE, shows the most suitable thermal capacity and insulation for different conditions. Where practicable, walls and roofs of buildings should be shaded without obstructing ventilation. In general, dark surfaces should be avoided, as previously described. On the other hand, shiny, metallic surfaces reflect much of the heat radiated by the sun but do not readily release heat by their own radiation.

Air Movement

BRE Digest 302[1] describes how in warm, humid conditions, movement of air helps to improve comfort by cooling the skin. Hence buildings should be as open as practicable to offer minimum resistance to light winds, but screening may be necessary for protection and

Table 16.1 Thermal capacity and insulation requirements

Type of climate	Predominate use	Naturally ventilated	Air-conditioned
Warm humid. Small diurnal temperature range: mean wet bulb temperature exceeds 24°C	Night	Low to moderate thermal capacity, insulation only secondary	Low thermal capacity. Insulation not as important as vapour barrier towards outer skin
	Day	Thermal capacity not important. Roof insulation useful; ventilated double roof better	Moderate thermal capacity; good insulation; vapour barrier
Hot dry. Large diurnal temperature range: mean wet bulb temperature below 24°C	Night	Low thermal capacity unless night temperature low — then moderate insulation only secondary	Low thermal capacity; insulation not important
	Day	High thermal capacity; good insulation, best towards outside of structure	High thermal capacity; very good insulation

Source: BRE Digest 302[1]

privacy. In late afternoon and evening, plenty of fresh air is needed to remove heat gained during the day. The upper floors of high buildings are likely to be cooler where the winds are stronger, except possibly the top floor, where solar heat may be transmitted through the roof.

When considering orientation for both sun and wind, sun should be the prime criterion and the buildings should be sited as near east-west as practicable, unless the sky is normally overcast. Ventilation should be designed to take advantage of prevailing winds and is more effective if openings are set at an angle to the prevailing wind rather than at right angles to it. Maximum air movement can be obtained with cross-ventilation. Some improvement of comfort will result from mechanical air circulation using large diameter ceiling fans. Ventilation as near to the floor as possible provides welcome air movement around persons who

are seated or lying on beds. In conditions of still air, movement will be increased by 'stack-effect', where ventilators are positioned at both low and high levels.

In hot, dry climates, ventilation must be controlled to exclude heat and dust, but an abundance of fresh air is needed after sunset to cool the building. At night, the indoor temperature is often 3°C or more higher than that outside. Roof level extract fans can be installed to assist in drawing in cooler night air.

BRE Overseas Building Note 158[3] gives guidance on the design of ventilators in low cost housing and schools, and recommends the following action to control ventilation

(1) No breeze: low and high vents open, windows closed.

(2) Little breeze: all vents and windows open.

(3) Hot dust-laden winds: all vents and windows closed on windward side.

(4) Cooler conditions in mountains: vents and/or windows closed in cooler seasons as required. If necessary, fly screens can be provided.

For many years there was a widespread, but largely erroneous, belief that ceilings should be high for health and comfort in warm climates. Furthermore, building regulations in some tropical countries required minimum ceiling heights of 2.75 m, 3.05 m and 3.65 m in habitable rooms. By comparison with a more realistic ceiling height of 2.50 m, wall areas were increased by 12 to 30 per cent, resulting in more expensive buildings.

Artificial Cooling and Air Conditioning

BRE Digest 302[1] describes how in a very dry climate, it may be possible to cool a building below the lowest shade temperature with evaporative (desert) coolers, or alternatively cooling can be carried out by refrigeration (by drawing filtered air over a chilled coil). Part of the moisture in the air is condensed out. However, the system is expensive to operate, except where electricity is available very cheaply, and this is likely to remain the exception in many tropical countries. To minimise running costs, the cooling load should be reduced by ample shading and a good standard of insulation and, in a humid climate, by an effective vapour-barrier outside the insulation or an impervious external skin. The building volume should be kept to a minimum consistent with essential needs to reduce the cost of installing and operating the plant. Any sources of heat and moisture within a building should receive separate treatment to avoid additional load on the air conditioning system.

BRE Overseas Building Note 158[3] emphasised how a complete air conditioning system could be designed to provide the required internal environment, but it had the major disadvantage of high initial cost and subsequent running and maintenance expenses. Moreover, mechanical plant is subject to failure and skilled maintenance staff and spare parts are not always available. Buildings may also be required where there is no electricity. If plant is to be installed, its size and running costs must be minimised by giving careful thought to the siting and construction of the building. In more prosperous countries very sophisticated plant may be used, such as the computerised air conditioning plant in the main Standard Chartered Bank building in Singapore.

CONSTRUCTION OF BUILDINGS IN WARM CLIMATES

Foundations

BRE Overseas Building Note 179[4] describes how soils susceptible to expansion and shrinkage with changes of moisture content are a constant source of trouble in the design and construction of foundations. Such soils, popularly known as 'black cotton soils', occur extensively in central, southern and western parts of India. Buildings constructed on such soils, adopting the types of foundation commonly employed for other types of soil strata, are observed to crack extensively within a short period of construction, despite every reasonable precaution being taken. Effective solutions to this problem are described later in the section on Special Structural Problems.

Walls

Walls may be built of bricks and blocks of many types or of *in situ* concrete or timber and other claddings. In most climates, some types of bricks and blocks are suitable only for internal use, suitably protected from the weather; some are suitable for internal or external use, as long as they are not exposed to freezing conditions when very wet; and some will withstand all conditions. Some types of clay brick may themselves be subject to small progressive expansion; others, containing soluble salts, are liable to cause expansion of bedding mortars in some conditions of use. It is always advisable to specify only bricks or blocks that are suitable for the intended location and use, and to provide in the design for any movement that may be expected. In very arid climates, all types of brick or block may be used internally or externally, and even unburnt earth blocks have been used successfully in the Sudan, as described in BRE Overseas Building Note 165,[5] and in Egypt.

Although soil may be sufficiently strong for building the walls of dwellings in dry, arid climates, it is not very durable and has little resistance to moisture. The effect of its poor weathering performance can be seen in many developing countries where rural houses must be maintained regularly or even completely rebuilt at intervals, especially in areas with moderate rainfall. BRE Overseas Building Note 176[6] describes how this type of house wall is traditionally thick, partly because the dried mud used in construction does not have a high strength and also because the thick walls give better insulation against intense heat. Houses in the Middle East were often built around a central courtyard, to give the family a cool refuge when the heat of the day penetrated the fabric of the house.

For this reason methods have been developed to improve the natural durability and strength of the soil, as described in BRE Overseas Building Note 184,[7] and commonly referred to as soil stabilisation. There are numerous examples of how these blocks can be used to build inexpensive but durable houses in many parts of the developing world. The best and most widely used method is to add cement to the soil while it is still wet, and this binds together the soil particles or clusters of particles, to resist moisture movement and improve strength. Other suitable cementitious materials which can be used include hydraulic lime and lime-pozzolana mixes. Other types of stabilisers, such as bitumen and asphalt, act as waterproofing agents by providing a physical barrier to the passage of water. The grading of the soil will determine how effectively it can be compacted and stabilised. A number of simple presses have been developed specifically for making soil blocks, comprising constant pressure or constant volume, and being either manually operated or preferably power-driven.

BRE Overseas Building Note 184[7] recommends that

buildings constructed of stabilised blocks should, whenever possible, be designed in accordance with the relevant Codes of Practice for concrete blockwork. Concentrated loads on walls should be minimised by using lintels over door and window openings and providing a wall plate to distribute loads. The wall layout should be simple with few breaks and avoiding slender sections of walling between windows and doors. It is good practice to construct the base of the wall with concrete or stone for a height of at least 250 mm or to protect the soil wall for the same height with a rendering. Walls should be protected as much as possible from the erosive effects of rain and generous roof overhangs of at least 1 m should be provided in wet areas.

Reinforced concrete is used extensively in the frameworks and cladding of buildings, particularly those of more than two storeys. BRE Overseas Building Note 139[8] describes how loss of strength and chemical resistance of the set concrete can occur when the temperature reaches about 30°C in the presence of moisture, although the risk can be reduced by using low water/cement ratios. In desert regions the choice and supply of aggregates may be restricted or they may be of indifferent quality or contaminated with sulphates, chlorides or saline water. In arid countries it is often necessary to use sea water or brackish well water for mixing and curing concrete. In these circumstances corrosion of steel reinforcement is likely and additional precautions should be taken, such as increased cover to the steel or richer mixes of concrete.

The following procedures can be used for lowering the temperatures of aggregates in hot climates

(1) Shading the stockpiles from the direct rays of the sun.
(2) Spraying the stockpiles with water and keeping them moist and making provision for drainage and the removal of unwanted silt.
(3) Stockpiles cooled by refrigerated water or by passing refrigerated air through them.
(4) Use of ice.

Concreting at high temperatures presents special problems, as the quality of concrete is adversely affected by high temperatures during mixing, placing and curing. The concrete must be adequately cured by the use of absorbent materials kept constantly damp, or be sheeted or sprayed with an impervious film.

BRE Overseas Building Note 146[9] describes some of the main problems connected with the use of timber structures in tropical countries. The section of a timber structure which requires most protection is the point of ground support. This area sustains maximum exposure to the elements with minimum protection from the structure or surroundings. Hence a permanent timber structure should be isolated from the ground, usually by means of a steel shoe, steel plate, steel posts or bars with a concrete base. Whenever possible there should be an air space between the end grain of timber and the shoe or plate; if not the abutting area should be minimised and the end grain sealed. A ground level problem peculiar to the tropics and sub-tropics is the probability of attack by subterranean termites, which is further described later in the section on Behaviour of Building Materials. The design must effectively isolate the wood from the ground, so that termite activity is readily seen, and also incorporate physical barriers, such as termite shields which are usually of copper.

Roofs

Roof design requires special attention in warm climates, and with due regard to thermal performance. In the tropics the roof receives the greatest proportion of the sun's radiation because of its inclination in relation to solar altitude and its area as compared with the external walls. The roof also loses the greatest amount of stored heat by long wave radiation to the night sky. Apart from the requirements of structural safety and protection from the direct sun and rain, the roof construction plays a major role in attaining thermal comfort, the capital and running costs of the building, and the capacity and size of the air conditioning plant in an air conditioned building.

BRE Overseas Building Note 182[10] describes how both flat and pitched roofs are used extensively in Northern Sudan. For example, solid reinforced concrete slabs are used in the Khartoum area with various forms of insulation and protection from the sun. Roof shading is used in some public buildings and is generally formed of corrugated galvanised steel or asbestos cement sheets. Both materials are imported and expensive and the asbestos cement sheets involve a greater risk of damage from extreme temperatures and high speed winds.

Light or sheet roofs are mainly covered with corrugated galvanised steel and asbestos cement sheets, and are used extensively in Sudanese government buildings and housing. Corrugated galvanised steel sheeting is the most popular laid to about a 10° pitch, coupled with a simple structure in most dwellings concealed by parapet walls. Most houses have no ceilings and the internal temperatures are well in excess of the outdoor air temperature during day-time hours. Problems also arise from the noise of rain falling on the roof and the thermal movements of the sheets.

In rural areas, traditional roofs are the most common, using timber from the palm trees and earth finished with 'Zibala', which is a mixture of animal dung, straw and mud for rendering mud walls and roofs. Initial costs are low but annual maintenance is necessary, usually comprising another layer of 'Zibala'.

Experiments by Mukhtar[10] showed that a concrete roof slab with a suitable layer of insulation on top, or shading with a light reflective material like galvanised steel, gives minimum heat transfer, together with a delay in reaching the maximum internal temperature. The combination is simple and can be achieved at low cost if local materials are used for insulation and shading. The research also showed that the thermal performance of roofs can be improved by any of the following methods, used singly or in combination.

(1) applying a coat of white paint to the external surface of the roof;

(2) ventilating during the hours when the external air temperature is lower than the corresponding internal air temperature; and

(3) allowing the shaded roof construction to lose most of its gained heat to the cool night sky by folding or rolling away the shading device during night time.

In warm climates, uninsulated reinforced concrete roofs are unlikely to provide adequate protection for residential buildings. Whitewashing is not a viable proposition for heavy flat roofs, as in mainly dry areas it is quickly rendered ineffective by the accumulation of dust and dirt. A layer of light-coloured stone gives better results. BRE Overseas Building Note 164[11] describes how the use of reed panels as a shading device has reduced the maximum ceiling temperature by about 5°C, prolonged the time lag by one hour and reduced the amplitude of the internal surface wave by 6°C, but they require renewal about every two years. However, for residential buildings, the generally recommended time lag of over 8 hours is difficult to achieve without the addition of insulation layers.

There is often a problem with accentuated thermal movement leading to failure of joints and membranes, which can be accentuated by incorporating insulating materials. The use of inverted membrane (upside-down) systems may be beneficial for flat roofs, and BRE Digest 302[1] describes how an additional open 'parasol' roof to provide shade is very effective. In hot, humid climates, lightweight roofs, with good ventilation of roof spaces, generally provide the best solution. Where air conditioning is used the external envelope of the roof should be water vapour tight and sealed to the wall structure, which should also be designed to prevent the entry of warm, humid, external air.

Probably the most important example of protection by design is a large roof overhang at the eaves. In most climates an important effect of the eaves overhang is the significant reduction in the flow of rainwater over a wall, thus reducing significantly the risk of decay in external joinery. Of greater consequence in the tropics is the shading from direct sunlight which avoids high surface temperatures and excessively steep moisture and temperature gradients within the woodwork. In high rainfall areas of the tropics, metal flashings may be used to protect woodwork when eaves, reveals and canopies are not considered appropriate or adequate. Direct nailing of metal flashings to wood should be avoided, as unequal thermal expansion/contraction will result in consequent buckling and possible entrapment of water.

In some tropical areas, roof design must also take into account the possibility of earthquakes and/or hurricanes. Not only must roofs be securely held together and firmly tied to the wall structure, but they must also be built so that the materials used in them will be unlikely to cause injury by falling on people either within or outside the building.

Solar Energy Applications

The use of solar energy in warm countries for heating, cooking, refrigeration systems, power generation and water pumping is extensive. The applications are particularly relevant for families in both urban and rural areas in developing countries. BRE Overseas Building Note 192[12] describes how when capital for investment is readily available the choice of options is wide but where capital is scarce, simpler home-based equipment should be encouraged. Solar data in the form of maps or tables are invaluable in making an initial assessment of the possible effectiveness of a particular solar application.

Some systems may require certain imported raw materials or equipment, but these should be kept to a minimum. This section is, however, primarily concerned with systems which can be produced mainly from local resources and they include solar cookers and ovens; solar driers for agricultural produce; domestic hot water systems; and solar powered water pumping systems. In many cases the availability of water may be vital and solar powered water pumping is therefore often needed in order to use the other systems.

Solar cookers to be widely used should be as simple and inexpensive as possible; be capable of being made locally; and should need little or no positional adjustment during the day or year. They should also be durable and need minimal maintenance. The solar collector and cooking utensils should be separate items and the cooker or oven should operate without the need for constant attendance. Their effective use is however restricted to periods of direct solar radiation. The simplest form of cooker, an insulated rectangular container with a transparent lid, does absorb indirect radiation but its effect will not be sufficient alone to boil water or maintain water at boiling point. The basic box type of cooker can be improved by adding a reflector to increase the direct radiation into the cooker and by providing effective insulation to reduce heat losses from the cooker.

Solar powered domestic hot water heating systems are simple in principle. They may be constructed from basic materials and will have a performance which approaches that of systems using expensive manufacturing techniques and materials. Some basic details of a typical collector assembly are illustrated in BRE Overseas Building Note 192.[12] The collector frame and case should be constructed of rigid materials which will last for some years without deterioration due to climatic conditions. They should not corrode through the action of rain or condensation; steel, suitably protected or painted to withstand local conditions, may be used. Insulation should be 75 to 100 mm thick and be kept dry. Suitable insulants for this equipment and cookers and ovens include infills of rice or corn husks, dried grass, mineral wool or fibre-glass covered by solid non-soluble material.

Collectors are usually made from steel, copper or aluminium, although the latter is more difficult to fabricate. A commonly used type incorporates two metal sheets, 1 to 2 mm thick, joined along the edges by soldering, welding or pressing. The edges are shaped to leave a gap between the two sheets approximately 6 mm wide and to permit the connection of 20 to 25 mm diameter pipework. Another type of collector uses a flat sheet of metal to which serpentined pipework, normally 12 to 15 mm diameter, is soldered or welded. Both types of collector should be painted matt black on their upper surfaces. Multiple collectors may also be used. Glazing may be glass or any transparent plastic which does not degrade in sunlight, while storage vessels are of metal and contain a heating coil or annulus.

Sewage Treatment

BRE Digest 302[1] emphasises that efficient sanitation is an important feature of many overseas construction projects. Where water supplies are severely restricted, various on-site systems are used. In urban areas, some arrangement for collection such as cesspits or emptiable pit latrines or even a full sewerage system may be provided. The former requires good organisation and the latter involves considerable investment. At high ambient temperatures, sewage becomes septic more quickly than in temperate zones, while in dry regions, limited water supplies can aggravate the problem, as the sewage is stronger and is flushed away at a slower rate. The consequent increase in acidity can lead to rapid corrosion of cement-based materials, and in the Middle East, in particular, there is considerable use of non-corrosive internal coatings of plastics (PVC-U and GRP) for sewer pipes.

BRE Overseas Building Note 174[13] describes how conventional primary and secondary sewage and sludge treatment processes of the kind commonly encountered in the United Kingdom, Europe and North America are rarely used in hot climates, mainly on grounds of costs, maintenance aspects and problems of operation. Hence sewage treatment processes in hot countries are often restricted to waste stabilisation ponds (also called oxidation ponds); aerated lagoons; and oxidation ditches.

While BRE Overseas Building Note 187[14] describes how in low density, high class residential areas throughout the world, septic tanks are the most common method of providing water-carried sanitation where there is no municipal sewerage system, for the household a flush WC connected to a well designed septic tank with an effective effluent disposal system has virtually all the advantages of a sewer connection. In a septic tank system excreta are conveyed through pipes from water closets and are retained and partially treated in the tank. Sullage (used water from bathing, clothes washing and kitchens) may also pass to septic tanks, but in developing countries it is often discharged directly to the stormwater or monsoon drainage system, to separate soakaways, or on to gardens or agricultural land.

If water is scarce as, for instance, where it is obtained from a public standpipe or a single household tap, a small septic tank can be provided under the latrine. Excreta fall directly into the tank, from a squatting plate or seat above it, without passing through a water seal, and this system is known as the 'aqua-privy'. Individual household septic tanks and aqua-privies are not low cost solutions to the problem of providing satisfactory excreta disposal. They are therefore not appropriate for the poorest groups of people in developing countries. Where communal latrines are socially acceptable and are kept clean, septic tanks and aqua-privies can provide a satisfactory form of sanitation, and there are many good examples in West Africa and India.

Most aqua-privies are built under a squatting slab and a chute is formed by a vertical pipe, which is often 100 mm diameter. It is important that the pipe extends at least 75 mm below the water level in the tank. The pipe may be integrally cast with the squatting plate. There are two main variations to the system: (1) aqua-privies with seats where the local practice is to sit rather than to squat as in the West Indies and Botswana; and (2) pour-flush water seals can be provided where the local practice is to use water for cleaning as in the Calcutta type.

In either a septic tank or an aqua-privy system, the waste material entering the tank receives primary treatment, solids separate out to form sludge and scum, and a partially treated effluent is discharged. In most developing countries, the tank has a concrete floor, rendered blockwork walls and a reinforced concrete cover slab. T-junctions are normally used for inlets and outlets to ensure minimum disturbance of settled sludge and scum. A ventilating pipe should be pro-

vided for the emission of foul gases.

The second stage of treatment is the biological breakdown of the effluent, which usually takes place as it soaks into soil from a soakpit or percolation trench. Alternatively, the effluent from a large septic tank, such as one serving an institution or a group of houses, may be treated in a percolating or trickling filter before being discharged to a watercourse or irrigation area.

BRE Overseas Building Note 189[15] emphasises that human excreta can be a source of infection from pathogenic bacteria, viruses and eggs of parasitic worms. Hence great care is necessary in siting possible sources of pollution, such as pit latrines and soakage systems from septic tanks and aqua-privies, to avoid pollution of the ground and water supplies by organisms and chemicals that are harmful to health.

The same paper describes in detail the diverse sanitation systems in use throughout the world and these are now listed.

(1) Nightsoil bucket and collection; high operating cost and health hazard.

(2) Overhung latrine; low cost but high health hazard.

(3) Pit latrine; low cost, medium health hazard and not permanent.

(4) Bored hole latrine; low cost, medium health hazard and not permanent.

(5) Ventilated improved pit; low cost and health hazard, not permanent.

(6) Permanent improved pit; low cost and health hazard; small double emptiable chambers.

(7) Compost latrine; low cost, medium health hazard and needs extra user care.

(8) Pourflush latrine; medium cost and health hazard, and needs effective soakaway.

(9) Vault and vacuum tanker collection; high operating cost, medium health hazard, and requires efficient management.

(10) Septic tank and soakaway; high cost, low health hazard and requires effective soakaway.

(11) Aqua-privy and soakaway; medium cost, low health hazard and requires effective soakaway.

(12) Sewered self-topping aqua-privy; very high initial cost, low health hazard and treatment required as for sewerage.

(13) Full sewerage; very high cost and very low health hazard.

Village Water Supplies

BRE Overseas Building Note 178[16] postulates that at least one in four of the rural water supplies in most developing countries is out of order and has frequently been broken down for some time. It is important that works of water supply shall be able to operate under prevailing conditions with the resources available. It is advisable to find a source of good quality water and to protect it from pollution. Pumps often break down or fall into disuse in rural areas and motorised pumps should only be installed where there is adequate finance to cover the running costs. It is preferable to select a source of water high enough to permit a gravity-fed supply of a fail-safe character.

Rainwater from metal roofs is relatively pure but requires large and expensive storage tanks, whereas surface water is often of inferior quality. Springs make ideal sources of community water supply but must be suitably protected. Wells can be sunk in a variety of ways and require protection from pollution at the surface by a concrete apron, open hand dug wells being the most vulnerable. There are many methods of lifting water but some may be too expensive to install and operate. The simplest methods are often the cheapest, and can be more easily made and repaired with local materials. For example, shadufs, consisting of bamboo poles with a balancing device, are used extensively along the banks of the Nile. For tube wells or very deep hand dug wells, a simple hand pump made locally from wood or plastics may suffice.

Adequate storage tanks may be built of local building materials such as bricks or masonry, especially if galvanised wire is laid between courses to provide horizontal reinforcement. Care should be taken to prevent them becoming breeding places for malarial mosquitoes, especially in seasonally arid areas, by covering them and screening ventilation pipes with mosquito-proof mesh. Storage for 48 hours with a slow and even movement of the water will permit some silt to settle out and some disease organisms to die off, and helps to clarify water for treatment by filtration or chlorination. Ideally, water should be provided in each house, and to economise in the use of water the tap can be fitted with a special valve to discharge a fixed volume of water, usually a litre, each time it is operated.

BEHAVIOUR OF BUILDING MATERIALS

General Factors

BRE Overseas Building Note 145[17] emphasises that when considering the use of building materials in tropical developing countries, two factors are of paramount importance, namely availability and performance. Indigenous products should be used whenever practicable to avoid foreign exchange expenditure and to secure the economic use of locally produced materials or surplus products from industrial or agricultural processes, as well as the maximum employment of local labour.

BRE Digest 302[1] describes how warm regions have high levels of solar radiation as well as high temperatures. Where rainfall and humidity are high, insects and micro-organisms are generally very active and organic materials, such as timber and building boards, are much more vulnerable than in a cool climate. Prolonged high humidity, as experienced in Singapore and Malaysia, encourages mould growth and decay, although treatment with preservative can greatly reduce the problem. Prolonged low humidity causes some materials, particularly concrete, mortar and plaster mixes, to dry out too quickly. Adequate precautions must be taken to avoid this with special emphasis on curing. In some regions there are significant changes in humidity which affect the moisture content of certain building materials and cause dimensional change. A typical example is timber, which may warp or split.

Severe climatic conditions can have adverse effects on many building materials. In particular, organic materials, which include paint and plastics, perform satisfactorily in temperate climates but are much less durable in the tropics, because of the combination of high temperatures, ultra-violet light levels and, possibly, high ambient humidities, and this is further examined in BRE Digest 382.[26] Table 16.2 shows the effect of a variety of factors or weathering agencies on various materials. This table will help in selecting the most appropriate materials for storage, use and maintenance. Many of the effects, although similar to those experienced in the United Kingdom, are more severe in the tropics.

Thus environmental factors have an important influence on the durability of materials. For example, in areas of continuous high humidity, some materials may retain sufficient moisture to have deleterious effects,

Table 16.2 Factors affecting the performance of building materials in warm climates

Agent	Bitumen felt	Concrete	Elastomers and rubbers	Fibre-reinforced cement composites	Lime
Moisture	Felts may blister. Rain removes water-soluble material, and washes off surface protective granules in time	Premature hydration of cement during storage reduces performance. Moisture aids curing of set concrete		Excessive cyclic moisture movements lead to cracking. High humidity encourages deterioration of natural organic fibres, and steel fibres used in composites	Premature set during storage of hydraulic limes. Slaking of bagged quicklime. In very dry conditions, non-hydraulic limes carbonate and harden slowly
Solar radiation (photochemical action)	Ultra-violet radiation causes embrittlement, cracking and formation of water-soluble compounds		May lead to surface oxidation, crazing and embrittlement	Degradation of natural and synthetic organic fibres	
Solar radiation (thermal action)	Softening and flow. Expansion of entrapped air or moisture causes blistering	Premature evaporation of water and rapid setting lead to surface cracking and poor curing of concrete	Softening, possible dimensional change, shrinkage when warm; longer-term chemical hardening	Rapid evaporation may lead to poor curing of matrix; drying shrinkage resulting in warping and cracking	Premature evaporation of water causes poor hardening of hydraulic limes
Pollution – atmospheric gases, salt (airborne or in ground) and grit		Chlorides in sea spray or aggregates affect alkalinity, lead to corrosion of reinforcement. Carbon dioxide causes carbonation. Sulphates in ground cause swelling and disruption	Ozone causes embrittlement, hardening and cracking	Carbon dioxide affects cement matrix	
Insects, micro-organisms	Mould growth on surfaces may lead to softening and cracking. Not immune to termites	Surface mould-growth can cause blackening	Surface mould-growth possible in humid environment	Surface mould-growth in humid environment	
General	High winds may cause felts to crack and tear	Special problems in hot, dry climates. Cement storage needs attention in humid climates		Rapid loss of fracture toughness in humid conditions of glass-fibre cement composites. Care necessary in production and handling, especially of asbestos-fibre reinforced materials	Storage is chief problem in wet climates

Source: BRE Digest 302[1]

Table 16.2 continued

Agent	Metals	Paints	Plastics	Sealants	Timber and wood products
Moisture	High humidity encourages corrosion and slows evaporation from wet surfaces. Rain can reduce corrosion by removing contaminating salts	Movement leads to loss of adhesion, crazing and cracking. Condensation on a drying film causes loss of gloss, pitting, slow drying and general weakening of film	Moisture may contribute to surface breakdown of glass reinforced plastics	May affect adhesion	Large moisture movement can cause checking, splitting, warping and raising grain. Swelling and delamination of wood-based boards
Solar radiation (photochemical action)		Colour change and chalking of pigments and embrittlement of binder accelerated	Oxidative degradation, leading to embrittlement, crazing and discolouration	Surface hardening, reduced resilience and adhesion	Embrittlement and discolouration of surface
Solar radiation (thermal action)		Heat polymerisation and loss of volatile components causes cracking and flaking. Movement may cause cracking. Difficulty of application on hot surface. Drying and hardening period is shortened	Accelerates any breakdown reaction; loss of strength and elasticity. Loss of plasticiser causes shrinkage and embrittlement	Hardening, loss of resilience and failure in adhesion or cohesion. With some softening and slumping	Thermal movement slight, may be masked by moisture movement
Pollution – atmospheric gases, salt (airborne or in ground) and grit	Salts and sulphur gases promote corrosion	Wind-borne grit may cause abrasion; dirt adheres to softer coatings	Wind-borne grit causes abrasion of some glossy and transparent surfaces	Dirt adheres to surfaces	
Insects, micro-organisms		Fungal and algal growth on film and on surface dirt, particularly when film remains wet, may cause film breakdown. Not immune to termites	Growth on organic fillers and plasticisers; seldom serious	Surface growth possible, not serious	Attacked by many insects and fungi, depending on timber type and moisture content
General	Contaminants (salt and sulphur gases) cause active corrosion. Careful protection may be required	Best stored in cool (not cold) conditions; warmth reduces shelf-life	Components for exterior use must be properly stabilised and employ light-fast pigments. Components must be protected during transport and site storage	Important to retain adhesion and elasticity	Rain-washing may leach out poor preservatives. Design detail important in avoiding rot

Source: BRE Digest 302[1].

while in drought zones they may deteriorate or fail to develop their potential properties because of dehydration. Moist conditions encourage fungal and insect attack of organic materials and the corrosion of metals and, in addition, impair the thermal properties of porous cladding materials. A very dry atmosphere, on the other hand, can result in the premature drying of cement products before curing is complete.

The shade temperature in some tropical regions may reach 60°C and surfaces exposed to direct sunlight can approach 100°C. Diurnal range of air temperature can exceed 50°C. A change in temperature can cause reversible changes in physical properties such as hardness, rigidity and strength, and can also result in permanent changes in properties from chemical degradation. For example, the degradation reactions responsible for the breakdown of plastics sheets are initiated by ultra-violet radiation, but the rate of deterioration is largely dependent on temperature. Temperature variations also cause dimensional changes, leading to splitting, warping or crazing of components. Deterioration can also result from high winds, impinging rain, salt, sand or dust upon exposed surfaces.

Concrete

Concrete is the most widely used structural material as it can offer a good performance over long periods, both with and without reinforcement. In warm, humid climates there are usually no major problems in producing good concrete, provided the cement is fresh, aggregates and water are clean, and satisfactory designs are used. Surface mould growth is likely to darken surfaces, especially if they remain wet, but it is appearance that is mainly affected rather than strength or durability.

In hot, dry climates, particularly near to coasts where the water table is high and the ground is contaminated with salts, difficulties may arise. Sulphate-resisting Portland cement is usually required below ground level. Clean aggregates may not be readily available, nor the water for washing dirty ones, but it is important to avoid salt-contaminated materials, especially for reinforced concrete. Chlorides aggravate the corrosion of reinforcement, with resultant expansion, and the concrete cracks and spalls. It is essential to keep the temperatures of all concrete materials as low as possible during casting to control rates of setting and curing and to minimise thermal expansion. After casting, surfaces should be protected and kept moist sufficiently long to avoid surface cracking and to ensure proper curing. As a final precaution, BRE Digest 302[1] recommends the use of surface coatings to protect the concrete from carbonation and possible subsequent reinforcement corrosion.

Brickwork and Blockwork

BRE Overseas Building Note 177[18] describes how defects in brickwork and blockwork, one of the most traditional types of construction throughout the world, are almost as common with the newer types of bricks and blocks. They arise mainly from inadequacies in the quality of the bricks and blocks, or the bedding mortar, from faults in design making insufficient provision for movement of the building or its components, or from unsatisfactory workmanship. Mortars generally have to be made from local sand and serious defects are only likely to occur where it contains soluble salts, which cause expansion, or organic matter, which reduces its strength.

Timber

In many areas, particularly in the developing world, timber is the most important structural material. Wood is potentially very durable but, being organic, is affected by climatic and biological factors and in the tropics the effects can be severe. Moisture is the most important factor, as variations in moisture content give rise to dimensional changes, particularly in variable conditions, and often lead to distortion and cracking. Moisture must be present to create the right conditions for biological action. Heat contributes directly to the thermal expansion of timber and it may also cause movement of moisture within the wood, resulting in distortion. In savannah areas the variation in equilibrium moisture content between seasons can be very large and poses special problems. Exposure to sunlight and weather causes colour changes and degradation in timber and leads to deterioration and loss of strength. In the appropriate conditions of temperature and moisture, moulds cause discolouration, fungi attack and destroy wood, and insects feed on it.

It is important to select an appropriate type and

grade of timber and to ensure suitable treatment with chemicals and surface coatings in exposed conditions in hot climates. In the humid tropics only glues of WBP (weather and boil proof) type as specified in BS 1203 and BS 1204 are suitable. For drier climates a BR (boil resistant) type may suffice. Similarly, fastenings when used in the humid tropics need to be more heavily plated than is usual or be of corrosion resistant metal.

In the tropics, particularly for structural work or where the timber is inaccessible, pressure wood preservative treatment is required to ensure deep penetration. Any sapwood must be thoroughly impregnated. BRE Overseas Building Note 183[19] lists the preservation treatments appropriate to various degrees of risk, related to the type of member and its location. Pressure and open tank treatments are effective in most situations.

Termites

BRE Overseas Building Note 170[20] describes how there are about 1900 known species of termites, distributed mainly throughout the tropics with a few species extending into temperate zones. The two main categories are drywood termites and subterranean or soil termites. They constitute the major threat in warmer countries to wood and wood-based materials used in building.

Drywood termites make their homes within the attacked timber and need no contact with the ground, and the protective measures used are similar to those employed for wood-boring beetles. They may fly into the building or be carried there in previously infested timber, and are generally restricted to coastal areas and very damp inland areas, as they require fairly high relative humidity.

Subterranean or soil termites are more numerous and widespread but all need to maintain contact with the ground. They can be physically excluded from buildings by precautionary measures during the building process and by vigilance on the part of the occupier or maintenance workers.

Building sites should be well drained and cleared of all wood, leaves and debris. Any termite mounds on the site should be levelled or removed and the soil replaced and well compacted before soil poisoning is carried out. Poisoning the soil produces a barrier through which it is difficult for the termites to penetrate. The poisons dieldrin, aldrin and chlordane, ap-

plied as 0.5 or 1.0 per cent emulsions in water, have been found to be the most reliable and by their use it is possible to obtain protection for periods of at least ten years, and often much longer in the warm humid regions close to the equator where the risk of termite attack is very high and the durability of soil poisons is poorest.

Buildings are usually constructed with slab-on-ground construction, preferably of the reinforced monolithic type, where the floor slab and foundations are poured in one continuous operation. After the ground has been filled, the hardcore and soil are treated as follows

(1) Before casting the slab, poison is applied at the rate of 5 ℓ/m^2 over the whole area to be covered by the building, or 7.5ℓ/m^2 if the fill is gravel, when the soil is fairly dry.

(2) Poison is applied along the inside of the foundation walls, around plumbing, and in wall voids at 6 ℓ/m.

(3) After the floor slab has been constructed and all levelling completed, a trench is dug around the outside of the foundations and sprayed with poison at the rate of 6 ℓ/m. The replaced excavated soil is similarly treated.

With hollow block walls or cavity walls, the bottom two or three courses should be built solid in poisoned mortar. All timber should be treated with preservative. Openings for ventilation should be protected by insect screens, which also serve to keep out other pests including drywood termites.

Metals

BRE Overseas Building Note 171[21] describes how the tropics are usually considered to present a highly corrosive environment, but this is not true of the hot dry regions. Even the hot humid zones are rarely as aggressive as many temperate industrial areas. Severe corrosivity is a very localised phenomenon, depending on special conditions rather than general climate, and especially the presence of pollutants such as salt, acids or sulphur dioxide. However, wind-driven sand can abrade both metal and protective coatings and lead to increased corrosion.

A relative humidity greater than 70 per cent is usually considered necessary for atmospheric corrosion to proceed, and the higher the humidity and the longer the period at which it exceeds this level, the greater the corrosion to be expected. The critical humidity may be below 70 per cent where sea salt is deposited from spray, since the sea salt is hygroscopic. For a given relative humidity, the corrosion rate will rise with increasing temperature.

Asphalt and Bitumen

On roofs and other exposed places, the viscoelastic properties of bituminous materials are seriously affected by repeated wetting and drying, while heat produces softening and blistering. Ultra-violet radiation leads to embrittlement and the production of water-soluble materials which can be leached out by rain; felt fibres may rot or deteriorate as a result of moulds, or be attacked by termites. High winds cause felts to crack and tear, and they can be abraded by sand storms. Modern roof coverings are sophisticated systems using bituminous felt, plastics sheet or film, metal foil, asbestos and glass fibre reinforcement and mineral granules in various combinations. Such systems are prone to weathering to a greater or lesser extent and need careful evaluation.[17]

Plastics and Paints

Plastics are used in many developing countries in flooring, water supply and drainage, thermal insulation, rigid sheeting for roof and wall cladding, and flexible sheets and films for waterproofing. Their more extensive use is inhibited by their combustibility and doubtful weathering properties. They are also more prone to degradation by ultra-violet radiation in the tropics, and this process is accelerated by high temperatures and tropical storms.

Paints are subject to degradation by solar radiation, binders and plasticisers may decompose or be leached out by rain, leading to cracking of the paint film, and pigments tend to fade. Hence they require careful selection and application.[17]

SPECIAL STRUCTURAL PROBLEMS

General Factors

BRE Digest 302[1] points out how designers may have to face one or more of three special structural problems

which, while not unique to warm climates, are certainly more prevalent in these regions. These hazards comprise the problem of building on expansive clay (heaving or black cotton) soils, and the severe effects of earthquakes and cyclonic storms, such as typhoons and hurricanes. Where these problems are encountered, the design of large buildings, in particular, calls for a sophisticated approach and the services of a specialist consultant are often required. Small buildings are best constructed using rigid construction.

Expansive Clay

This soil is sometimes referred to as heaving or black cotton soil. Volume change movements in expansive clays under foundations is a serious worldwide problem and can cause distortion and cracking of low rise buildings. Damage tends to be most severe in areas where the climate is either arid or has pronounced wet and dry seasons. Swelling of the clay results from infiltration of water from rainfall or domestic sources, while shrinkage of the clay is caused by evaporation and transpiration. The scale of damage and consequent financial loss is very high, and has been estimated at more than £2b annually. In fact expansive soils cause more damage than the combined effects of earthquakes, tornados, floods and landslides. BRE Overseas Building Note 191[22] has shown that expansive clays still cause widespread damage to buildings, mainly because of the difficulty of designing foundations which are both economical and suitable for local ground conditions.

The Building Research Establishment, in collaboration with the Building Research Center of the Royal Scientific Society, conducted a four year study of damage to buildings constructed on expansive clays in the highland area of northern Jordan. This informative study found that various types of foundations have been used, often with little understanding of the expansive behaviour of the clays. In consequence, foundation movements have severely damaged many low rise buildings, including dwellings, hospitals and schools.

In the Mediterranean climate of the northern highlands of Jordan the clay shrinks substantially in the long hot, dry summers (average maximum temperature 32°C and nil rainfall), swelling back by similar amounts in the cool wet winters (average maximum temperature 15°C and rainfall 350 to 700 mm). The magnitude of the changes and depth of ground affected is also influenced by the proximity of environmental features, such as trees. Open ground is subject to seasonally cyclic movements, decreasing in magnitude with depth, from about 50 mm at the surface to negligible amounts below about 2.5 m. In contrast, in locations close to trees ground movements can extend down to at least 5 m. Ground close to sources of water infiltration, such as leaking pipes, often show an opposite trend of long term movement, with progressive swelling superimposed on the seasonal movement.

The damage most commonly found was cracking of internal and external walls. About ten per cent of the buildings examined suffered severe or very severe damage. The most common crack pattern was one arising from relative downward movement of the corners of the buildings, with the cracks starting at ground level near the centre of walls and rising diagonally to the top of a corner. The cracks generally tended to be zig-zag, picking out weaknesses in the structure, particularly mortar joints and window and door openings.

Three types of foundation, in particular, often showed building damage, namely

(1) strip foundations of reinforced concrete at depths ranging from 1 to 3 m;

(2) pier foundations and ground beams, the piers being 1 to 2 m in section and 2 to 5 m deep, with the ground beams resting directly on clay at shallow depths; and

(3) pad foundations with columns and ground beams, the pads being of reinforced concrete 1 to 2 m square and up to 4 m deep, while the ground beams are seated directly on clay near ground level.

These three types of foundation have a major design deficiency: all have parts of the foundation seated at relatively shallow depths on clay affected by ground movement.

Apart from the cracking of walls, the most common damage found in the study was the distortion of floors, mainly resulting from inadequate compaction of clay fill under the floors. Subsequent absorption of ground moisture causes it to settle over a period of years. The problem can be minimised by using non-cohesive backfill, or avoided by constructing suspended floors.

Damage to low rise buildings can be avoided either by using foundations which penetrate through the zone of ground movement, or by constructing foundations and superstructures which are tolerant of ground movement.

However, in situations where there is likely to be deep seated ground movement, designs which completely eliminate damage can be prohibitively expensive in relation to the value of the building, and a compromise approach may be necessary which restricts damage to minor cracking. In these cases the building and its foundations should be designed to be safe against shallow seasonal ground movements, and environmental control measures implemented to minimise the long term deep seated changes of moisture content and resulting ground movement.

Two effective designs are very stiff raft foundations and rigid superstructures, which move as a unit, minimising distortion and cracking when subject to differential ground movement; and flexible foundations and superstructures to accommodate ground movement without damage, incorporating carefully designed movement joints in the structure. The most cost-effective way of preventing building movement and subsequent damage is to provide foundations deep enough to penetrate the zone affected by clay movement to stable ground below. Away from trees, and in the absence of excessive water infiltration near the building, foundation depth should be about 3 m. If large trees are present or are likely to be planted in the future, foundations must be substantially deeper (at least 5 m).

BRE Overseas Building Note 191[22] recommends four types of foundation design which satisfy the criterion of isolating the building from the zone of ground movement

(1) A basement approximately 3 m deep over the whole area of the building. This approach is expensive and is rarely suitable for low cost housing.

(2) Pile and beam foundations have been used successfully in many parts of the world. The piles are small diameter bored piles usually drilled by a mechanical auger, and extending to a depth well below the zone of clay movement by inserting a reinforcing cage and concreting to the ground surface. The reinforced concrete ground beams must be kept well clear of the ground so that ground swelling pressure is not transferred to the beams.

(3) Pad and beam foundations consisting of pad foundations on stable ground below the movement zone, a minimum of 3 m deep, from which reinforced concrete columns support suspended ground beams.

(4) Pier and beam foundations are similar to those

described in (3) but use mass concrete piers supporting the suspended ground beams instead of pads and columns. With deep foundations, this entails the use of large quantities of concrete.

BRE Overseas Building Note 191[22] emphasises that in all cases excessive moisture penetration must be prevented by

(1) Ensuring that all water supply and drainage pipes are watertight and have flexible joints to accommodate ground movement.

(2) Ensuring that all water tanks, septic tanks and the like are reinforced to minimise cracking and have flexible waterproofing.

(3) Providing a watertight drainage system which collects rainwater falling on roofs and paved areas and carries it well away from the foundations.

(4) Providing an outward sloping paved area around the building to minimise seasonal moisture changes near the wall foundations. The recommended width is about 2 m with a slope of 1:15. It should be made completely waterproof by laying it on thick polythene or PVC sheeting. A watertight sliding joint should be provided where the paving meets the building.

Earthquakes

BRE Overseas Building Note 190[23] describes how in recent years many regions in the Middle East experienced an unparallelled increase in construction activity, resulting from the receipt of very large oil revenues, and it was considered vital that the numerous large buildings erected should remain largely undamaged in the event of an earthquake. One of the first essential preliminary steps was to prepare maps showing zones in which buildings must be designed to resist earthquakes. The seismic zones are based on an examination of the tectonic structure of the area, seismograph records of earthquakes in the last 50 to 100 years, and on historical accounts of earthquakes dating back as far as 1500 BC.

An earthquake is defined by four parameters

(1) the origin time of the event;
(2) the co-ordinates of the epicentre;
(3) the focal depth of the source; and
(4) the size of the event, usually expressed on a magnitude scale.

In general the intensity of an earthquake shock experienced at any point on the surface varies inversely with the depth of the source fracture. A shallow earthquake of medium magnitude can thus be more potentially devastating than an event of larger magnitude at a deeper focus. One of the most devastating earthquakes of recent times was that at Agadir, which had a magnitude of 5.7 but a focal depth of no more than 2 or 3 km.

The design and construction of buildings in vulnerable areas has to allow for the effects of earthquakes in order to minimise the risk of total collapse, loss of life, and essential services being out of use. BRE Overseas Building Note 143[24] estimated that to erect an earthquake-resisting building instead of a normal one of good quality would add from 2 to 7 per cent to its cost.

Widespread experience shows that buildings made with good materials, soundly constructed and with all component parts thoroughly tied to one another generally survive. The following useful guidelines are contained in BRE Overseas Building Note 143[24]

(1) Wood framed buildings usually perform well because they are both strong and slightly flexible, provided the wood frames are properly braced.

(2) The effectiveness of brick or masonry walls depends largely on the type of mortar. For example, earth or lime mortars generally lead to failures. Reinforced brickwork or masonry performs well in most instances.

(3) Mud block walls almost invariably collapse.

(4) Unreinforced piers or columns of brick or masonry are dangerous.

(5) In framed buildings the panel walls are apt to fall away from the frame if they are not well tied to it.

(6) Soundly constructed foundations are important and in vulnerable areas they should be tied together with continuous reinforcement. Rigid buildings founded on rock or hard ground perform much better than those on alluvium or fill.

(7) Parapet walls, ornamental features, untied walls and gable ends are liable to collapse.

(8) Buildings with a heavy ground floor and a light upper storey, where applicable, and a light roof perform better than those with light walls and a heavy roof.

(9) Arches are a source of potential weakness.

BRE Overseas Building Note 143[24] further recommends that

(1) Small or medium-sized buildings should ideally take the form of a rigid box strengthened by cross walls which are well tied in and spaced as symmetrically as possible. Large openings should be avoided, particularly in external walls near the corners of the buildings. Bay windows are always a source of weakness.

(2) When siting a building in a seismic area the foundations should not span both hard and soft ground as this could result in a fracture where the strata meet. Foundations should not be stepped on sloping ground, but carried down to the same depth all round the building, and preferably tied together with continuous reinforcement.

(3) Records show that single storey dwellings built of blocks or burnt bricks with cement: sand or cement: lime: sand mortar are generally satisfactory, particularly where the roofs are lightweight. In minor seismic zones the best strengthening measure is to incorporate a continuously reinforced concrete ring beam on top of all the walls and immediately below the wall plate. In zones of greater severity the next precaution is to reinforce the walls at all corners and wall intersections, and where walls are built of hollow blocks to fill some of the holes with reinforcing bars and concrete. In the worst zones it is also advisable to provide horizontal reinforcement. In all cases any brick, block and masonry chimneys should be well tied to the walls with ample reinforcement.

(4) In certain climates houses are erected on piers to permit air to flow under the floor and cool the building. This type of construction needs special care in earthquake areas, because the piers are liable to fail by shear at the top or bottom during a tremor.

(5) For medium-sized buildings it is often best to use a reinforced concrete frame with cross and external walls tied to the frame. These walls may be of reinforced concrete or reinforced blockwork, brickwork or masonry.

(6) Much damage and loss of life has been caused by appendages falling from buildings, such as parapets, balconies, cornices, chimneys and other projecting features. Gable walls also need strengthening.

(7) Medium-sized buildings in the shape of an L are best designed as two and those in the shape of a U in three separate blocks. The external walls of the blocks can be joined with a 'crumple section', possibly consisting of a veneer of plaster which readily crushes when there is differential movement between the blocks. Afterwards the plaster can be cheaply restored.

The crumple section should be at least 25 mm wide for a building 6 m high plus an extra 12 mm for each additional 3 m or part thereof in height. For practical purposes it will probably be easier to make it at least 150 mm wide. Care must also be taken to enable the two roofs to move relative to one another.

(8) Similarly towers on buildings need carefully designing as they are liable to shear at the base.

(9) Water tanks should not be installed in the roofs of buildings unless unavoidable.

(10) Roof trusses should be bolted at junctions and all members of wood framed buildings should be adequately strapped together.

(11) Sanitary appliances and pipes should have flexible joints.

(12) The normal requirements for chimneys and flues must be carefully observed and, in particular, no woodwork must be built into the chimney.

(13) A fire nearly always follows a severe earthquake, hence

    (i) steel frames must be well encased;
    (ii) the ends of reinforcing bars should be hooked rather than left straight with reliance on the bond with concrete; and
    (iii) the fire precautions should exceed the statutory minimum, because the fire service may not be able to assist for some time.

(14) In the USSR the distance between buildings required for fire precautions is increased by 20 per cent in areas liable to severe earthquakes.

Tropical Windstorms

BRE Overseas Building Note 188[25] categorises the formation and development of windstorms in the following way

(1) Tropical disturbance: the formative stage, when the weather is unsettled and disturbed.

(2) Tropical cyclone: the start of a closed circulation over tropical oceans, with the energy being derived from warm seas where the temperature exceeds 27°C.

(3) Tropical depression: a larger and more intense tropical cyclone, with mean speeds up to 17 m/s.

(4) Tropical storm: a larger and more intense tropical depression, with mean wind speeds greater than 17 m/s but not exceeding 33 m/s.

(5) Hurricane: a very intense tropical storm with mean wind speeds exceeding 33 m/s, and gust speeds will be much greater.

(6) Dissipation stage: the stage when there is decrease in wind speed and an increase in the barometric pressure.

Countries which experience the highest gust speeds (70 to 90 m/s) are Rodriguez (South Indian Ocean), Hong Kong and Taiwan.

Every year, many hurricanes strike islands or populated coastal zones of continents, causing major structural damage to buildings, making many people homeless and causing loss of life. Throughout the world on average 23 000 people are killed and 2.6 million are injured or left homeless each year as a result of windstorms and the associated flooding from storm surges.

Some buildings are damaged because of their siting and position; for example wind speeds are greater near the top of a hill. Conversely, many buildings resist hurricanes very well because they are sheltered by trees or other surrounding buildings. A building may be destroyed or have a large part, such as a complete roof, removed largely because of its particular geometrical shape. Typical examples are tall buildings and projections to buildings such as parapets. Large roof overhangs present an easy target for strong winds, and flat or low pitched roofs will experience a much greater uplift than more steeply pitched roofs.

By far the majority of buildings fail because of poor constructional details, resulting from inferior design and/or construction on site, with little appreciation of the forces that can be exerted during a hurricane. The majority of buildings in the world are low rise houses and most of them are not designed at all. They are often built by house owners or small contractors, using traditional methods and without any specialist constructional knowledge. This often results in vital connections, such as those required between walls and roofs, being inadequate or omitted altogether with disastrous consequences in a hurricane. An important factor in designing buildings to prevent wind damage is a knowledge of the likely exposure to storms. There are two components to this assessment, namely the meteorological data and probabilities of a given wind speed, and the nature of the local terrain and surroundings which will affect factors such as the likelihood of storm surge. The likely wind flow patterns need to be understood and steps taken to counteract them.

BRE Overseas Building Note 188[25] points out that in most countries there is little engineering design input into those aspects of buildings which relate to wind resistance. If the amount of damage that occurs because of hurricanes is to be reduced, it is essential that the level of this engineering input is increased. This will require a greater and more widespread understanding of the structural principles involved in resisting the pressures and forces. It is particularly important to realise that pressures act at right angles to a surface, and in the case of low pitched roofs the uplift forces are extremely large.

REFERENCES

1. *BRE Digest 302*: Building overseas in warm climates (1985)
2. *BRE OBN 160*: Low cost housing in urban and peri-urban areas (1975)
3. *BRE OBN 158*: Building for comfort (1974)
4. *BRE OBN 179*: Foundations in poor soils including expansive clays (1978)
5. *BRE OBN 165*: Buildings and the environment (1975)
6. *BRE OBN 176*: Building materials in the Arabian Gulf – their production and use (1977)
7. *BRE OBN 184*: Stabilised soil blocks for building (1980)
8. *BRE OBN 139*: Problems of concrete production in arid climates (1971)
9. *BRE OBN 146*: Timber in tropical building (1972) (Revised in OBN 199: 1993)
10. *BRE OBN 182*: Roofs in hot dry climates, with special reference to northern Sudan (1978)
11. *BRE OBN 164*: The thermal performance of concrete roofs and reed shading panels under arid summer conditions (1975)
12. *BRE OBN 192*: Solar energy applications (1985)
13. *BRE OBN 174*: Sewage treatment in hot countries (1977)
14. *BRE OBN 187*: The design of septic tanks and aquaprivies (1980)
15. *BRE OBN 189*: Sanitation for developing communities (1982)
16. *BRE OBN 178*: Village water supplies (1978)
17. *BRE OBN 145*: Durability of materials for tropical buildings (1972)

18. *BRE OBN 177*: Avoiding faults and failures in buildings (1977)
19. *BRE OBN 183*: Preservation of timber for tropical building (1979)
20. *BRE OBN 170*: Termites and tropical building (1976)
21. *BRE OBN 171*: Protection of steelwork in building (1976)
22. *BRE OBN 191*: Guidelines for foundation design of low rise buildings on expansive clay in northern Jordan (1984)
23. *BRE OBN 190*: Earthquake risk to buildings in the Middle East (1982)
24. *BRE OBN 143*: Building in earthquake areas (1972)
25. *BRE OBN 188*: Buildings and tropical windstorms (1981)
26. *BRE Digest 382*: New materials in hot climates (1993)

APPENDIX: METRIC CONVERSION TABLE

Length

1 in = 25.44 mm (approximately 25 mm)

then $\frac{mm}{100} \times 4$ = inches

1 ft = 304.88 mm (approximately 300 mm)
1 yd = 0.914 m (approximately 910 mm)
1 mile = 1.609 km (approximately $1\frac{3}{5}$ km)
1 m = 3.281 ft = 1.094 yd (approximately 1.1 yd)
(10 m = 11 yd approximately)
1 km = 0.621 mile ($\frac{5}{8}$ mile approximately)

Area

1 ft^2 = 0.093 m^2
1 yd^2 = 0.836 m^2
1 acre = 0.405 ha (1 ha or hectare = 10 000 m^2)
1 mile^2 = 2.590 km^2
1 m^2 = 10.746 ft^2 = 1.196 yd^2 (approximately 1.2 yd^2)
1 ha = 2.471 acres (approximately $2\frac{1}{2}$ acres)
1 km^2 = 0.386 mile^2

Volume

1 ft^3 = 0.28 m^3
1 yd^3 = 0.765 m^3
1 m^3 = 35.315 ft^3 = 1.308 yd^3 (approximately 1.3 yd^3)
1 ft^3 = 28.32 litres (1000 litres = 1 m^3)
1 gal = 4.546 litres
1 litre = 0.220 gal (approximately $4\frac{1}{2}$ litres to the gallon)

Pressure

1 lbf/in^2 = 0.007 N/mm^2
1 lbf/ft^2 = 47.88 N/m^2
1 tonf/ft^2 = 107.3 kN/m^2

Mass

1 lb = 0.454 kg (kilogramme)
1 cwt = 50.80 kg (approximately 50 kg)
1 ton = 1.016 tonnes (1 tonne = 1000 kg = 0.984 ton)
1 kg = 2.205 lb (approximately $2\frac{1}{5}$ lb)

Density

1 lb/ft^3 = 16.019 kg/m^3
1 kg/m^3 = 0.062 lb/ft^3

Velocity

1 ft/s = 0.305 m/s
1 mile/h = 1.609 km/h

Energy

1 therm = 105.506 MJ (megajoules)
1 Btu = 1.055 kJ (kilojoules)

Thermal conductivity

1 Btu/ft^{2}h°F = 5.678 W/m^2°C

(where W = Watt)

Temperature

$x°F = \frac{5}{9}(x - 32)°C$
$x°C = \frac{9}{5}x + 32°F$
0°C = 32°F (freezing)
5°C = 41°F
10°C = 50°F (rather cold)
15°C = 59°F
20°C = 68°F (quite warm)
25°C = 77°F
30°C = 86°F (very hot)

INDEX